Managing Risk in Agriculture: A Development Perspective

Managing Risk in Agriculture: A Development Perspective

Edited by

Ashok K. Mishra

Kemper and Ethel Marley Foundation Chair in Food Management
Morrison School of Agribusiness
W. P. Carey School of Business
Arizona State University, Tempe, Arizona, USA

Subal C. Kumbhakar

Distinguished Professor of Economics
Department of Economics
State University of New York at Binghamton, New York, USA
&
Department of Economics, Faculty of Economics and Management, Czech University of Life Sciences Prague, Prague, Czech Republic

and

Gudbrand Lien

Professor in Business Economics
Inland School of Business and Social Sciences,
Inland Norway University of Applied Sciences, Lillehammer, Norway
&
Senior Researcher within Agricultural Economics
Norwegian Institute of Bioeconomy Research (NIBIO), Ås, Norway

CABI

CABI is a trading name of CAB International

CABI
Nosworthy Way
Wallingford
Oxfordshire OX10 8DE
UK

Tel: +44 (0)1491 832111
E-mail: info@cabi.org
Website: www.cabi.org

CABI
200 Portland Street
Boston
MA 02114
USA

Tel: +1 (617)682-9015
E-mail: cabi-nao@cabi.org

A catalogue record for this book is available from the British Library, London, UK.

ISBN-13: 9781800622265 (hardback)
9781800622272 (ePDF)
9781800622289 (ePub)

DOI: 10.1079/9781800622289.0000

Commissioning Editor: David Hemming
Editorial Assistant: Lauren Davies
Production Editor: Rosie Hayden

Typeset by Straive, Pondicherry, India
Printed and bound in the UK by Severn, Gloucester

While the publisher has attempted to ensure that the full-text version of this eBook is accessible to all readers, rendering of native MathML is not fully functional in all support scenarios.

Dedication

This book is dedicated to J. Brian Hardaker
for his pioneering work in agricultural risk management

Brian was one of the authors (together with Jock R. Anderson and John L. Dillon) of the pathbreaking book *Agricultural Decision Analysis* back in 1977 and is the first author of the book *Coping with Risk in Agriculture* from 2015 (3rd edition). Brian has published more than 100 scientific articles, with important contributions to studies of risk in agriculture in developed and developing countries.

Contents

Contributors

Tahirou Abdoulaye, *Agricultural Economist and Principal Scientist, International Institute of Tropical Agriculture, Bamako, Mali.*

Rachidi Aboudou, *Agricultural Economist, Research Assistant on Quantitative Impact Assessment, Africa Rice Center (AfricaRice), Abidjan, Côte d'Ivoire.*

Essossinam Ali, *Assistant Professor, University of Kara, Kara, Togo.*

Aminou Arouna, *Program Leader & Impact Assessment Economist, Africa Rice Center (AfricaRice), Abidjan, Côte d'Ivoire.*

Victor Barrera, *Senior Economist, Department of Agricultural Economics, National Institute for Agricultural Research of Ecuador (INIAP), Quito, Ecuador.*

Toritseju Begho, *Researcher, Rural Economy, Environment & Society, Scotland's Rural College, Edinburgh, UK.*

Pratap S. Birthal, *Director, ICAR-National Institute of Agricultural Economics and Policy Research, New Delhi, India.*

Martina Bozzola, *Senior Lecturer, Queen's University Belfast, Belfast, UK; ZHAW Zurich University of Applied Sciences, Winterthur, Switzerland.*

Selina Bruns, *Postdoctoral Researcher at the Department of Agricultural Economics and Rural Development, University of Göttingen, Göttingen, Germany. Visiting Researcher at the School of International Development, University of East Anglia, Norwich, UK.*

Simone Cerroni, *Associate Professor, Department of Economics and Management and C3A, University of Trento, Trento, Italy.*

Jean-Paul Chavas, *Professor Emeritus, University of Wisconsin, Madison, USA.*

Juan Dominguez, *Associate Dean of Research, Graduate School of Management, Escuela Superior Politécnica del Litoral (ESPOL), Guayaquil, Ecuador.*

Marcella D'Souza, *Director, Centre for Resilience Studies, Watershed Organisation Trust, Pune, India.*

Hongli Feng, *Assistant Professor, Department of Economics and Center for Agricultural and Rural Development (CARD), Iowa State University, Ames, USA.*

Ricardo Fort, *Senior Researcher, Group for the Analysis of Development (GRADE), Lima, Peru.*

J. Brian Hardaker, *Professor Emeritus, University of New England, Armidale, Australia.*

Jaweriah Hazrana, *Post-Doctoral Fellow, Ohio State University, Columbus, USA.*

David A. Hennessy, *Professor, Department of Economics and Center for Agricultural and Rural Development (CARD), Iowa State University, Ames, USA.*

Maman H. Karmana, *Professor, Faculty of Agriculture, Universitas Padjadjaran, West Java, Indonesia.*

Aditya R. Khanal, *Associate Professor, Department of Agricultural and Environmental Sciences, College of Agriculture, Tennessee State University, Nashville, USA.*

Abel-Gautier Kouakou, *Agricultural and Environmental Economist, International Institute of Tropical Agriculture, Bamako, Mali.*

Subal C. Kumbhakar, *Distinguished Professor of Economics, Department of Economics, State University of New York at Binghamton, New York, USA; Department of Economics, Faculty of Economics and Management, Czech University of Life Sciences Prague, Prague, Czech Republic.*

Alfons G.J.M. Oude Lansink, *Professor, Business Economics, Wageningen University & Research, Wageningen, Netherlands.*

Seo Woo Lee, *PhD candidate, Department of Agricultural, Food, and Resource Economics, Michigan State University, East Lansing, USA.*

Sergio H. Lence, *Professor and Marlin Cole Chair of International Agricultural Economics Department of Economics, Iowa State University, Ames, USA.*

Gudbrand Lien, *Professor in Business Economics, Inland School of Business and Social Sciences, Inland Norway University of Applied Sciences, Lillehammer, Norway; Senior Researcher within Agricultural Economics, Norwegian Institute of Bioeconomy Research (NIBIO), Ås, Norway.*

Lu Liu, *Graduate student, Institute of Agricultural Economics and Development, Chinese Academy of Agricultural Sciences, Beijing, China.*

Fulco Ludwig, *Water Systems and Global Change Group, Wageningen University & Research, Wageningen, Netherlands.*

Wanglin Ma, *Associate Professor, Department of Global Value Chains and Trade, Faculty of Agribusiness and Commerce, Lincoln University, Christchurch, New Zealand.*

Hui Mao, *Northwest Institute of Historical Environment and Socio-economic Development, Shaanxi Normal University, Xi'an, China.*

Miranda P.M. Meuwissen, *Professor, Business Economics, Wageningen University & Research, Wageningen, Netherlands.*

Ashok K. Mishra, *Kemper and Ethel Marley Foundation Chair in Food Management, Morrison School of Agribusiness, W. P. Carey School of Business, Arizona State University, Tempe, USA.*

Kehinde Odeniyi, *Lecturer II, Department of Agricultural Economics and Agribusiness Management, Osun State University, Osogbo, Nigeria.*

Asis Kumar Senapati, *Assistant Professor, Department of Economics and Development Studies, Central University of Jharkhand, Jharkhand, India; Department of Economics, Ravenshaw University, Cuttack, India.*

Ariel Singerman, *Assistant Professor, Food and Resource Economics Department, Citrus Research and Education Center, University of Florida. Gainesville, USA.*

Arjuna Srinidhi, *PhD candidate, Water Systems and Global Change Group, Wageningen University & Research, Wageningen, Netherlands; Watershed Organisation Trust, Pune, India.*

Ruggiero Rippo, *PhD student, C3A, University of Trento, Trento, Italy.*

Nicholas Tyack, *Assistant Professor, Department of Agricultural & Resource Economics, University of Saskatchewan, Saskatoon, Canada.*

Marrit van den Berg, *Professor, Development Economics Group, Wageningen University, Wageningen, Netherlands.*

Puneet Vatsa, *Senior Lecturer, Department of Global Value Chains and Trade, Faculty of Agribusiness and Commerce, Lincoln University, Christchurch, New Zealand.*

Alexis H. Villacis, *Assistant Professor, Morrison School of Agribusiness, W. P. Carey School of Business, Arizona State University, Tempe, USA.*

H. Holly Wang, *Professor, Department of Agricultural Economics, Purdue University, West Lafayette, USA.*

Saskia E. Werners, *Head of Vulnerability Assessment, Risk Management & Adaptive Planning (VARMAP), United Nations University, Bonn, Germany; Water Systems and Global Change, Wageningen University & Research, The Netherlands.*

Tesfamicheal Wossen, *Economist, International Institute of Tropical Agriculture, Nairobi, Kenya.*

Eliana Wulandari, *Associate Professor, Faculty of Agriculture, Universitas Padjadjaran, West Java, Indonesia.*

Xiaoheng Zhang, *College of Economics and Management, Nanjing University of Aeronautics and Astronautics, Nanjing, China.*

Hongyun Zheng, *College of Economics and Management, Huazhong Agricultural University, Wuhan, China.*

Foreword

Barry K. Goodwin
William Neal Reynolds Distinguished Professor, North Carolina State University, Raleigh, USA

Since nearly all producers operate in a dynamic world of uncertainty, risk is an omnipresent characteristic of production. This is particularly true in agriculture, where biological constraints place significant lags on production, leading to uncertain economic outcomes. The agricultural economics field courses at my home institution, North Carolina State University, consist of a course in Production Economics and a parallel course in Consumer Demand. I have often challenged my graduate students to reveal which of the two areas of study is more difficult. The standard response is that the demand side of the equation is the more difficult since one must deal with the vagaries of human behavior while the production side is often just an engineering question of inputs and outputs (or netputs, if you will). However, risk changes the simple answer since the manner in which economic actors respond to risk involves aspects of human behavior that are at least as difficult to conceptualize and measure.

Frank Knight famously made the distinction between risk and uncertainty in his classic 1921 text *Risk, Uncertainty, and Profit*. According to Professor Knight, risk pertains to situations where the outcome of an event is uncertain but the underlying probability-generating functions are well understood. Uncertainty, in contrast, applies to situations where neither the outcome nor the probabilities of different outcomes are known a priori. The management of risk has always been a core principle of business practices. The proper management of risk must first determine the probability-generating functions governing alternative outcomes. Without uncertainty, fields such as actuarial science and political polling sciences (and, indeed, much of economics), which largely involve efforts to measure the odds of different outcomes, would not exist. On a point which many of us may take issue with, Knight argued that 'true uncertainty is not susceptible to measurement.'

As I noted, risk is, by definition, a dynamic concept. If there are no lags in production or behavior, outcomes are fully known and there is no uncertainty or risk. Risk is especially important in agricultural markets because of the existence of fixed (within certain stochastic bounds) lags in production. In the US Corn Belt, for example, corn is usually planted in April or May and harvested in October. Broilers usually have a production cycle of 4 to 7 weeks, depending on their intended use. Steers are usually finished on feed for 5 to 6 months. These production cycles are determined by biology as well as by certain exogenous factors such as price, weather, specific hybrid or variety, and by other production practices. One does not know the outcomes at the time a crop is planted or an animal is placed on feed. The variability of these outcomes reflects the underlying risk associated with production.

A large number of mechanisms for managing risk in agriculture have been developed. These include subsidized (and unsubsidized) crop insurance, futures and options markets, specific production practices including the choice and application of productive inputs, and so forth. It is the application and use of such mechanisms that gives rise to the questions typically addressed by economists studying risk and risk management in agriculture. The questions are endless and the policies and market environment are constantly changing. In recent years, risk management research has focused on climate change as an important factor that is evolving and directly impacting agricultural production. Policy is a critical factor in the risk faced by farmers. Almost all farmers in the US participate in the federal crop insurance program, which has become the main support mechanism for US farmers.

Risk management often assumes a different role in developing economies. Credit markets may be less developed, and the lack of liquid capital for farmers to borrow is often a barrier to the adoption of new technologies. Insurance, which plays a very prominent role in the agricultural economy of the US, as well as many other developed countries, may be far less developed in these economies. Insurance and credit are strongly linked since insurance provides collateral to back the loans that many farmers need access to in order to adopt new technologies or to expand to take advantage of economies of scale.

I was very pleased (and flattered) when I was asked to provide this foreword for what I consider to be an outstanding and important book. *Managing Risk in Agriculture: A Development Perspective* (edited by Ashok K. Mishra, Subal C. Kumbhakar and Gudbrand Lien) provides a comprehensive review of various issues related to the management of risk in agriculture, including efforts to measure what Knight called 'true uncertainty.' The text fills a substantial void in the professional literature on the measurement and management of risk in agriculture. Risk and risk management has become the most researched topic in agricultural economics. Developments in the professional literature are occurring at a rapid pace, making it difficult to stay abreast of the most recent developments. This valuable book, which includes chapters by the profession's leading researchers, conveys the current state of the art in the study of agricultural risk.

The individual chapters cover a wide range of topics important to agricultural risk. Valuable surveys of the existing literature addressing risk issues in agriculture are presented in the text. Hardaker and Lien (Chapter 1) present an overview of research on agricultural decision analysis. From an interesting perspective, they argue that risk attitudes have been overly emphasized in explaining behavior under risk. Kumbhakar (Chapter 2) surveys the current state of research on risk preferences and price and production risk. Bruns (Chapter 8) surveys the measurement of risk elicitation methods in rural households. Cerroni and Rippo (Chapter 3) survey the literature investigating the role of subjective probabilities in determining farmers' decision-making under uncertainty in developing countries.

Singerman and Lence (Chapter 4) summarize the role of strategic risk with a particular focus on pest management. Strategic risk arises when agents' behaviors depend on their beliefs about the actions of other agents. Chavas (Chapter 9) provides an overview of the economics of risk in agriculture. He develops a conceptual model of farm-level agricultural risk with a focus on price and production risk. He makes several interesting points regarding the downside of risk (e.g. low prices or yields).

These surveys are very valuable to teachers and their students. I have found such surveys to be invaluable in my own graduate classes since much of the material involves a survey of the existing state of the art in research. These chapters certainly fit the bill – they offer outstanding summaries of the contemporary state of knowledge.

A number of other chapters report on case studies of risk in agriculture. Khanal, Mishra and Lien (Chapter 5) adopt a semi-parametric smooth-coefficient framework to evaluate risk-reducing actions among smallholder onion producers in India. Begho and Odeniyi (Chapter 7) investigate the role of risk attitudes in decision-making in India and Nepal. Cerroni, Bozzola and Rippo (Chapter 6) report on an investigation that elicited risk preferences for horticultural producers in Zimbabwe. They adopt a methodology that is based on incentivized and incentive-compatible mechanisms for eliciting beliefs and preferences. Arouna, Tyack and Aboudou (Chapter 10) elicit risk preferences from smallholder rice farmers and consider how these preferences affect the value assigned to having access to seed of farmers' varieties of rice from the gene bank of the Rice Biodiversity Center for

Africa. Hazrana and Birthal (Chapter 19) consider how crop diversification influences productivity and resilience in Indian agriculture.

As I have noted, the role of credit is essential to any consideration of risk management. Credit markets provide farmers with a means of withstanding negative shocks to production and prices. Credit and risk are intimately tied to one another. Higher risk means more expensive terms of borrowing. Risk management mechanisms also influence the terms of credit, with farmers who adopt mechanisms to manage risk generally being offered more favorable borrowing terms. Wulandari, Meuwissen, Karmana and Lansink (Chapter 14) consider borrowing from different finance providers and contract fulfillment rates for Indonesian chili farmers. Vatsa, Ma, Mao and Zhang (Chapter 20) evaluate linkages between extreme weather events and apple farmers' demand for credit in six counties of China's Shaanxi province. Their analysis uses a double-hurdle model to investigate the likelihood of borrowing and the level of credit obtained.

A central focus of many chapters and, if one wanted to define such, a theme of the volume is climate change. I'm not sure that there is too much more of a debate about whether climate is changing. It is. The reasons for the change are not entirely known by scientists, and approaches to reversing or mitigating the change are even more opaque. Several chapters address risk issues that are connected in some measure to a changing climate. In much of the existing literature, researchers use measurements of technology that may be 50 years old to extrapolate 50 or more years into the future. I find this approach very suspect and am glad to see that this approach is not adopted in the chapters presented in this text.

Villacis, Mishra, Barrera and Dominguez (Chapter 16) examine Ecuadoran cacao farmers' perceptions of climate change. They find that farmers' views about risk are related to perceptions about climate change. In their empirical application, they find that potential agronomic mitigation activities occur only for fertilizer. Ma, Zheng and Vatsa (Chapter 18) investigate risk exposure and adaptation to climate change for wheat farmers in China.

Lee, Feng and Hennessy (Chapter 21) examine how beliefs about climate change evolve after experiencing an extreme weather event. They also use survey data collected before and after a severe drought to examine how views about climate change evolve after experiencing a severe weather event. Van den Berg and Fort (Chapter 17) examine the impact of a major natural disaster – an earthquake – on farm households' behavior. They found that risk aversion increased significantly following the earthquake but that famers did not change their preferences for crop insurance. Srinidhi, Werners, Ludwig, D'Souza and Meuwissen (Chapter 15) examine the climate resilience of semi-arid farming in India.

As noted above, agricultural insurance has become the primary program for supporting farmers in the US and has also expanded around the world. Wang and Liu (Chapter 11) consider small farmers' willingness to pay for crop insurance in China. They consider different forms of crop insurance and examine preferences for the alternatives. Abdoulaye, Essossinam, Kouakou and Wossen (Chapter 13) examine the willingness to pay for weather index insurance and drought-tolerant maize in West Africa. Senapati (Chapter 12) examines the role of risk from extreme weather shocks and its relationship to crop insurance purchases by Indian paddy rice growers.

In all, I found this text to be a valuable resource and a significant reference point for characterizing the latest in research into risk and its impacts on agriculture. Many of the chapters make a distinction in risk management and attitudes toward risk between developed and developing countries. Climate change is a central theme of many chapters – and rightfully so. For agriculture, climate change is central to any consideration of risk management. The text is an excellent collection of chapters written by the leading researchers who are addressing risk and risk management questions in agriculture. The text is a valuable resource for those involved in teaching production economics. I commend the authors and editors for producing such an excellent collection of chapters.

Acknowledgments

Any large project involves an equally large number of individuals, and this book is no exception. As editors of this volume, we are grateful to a significant number of people who collaborated with us, supported us and assisted us in preparing the chapters of this book, as well as with the activities that underlie them. First, we appreciate David Hemming, who approached us with the idea for this book several years ago. Over the years, the editors became excellent friends, with similar educational attainment, backgrounds and research interests, though a continent apart. Second, we would like to thank the chapter writers who willingly and efficiently revised or prepared materials to be utilized in this form. Third, country research teams organized and conducted studies in Ecuador, Peru, India, Ghana, Nigeria, China, Indonesia, the USA and other developing and emerging economies.

Prof. Mishra would like to thank the Marley Foundation for supporting this book. We sincerely appreciate the sacrifices made by our families since the time spent editing this book was time that would have otherwise been spent at home and with family members. Their sacrifice and understanding enabled the editors to spend the time required for this volume to become a reality.

The editors wish special thanks to Ms. Jane Larson for helping to edit the book. Without her help, this volume would not have become a reality. Finally, the views expressed are purely those of the authors and may not in any circumstances be regarded as stating an official position of the university and/or any other institution with which they are associated.

Ashok K. Mishra
Subal C. Kumbhakar
Gudbrand Lien

Introduction and Overview

Ashok K. Mishra,[1] Subal C. Kumbhakar[2,3] and Gudbrand Lien[4,5]
[1]*Arizona State University, Tempe, USA;* [2]*State University of New York at Binghamton, New York, USA;* [3]*Czech University of Life Sciences Prague, Prague, Czech Republic;* [4]*Inland Norway University of Applied Sciences, Lillehammer, Norway;* [5]*Norwegian Institute of Bioeconomy Research (NIBIO), Ås, Norway*

A comprehensive risk analysis of a business such as farming entails questions about the importance of risk perceived by farmers and how to deal with it (Wauters *et al*., 2014). These perceptions and farmers' experiences with risks influence their risk responses (Flaten *et al*., 2005). Many scholars have studied farmers' behavior in coping with agricultural risks and the management measures taken. Generalization is complicated since the risks farmers face and the coping strategies they subsequently adopt vary by farming system, agro-climatic region, local policy and institutional setting. The assessment of farmers' perceptions and how they react to risk is crucial to better understanding their decision-making when faced with risk (Flaten *et al*., 2005). It contributes to formulating more appropriate policies to help overcome the risk associated with farming.

Another risk facing farming businesses is climate change or weather risk. Exposure to adverse weather and climate change increasingly affects the livelihoods of the rural population. It challenges their food security and economic well-being (Bobojonov and Sommer, 2011), especially farmers who depend on subsistence agriculture and lack access to alternative livelihoods. Beyond weather variability and climate change, other factors, such as market conditions, seed availability and labor supply, also play a decisive role in shaping farmers' responses to risks associated with farming (Mertz *et al*., 2009). Increased adverse weather conditions such as droughts and floods likely exacerbate existing gaps, making the poor more vulnerable and leaving many rural farmers in a vicious cycle of poverty. Moreover, agricultural risks are the principal cause of transient food insecurity, creating a poverty trap for millions of households across the developing world that enforces a vicious cycle of shock and recovery. Risks can have potentially significant implications for stakeholders, investments and development in the agricultural sector (Giertz *et al*., 2015). This linkage to policy-making is critical considering the increase in risk due to climate changes, risk perceptions and risk attitudes. Governments can facilitate farmers' greater access to risk management strategies, improve farm management skills, engage private enterprises in managing risks and climate information and enable farmers and smallholders to adapt to long-term climate change.

How farmers deal with risk has been an essential topic in agricultural economics. However, interest has developed more recently in how risk issues (risk preferences and risk attitudes) impact technology adoption, the adoption of risk management strategies, crop portfolios, food security, responses to climate change, conservation and environmental management choices. Although risk itself is critically important, even more important is risk perception, and risk response in developing and emerging economies

(Elbers *et al.*, 2007; Dercon, 2008) is paramount. But there is little evidence on probabilistic risk formation in developing and emerging countries (Attanasio, 2009; Delavande, 2014), as most research on the topic looks at professional forecasters.

There is scant empirical evidence from farm households to suggest that heterogeneity in subjective probabilities is driven by farmers' risk attitudes, preferences, household- and farm-specific characteristics and access to coping mechanisms (Giné *et al.*, 2009). To this end, experimental economics methods have proven helpful in collecting information on individual risk preferences, attitudes, risk management strategies and responses to climate risks. Applied economists and sociologists have conducted experiments in several African and Asian countries to collect timely information on risk in agriculture.

The Rationale for Managing Risk in Agriculture: A Development Perspective

Risk attitude also is proven to be context-specific (Pennings and Garcia, 2001). Although risk perception and risk attitude are described independently in the literature, much less research shows the interaction between risk perception and risk attitude and how they collectively guide risk behavior. Farming, especially in developing and emerging economies, offers a fascinating case study investigating risk behavior. Agriculture is increasingly confronted with risk and uncertainty from various sources, such as production risk, price volatility, personal risks and policy changes (Hardaker *et al.*, 2015). Farming decisions are made primarily by a single person who aims to maximize production and profit and sustain farming as an occupation. The decision-making process regarding risk management strategies and the role of risk attitude enhances the understanding of farmers' risk behavior, acreage allocation, yields and technology adoption. Smallholders may not always understand that their choices differ from those of other smallholders due to differences in risk perception and risk attitudes, rather than being driven by external influences and structural barriers.

Many applications of risk frameworks to agricultural decision processes have used an expected utility (EU) framework (e.g. Pennings and Garcia, 2001; Hardaker *et al.*, 2015). Decision makers are expected to maximize their EU of returns when making decisions. The relevant utility function depends on whether decision makers are risk-averse, risk-neutral or risk-loving. Consistent with classical definitions, outcomes can be characterized by predictable randomness (risk) and unpredictable randomness (uncertainty), which typically are bundled under the rubric of risk from the decision makers' point of view. Applying the classic 1954 Savage model of subjective EU allows for variations between individual choices to the extent that their subjective risk assessments differ.

The challenge of analysis is to manage the interplay between the actual risks and uncertainties involved, the perceptions of risks held by a landholder and how risk factors are incorporated into individual utility functions. Recent decades have seen a growing acceptance that EU models are not always appropriate, and several non-expected utility (NEU) models, such as prospect theory (Kahneman and Tversky, 1979), have emerged to explain behaviors such as loss aversion and probability weighting. These models primarily focus on addressing the 'fourfold pattern of risk attitudes' (Tversky and Kahneman, 1992), which states that when faced with a risky prospect, people (in general) will be: (i) risk-seeking over low-probability gains; (ii) risk-averse over high-probability gains; (iii) risk-averse over low-probability losses; and (iv) risk-seeking over high-probability losses (Harbaugh *et al.*, 2010).

EU and NEU models continue to be widely applied in agricultural economics to explain farmers' choices under risky conditions. The EU models continue to be widely used because they support more specific representative functions, are more tractable in analysis and results are relatively less ambiguous than with other approaches (Ghadim and Pannell, 2003; Gregg and Rolfe, 2017). Adopting an EU framework, analysts can assess decision makers' risk preferences and then essentially assume that these have some proportional relationships to the risk profiles of alternate enterprise outcomes. In other words, risk-averse farmers will have a much lower EU for risky outcomes than risk-loving farmers.

Thus, different theories on decision-making under uncertainty coincide in characterizing decisions due to preferences and the (subjective) probabilities of different risky states of nature. The process can be thought of as having two steps (see Fox and Tversky, 1998). The first step is for individuals to objectively assess the probability of a rare random event. In the second step, farmers decide on farming and crop portfolios based on this probability assessment and considering their preferences. Recently it has been recognized that it is necessary to evaluate farmers' subjective probabilities when analyzing their preferences regarding risk and uncertainty (Cerroni, 2020).

In sum, this book critically examines the currently held beliefs about risk preference, attitudes and empirical estimation of risk management strategies, emphasizing developing and emerging economies. The book adds to the discussion of where knowledge gaps exist. Readers will come away from the book understanding risk preferences, attitudes toward risk and risk-managing tools used by farmers in developing and emerging economies. However, we do not propose providing easy answers, as some more popular books seem to do. Instead, a discussion of how these decisions are rooted in rational choices made by profit-maximizing, risk-minimizing individuals will show that often-cited risk management strategies may not be profitable for smallholders in developing and emerging economies.

Policymakers, especially in emerging economies, need a better understanding of farmers' attitudes toward risk and choices of risk management strategies when designing policies to support agricultural production. This book lays the foundation for empirical research by first addressing and documenting farmers' risks in developing and emerging economies. Second, it draws lessons from experimental economics on measuring risk preferences, attitudes, gender differences in managing risks and risk management strategies in countries across Africa and Asia.

This book will guide academics, researchers, students, legislative liaisons, non-profit associations and think-tank groups. It includes chapters on three themes: (i) understanding risk, attitudes and preferences; (ii) using experimental economics to measure risk, preferences and risk management strategies; and (iii) understanding climate change, risk and risk management. Based on these studies, a wealth of knowledge on strategies and contexts is developed. The concluding chapter identifies critical knowledge gaps and future priorities. Where feasible, the book draws implications for policy development and practice based on the available evidence.

References

Attanasio, O.P. (2009) Expectations and perceptions in developing countries: their measurement and their use. *American Economic Review* 99, 87–92.

Bobojonov, I. and Sommer, R. (2011) Alternative insurance indexes for drought risk in developing countries. Paper prepared for presentation at the EAAE 2011 Congress ETH Zurich. Zurich, Switzerland.

Cerroni, S. (2020) Eliciting farmers' subjective probabilities, risk, and uncertainty preferences using contextualized field experiments. *Agricultural Economics* 51, 707–724.

Delavande, A. (2014) Probabilistic expectations in developing countries. *Annual Review of Economics* 6, 1–20.

Dercon, S. (2008) Fate and fear: risk and its consequences in Africa. *Journal of African Economies* 17(Suppl. 2), ii97–ii127.

Elbers, C., Gunning, J.W. and Kinsey, B.H. (2007) Growth and risk: methodology and micro evidence. *The World Bank Economic Review* 21, 1–20.

Flaten, O., Lien, G., Koesling, M., Valle, P.S. and Ebbesvik, M. (2005) Comparing risk perceptions and risk management in organic and conventional dairy farming: empirical results from Norway. *Livestock Production Science* 95, 11–25.

Fox, C.R. and Tversky, A. (1998) A belief-based account of decision under uncertainty. *Management Science* 44, 879–895.

Ghadim, A.K. and Pannell, D.J. (2003) Risk attitudes and risk perceptions of crop producers in Western Australia. In: Babcock, B.A., Fraser, R.W. and Lekakis J.N. (eds) *Risk Management and the Environment: Agriculture in Perspective*. Kluwer, Dordrecht, Netherlands, pp. 113–133.

Giertz, A., Mudahar Mohinder, S., Gray, G., Rubaiza, R., Garpelin, D. *et al.* (2015) *Rwanda Agricultural Sector Risk Assessment*. World Bank, Washington, DC.

Giné, X., Townsend, R. and Vickery, J. (2009) Forecasting when it matters: evidence from semi-arid India. Mimeo. Available at: https://www.tse-fr.eu/sites/default/files/TSE/documents/sem2015/developpement/gine.pdf (accessed February 7, 2023).

Gregg, D. and Rolfe, J. (2017) Risk behaviours and grazing land management: a framed field experiment and linkages to range land condition. *Journal of Agricultural Economics* 68, 682–709.

Harbaugh, W., Krause, L. and Vesterlund, L. (2010) The fourfold pattern of risk attitudes in choice and pricing tasks. *The Economic Journal* 120, 595–611.

Hardaker, J.B., Lien, G., Anderson, J.R. and Huirne, R.B.M. (2015) *Coping with Risk in Agriculture, Applied Decision Analysis*, 3rd edn. CAB International, Wallingford, UK, p. 276.

Kahneman, D. and Tversky, A. (1979) Prospect theory: an analysis of decision under risk. *Econometrica* 47, 263–291.

Mertz, O., Mbow, C., Reenberg, A. and Diouf, A. (2009) Farmers' perceptions of climate change and agricultural adaptation strategies in rural Sahel. *Environmental Management* 43, 804–816.

Pennings, J. and Garcia, P. (2001) Measuring producer risk preferences: a global risk-attitude construct. *American Journal of Agricultural Economics* 83, 993–1009.

Tversky, A. and Kahneman, D. (1992) Advances in prospect theory: cumulative representation of uncertainty. *Journal of Risk and Uncertainty* 5, 297–323.

Wauters, E., Van Winsen, F., De Mey, Y. and Lauwers, L. (2014) Risk perception, attitudes towards risk and risk management: evidence and implications. *Agricultural Economics* (Czech Republic) 60, 389–405.

Section I

Understanding and Modeling Risk, Attitudes and Preferences

The first section of this book includes five chapters outlining the measurement of risk perceptions and risk attitudes and the modeling of risk in agricultural production. It illustrates the stock of knowledge on subjective measurements of risk and risk preferences and of estimation methods used to explain the direct and indirect channels through which risk-reducing strategies regarding inputs can influence productivity.

Chapter 1 discusses the importance of risk aversion and how it generally has been exaggerated relative to the task of finding better ways to deduce relevant and reliable probabilities. The authors aim to identify areas of difficulty and possible confusion and suggest some principles of good practice based on pertinent literature findings and reasoned arguments. The authors conclude that it is time for our profession to break down the facade of supposedly respectable 'objectivity' and to be more honest about the many subjective judgments that we inevitably must make in risk analysis. The authors note that researchers and leaders should recognize that thoughtful and transparent subjectivity is the only sensible basis for analyzing many important and risky economic decisions in agriculture.

Chapter 2 presents the modeling of risk preferences when producers face uncertainties in output and input prices, in addition to production risk. The author derives risk preference functions associated with these uncertainties without assuming a specific form of the utility function. The author carefully specifies production risk and risk aversion behavior so researchers and applied economists can test for different risk aversion behavior.

Chapter 3 reviews studies exploring the role of subjective probabilities in explaining farmers' decision-making under risk and uncertainty in developing countries. The authors note that the number of studies looking at farmers' subjective probabilities in developing countries is small but growing. Most studies focused on expectations regarding future climate variability, future yields and price fluctuations. All but two studies elicited subjective probabilities using direct methods that, by definition, are not compatible with incentives and hence may not be able to provide sincere beliefs. The authors believe there is an opportunity to test whether indirect approaches could be used to elicit farmers' subjective probabilities in developing countries.

In Chapter 4, the authors discuss the key issues arising from strategic risk among agricultural producers by looking at the case of voluntary pest control. The authors focus on collective action because the agricultural economics literature does not seem to have sufficiently emphasized the uncertainty underlying the aggregate success of collective action efforts or individuals' associated opportunity costs of collective action.

Chapter 5 introduces applied economists to a semi-parametric smooth-coefficient (SPSC) framework to explain the direct and indirect channels through which applying risk-reducing strategies to inputs can influence productivity. The authors examine four broad types of risk-reducing strategies: farm diversification, the use of risk-reducing market-based instruments, the adoption of good agricultural practices and access to off-farm income sources.

1 Principles of Good Practice in Dealing with Risk in Agriculture: A Review of Agricultural Decision Analysis

J. Brian Hardaker[1] and Gudbrand Lien[2,3]*

[1]University of New England, Armidale, Australia; [2]Inland Norway University of Applied Sciences, Lillehammer, Norway; [3]Norwegian Institute of Bioeconomy Research (NIBIO), Ås, Norway

Abstract

Because the analysis of risky choice in agriculture is important but difficult, agricultural economists need both suggestions on how they can do decision analysis better and some agreed-on principles. In this chapter, we argue that the importance of risk aversion has generally been exaggerated relative to the task of finding better ways to deduce relevant and reliable probabilities. Getting better probabilities demands careful thought, drawing on what is known about the pitfalls and on evolving insights into better methods. Our aim is to stimulate a debate leading to a clearer consensus about better practice in these matters.

1.1 Introduction

Economic risk is difficult to deal with consistently because it is nearly always about some future outcome or outcomes that are therefore nearly always wholly imaginary. Finding a way to deal rationally with such imaginary problems is difficult and leaves scope for a range of approaches, some more sensible than others. The first person to suggest a rational approach to risky choice was Daniel Bernoulli in 1738. Bernoulli, who wrote in Latin, proposed that, faced with risky choices, decision makers should act consistently with their beliefs about the chances of occurrence of different outcomes and with their relative preferences for those outcomes. In what follows, we call Bernoulli's logical approach 'decision analysis,' which we view as a rational or normative way of coping with risk. We neither imply nor expect that all decision makers will adopt this form of analysis. Indeed, there is ample evidence that many people do not (e.g. Ellsberg, 1961; Kahneman and Tversky, 1979; Allais, 1984; Schmeidler, 1989; Quiggin, 1993; Rabin, 2000; Rabin and Thaler, 2001; Kahneman, 2011; Bocquého *et al.*, 2014; Machina and Siniscalchi, 2014). Nevertheless, in what follows we seek to show how decision analysis can be applied successfully to agriculture, along with some suggestions on how current practices might be improved or refined. In doing so, we generally ignore the often-strange behaviors that have been observed in studies of risky choice.

*E-mail: gudbrand.lien@inn.no

DOI: 10.1079/9781800622289.0001

Agricultural production is risky, and decision analysis and risk management tools have become increasingly popular. Unfortunately, some risk analysis has not been done well. Given the complexity of the task, it is not surprising that some mistakes have been made and that there is scope for disagreement on how to proceed. With an econometric focus, Just (2000, 2003) and Just and Pope (2001) have assessed principles of good practice for research on agricultural risk. In this chapter, we focus on prescriptive decision analysis. We aim to identify areas of difficulty and possible confusion and to suggest some principles of good practice, based on relevant findings in the literature or on reasoned argument.

1.2 General Principles of Modeling

1.2.1 The importance of accounting for risk

Despite efforts over the centuries by agricultural decision makers to find ways to manage risk, and notwithstanding government interventions over many decades to try to help, risk remains unavoidable. But does risk in farm and policy decision-making really matter? Clearly, some risks, such as the possibility of an outbreak of a highly contagious animal disease or an environmental catastrophe, must be taken seriously (e.g. Kunreuther, 2002). In farming, risks such as a temporary feed shortage on a grazing farm or low returns from a particular cash crop usually can be 'ridden out' by a reasonably solvent and competent manager. Similarly, some resource management risks may be so limited in their impacts as to be of small importance. Evidently, only 'important' risks matter – a point not always reflected in the literature – and these important risky choices faced by policymakers, farmers and other resource managers do warrant systematic analysis to assess the 'best-bet' actions.

Accounting for risk can be important because, in a non-linear system, even setting input variables at their mean values will give a biased (and usually overestimated) value of the consequences (Anderson, 1976; Hardaker *et al.*, 2015, Chapter 1). This effect, called downside risk, occurs whether the decision maker is risk-averse or not. Risk analysis is needed to deal with it.

In addition, if a decision maker is risk-averse, as most almost certainly are, analysis based on maximizing expected money value generally will not lead to the most preferred choice (Arrow, 1951; Robison and Barry, 1987). The size of the bias depends on the riskiness of the particular decision and on the decision maker's degree of risk aversion. These considerations lead to our first principle:

Principle 1. All decisions are risky, some more than others. For choices with significant risk, a decision analysis that ignores risk and/or risk aversion is likely to be flawed and hence misleading, perhaps seriously so.

Because systematic risk analysis requires more information and judgments about the uncertainties to be faced than deterministic analysis requires, it should lead to an estimate of expected consequences that is at least no worse than that obtained by deterministic analysis. Moreover, by having information about the distribution of the consequences, a better assessment of the risks can be made, leading to a better choice.

Practitioners of deterministic analyses often claim that sensitivity analysis accommodates 'risk.' Yet, at least as conventionally applied, sensitivity analysis is a poor substitute for a proper risk analysis, although perhaps better than nothing. However, as Reutlinger (1970) argued, varying selected uncertain coefficients is of little use unless accompanied by thoughtful specification of their probability distributions, or at least of their feasible ranges. Moreover, the usual one-by-one variation of uncertain coefficients tells nothing about possibly crucial stochastic dependencies or other interactions, such as complementarities between variables.

Anderson (1976) argued that full disclosure of information and its quality is important in modeling. In analyses under assumed certainty, disclosing uncertainty about key coefficients is rare and may even be seen as an admission of weakness. By contrast, risk analysis requires careful consideration and explicit modeling of the main uncertainties. Provided the data, assumptions and results are presented informatively, risk analysis therefore provides support for more transparent decisions.

Many people balk at the inevitable subjectivity of most risk analyses. Certainly, there is ample evidence of the fallibility of human judgment. Yet risk analysis exposes the subjectivity, while analysis under assumed certainty typically hides it. Systematic risk analysis does not overcome human fallibility, but the need for analysts to think deeply about the uncertainty associated with the problem likely reduces the bias inherent in single-valued estimates (Morgan and Henrion, 1990, p. 44).

Finally, surveys have shown that farmers view some risks as worrisome (e.g. Meuwissen *et al.*, 2001; Komarek *et al.*, 2020), suggesting a need for decision support to help them cope. Komarek *et al.* (2020), in their literature review of types of risk in agriculture, found that earlier research often focused on only one type of risk. For the future, they ask for a shift in research toward the analysis of multiple types of risk.

1.2.2 The difficulty of risk analysis

Despite the need, evaluating risky choices can be hard. The difficulty at the farm level arises partly because agricultural businesses often are best modeled in a system context. A system view includes dynamic, stochastic, biological, technical, financial and human factors that interact (Pannell *et al.*, 2000). Further, all possible alternatives and all the risks bearing on them should ideally be considered as elements in a portfolio of risky choices that the decision maker faces (e.g. Anderson, 1982).

Even in a marginal analysis, there are usually several uncertain quantities to consider, and in a whole-farm analysis there will be many. Consequently, as discussed below, stochastic dependencies between variables may be important yet difficult to handle (e.g. Reutlinger, 1970; Taylor, 1990; Hardaker *et al.*, 2015, pp. 64–79).

Evidently, comprehensive modeling is likely to be difficult, perhaps impossible. Even if feasible, too much complexity makes a model difficult to build, debug and use, and may give results little better than those that could have been obtained from a simpler representation. Morgan and Henrion (1990, Chapter 12) argue that a complex model often will do a worse job than a simpler one. It is therefore best to keep the model as simple as is judged reasonable. The intention is not to give exact answers but to highlight relative consequences of alternatives and develop insight and understanding. The analysis need be only as sophisticated as is necessary to provide the decision maker with a good yet timely answer to the problem. Such an analysis is essentially 'artistic,' just as a portrait painter seeks to capture the most important features of a face or personality. Hence, our second principle:

Principle 2. Risk analysis is 'the art of the possible.'

The difficulty and unavoidably artistic nature of decision analysis should not deter an analyst from attempting some modeling. Only exceptional people have the capacity to make wise decisions intuitively. For most of us, some decision analysis is likely to be better than none, at least for important risky choices (Arrow, 1951).

By the nature of artistry, there can be few firm guidelines on how to do decision analysis well. However, Clemen (1996 Chapters 2 and 3), for example, describes the basic steps, including preliminary structuring steps that force the analyst to understand all aspects of the problem. It is in these steps that 'artistic' choices must be made about what to do and how to do it, e.g. deciding which uncertainties should and should not be represented by probability distributions (Morgan and Henrion, 1990, pp. 50–56). Vose (2000, pp. 6–12) suggests doing a preliminary informal analysis to exclude uncertainties that have little impact on the riskiness of the consequences.

Tools useful in problem specification as well as those suited to the subsequent decision analysis are described in several texts (e.g. Clemen, 1996; Hardaker *et al.*, 2015). The capacity for risky decision analysis has increased greatly due to the increased availability of powerful, special-purpose software, the ease of use of which creates the danger that analysts may rely too much on the software and give insufficient thought to the modeling task. In recent years, with increased data availability, machine learning has gained increased attention in decision analysis and farm risk management. Without doubt, it has a large potential, but sound practice and good modeling are also highly important in that field (Ghaffarian *et al.*, 2022).

Every decision both closes and opens doors to future choices. Particularly important in structuring the analysis is the need to think carefully about how the decision maker can adapt to bad or good outcomes as they unfold. Imaginative thinking may be needed to find realistic tactics to deal especially with bad outcomes (see e.g. Kingwell *et al.*, 1993). Often, of course, there may be little that can be done about bad outcomes due to irreversibilities such as sunk costs, soil erosion, or species extinction. Then, keeping options open may be the best way to limit downside risk. Keeping options open and postponing choice is not the only option to consider. In principle, all options influenced by the current choice ought to be considered. Real option valuation takes account of the uncertainty of predicted cash flows and of the decision maker's opportunities to react to changed circumstances. Abandoning a project or selling some of the assets if the project is no longer profitable may be possible. Similarly, there may be scope to expand or contract the scope of a project, to vary the methods or pace of implementation, or to opt out and terminate a failing project. Several ways of incorporating real options into decision analysis have been proposed (e.g. Dixit and Pindyck, 1994; Hardaker *et al.*, 2015). Examples of agricultural applications include Tegene *et al.* (1999), Luong and Tauer (2006) and Kim *et al.* (2020).

Although artistry is unavoidable in the analysis of risky choice, validation of the model developed is important to check that it provides a reasonable representation of the actual problem (van Horn, 1971). Often this means reviewing and refining what was done. Ultimately, of course, the decision maker(s) must find the analysis credible – an essentially subjective judgment.

1.3 Expected Utility and Risk Aversion

1.3.1 Choice of the best theory

The subjective expected utility (SEU) hypothesis, also known as Bernoulli's principle, has a long history (e.g. Bernoulli, 1738; von Neumann and Morgenstern, 1944; Arrow, 1951; Savage, 1954). Basically, the hypothesis is that rational people will seek to make risky choices consistent with what they believe, as measured by their subjective probabilities for outcomes, and with what they prefer, as evaluated via their utility functions for consequences. The decision maker's utility function for outcomes is needed because the shape of the utility function reflects that person's attitude to risk (Anderson *et al.*, 1977, pp. 66–69). The SEU hypothesis states that the utility of a risky alternative is the decision maker's expected utility, meaning the probability-weighted average of the utilities of outcomes.

The SEU hypothesis has been criticized because it has long been recognized that many people do not act consistently with the theory in some risky choice situations (e.g. Allais, 1984; Kahneman, 2011; Bocquého *et al.*, 2014). Rabin (2000) has shown that the typical aversion to a single risky prospect with small losses is so great as to be inconsistent with any utility function expressed in terms of the utility of wealth. Such *loss aversion* implies *failure in asset integration*, meaning that people seemingly do not regard small gains and losses as changes in wealth (see e.g. Binswanger, 1981). Prospect theory (Kahneman and Tversky, 1979), as an alternative to SEU, has potential to give good insights to describe farmers' behavior in different risky environments. The SEU hypothesis is flawed as a behavioral theory of choice (Rabin and Thaler, 2001). Yet these criticisms miss an important point. As is clear from Bernoulli's original proposition, and as explicitly stated by von Neumann and Morgenstern in their early axiomatic treatment of the hypothesis, the hypothesis is essentially normative, offering a way to make better, more logical risky choices. It never was intended as a model of actual behavior in the face of risk, although the hypothesis sometimes may provide insights into responses to risk. We therefore argue that we should stick with the SEU hypothesis if we want to act sensibly, which leads to our third proposed principle:

Principle 3. Despite some objections that have recently been given renewed prominence by Nobel Laureate Daniel Kahneman (2011), the SEU hypothesis has yet to be displaced as the best model for normative decision analysis.

For support of this principle, see Hardaker *et al.* (2015, pp. 93–94), Machina (1987), Edwards (1992), Meyer (2001) and Just (2003).

1.3.2 Difficulties in applying the SEU hypothesis

While we believe the hypothesis to be valid for prescriptive decision analysis, there are difficulties in applying it. Attempts to elicit relevant utility functions from decision makers to implement the approach in the analysis of risky alternatives in agriculture have often been rather unconvincing (King and Robison, 1984; Anderson and Hardaker, 2003). Many people apparently lack the introspective capacity to respond consistently to hypothetical questions about their risk attitudes (see e.g. Iyer *et al.*, 2020). Even when real money prizes have been offered (e.g. Binswanger, 1981), loss aversion has led to unsatisfactory results.

If direct elicitation of risk attitudes has proved tricky, attempts to elicit risk attitudes from observed behavior also have problems. First, if based on historical data, such studies may not describe decision makers' current risk attitudes. Second, errors in model specification tend to cause overestimates of risk aversion coefficients. For example, omission of some constraints in whole-farm programming models may lead to attributing diversification to risk aversion when it is really a response to technical considerations such as the need for crop rotations. Similar bias may affect the results of econometric studies in which such constraints are unrecognized. Third, it is usually assumed that the decision maker's probabilities about the uncertain events of concern were the same as those of the analyst, the latter typically being drawn from some historical data that may not match the decision maker's experience and expectations. For such reasons, some studies deducing risk aversion measures from observed behavior have found coefficients of risk aversion too high to credit.

These considerations lead to our fourth principle:

Principle 4. Measuring decision makers' risk attitudes is difficult and will give misleading results if based on inappropriate hypothetical or real-choice situations or on inappropriate analyses of observed behavior.

Note that we are not asserting that relevant measures of risk aversion cannot be obtained – it is often possible to elicit a reliable utility function for wealth from well-motivated and well-briefed decision makers. Rather, we flag the need for care in elicitation efforts and for caution in interpreting results.

Stochastic efficiency criteria have been developed partly to avoid the need to elicit a specific utility function (or a single value for a coefficient of risk aversion). They are useful for a decision maker whose preferences are not known precisely and when there is more than one decision maker, such as in formulating recommendations for a group of farmers. Stochastic efficiency methods, and particularly stochastic efficiency analysis with respect to a function (SERF) (Hardaker *et al.*, 2004), are consistent with the SEU hypothesis and narrow down the range of choice to an efficient set. Moreover, these methods can be used in most forms of risk analysis. That surely should quiet those critics of decision analysis who see problems in utility function elicitation as a stumbling block.

Because there is a limit to what can be achieved by stochastic efficiency methods, a need remains for work on how risk attitudes can be assessed and rationalized more reliably. Just (2003) noted that elicitation of risk attitudes has neglected intertemporal risk. Multi-attribute utility functions have been similarly neglected, despite evidence of the diversity of farmers' goals (e.g. Gasson *et al.*, 1988; Bergevoet *et al.*, 2004). Such functions may be particularly useful for risky choices in resource and environmental management, perhaps offering greater flexibility and precision in modeling preferences than methods such as conjoint analysis and choice modeling (e.g. Duke and Ilvento, 2004).

1.3.3 The importance of risk aversion in agriculture

Given the difficulties in applying utility theory, how important is decision makers' risk aversion in agricultural decision analysis? Some farm-level studies have implied or assumed (often without explicit justification) that farmers are very risk-averse, even for quite marginal decisions. Yet it has been shown that the cost of ignoring risk may be small, at least for farmers' partial or short-term decisions (e.g. Pannell *et al.*, 2000).

Reasons why the importance of risk aversion has been exaggerated include:

1. Too narrow analysis that ignores interactions with other choice options on- and off-farm, including risk-sharing opportunities and possibilities for tactical or sequential adjustment (Antle, 1983; Pannell *et al.*, 2000).
2. Overestimates of decision makers' degree of risk aversion, as discussed above.
3. Erroneous use of risk aversion measures, such as the use of coefficients of relative risk aversion for wealth in the analysis of payoffs expressed in terms of losses and gains (Anderson and Hardaker, 2003).
4. At least in more developed countries where access to credit is easy, farmers with reasonable equity can ride out normal variations in annual income (Just, 2003).

Assuming asset integration makes sense, the decision maker's utility function for wealth is appropriate for prescriptive analysis. Risk aversion can be described by the absolute and relative risk aversion functions for wealth (Pratt, 1964; Arrow, 1965). Arrow (1965) suggested that the value of the relative risk aversion function for wealth is positive and close to 1, and it seems likely that it is seldom above about 4 (Anderson and Dillon, 1992). Hence, a plausible range of absolute risk aversion coefficients can be derived for a range of assumptions about relative risk aversion and wealth. Using such a range, Hardaker *et al.* (2015, pp. 99–102) show that, for capitalist farmers, risk aversion, though not negligible, is likely to be unimportant for most of their partial and short-term decisions.

Support for the view that risk aversion's importance often is overestimated may be found in e.g. Pannell *et al.* (2000), Meyer (2001) and Just (2003). Hence, we propose the following principle:

Principle 5. The importance of risk aversion has sometimes been exaggerated, especially in the context of farmers' partial or short-term decisions in more developed countries.

The implication is that, for many farm decisions, basing choice on maximizing expected money value is adequate. When that seems too strong an assumption, the handling of risk aversion depends on how the problem is cast and hence on how the consequences are measured (Hardaker *et al.*, 2015, pp. 96–98). However, we do not assert that risk aversion never matters in farm-level decision-making. It will be important for asset-threatening decisions, e.g. major investments. For poor farmers in less developed countries, even moderate risks may matter a lot. The general rule, however, is that risk aversion matters for the risks that threaten the viability of a farm or the sustainability of a family, a community, or an ecosystem (Just, 2003).

1.3.4 Risk aversion and public choice

Dealing with risk in public choice, in principle, merely involves applying the same methods as for private choice. Thus, analysis for support for public decisions should take into account any possible downside risk and should include consideration of ways to deal with bad consequences, if and when they occur. But what utility function should be used in such analyses? When all the consequences of the policy choice are adequately measured in money units, via some form of cost–benefit analysis, leading to a distribution of a measure of worth such as net present value, Arrow and Lind (1970) argued persuasively that society is able, at least potentially, to pool its risks across the whole population. Consequently, society should usually be neutral toward risk, a view supported by Samuelson and Vickrey (1964). So, we present our sixth principle:

Principle 6. For policy decision-making, it is usually appropriate to assume indifference to risk because society can potentially share bad consequences among the large number of citizens.

Cases that are exceptional relate mainly to risk in project appraisal in less developed countries. Little and Mirrlees (1974, p. 316) argued that some accounting for risk aversion is appropriate when a public project is large relative to national income, when project returns are highly correlated with such income, or when a particular disadvantaged group is involved. For a discussion of the appropriate risk deduction in such cases, see Anderson (1989).

It seems that the above principle is seldom applied to choices relating to environmental and food safety issues. For example, Brunton (1995) argued for risk-averse decisions in dealing with environmental issues. Similarly, in these fields it is not unusual to find decision makers or analysts invoking such notions as the *precautionary principle* (Ricci *et al.*, 2003) or a *safe minimum standard* (Crowards, 1998). Depending on circumstances, such rules can imply an extreme but undisclosed degree of risk aversion, which may be inappropriate. Although there are indeed cases in which extreme caution is justified, it may be better (at least for reasonably well understood risks) not to apply such arbitrary rules but to use a conventional decision analysis in which judgments about probabilities are separated from assessments of the importance of alternative consequences – including potentially catastrophic outcomes (Morris, 2002).

1.4 State-Contingent Approach

The state-contingent (SC) approach was pioneered by Debreu (1952) and Arrow and Debreu (1954). They showed that, if uncertainty is represented by a set of possible states of nature, production uncertainty can be brought within the ambit of conventional production theory, with output in different states treated in much the same way as different types of output. After languishing for many years, at least in agricultural decision analysis, the SC approach was revived in an important book by Chambers and Quiggin (2001). Their treatment of the approach was theoretically rigorous, but it was hard to understand for many practitioners of decision analysis. Clarifying papers by Chambers and Quiggin (2004), Quiggin and Chambers (2006), Rasmussen (2003, 2011) and Hardaker *et al.* (2015, Chapter 8), among others, have largely dispersed the mists so that several agricultural applications have appeared.

In the context of the SC approach, the word 'contingent' means 'depending upon.' As indicated, central to the approach is the notion of consequences of risky choices being analyzed as depending upon which of the possible uncertain states of the world occurs. It implies that the analysis of production under uncertainty reveals that risk can be accommodated within the ambit of existing production economics theory. It makes it clear that risk should not be regarded as a limitation of that theory. The method also has appeal as providing, at least potentially, a more satisfactory way of solving risky choices about production. Moreover, the scope is wider than just the analysis of a stochastic production function problem (as were the first applications with this approach within agricultural economics). The same methods work with cost functions, profit functions, etc. Indeed, almost any problem in decision analysis can be framed in an SC context. Note also that methods such as decision trees, payoff tables and other forms of analysis account for risky consequences contingent upon which state occurs (Hardaker *et al.*, 2015, Chapter 8).

We propose the following principle:

Principle 7. The SC approach to the representation of production under uncertainty is superior to the stochastic production function approach whenever the available data and estimation methods allow its use.

We offer this principle somewhat tentatively since it will not be until more empirical applications appear that the approach's advantages and limitations become more apparent. The limited numbers of studies using the SC approach in agriculture include O'Donnell and Griffiths (2006), Chavas (2008), O'Donnell *et al.* (2010) and Nauges *et al.* (2011).

1.5 Probabilities for Risk Assessment

Our treatment of this topic is founded on the view that all probabilities are subjective statements of degrees of belief in propositions about the world – a view not accepted by many, perhaps most, agricultural economists. However, this is not the place to try to convert the unbelievers about subjective probabilities. Instead, we merely offer the following observations:

1. The notion of probabilities as subjective has been supported by some distinguished thinkers. The theory is based on plausible axioms and sound logic, and there are well-tried methods of implementation (e.g. Savage, 1954, 1971; de Finetti, 1974).
2. Deriving probabilities from relevant, reliable and reasonably abundant relative frequency data is entirely consistent with the subjectivist view.

3. Rejection of subjective probabilities implies that no systematic analysis is possible to support most of the important risky choices that farmers face.

The need for, and appropriateness of, subjectivity in decision analysis has found wide acceptance (e.g. Raiffa, 1968; Anderson *et al.*, 1977; Morgan and Henrion, 1990; Clemen, 1996). As long ago as 1951, Arrow wrote that 'the uncertainty of the consequences ... is basically that existing in the mind of the chooser.' More recently, the authors of this chapter also focused on the importance of reliable probabilities for decision analysis in agriculture (Hardaker and Lien, 2010).

1.5.1 Sparse or zero data situations

When there are abundant data that are relevant and reliable, the subjective and relative frequency views of probability usually will coincide. However, it is not always recognized that accepting probabilities drawn from historical data involves a subjective belief that the historical data are relevant to the future, i.e. using probabilities based on historical data involves a seldom-examined assumption of *stationarity*. The subjectivist is more likely to question this assumption, which the objectivist usually will be obliged to adopt willy-nilly.

The difference between the two main views of probability becomes sharper when data are sparse or absent. Then, rejection of the subjective view can lead to some unfortunate mistakes, e.g. use of inappropriate but available data, ignoring the most important problems or variables for lack of data, use of cross-section data to reflect uncertainty over time (see Taylor and Zacharias, 2001; Just, 2003, p. 153). Yet the all-too-familiar situation is one in which abundant and relevant data are neither available nor obtainable at reasonable cost.

If data are sparse or absent, the subjectivist will seek the best-obtainable probability judgments about important uncertain variables to allow analysis to proceed, whereas the relative frequentist can do nothing, which is surely unsatisfactory. Hence, we propose:

Principle 8. When abundant, comprehensive and relevant data are not available or obtainable for important risky agricultural decisions, analysis must be based on carefully assessed subjective probabilities.

The implications of accepting this proposition are far-reaching. However, we have sought to draw out below only the main implications for applied decision analysis.

1.5.2 Making good probability judgments

Of course, subjective probabilities must obey the usual rules of probability theory. Moreover, they should be consistent with what the decision maker (or the decision analyst acting for that person) really believes about the uncertainty to be faced. Thinking rationally about probabilities means striving for consistency in one's whole network of beliefs. So, if some information is believed to be relevant to assessing the probability of some uncertain event, the assigned probability should reflect that belief in that information. Hence, we propose the following principle:

Principle 9. All probabilities for decision analysis should be thoughtfully and transparently obtained using the best available information, advice and well-considered and disclosed methods.

This means that possible sources of information should be considered and evaluated as to their suitability in supporting the formation of probability judgments. Sometimes, there are no existing useful data and none can be collected, as with risks that have not yet been experienced. Then, probabilities must be wholly subjective judgments by the decision maker, or those assisting that person. If some data are already available, their relevance, reliability and applicability to the assessment task at hand must be assessed carefully. Particularly when data are sparse or suspected to be biased, they will need to be combined with some subjective judgments before use. Below we deal in more detail with the two cases of no data and some data.

Morgan and Henrion (1990, Chapter 3) have emphasized the need to make probability assessments transparent. This need in public decision analysis is obvious, but even for private decisions, documenting what was done and assumed facilitates later review and offers better opportunities to learn from past failures and successes. The view that probabilities must be relative frequencies seems to have led to too many studies in which probability assumptions are inadequately documented and justified, or in which probabilities have been based on inappropriate data.

1.5.3 Subjective elicitation of probabilities

The subjective assessment of the probabilities usually needs to start with clarifying the nature of the uncertainty to be faced. This process, which is often demanding, may be assisted by use of influence diagrams or probability trees (Clemen, 1996, Chapters 4 and 7). Thereafter, several techniques are available for eliciting subjective probabilities, such as the equivalent lottery, visual impact and fractile methods (e.g. Clemen, 1996, Chapter 10; Hardaker *et al.*, 2015, Chapters 3 and 4).

The chief difficulty in eliciting probabilities is bias in the assessments. The psychological literature describes the various types of error that people typically make in such assessment tasks (see e.g. Plous (1993), summarized in Hardaker *et al.* (2015, pp. 33–35)). These include overconfidence in one's judgment (leading to too tight distributions), avoidance of uncertainty, misconception of chance, anchoring problems and problems due to motivational incentives. Yet there is rather little research (and almost none in agricultural economics) on how to overcome these problems.

Asking subjects for reasons and getting them to construct careful arguments in support of their judgments appear to reduce bias and improve the quality of the assessments (Morgan and Henrion, 1990, Chapter 6). Training assessors in advance of the main assessment also can help. A *proper scoring rule* is used to reward assessors who assign distributions to almanac-type questions so that their total expected scores are maximized only if they act consistently with what they really know about the questions posed (e.g. Savage, 1971). Offerman *et al.* (2009) and Hossain and Okui (2013) propose extension of the proper scoring rule (which requires risk-neutral respondents), by introducing scoring rules that allow for beliefs that deviate from risk neutrality, and Trautmann and van de Kuilen (2015) in an experiment compare different scoring rule methods.

With so little empirical evidence on how to minimize bias in subjective probability assessments, what is best is not clear. However, we propose the following general principle:

Principle 10. In seeking subjective probabilities for risk analysis, analysts should encourage assessors to think carefully about the nature of the uncertainties to be faced and should look for ways to motivate them to minimize bias in their judgments.

In this context, more research is needed to calibrate probability distributions obtained from decision makers using different elicitation methods to be able to improve these methods. Some bias in probability assessment appears to arise from failures of imagination, particularly about low probability events. It may help to brainstorm to bring into mind a more complete set of possible futures (although Kunreuther (2002) has suggested that people may not even want information about the risk of extreme events occurring). Moreover, if an assessor has had no experience with a particular outcome, that person may have difficulty assigning a meaningful probability to it. People also have difficulty discriminating between very small probabilities. A chance of 0.001 may seem not much different from 0.000001, yet the difference for risk management can be great. People can better grasp extreme probabilities if they are related to familiar scenarios described in realistic detail (Kunreuther *et al.*, 2001).

1.5.4 Use of experts

Normally the decision maker's personal probabilities should be used in decision analysis, but when that person is not well informed about the risk to be faced, seeking expert advice makes sense (Cooke, 1991). Use of experts may be particularly apposite in public decision-making. Moreover, if one expert is consulted, it might seem better still to consult several, provided each brings some unique insights to the assessment. Our next principle, therefore, is:

Principle 11. In cases in which good data are sparse or absent, it will be wise to seek probability assessments from people with relevant expert knowledge. Moreover, provided the assessments are made independently, using multiple experts is likely to give better results than consulting only one.

The proviso about the need for independence gives a clue to how experts should be selected and used. The aim should be to pick people who are knowledgeable about the risk to be faced but who have different professional backgrounds and experiences. Moreover, there will be diminishing returns from expanding the number of experts because the criterion of reasonable independence will become harder to fulfill.

It usually will be necessary to combine the assessments of the different experts into the single probability distribution(s) needed for analysis. A number of methods have been proposed, but it is not clear which is best (Clemen and Winkler, 1999). It usually makes sense, when possible, to start by pooling information among the experts to try to reduce differences between them in their perceptions. On the other hand, getting the experts together can be dysfunctional. For example, sometimes the group dynamics work to produce a more extreme and more overconfident assessment than any of the individuals would have suggested alone. The dynamics of a group can carry doubters with them, especially if one strong-willed or high-status (but ill-informed) individual is able to dominate the group. An infamous case of group dysfunction occurred in 2003 when a group of men around US President George Bush concluded that an invasion of Iraq was necessary because Saddam Hussein was believed to be in possession of weapons of mass destruction. The decision to invade was made despite the lack of evidence that such weapons were present in Iraq. No such weapons were found, yet the war cost many lives and led to major problems in Iraq over subsequent years (Badie, 2010).

Such evidence of possible dysfunction of groups led to the development of forms of information-sharing followed by reassessment in which the anonymity of group members is preserved, the best known and most widely used of which is the Delphi method (Linstone and Turoff, 2002).

These considerations lead to our next proposed principle:

Principle 12. Sharing information among assessors makes sense, so some form of feedback and reassessment has the potential to yield better probability assessments. However, analysts generally should seek to use group methods with feedback that are less likely to be dysfunctional, such as the Delphi method.

1.5.5 The consensus problem

Getting experts to share information can be expected to lead to some convergence of views, but a complete consensus is unlikely unless the members meet and resolve their differences. As noted, that is usually not a good idea. Consequently, the analyst must decide how to combine different probabilities (or probability distributions) from the different experts. For this purpose, there are two main options. First is some form of weighted average of the individual distributions. Different weighting schemes have been proposed, with the best usually being weights based on the past performance of the experts in similar assessment tasks (e.g. Clemen and Winkler, 1999). Second is use of Bayesian methods. Although these seem logical, there are operational problems relating chiefly to obtaining appropriate likelihoods. Although Jouini and Clemen (1996) have proposed an innovative solution in which dependence among sources is encoded into a copula, experimental evidence suggests that weighted averages often work about as well as more sophisticated methods (Clemen and Winkler, 1987). Hence, by recourse to Occam's razor, that seems the best choice for now.

Since the consensus problem is still not fully resolved, we have not proposed any principle relating to it. Instead, we urge analysts to become acquainted with the pros and cons of different existing methods, as reviewed by Clemen and Winkler (1999), and to watch for new developments.

1.5.6 Feedback to improve probability assessment

For events that occur repeatedly, it is possible to match an assessor's subjective probability judgments against observed relative frequencies. For example, if, on a number of occasions, an economist has specified a 30% chance that the world price of wheat will increase, wheat prices should have subsequently risen on about 30% of those occasions for the forecasts to be well calibrated. Poorly calibrated (unreliable) probability assessments imply some significant bias that should be corrected (Morgan and Henrion, 1990, Chapter 6). We therefore propose the following principle:

Principle 13. Whenever possible, subjective probabilities should be calibrated against actual outcomes so that assessors can be given feedback on their performance with a view to improving the reliability of their assessments.

Although feedback and training of assessors can be arduous, it should be useful, even essential, in fields such as economic and weather forecasting (Murphy and Winkler, 1977). Moreover, the more such studies are reported, the more we can learn about types and sources of bias in probability assessment, increasing the opportunities to mitigate these problems.

1.5.7 Using data in probability assessment

We now turn to the situation in which some data are available upon which to base the assessment of probabilities. Here it is important first to consider the reliability and the temporal and spatial relevance of the data to the assessment of the uncertainty at hand. If the data are likely to be biased, ways to try to correct the bias should be considered and applied. In the worst cases, it may be best to discard misleading information and to use more subjective assessments as described above.

Basing probability assessment on dubious data alone is clearly irrational. Hence, we suggest the following principle:

Principle 14. Sparse or suspect data should only be used for decision analysis in combination with a careful assessment of their relevance and of how to accommodate their imperfections.

Any measurement can be regarded as a sum of the true value, systematic error (measurement bias and sampling bias) and sampling error (e.g. Schlaifer 1959, Chapter 31). It is easy to underestimate the effect of little-known sources of error so that systematic error may not be recognized (Morgan and Henrion, 1990, Chapter 4). The sample selection method is crucial to minimize the systematic error. An example of potential sampling bias is use of experimental yield data for farm analysis. Performance under experimental conditions usually will significantly exceed that achieved under workaday farm conditions (Dillon and Anderson, 1990). A promising avenue for the future is, to a larger extent than available today, to use satellite data to estimate, e.g. yield forecast and uncertainties (e.g. Lobell *et al.*, 2007; Kamir *et al.*, 2020).

Agricultural production is fragmented and heterogeneous, yet many statistics are available only at some level of aggregation. As Just (2003) stresses, farm-level analyses should be based on micro-level data, so there is a need to collect panel data. Such data are needed for deriving probabilities for the risks that farmers face. Presumably, the lack of such micro-level time-series data has led some researchers into the error of using aggregated data to assess risks facing individual farmers (Just and Weninger, 1999).

1.5.8 Smoothing of empirical distribution

In cases with sparse data, or even when the data are relatively abundant, irregularities are usually present due to sampling errors. It is almost always reasonable to smooth out these irregularities in fitting a distribution (Schlaifer, 1959, Chapter 31; Anderson, 1974; Anderson *et al.*, 1977). The smaller the number of observations, the greater the need for smoothing (Whittle, 1957).

Before smoothing is attempted, all supplementary information that can make the process more trustworthy should be considered, especially when data are sparse. For example, it will usually be reasonable to assume that the upper and lower bounds of a distribution will be more extreme than those observed in a small dataset. Often, expert judgments can be used to get estimates on such bounds and perhaps also to estimate the modal value (Vose, 2000, Chapter 10).

With some exceptions, it is usually safe to assume that the marginal cumulative distribution function (CDF) of some continuous uncertain quantity is a smooth sigmoidal curve (Anderson, 1974; Anderson *et al.*, 1977, Chapter 2; Clemen,

1996, Chapter 10; Hardaker *et al.*, 2015, pp. 53–64). To fit a CDF, one option is to plot the data points using the fractile rule, and to draw a curve approximating these points by hand (Schlaifer, 1959, Chapter 31; Hardaker *et al.*, 2015, pp. 55–57). Hand smoothing gives the opportunity to incorporate additional information about the shape and location of the distribution. Non-parametric methods also can be used, such as spline and kernel methods (e.g. Whittle, 1957; Silverman, 1986). Or a theoretical distribution can be fitted to the data (e.g. Vose, 2000, Chapter 9; Hardaker *et al.*, 2015, Chapter 4), perhaps using software such as ModelRisk or Simetar. However, it may be unsafe to assume that an uncertain quantity conforms adequately to some convenient parametric distributional form. Especially if the risk is concentrated in one tail of the distribution, some tests of goodness of fit may be unsatisfactory.

1.5.9 Collecting more information to supplement sparse data

When data for probability assessment are inadequate, it is wise to consider collecting more information. Two options are to collect more empirical data or to seek probability assessments from experts, as discussed above. Yet there is a human tendency to limit the amount of information sought or used in choice (Morgan and Henrion, 1990, Chapter 6). These considerations lead us to Principle 15:

Principle 15. When data for probability assessment are inadequate, more information should be collected, whenever this is possible and justified.

Before taking steps to collect more information, one should be sure that the necessary information is not already available from another source. If not, then the expected value of the new information should be compared with the likely costs of collection (including the costs of delay).

1.5.10 Revision of probabilities

New information for decision analysis should permit better probability judgments. Sometimes the new data can simply be added to the existing data and the analysis can proceed using a larger sample. However, in other cases, the analyst is faced with using data of different types or from different sources. The evidence is that most people exhibit *conservatism* in this task and do not revise their prior probabilities sufficiently in light of the new evidence (e.g. Peterson and Beach, 1967). Sometimes Bayes' theorem can be used to overcome this problem, since it ensures that full weight is given to any additional information (Anderson *et al.*, 1977). Hence, we propose the following principle:

Principle 16. When possible and appropriate, Bayesian methods should be used to update prior probabilities in light of new evidence.

Bayes' theorem provides a logical mechanism for consistent processing of additional information. Brandford and Kelejian (1978) have shown how Bayesian updating can improve the quality of information used to construct forecasts of wheat prices. Kristensen (1993) illustrated how Bayesian updating can be applied in dynamic programming. However, an issue in Bayesian updating is how to get the likelihood probabilities that measure the reliability of the additional information. If there is no sound basis for the likelihoods, they must be subjective judgments. In that case, it may be as easy to assess the posterior probabilities directly, although a Bayesian calculation still may be useful as a cross-check.

1.5.11 Dealing with stochastic dependency

Most decision problems involve more than one stochastic variable. If stochastic dependency between two or more variables is ignored for ease in modeling, the distribution of the output variable(s) may be seriously astray. Hence, we propose the following principle:

Principle 17. Stochastic dependency among uncertain input variables will affect the distribution of consequences. Omitting such dependency in analysis may lead to erroneous results.

Stochastic dependency is seldom easy to assess and can be even harder to model effectively. Hence, it is common to find that dependency has been assumed away or, if accounted for, linear correlations have been assumed to represent the dependency, usually applied in conjunction with the assumption of a multivariate normal distribution. In reality, the dependency between uncertain variables may differ at different levels of those variables, and the normal distributions will often not represent agricultural phenomena well (e.g. Anderson, 1982; Ramirez *et al.*, 2003). Assuming normal distributions inappropriately may lead to appreciable errors in the tails of the distributions where the most serious risks often occur (e.g. insurance and derivative instruments, food safety risk). If the marginal distributions are not assumed to be normal (or lognormal), it is usual to draw correlated samples in Monte Carlo simulation accounting only for rank correlations. Again, because rank correlations do not tell the whole story (also measuring only linear dependency), the sampled values may be a poor representation of the 'true' stochastic dependency.

There are, of course, circumstances in which dependencies can legitimately be ignored, as when the dependency is weak, or is between variables with little influence on the outcomes of interest (e.g. Vose, 2000, Chapter 11).

When dependency must be addressed, several methods are available. If abundant relevant data are available, they can be allowed to 'speak' by using the historical information directly. Alternatively, appropriate statistics can be estimated from the data that summarize the joint distribution, including the dependencies.

If dependency is not a statistical artifact, there must be underlying causal factors. If these can be identified and their distributions and effects quantified, the processes causing dependency can be modeled. Burns and Clemen (1993) suggest the use of influence diagrams for this purpose. Hardaker *et al.* (2015, pp. 66–68) call the method the *hierarchy of variables* approach. Of course, a common problem with this approach is lack of data. For example, crop yield may be a function of a range of soil and weather variables that interact in complex and poorly understood ways.

The conditional distribution method has been used to model multivariate non-normal probability distributions for simulation and other risk analyses (e.g. Taylor, 1990). However, if one variable depends on many other variables (so that many conditional probabilities are required), the procedure is not operational.

Copula methods have become increasingly popular to describe and simulate multivariate distributions, especially in finance (Cherubini *et al.*, 2004). Given two or more marginal distributions, the copula function describes how they 'come together' to define the multivariate distribution. Any types of marginal distributions can be combined with an appropriate copula function. Moreover, the copula allows for Monte Carlo sampling from the implied joint distribution. If only sparse or no data are available, copula methods have also been proposed for the subjective assessment of joint probability distributions (e.g. Clemen and Reilly, 1999; Hardaker *et al.*, 2015, pp. 75–78).

There are many different copulas (Nelson, 1999). Some better account for the dependency in the tails of the stochastic variables than others. Copulas are estimated statistically from a dataset, so the method usually can be applied only when abundant data are available.

In the agricultural and resource economics literature, application of copula methods includes Taylor (1990), Richardson *et al.* (2000) and Goodwin and Hungerford (2015). We believe there is considerable scope for more research on the use of copulas in applied risk analysis in agricultural economics.

1.6 Conclusion

Our aim in this chapter has been to suggest how decision analysis in agricultural economics can be done better. Our list of principles is no doubt contentious because it challenges some established beliefs and practices. We suggest it is time for our profession to break down the facade of supposedly respectable 'objectivity' and to be more honest about the many subjective judgments that inevitably must be made in risk analysis. It is time to recognize that thoughtful and transparent subjectivity is the only sensible basis for the analysis of many important risky agricultural economic decisions.

Acknowledgments

We would like to thank David Hennessy for his comments, as they improved the chapter.

References

Allais, M. (1984) The foundations of the theory of utility and risk. In: Hargen, O. and Wenstrop, F. (eds) *Progress in Utility and Risk Theory*. Reidel, Dordrecht, Netherlands, pp. 3–131.

Anderson, J.R. (1974) Sparse data, estimational reliability and risk-efficient decisions. *American Journal of Agricultural Economics* 56, 564–572.

Anderson, J.R. (1976) Essential probabilistics in modelling. *Agricultural Systems* 1, 219–231.

Anderson, J.R. (1982) Agricultural economics, interdependency and uncertainty. *Australian Journal of Agricultural Economics* 26, 89–97.

Anderson, J.R. (1989) Reconsiderations on risk deductions in public project appraisal. *Australian Journal of Agricultural Economics* 33, 136–140.

Anderson, J.R. and Dillon, J.L. (1992) *Risk Analysis in Dryland Farming Systems*. Farming Systems Management Series No. 2, FAO, Rome.

Anderson, J.R. and Hardaker, J.B. (2003) Risk aversion in economic decision making: pragmatic guides for consistent choice by natural resource managers. In: Wesseler, J., Weikard, H-P. and Weaver, R. (eds) *Risk and Uncertainty in Environmental Economics*. Edward Elgar, Cheltenham, UK, pp. 171–188.

Anderson, J.R., Dillon, J.L. and Hardaker, J.B. (1977) *Agricultural Decision Analysis*. Iowa State University Press, Ames, Iowa.

Antle, J.M. (1983) Incorporating risk in production analysis. *American Journal of Agricultural Economics* 65, 1099–1106.

Arrow, K.J. (1951) Alternative approaches to the theory of choice in risk-taking situations. *Econometrica* 19, 404–437.

Arrow, K.J. (1965) *Aspects of the Theory of Risk-Bearing*. Yrjö Jahnssonin Säätiö, Academic Bookstore, Helsinki.

Arrow, K. and Debreu, G. (1954) Existence of an equilibrium for a competitive economy. *Econometrica* 22, 265–290.

Arrow, K.J. and Lind, R.C. (1970) Uncertainty and evaluation of public investment decisions. *American Economic Review* 60, 364–378. Comments by Mishan, E.J., McKean, R.N., Moore, J.H. and Wellington, D., reply by Arrow, K.J. and Lind, R.C. *American Economic Review* 62, 161–172.

Badie, D. (2010) Groupthink, Iraq, and the war on terror: explaining US policy shift toward Iraq. *Foreign Policy Analysis* 6, 277–296.

Bergevoet, R.H.M., Ondersteijn, C.J.M., Saatkamp, H.W., van Woerkum, C.M.J. and Huirne, R.B.M. (2004) Entrepreneurial behaviour of Dutch dairy farmers under a milk quota system: goals, objectives and attitudes. *Agricultural Systems* 80, 1–21.

Bernoulli, D. (1738) Specimen theoriae novae de mensura sortis, St Petersburg. *Commentarii Academiae Scientiarum Imperialis Petropolitanae* 5, 175–192. Translated by Somer, L. (1954) in: Exposition of a new theory on the measurement of risk. *Econometrica* 22, 23–36.

Binswanger, H.P. (1981) Attitudes towards risk: theoretical implications of an experiment in rural India. *Economic Journal* 91, 395–407.

Bocquého, G., Jacquet, F. and Reynaud, A. (2014) Expected utility or prospect theory maximisers? Assessing farmers' risk behaviour from field-experiment data. *European Review of Agricultural Economics* 41, 135–172.

Brandford, D.F. and Kelejian, H.H. (1978) The value of information for crop forecasting with Bayesian speculators: theory and empirical results. *The Bell Journal of Economics* 9, 123–144.

Brunton, R. (1995) We must adopt a risk-averse approach and always err on the side of caution when dealing with environmental issues. In: Bennett, J. (ed.) *Tall Green Tales*. Institute of Public Affairs, Perth, Australia.

Burns, W.J. and Clemen, R.T. (1993) Covariance structure models and influence diagrams. *Management Science* 39, 816–834.

Chambers, R.G. and Quiggin, J. (2001) *Uncertainty, Production, Choice, and Agency: The State-Contingent Approach*. Cambridge University Press, Cambridge, UK.

Chambers, R.G. and Quiggin, J. (2004) Technological and financial approaches to risk management in agriculture: an integrated approach. *Australian Journal of Agricultural and Resource Economics* 48, 199–223.

Chavas, J.P. (2008) A cost approach to economic analysis under state-contingent production uncertainty. *American Journal of Agricultural Economics* 90, 435–446.

Cherubini, U., Luciano, E. and Vecchiato, W. (2004) *Copula Methods in Finance*. Wiley Finance, Chichester, UK.
Clemen, R.T. (1996) *Making Hard Decisions: An Introduction to Decision Analysis*, 2nd edn. Duxbury Press, Belmont, California.
Clemen, R.T. and Reilly, T. (1999) Correlation and copulas for decision and risk analysis. *Management Science* 45, 208–224.
Clemen, R.T. and Winkler, R.L. (1987) Calibration and combining precipitation probability forecasts. In: Viertl, R. (ed.) *Probability and Bayesian Statistics*. Plenum, New York, pp. 97–110.
Clemen, R.T. and Winkler, R.L. (1999) Combining probability distributions from experts in risk analysis. *Risk Analysis* 19, 187–203. See also the Letter to the Editor by Kaplan, S. (2000) *Risk Analysis* 20, 155–156.
Cooke, R.M. (1991) *Experts in Uncertainty: Opinion and Subjective Probability in Science*. Oxford University Press, Oxford, UK.
Crowards, T.M. (1998) Safe minimum standards: cost and opportunities. *Ecological Economics* 25, 303–314.
Debreu, G. (1952) A social equilibrium existence theorem. *Proceedings of the National Academy of Sciences of the United States of America* 38, 886–893.
de Finetti, B. (1974) *Theory of Probability*. Wiley, New York.
Dillon, J.L. and Anderson, J.R. (1990) *The Analysis of Response in Crop and Livestock Production*, 3rd edn. Pergamon, Oxford, UK.
Dixit, A. and Pindyck, R. (1994) *Investment under Uncertainty*. Princeton University Press, Princeton, New Jersey.
Duke, J.M. and Ilvento, T.W. (2004) A conjoint analysis of public preferences for agricultural land preservation. *Agricultural and Resource Economics Review* 33, 209–219.
Edwards, W. (ed.) (1992) *Utility Theories: Measurements and Applications*. Kluwer, Boston, Massachusetts.
Ellsberg, D. (1961) Risk, ambiguity, and the Savage axioms. *The Quarterly Journal of Economics* 75, 643–669.
Gasson, R., Crow, G., Errington, A., Huston, J., Marsden, T. *et al.* (1988) The farm as a family business: a review. *Journal of Agricultural Economics* 39, 1–41.
Ghaffarian, S., van der Voort, M., Valente, J., Tekinerdogan, B. and de Mey, Y. (2022) Machine learning-based farm risk management: a systematic mapping review. *Computers and Electronics in Agriculture* 192, 106631.
Goodwin, B.K. and Hungerford, A. (2015) Copula-based models of systemic risk in US agriculture: implications for crop insurance and reinsurance contracts. *American Journal of Agricultural Economics* 97, 879–896.
Hardaker, J.B. and Lien, G. (2010) Probabilities for decision analysis in agriculture and rural resource economics: the need for a paradigm change. *Agricultural Systems* 103, 345–350.
Hardaker, J.B., Richardson, J.W., Lien, G. and Schumann, K.D. (2004) Stochastic efficiency analysis with risk aversion bounds: a simplified approach. *Australian Journal of Agricultural Economics* 48, 253–270.
Hardaker, J.B., Lien, G., Anderson, J.R. and Huirne, R.B.M. (2015) *Coping with Risk in Agriculture, Applied Decision Analysis*, 3rd edn. CABI International, Wallingford, UK.
Hossain, T. and Okui, R. (2013) The binarized scoring rule. *Review of Economic Studies* 80, 984–1001.
Iyer, P., Bozzola, M., Hirsch, S., Meraner, M. and Finger, R. (2020) Measuring farmer risk preferences in Europe: a systematic review. *Journal of Agricultural Economics* 71, 3–26.
Jouini, M.N. and Clemen, R.T. (1996) Copula methods for aggregating expert opinions. *Operations Research* 44, 444–457.
Just, R.E. (2000) Some guiding principles for empirical production research in agriculture. *Agricultural and Resource Economics Review* 29, 138–158.
Just, R.E. (2003) Risk research in agricultural economics: opportunities and challenges for the next twenty-five years. *Agricultural Systems* 75, 123–159.
Just, R.E. and Pope, R.D. (2001) Past progress and future opportunities for agricultural risk research. In: Just, R.E. and Pope, R.D. *A Comprehensive Assessment of the Role of Risk in U.S. Agriculture*. Kluwer, Boston, Massachusetts, pp. 557–577.
Just, R.E. and Weninger, Q. (1999) Are crop yields normally distributed? *American Journal of Agricultural Economics* 81, 287–303.
Kahneman, D. (2011) *Thinking, Fast and Slow*. Penguin, London.

Kahneman, D. and Tversky, A. (1979) Prospect theory: an analysis of decisions under risk. *Econometrica* 47, 263–291.
Kamir, E., Waldner, F. and Hochman, Z. (2020) Estimating wheat yields in Australia using climate records, satellite image time series and machine learning methods. *ISPRS Journal of Photogrammetry and Remote Sensing* 160, 124–135.
Kim, Y., Yu, J. and Pendell, D.L. (2020) Effects of crop insurance on farm disinvestment and exit decisions. *European Review of Agricultural Economics* 47, 324–347.
King, R.P. and Robison, L.J. (1984) Risk efficiency models. In: Barry, P.J. (ed.) *Risk Management in Agriculture*. Iowa State University Press, Ames, Iowa, pp. 68–81.
Kingwell, R.S., Pannell, D.J. and Robinson, S.D. (1993) Tactical responses to seasonal conditions in whole-farm planning in Western Australia. *Agricultural Economics* 8, 211–216.
Komarek, A.M., De Pinto, A. and Smith, V.H. (2020) A review of types of risks in agriculture: what we know and what we need to know. *Agricultural Systems* 178, 102738.
Kristensen, A.R. (1993) Bayesian updating in hierarchic Markov processes applied to the animal replacement problem. *European Review of Agricultural Economics* 20, 223–239.
Kunreuther, H. (2002) Risk analysis and risk management in an uncertain world. *Risk Analysis* 22, 655–664.
Kunreuther, H., Novemsky, N. and Kahneman, D. (2001) Making low probabilities useful. *Journal of Risk and Uncertainty* 23, 103–120.
Linstone, H.A. and Turoff, M. (eds) (2002) *The Delphi Method: Techniques and Applications*. New Jersey Institute of Technology, Newark, New Jersey.
Little, I.M.D. and Mirrlees, J.A. (1974) *Project Appraisal and Planning for Developing Countries*. Heinemann, London.
Lobell, D.B., Ortiz-Monasterio, J.I. and Falcon, W.P. (2007) Yield uncertainty at the field scale evaluated with multi-year satellite data. *Agricultural Systems* 92, 76–90.
Luong, Q.V. and Tauer, L.W. (2006) A real options analysis of coffee planting in Vietnam. *Agricultural Economics* 35, 49–57.
Machina, M.J. (1987) Choice under uncertainty: problems solved and unsolved. *Journal of Economic Perspectives* 1, 121–154.
Machina, M.J. and Siniscalchi, M. (2014) Ambiguity and ambiguity aversion. In: Machina, M.J. and Viscusi, W.K. (eds) *Handbook of the Economics of Risk and Uncertainty, Vol. 1*. North-Holland, Kidlington, UK, pp. 729–807.
Meuwissen, M.P.M., Huirne, R.B.M. and Hardaker, J.B. (2001) Risk and risk management: an empirical analysis of Dutch livestock farmers. *Livestock Production Science* 69, 43–53.
Meyer, J. (2001) Expected utility as a paradigm for decision making in agriculture. In: Just, R.E. and Pope, R.P. (eds) *A Comprehensive Assessment of the Role of Risk in U.S. Agriculture*. Kluwer, Boston, Massachusetts, pp. 3–19.
Morgan, M.G. and Henrion, M. (1990) *Uncertainty: A Guide to Dealing with Uncertainty in Quantitative Risk and Policy Analysis*. Cambridge University Press, New York.
Morris, J. (2002) The relationship between risk analysis and the precautionary principle. *Toxicology* 181–182, 127–130.
Murphy, A.H. and Winkler, R.L. (1977) Reliability of subjective probability forecasts of perceptions and temperature. *Applied Statistics* 26, 41–47.
Nauges, C., O'Donnell, C.J. and Quiggin, J. (2011) Uncertainty and technical efficiency in Finnish agriculture: a state-contingent approach. *European Review of Agricultural Economics* 38, 449–467.
Nelson, R.B. (1999) *An Introduction to Copulas*. Springer, New York.
O'Donnell, C.J. and Griffiths, W.E. (2006) Estimating state-contingent production frontiers. *American Journal of Agricultural Economics* 88, 249–266.
O'Donnell, C.J., Chambers, R.G. and Quiggin, J. (2010) Efficiency analysis in the presence of uncertainty. *Journal of Productivity Analysis* 33, 1–17.
Offerman, T., Sonnemans, J., Van de Kuilen, G. and Wakker, P.P. (2009) A truth serum for non-Bayesians: correcting proper scoring rules for risk attitudes. *The Review of Economic Studies* 76, 1461–1489.
Pannell, D.J., Malcolm, B. and Kingwell, R.S. (2000) Are we risking too much? Perspectives on risk in farm modelling. *Agricultural Economics* 23, 69–78.
Peterson, C.R. and Beach, L.R. (1967) Man as an intuitive statistician. *Psychological Bulletin* 68, 29–46.
Plous, S. (1993) *The Psychology of Judgment and Decision Making*. McGraw-Hill, New York.
Pratt, J.W. (1964) Risk aversion in the small and in the large. *Econometrica* 32, 122–136.

Quiggin, J. (1993) *Generalized Expected Utility Theory: The Rank Dependent Model*. Kluwer, Boston, Massachusetts.
Quiggin, J. and Chambers, R.G. (2006) The state-contingent approach to production under uncertainty. *Australian Journal of Agricultural and Resource Economics* 50, 153–169.
Rabin, M. (2000) Risk aversion and expected utility theory: a calibration theorem. *Econometrica* 68, 1281–1292.
Rabin, M. and Thaler, R.H. (2001) Anomalies: risk aversion. *Journal of Economic Perspectives* 15, 219–232.
Raiffa, H. (1968) *Decision Analysis*. Addison-Wesley, Reading, Massachusetts.
Ramirez, O.A., Misra, S. and Field, J. (2003) Crop-yield distributions revisited. *American Journal of Agricultural Economics* 85, 108–120.
Rasmussen, S. (2003) Criteria for optimal production under uncertainty: the state-contingent approach. *Australian Journal of Agricultural and Resource Economics* 47, 447–476.
Rasmussen, S. (2011) *Optimisation of Production under Uncertainty: The State-Contingent Approach*. Springer, Heidelberg, Germany.
Reutlinger, S. (1970) *Techniques for Project Appraisal under Uncertainty*. John Hopkins Press, Baltimore, World Bank Occasional Paper No. 10.
Ricci, P.F., Rice, D., Ziagos, J. and Cox Jr., L.A. (2003) Precaution, uncertainty and causation in environmental decisions. *Environmental International* 29, 1–19.
Richardson, J.W., Klose, S.L. and Gray, A.W. (2000) An applied procedure for estimating and simulating multivariate empirical (MVE) probability distributions in farm-level risk assessment and policy analysis. *Journal of Agriculture and Applied Economics* 32, 299–315.
Robison, L.J. and Barry, P.J. (1987) *The Competitive Firm's Response to Risk*. Macmillan, New York.
Samuelson, P.A. and Vickrey, W. (1964) Discussion – principles of efficiency. *American Economic Review* 54, 88–96.
Savage, L.J. (1954) *Foundations of Statistics*. Wiley, New York.
Savage, L.J. (1971) Elicitation of personal probabilities and expectations. *Journal of the American Statistical Association* 66, 783–801.
Schlaifer, R. (1959) *Probability and Statistics for Business Decisions*. McGraw-Hill, New York.
Schmeidler, D. (1989) Subjective probability and expected utility without additivity. *Econometrica* 57, 571–587.
Silverman, B.W. (1986) *Density Estimation and Data Analysis*. Chapman and Hall, New York.
Taylor, C.R. (1990) Two practical procedures for estimating multivariate nonnormal probability density functions. *American Journal of Agricultural Economics* 72, 210–217.
Taylor, C.R. and Zacharias, T.P. (2001) Programming methods for risk-efficient choice. In: Just, R.E. and Pope, R.D. (eds) *A Comprehensive Assessment of the Role of Risk in U.S. Agriculture*. Kluwer, Boston, Massachusetts, pp. 193–211.
Tegene, A., Wiebe, K. and Kuhn, B. (1999) Irreversible investment under uncertainty: conservation easements and the option to develop agricultural land. *Journal of Agricultural Economics* 50, 203–219.
Trautmann, S.T. and van de Kuilen, G. (2015) Belief elicitation: a horse race among truth serums. *The Economic Journal* 125, 2116–2135.
van Horn, R.L. (1971) Validation of simulation results. *Management Science* 17, 247–258.
von Neumann, J. and Morgenstern, O. (1944) *Theory of Games and Economic Behavior*. Princeton University Press, Princeton, New Jersey.
Vose, D. (2000) *Risk Analysis: A Quantitative Guide*, 2nd edn. Wiley, Chichester, UK.
Whittle, P. (1957) Curve and periodogram smoothing. *Journal of the Royal Statistical Society* 19, 38–63.

2 Risk Preferences, Price Uncertainties and Production Risk: A Survey[1]

Subal C. Kumbhakar[1,2*]

[1]*State University of New York at Binghamton, New York, USA;* [2]*Czech University of Life Sciences Prague, Prague, Czech Republic*

Abstract

This chapter deals with modeling of risk preferences when producers face uncertainties in output and input prices, in addition to production risk. All these uncertainty components are modeled assuming that the producers maximize expected utility of anticipated profit. Risk preference functions associated with these uncertainties are derived without assuming a specific form of the utility function. Moreover, no distributional assumptions are made on the distributions of the random variables representing price and production uncertainties. A multi-stage estimation procedure is discussed to estimate the parameters of the production function and risk preference functions associated with output price uncertainty, input price uncertainty and production risk. Production risk is specified in such a way that one can identify inputs with increasing, decreasing and constant production risks. Similarly, risk aversion behavior is specified in such a way that one can test for different types of risk aversion behavior. Different modeling strategies that have been used in the past are also discussed. We also consider the certainty equivalent approach.

2.1 Introduction

Risk is inherent in any production activities, more so in agriculture. United States Department of Agriculture (USDA) identifies different types of risk in agriculture. Among many of them production and price risks are dominant. 'Production risk derives from the uncertain natural growth processes of crops and livestock. Weather, disease, pests, and other factors affect both the quantity and quality of commodities produced. Price or market risk refers to uncertainty about the prices producers will receive for commodities or the prices they must pay for inputs. The nature of price risk varies significantly from commodity to commodity' (USDA, 2023). There are many other risks as well. In this chapter we focus on the two above-mentioned risks.

In the literature, there is a distinction between risk and uncertainty associated with any event. Following Knight, 'Randomness that is "susceptible of measurement" by statistical calculation, historical experience, or deduction is called risk. Uncertainty is defined residually as randomness that is not "susceptible of measurement" by statistical calculation, historical experience, or deduction. Practically speaking, risk involves randomness to which objective probability assessments can be attached, and uncertainty involves randomness to which objective probability assessments cannot be attached' (Chambers, 2022). However, here we

[1]This survey draws materials from some of the papers by the author and various co-authors.

*E-mail: kkar@binghamton.edu

 Managing Risk in Agriculture: A Development Perspective (eds A.K. Mishra, S.C. Kumbhakar and G. Lien).

DOI: 10.1079/9781800622289.0002

will not me making a strict distinction, although the situations described here are close to uncertainty according to Knight (1921).

Risk in production can be analyzed under (i) price uncertainty and (ii) production uncertainty (commonly known as production risk). Price uncertainty can further be addressed under either output price uncertainty (which is widely discussed in the theoretical literature; for example, see Sandmo (1971), Chambers (1983), Robison and Barry (1987)). Output price can be uncertain due to a variety of reasons. Perhaps the most important factor is the presence of a time lag between use of inputs and output produced. That is, produced output which is sold at a later date is likely to be different from the date when the production plan was made. Although no such lags exist in inputs, input prices might not be known in advance with certainty, especially when input markets are non-competitive and the quality of inputs differs. Finally, input quantities not only determine amount of output produced, but might also affect variability of output (often interpreted as risk). For example, some inputs might be risk augmenting while others might decrease output risk. Here we analyze the implications of the types of risk mentioned above under the assumption that producers maximize expected utility of anticipated profit. In particular, we examine input choice (input allocation decision) in the presence of price uncertainty (both output and inputs) and production risk. We also consider the certainty equivalent approach where producers choose input quantities by maximizing certainty equivalent profit. Since input demands (and therefore output supply) as well as own and cross-price elasticities and returns to scale are affected by the presence of uncertainty, no matter how uncertainty is modeled, it is desirable to take uncertainty into account.

One question that often arises is whether the production technology differs with risk. That is, whether there is a single technology irrespective of the level of risk or whether production risk alters the technology. We will be assuming that the mean technology is not risk-specific (as in the state-contingent production model) but the technology *ex-ante* depends on risk.

Our focus in this chapter is on agricultural production in which production risk plays an important role because of the risky environment. Econometric studies within agriculture focusing on risk and using a conventional productivity framework include Just and Pope (1979), Griffiths and Anderson (1982), Antle (1983), Antle and Goodger (1984), Wan and Anderson (1990), Kumbhakar (1993), Traxler *et al.* (1995), Battese *et al.* (1997), Regev *et al.* (1997), Di Falco *et al.* (2007), Shankar *et al.* (2008), Serra *et al.*, (2009), Tveterås *et al.* (2011) and Mishra *et al.* (2019).

First proposed by Sandmo (1971) and refined by Just and Pope (1979), the conventional formulation allows inputs to increase or decrease the variance of production or agricultural output. That is, riskiness is modeled in the variance of the error term in the production function. A weakness with this approach is that, since risk is defined in terms of the variance in production which is a function of the inputs, it is measured *ex-post*. But risk is *ex-ante* (something that has not happened yet), and uncertainty about outcomes may well influence input choices and hence *ex-post* output. So the observed (actual) output depends on the producer's attitude toward risk, which means input use decisions and therefore output/productivity are endogenous. The Just and Pope type of production function formulation does not take producers' attitudes to risk into account, since the mean output is not affected by the degree of risk perceived by a producer. The model does not consider endogeneity of inputs and therefore inputs are not affected by risk perceptions, which is clearly not realistic.

Chambers and Quiggin (2000), drawing on the work of Arrow and Debreu (1954), introduced the state-contingent production function. In this framework, output is contingent on the state of nature that eventuates, given the inputs applied. This is not a property of the conventional neoclassical production function, in which the role that inputs play remains the same regardless of which state occurs. Examples of empirical implementation of the state-contingent approach within agriculture include O'Donnell *et al.* (2010), Nauges *et al.* (2011), Hardaker and Lien (2014), Serra and Lansink (2014), and Serra *et al.* (2014). We will not survey this approach in this chapter.

There are several alternative ways, both to the Just and Pope framework and to the state-contingent production function framework, to model how risk-related variables affect productivity. These variables explain risk and are often labeled as the Z variables. One approach is to assume that the agents seek to maximize expected utility of profit in choosing levels of inputs and output. In this framework, one assumes a parametric profit function along with a production function to derive the input demand and output supply functions. This is the approach used in Ballivian and Sickles (1994). They estimated a system consisting of the input demand and output supply functions. This approach requires data on input and output prices, because the input demand and output supply functions depend on them. An alternative to the system of input demand and output supply equations is to consider the system consisting of the production function and the first-order conditions (FOC) of expected utility of profit maximization (Kumbhakar and Tveterås 2003).

Yet another alternative way is to treat the z variables similar to x and write the production function as y. An extension of this is to let risk indices (parametric functions of z variables) enter as inputs in the production function $(y = f(x, g(z)))$, as in Lien *et al.* (2017). But, a problem with both these approaches is that z is, mathematically speaking, similar to x and one cannot separate x from z.

Although the theoretical literature on risk is huge, there are relatively fewer empirical studies devoted to analyzing different types of risks. To examine producers' behavior under risk, some parametric forms of the utility function along with some specific distribution of the error term representing risk are considered in the existing literature. Here we consider a framework, which avoids both. Instead of specifying the utility function directly, we specify the absolute risk (AR) function, which in turn implies a specific utility function. The main advantage of working with the AR function is that it is not necessary to worry about the analytical solution for the utility function. Specific forms of the utility function are assumed, for example, in Love and Buccola (1991), Saha *et al.* (1994) and Chavas and Holt (1996) to derive the risk preference function. Choice of utility functions for which the risk preference function can be analytically derived is severely limited. In the present chapter the production function parameters, risk function (output risk) parameters and risk preference (associated with price and production uncertainties) of the producer (related to AR aversion function) are estimated either jointly or in a multi-step.

2.2 A Model with Only Production Risk

We start with the case with production risk and write the production technology as in the Just and Pope (1979) form, viz.,

$$y = f(x) + g(x)\varepsilon, \ \varepsilon : (0,1), \tag{2.1}$$

where y is output, x is a vector of variable inputs, ε is production risk and $f(x)$ is the mean output. Since output variance is represented by $g^2(x)$, the $g(x)$ function is labeled as the output risk function. An input j is said to be risk increasing (decreasing) if $\frac{\partial g(x)}{\partial x_j} > (<) 0$.

2.2.1 Exogenous inputs

An easy way of estimating the model is to assume x to be exogenous (Just and Pope, 1979; Tveterås, 1997). In this case, the model is simply heteroskedastic where the form of heteroskedasticity is dictated by the $g(x)$ function. To make it non-negative one can use an exponential form on it. The model can be estimated using a single or a two-step approach. In a two-step approach one estimates a parametric production function ignoring heteroskesdasticity. The absolute values of the residuals are then regressed on the $g(x)$ function to recover the parameters of it. Finally, the production function is re-estimated with heteroskedasticity correction. The estimated $g(x)$ can be used to calculate its derivative to find whether an input is risk increasing (decreasing).

2.2.2 Panel data model with exogenous inputs and inefficiency

Let the production process be characterized by

$$y = f(x;\alpha)e^{g(x;\beta)\varepsilon} \tag{2.2}$$

where y is output, x is a vector of inputs and ε is the error term. $f(x;\alpha)$ is the deterministic part of the production function. Production risk is captured by the $g(x;\beta)\varepsilon$ term. One important feature of the present model is that $f(x;\alpha)$ can be represented by a flexible functional form. For example, $\ln f(x;\alpha)$ can be a translog function which is linear in parameters although $f(x;\alpha)$ is not. Thus one can avoid using the non-linear estimation method in the first stage. Second, both $E(y)$ and $V(y)$ are affected by risk. Moreover, the specification allows positive as well as negative marginal risks because $g(.)$ is a linear function of the input variables.

Assuming that panel data is available, we specify ε as

$$\varepsilon_{it} = \tau_i + \lambda_t + v_{it} \qquad (2.3)$$

where i indexes firm ($i = 1,\cdots,N$) and t indexes time periods ($t = 1,\cdots,T$). τ is interpreted as time-invariant technical inefficiency, λ as time-specific effects and v as i.i.d. (independently and identically distributed) random noise specific to the firm and time period.

To derive the implications of the presence of risk on output, we rewrite Eqns 2.2 and 2.3 as

$$\ln y = \ln f(x;\alpha) + g(x;\beta)[\tau_i + \lambda_t + v_{it}] \qquad (2.4)$$

Note that the effect of technical inefficiency on log output is given by $g(x;\alpha)\tau_i$. Thus although τ is time-invariant, its effect on output is not. Similar is the case with λ_t: Second, the nature of heterogeneity across firms and over time depends on how one treats the error components and the form of the $g(.)$ function. Here we treat τ_i and λ_t as fixed (similar to the fixed effects (FE) model in the regular panel data models). Thus the only random component in the present model is v_{it}. Assuming v_{it} i.i.d. $N(0,1)$, the expected value of y, $E(y)$, and its variance, $V(y)$, are

$$E(y) = f(x;\alpha)e^{g^2(.)/2} \text{ and}$$
$$V(y) = f^2(.)e^{g^2(.)}[e^{g^2(.)/2} - 1] \qquad (2.5)$$

Thus $E(y) \geq f(x;\alpha)$ and marginal risk with respect to input i is

$$\partial V(y)/\partial x_i = 2f(.)\, e^{g^2(.)/2}\, [f_i(.)\, \{e^{g^2(.)} - 1\}$$
$$+ f(.)\, g(.)\, g_i(.)\, \{2\, e^{g^2(.)} - 1\}] \qquad (2.6)$$

where $f_i(.)$ and $g_i(.)$ are the partial derivatives of $f(.)$ and $g(.)$ functions with respect to x_i. It can be seen from Eqn 2.6 that marginal risk with respect to input i can be positive or negative depending on the sign and size of the $g(.)$ and $g_i(.)$ terms which vary with inputs, across firms and over time. The sign is unambiguously positive if $g(.)g_i(.) > 0$ and it is negative if $g(.)g_i(.) < 0$ and the second term under [.] is greater (in absolute value) than the first. (See Kumbhakar (1993) for estimation of this model).

2.2.3 Observed risk and exogenous inputs

If risk or risk perception variables are observed, one can include them in the model in several ways. We can include them in the production function directly, which means risk variables will be treated similarly to inputs treated as exogenous. If there are too many such variables, we can group them into fewer categories and use them in the production function as in Lien *et al.* (2017). The other alternative is to group the variables into several risk categories (labeled as risk indices) and use them in the production and/or input distance function. If there are Q such indices $T_q\left(z^1\right)$ defined as $T_q\left(z_i^q\right) = \left(\sum_{p=1}^{P_q} \varphi_p^q z_{pi}^q\right), \sum_{p=1}^{P_q} \varphi_p^q = 1 \forall q = 1,\cdots,Q$, these indices can be used in the production and/or input distance function and estimate the parameters in production and/or input distance function along with those in the risk indices using non-linear least squares. Note that in a translog function, the input elasticities and returns to scale will depend on the risk indices and therefore on the risk perception variables (see Lien *et al.* (2017) for details).

In another paper by Lien *et al.* (2022), risk perception and experience (z) are considered as environmental variables (facilitating inputs) which affect output via the inputs x. They first specify the production function as $y = A(z)f(x)$, where $A(z)$ is a non-parametric function which affects output neutrally. When the risk-related variables (z) are introduced in the intercept, they have a direct (neutral) impact on productivity, through a neutral shift in the production function. However, the z variables can affect y both neutrally and non-neutrally. For this, the z variables are introduced both in the intercept and in the slope parameters of the production function. The intercept captures the direct effect of z on y. The slope parameters capture the effect on y through the input elasticities – which constitute the indirect effect. This indirect effect results in a non-neutral shift of the production function and influences the marginal effects of the inputs.

The direct and indirect effects are captured in a semi-parametric smooth coefficient (SPSC) formulation (see Li *et al.* (2002) and Li and Racine (2010), for example). In a Cobb–Douglas (CD) model, $Y_i = X_i'\beta$, and the β parameters (the input elasticities) are non-parametric functions of Z. Note that the uppercase variables are in logarithms, except for some of the *z* variables. The CD SPSC model can be written as:

$$Y_i = \alpha(Z_i) + X_{1i}\beta_1(Z_i) + X_{2i}\beta_2(Z_i) + X_{3i}\beta_3(Z_i) + u_i, i = 1, \cdots, n, \tag{2.7}$$

where Y_i is the logarithm (log) of output (gross revenue) for farm i, X_{1i} is the log of land input, X_{2i} is the log of labor input, X_{3i} is the log of materials (materials) and Z_i is a vector of risk-related variables. In the Lien *et al.* (2017) paper three risk-related variables are used. These are: risk attitude (Z_1); perceptions of risk (Z_2); and risk management experience (Z_3). Finally, u_i is a random disturbance term, and the β parameters are smooth non-parametric functions to be estimated. Note that the model has a parametric structure in terms of the input variables. Such a model is very flexible because it allows the parameters to change with Z, making the parameters observation-specific if the Z variables are observation-specific. With this model, the overall marginal effect of the Z variables on output can be decomposed into a direct marginal effect and an indirect marginal effect. If all the β parameters are constants (i.e. $\beta_j(Z_i) = \beta_j, j = 1, \cdots, 3$), the model collapses to the partially linear model of Robinson (1988), where only the intercept, $\alpha(Z_i)$ is the non-parametric part of the model. On the other hand, if the β_j parameters, as well as the intercept, are linear functions of the Z variables, the model reduces to a restricted translog function (without the square and cross-products of the input and the Z variables). The model can take any parametric form of the $\beta_j(Z_i)$ and $\alpha(Z_i)$ functions, such as linear and quadratic (in log or in level). However, the main idea behind the SPSC model is not to assume any parametric form for them, although one can use any parametric form of $\beta_j(Z_i)$ and $\alpha(Z_i)$.

2.3 A Risk Model with Output Price Uncertainty and Production Risk

Now we consider a model that has production risk as well as uncertain output price. Following Kumbhakar (2002a) we assume that the objective of firms is to maximize expected utility of anticipated profit, $u(\pi)$, where $u(\pi)$ is a continuous and differentiable function of π. Anticipated profit, π, is defined as

$$\pi = p^e y - w \cdot x - C \tag{2.8}$$

where p^e is anticipated price of output, y. A firm's production technology is represented by $y = Af(x, z, t)$ in which x and z are variable and quasi-fixed input vectors, and t is the time trend variable. Heterogeneity in the production technology is introduced through the firm-specific parameter *A*. The parameters in $f(.)$ are common to all firms. The vector of variable input price is w and C denotes cost of fixed/quasi-fixed inputs. Following Zellner *et al.* (1966) we assume that $p^e = pe^\eta$, which implies that $E(p^e) = pE(e^\eta) = p$ when $E(e^\eta) = 1$ and p is the observed output price. Variance of e^η is assumed to be constant.

The FOC of expected utility of profit maximization can be written as

$$E(u'(\pi)\,\{Af_j(.)p^e - w_j\} = 0 \tag{2.9}$$

which can be expressed as

$$Af_j(.) = (w_j / p).(1/(1+\theta)) \equiv w_j / p^s \tag{2.10}$$

where $\theta = Cov(u'(\pi), e^\eta) / E(u'(\pi))$ is the risk preference function and $p^s = p(1+\theta)$. Note that the FOC in Eqn 2.10 are somewhat different from those derived and presented in Kumbhakar and Tveterås (2003). Although the risk preference function has the same interpretation, it is driven by production risk in Kumbhakar and Tveterås. Consequently, the effect of output price uncertainty and output risk on input demand is different. This difference can easily be seen by comparing the two sets of FOCs $f_j(.) = \frac{\partial f(.)}{\partial x_j}$. If producers are risk-neutral, θ would be zero. In such a case, maximization of expected utility of profit would be identical to maximization of expected profit.

2.3.1 Derivation of the risk preference function

Since we are interested in estimating the risk preference function, the most important task is to derive an algebraic form of θ that is easy to implement econometrically and imposes minimum restrictions on the structure of risk preferences on the individual producers. In general, the risk preference function θ defined above is a function of the parameters of the utility function, probability distribution of η, and data on inputs and output quantities. It is possible to derive an explicit form of θ if: (i) some particular forms of $u(.)$ are chosen or; (ii) some specific distributions on η are assumed. Choice between (i) and (ii) is very much limited in practice. For example, following Love and Buccola (1991), if one assumes $u(.)$ to be exponential and η normally distributed, or $u(.)$ expo-power and η follows the Weibull distribution (Saha *et al.*, 1994), it is possible to derive an analytical form for θ. Estimation, however, is a problem in both cases. The approach by Chavas and Holt (1996) does not require specification and estimation of θ directly. It does assume parametric utility and production functions. The method relies heavily on numerical methods, and therefore estimation can be a real burden if one wants to try more than one form of utility and production functions. Here we propose an alternative solution, which requires specification of neither any utility function (at least directly) nor any distributional assumption on η.

Proposition 1: If $u(.)$ is continuous and differentiable, and $u'(\pi)=u'(\mu_\pi+\sigma_\pi z)$ is approximated by a second order polynomial around $z=0$, then the risk preference function θ is:

$$\theta=\frac{-AR\beta^2 py+\frac{1}{2}DR\beta^3\gamma(py)^2}{1+\frac{1}{2}DR\beta^2(py)^2} \tag{2.11}$$

Kumbhakar and Tveterås (2003) derived a similar result based on a model that includes a Just and Pope (1979) type production risk. In the above formulation $\pi=pye^\eta-wx-C$ is rewritten as $\pi=\mu_\pi+\sigma_\pi z$, where the mean and standard deviation of profit are $\mu_\pi=\{py-wx-C\}$ and $\sigma_\pi=py\beta$. The random variable $z=(e^\eta-1)/\beta$ has a zero mean and unit variance where $\beta^2=var(e^\eta)$. AR is the Arrow–Pratt measure of AR aversion coefficient defined as $AR=-u''(\pi)/u'(\pi)$, and DR is the downside risk aversion coefficient (i.e. $DR=u'''(\pi)/u'(\pi)$). Finally, γ measures the skewness of e^η, i.e. $\gamma=E(z^3)$.

From an empirical point of view, the main advantage of the above result is that one needs to specify only the AR function to estimate θ (since $DR=\frac{\partial AR}{\partial\mu_\pi}+AR^2$). Consequently, a wide variety of parametric form on AR can be used without knowing the exact form of the underlying utility function. It is to be noted that all the parameters in θ can be identified, except when AR is a constant and DR ≈ 0.

Below we consider some important features of θ and several special cases that are used in the literature.

First, $\gamma=0$, which means that the distribution of output price is symmetric. Such a hypothesis can be empirically tested.

Second, constant AR aversion (CARA), decreasing AR aversion (DARA) and increasing AR aversion (IARA) hypotheses can be tested by specifying an appropriate form of AR. For example, if $AR=c_0+c_1\mu_\pi$ then $c_1=0$, positive and negative will imply CARA, IARA and DARA, respectively. Finally, if $c_1=c_0=0$ then producers are risk-neutral. Some other non-linear functional forms on AR can also be used to parameterize and test different forms of risk preferences. Although the AR function is linear the risk preference function is highly non-linear.

Third, since $DR=\frac{\partial AR}{\partial\mu_\pi}+AR^2$ it is not necessary to specify any functional form on DR, and no additional parameters are involved in DR. If AR is firm-specific, so is DR.

Fourth, the relative risk aversion function ($RR=AR.\mu_\pi$) is firm-specific with or without AR being firm-specific.

Fifth, all the parameters in θ are identified, including β and γ. This is also true even if $\gamma=0$. When DR is close to zero so that $\theta=-AR\beta^2py$, then β is not identified.

Sixth, it is possible to accommodate higher-order terms in the approximation of $u'(\pi)$ around $z=0$, and specify a variety of functional form for AR. Although a functional form on AR indirectly implies some functional forms for the utility function (Pratt, 1964), it is not necessary to have an analytical solution for the underlying utility function for every specification of AR.

If a firm is risk-averse, then $\theta<0$, which in turn implies that the relevant output price is less than the observed price (i.e. $p^s<p$). Consequently, the presence of output price uncertainty reduces demand for inputs and supply of output (Blair and Lusky, 1975) for a risk-averse firm. More generally, it can be seen from the FOC in Eqn 2.10 that input demand and hence output supply will depend on risk preferences, θ, as well as firm-effects, A. Thus, if firm-effects are assumed to be random, input variables in the

production function cannot be independent of the random firm components (in a one-way error component panel data model) of the production function. Consequently, the parameter estimates obtained from a single equation production function will be biased and inconsistent.

2.4 A Risk Model with Output Price and Input Price Uncertainties and Production Risk

Now we consider the case when both input and output prices are uncertain (Kumbhakar, 2001). We describe output price uncertainty by postulating anticipated price (p^e) as $p^e = pe^u$ (Zellner *et al.*, 1966) in which $E(e^u) = 1$ so that the expected value of p^e is the same as the observed price, p. The anticipated price differs from the observed price at a point in time because the production process is not always instantaneous, and the quantity of output cannot be perfectly predicted at the time the production decision is made. Similarly, input prices may not be known perfectly, especially if the input markets are non-competitive. Similar to anticipated output price, the anticipated input price vector (w^e) is assumed to be related to the observed input price vector w as $w^e = w.e^{\eta_j} \equiv (w_1 e^{\eta_1}, \ldots, w_J e^{\eta_J})$. Furthermore, we assume that $E(w^e) = w$, i.e. $E(e^{\eta_j}) = 1 \ \forall\ j$. For simplicity we assume that ε, u, are independent of η_j, and $E(\eta_j \eta_k) = 0 \ \forall\ j \neq k$.

To examine the effect of risk on optimal input quantities, we assume that producers maximize expected utility of anticipated profit $E(U(\pi^e))$. Define anticipated profit π^e as

$$\pi^e = p^e y - w^e.x = pf(x) - w.x + pf(x)\beta\left(\frac{e^u - 1}{\beta}\right) + pg(x)\sqrt{1+\beta^2}\left(\frac{e^u \varepsilon}{\sqrt{1+\beta^2}}\right) + \sum_j w_j x_j \gamma_j \left(\frac{e^{\eta_j} - 1}{\gamma_j}\right)$$
$$\equiv \mu_\pi + pf(x)\beta z_1 + pg(x)\sqrt{1+\beta^2} z_2 + \sum_j w_j x_j \gamma_j q_j \tag{2.12}$$

where $\mu_\pi = pf(x) - w.x$, $\beta^2 = Var(e^u)$, $\gamma_j^2 = Var(\eta_j)$, $z_1 = (\frac{e^u - 1}{\beta})$, $z_2 = (\frac{e^u \varepsilon}{\sqrt{1+\beta^2}})$, and $q_j = (\frac{e^{\eta_j} - 1}{\gamma_j})$. It is to be noted that z_1, z_2, and q_j, $j = 1, \ldots, J$ are (0,1) random variables. The FOC of expected utility of anticipated profit $E(U(\pi^e))$ maximization can be written as

$$E(U'(\pi^e)\{pf_j(x) - w_j + pf_j(x)\beta z_1 + pg_j(x)\sqrt{1+\beta^2} z_2 + w_j \gamma_j q_j\}) = 0 \tag{2.13}$$

Equation 2.10 can be expressed as

$$pf_j(x)\left(1 + \beta\frac{E(U'(\pi^e)z_1)}{E(U'(\pi^e))}\right) = w_j - g_j(x)p\sqrt{1+\beta^2}\left(\frac{E(U'(\pi^e)z_2)}{E(U'(\pi^e))}\right) - w_j\gamma_j\left(\frac{E(U'(\pi^e)q_j)}{E(U'(\pi^e))}\right)$$

We can further rewrite it to relate it to the case of output price uncertainty (e.g. Sandmo, 1971; Chambers, 1983) or production risk, viz.,

$$f_j(x)(1 + \beta\lambda) = \frac{w_j}{p} - g_j(x)\sqrt{1+\beta^2}\,\theta - \frac{w_j}{p}\gamma_j\psi_j \tag{2.14}$$

where

$$\lambda = \frac{E(U'(\pi^e)z_1)}{E(U'(\pi^e))}, \theta = \frac{E(U'(\pi^e)z_2)}{E(U'(\pi^e))}, \psi_j = \frac{E(U'(\pi^e)q_j)}{E(U'(\pi^e))} \tag{2.15}$$

The λ, θ and γ_j functions in Eqn 2.14 are risk preference functions associated with output price uncertainty, production risk and input price uncertainties, respectively. If producers are risk-averse then $\lambda < 0$ (i.e. an increase in u or z_1 increases π^e, which in turn reduces $U'(\pi^e)$ since $U''(\pi^e) < 0$ (utility function being concave)). A similar reasoning shows that $\lambda = 0$ when producers are risk-neutral (i.e. $U''(\pi^e) = 0$, which implies that utility function is linear), and if producers are risk-loving then $\lambda > 0$ (i.e. $U''(\pi^e) > 0$, which means that the utility function is convex). Finally, it can be shown, using similar arguments, that θ and ψ_j are negative if producers are risk-averse, positive for risk-loving producers and zero for risk-neutral producers.

The risk preference functions λ, θ and γ_j play an important role in input use decisions. This can be seen from Eqn 2.14 by expressing it as $f_j(x) = \frac{w_j}{p}\phi_j$, where $\phi_j = [1 - g_j(x)\theta\sqrt{1+\beta^2}\frac{p}{w_j} - \gamma_j\psi_j]/[1+\beta\lambda]$. For risk-averse producers $\phi_j < 1$, which means that producers do not equate marginal product (expected) of an input to the observed price of that input. If a producer is risk-averse then the value of (expected) marginal product of an input exceeds its price ($pf_j(x) > w_j$), thereby meaning that the producer is under-using the input relative to a risk-neutral producer ($\lambda = \theta = \gamma_j = 0$). Similarly, if producer A is more risk-averse than an otherwise identical producer B (i.e. ϕ_j for A is greater than ϕ_j for B), producer A will use less of input j than producer B. Thus, input demand functions (solution of x from Eqn 2.4 using the FOCs) will depend not only on observed prices but also on the risk preference functions. Consequently, anything that depends on the demand functions (e.g. own and cross-price elasticities, input productivity) will be affected by the presence of risk via λ, θ and γ_j. Since input demand functions are affected, output supply will also be affected, even if the producers share the same technology and face the same input and output prices.

Another implication of the first-order condition in Eqn 2.14 is that failure to take any one of the risk components is likely to make the model misspecified. For example, if only output price uncertainty is considered, as in Chambers (1983), while the producers also face input price uncertainty and production risk, the model represented by the first-order conditions in Eqn 2.14 will be imposing the restrictions $\theta = 0$ and $\psi_j = 0$. Imposing these restrictions without testing their appropriateness might make the model misspecified, and generate inconsistent parameter estimates. On the other hand, if one assumes only production risk, as in Eqn 2.1, the first-order conditions in Eqn 2.14 will also be misspecified because such a model imposes the restrictions $\beta = 0$ and $\gamma_j = 0$. Finally, the standard neoclassical model ignores all types of uncertainty ($\beta = \gamma_j = g_j(x) = 0$) by assuming $\phi_j = 1$. Consequently, the input demand functions derived from Eqn 2.14 by imposing these restrictions might be misleading, and the parameter estimates might be inconsistent.

2.5 Derivation of Risk Preference Functions with Production Risk, Input Output Price Uncertainties

Since we are primarily interested in estimating risk preferences, it is necessary to derive algebraic expressions for λ, θ and ψ_j in terms of observed data and unknown parameters. In doing so we try not to make the utility function restrictive, and at the same time try to avoid making distributional assumptions on z_1, z_2 and q_j. For example, if the utility function is quadratic in π^e, $U'(\pi^e)$ is linear in π^e and in such a case it is possible to derive analytical expressions for λ, θ and ψ_j. However, since flexible quadratic utility function implies $U''(\pi^e) = 0$, downside risk aversion cannot be captured by the quadratic utility function. Here we propose a solution, which does not require direct specification of the utility function.

Proposition 2: *If* $U'(\pi^e)$ is continuous and differentiable, and $U'(\pi^e) = U'(\mu_\pi + pf(x)\beta z_1 + pg(x)\sqrt{1+\beta^2}z_2 + \sum_j w_j x_j \gamma_j q_j)$ is approximated linearly around $z_1 = z_2 = q_j = 0$, then the risk preference functions λ, θ *and* Ψ_j are:

$$\lambda = \beta \frac{E(U'(\pi^e)z_1)}{E(U'(\pi^e))} = \frac{U''(\mu_\pi)}{U'(\mu_\pi)} pf(x)\beta = -AR(\mu_\pi)pf(x)\beta \tag{2.16}$$

$$\theta = \frac{E(U'(\pi^e)z_2)}{E(U'(\pi^e))} = \frac{U''(\mu_\pi)}{U'(\mu_\pi)} pg(x)\sqrt{1+\beta^2} = -AR(\mu_\pi)pg(x)\sqrt{1+\beta^2} \tag{2.17}$$

$$\psi_j = \frac{E(U'(\pi^e)q_j)}{E(U'(\pi^e))} = \frac{U''(\mu_\pi)}{U'(\mu_\pi)} w_j x_j \gamma_j = -AR(\mu_\pi)w_j x_j \gamma_j \tag{2.18}$$

where $AR(\mu_\pi) = -\frac{U''(\mu_\pi)}{U'(\mu_\pi)}$ is the Arrow–Pratt measure of AR aversion.

Proof: First-order Taylor series expansion of $U'(\pi^e)$ around $z_1 = z_2 = q_j = 0$ gives

$$\begin{aligned} U'(\pi^e) &= U'(\mu_\pi) + U''(\mu_\pi)pf(x)\beta z_1 \\ &+ U''(\mu_\pi)pg(x)\sqrt{1+\beta^2}z_2 \\ &+ U''(\mu_\pi)\sum_j w_j x_j \gamma_j q_j \end{aligned}$$

Thus,

$$\begin{aligned} E(U'(\pi^e)z_1) &= U''(\mu_\pi)pf(x)\beta, \\ E(U'(\pi^e)z_2) &= U''(\mu_\pi)pg(x)\sqrt{1+\beta^2}, \\ E(U'(\pi^e)q_j) &= U''(\mu_\pi)w_j x_j \gamma_j, \\ E(U'(\pi^e)) &= U'(\mu_\pi) \end{aligned}$$

which upon substitution in Eqn 2.15 gives Eqns 2.16, 2.17 and 2.18. One can include the second and higher-order terms in the Taylor series expansion to define λ, θ and ψ_j in a similar fashion.

Using these results, the FOC in Eqn 2.10 can be expressed as

$$f_j(x)[1 - ARpf(x)\beta^2] = \frac{w_j}{p}[1 + AR\gamma_j^2 w_j x_j] + g_j(x)(1+\beta^2)pg(x)AR \qquad (2.19)$$

A close look at the first-order condition in Eqn 2.10 shows that the focus of the problem has now been shifted from the utility function to the AR function. In addition to the mean production and risk functions, one needs to specify a functional form on AR which will define a system of J equations in J variable inputs (x) in Eqn 2.19. It is worth noting here that any specification of the AR function will indirectly imply some underlying utility function, such as $U = \int e^{-AR} d\mu_\pi$. Alternatively, the AR function gives all the information possessed by the utility function in Pratt (1964). The main advantage of working with the AR function is that one doesn't have to worry about: (i) what the underlying utility function is (which may not be always solvable analytically); (ii) derivation of λ, θ and ψ_j in Eqn 2.15 (which might not always give a closed-form solution); and (iii) the solutions of λ, θ and ψ_j using a specific utility function (which might not be easy to work with empirically). Furthermore, one can assume a functional form on AR that is flexible enough to test, first, whether producers are risk-neutral ($AR = 0$) or not. If not risk-neutral then we test for CARA, DARA and IARA hypotheses.

For testing hypotheses related to risk preferences, which can identify whether one or more sources of uncertainty are present in the data, one can estimate Eqn 2.19 and test singly or jointly parametric restrictions like $\beta = 0$, $\gamma_j = 0, \forall j$ and $g_j(x) = 0, \forall j$. If $\beta = 0$ then there is no uncertainty so far as output price is concerned. On the other hand, $\gamma_j = 0 \ \forall j$ would indicate no input price uncertainty. However, it is possible to have uncertainty in some but not in all input prices. This can be tested by specifying a subset of the $\gamma_j = 0$. Finally, absence of production risk can be tested by postulating $g_j(x) = 0 \ \forall j$ (which implies $g(x)$ is a constant), although it would be more interesting to test whether some inputs are risk increasing (decreasing) or not. If $\beta = \gamma_j = g_j(x) = 0$ then $f_j = w_j / p$ (i.e. $\phi_j = 1$) and we are back to the standard neoclassical case where input demand functions depend only on input and output prices. The same result is obtained if $AR = 0$. That is, if producers are risk-neutral then uncertainties in prices and output do not matter so far as input allocation is concerned.

Instead of specifying the AR function, one can work with the relative risk aversion function (RR), defined as

$$RR = -\frac{\partial \ln U'(\pi^e)}{\partial \ln \pi^e}\Big|_{\pi^e = \mu_\pi} = AR(\mu_\pi)\mu_\pi \qquad (2.20)$$

and rewrite Eqn 2.19 as

$$f_j = \frac{w_j}{p}\frac{[\mu_\pi + RR\gamma_j^2 w_j x_j]}{[\mu_\pi - RRpf(x)\beta^2]} + \frac{[RRg(x)g_j(x)p(1+\beta^2)]}{[\mu_\pi - RRpf(x)\beta^2]} \qquad (2.21)$$

Since the RR function gives as much information as the AR function, one could specify a parametric form of the RR function and use the system of equations in Eqn 2.21 to estimate all the parameters of the model.

To estimate the model outlined above, we need to assume a parametric form for the underlying production and risk functions, $f(.)$ and $g(.)$, respectively. The mean production function $f(.)$ is specified as linear quadratic and the risk function $g(.)$ as a Cobb–Douglas (see Kumbhakar and Tveterås (2003) for details), viz.,

$$\begin{aligned} f(\mathbf{x}) = {} & \alpha_0 + \sum_j \alpha_j x_j + 0.5\sum_i\sum_j \alpha_{ij} x_i x_j + \alpha_t t + \alpha_{tt} t^2 + \sum_j \alpha_{jt} x_j t \\ & + \alpha_{AGE} AGE + \alpha_{AGE} AGE^2 + \alpha_{AGE,t} AGE \cdot t + \\ & + \alpha_{q1} q_1 + \sum_j \alpha_{jq1} x_j q_1 + \alpha_{q1,t} q_1 t + \alpha_{q1,AGE} q_1 AGE \\ & + \alpha_{q2} q_2 + \sum_j \alpha_{jq2} x_j q_2 + \alpha_{q2,t} q_2 t + \alpha_{q2,AGE} q_2 AGE \end{aligned} \qquad (2.22)$$

In their paper, they used x_j = Feed (F), Fish (I), Capital (K), Labor (L), $g(\mathbf{x}) = (\prod x_j^{\beta_j}) \cdot AGE^{\beta_{AGE}} \cdot \exp(\beta_{q1} q_1 + \beta_{q2} q_2)$, $j = F, I, K, L$, $\alpha_0 = \sum_i \mu_i D_i$, where t is a time trend variable, AGE is farm age, D_i are farm dummies and q_1 and q_2 are labor quality dummy variables. The production function $Y = f(\mathbf{x}, \mathbf{q}, \mathbf{D}, t, AGE) + g(\mathbf{x}, \mathbf{q}, AGE)\varepsilon$ is rewritten as

$$Y / g(\mathbf{x}, \mathbf{q}, AGE) = f(\mathbf{x}, \mathbf{q}, \mathbf{D}, t, AGE) / g(\mathbf{x}, \mathbf{q}, AGE) + \varepsilon \qquad (2.23)$$

With the above functional forms for the $f(.)$ and $g(.)$ functions, the FOCs for feed and labor input (which are treated as variable inputs) are expressed as

$$\alpha_F + \alpha_{FF}x_F + \alpha_{FI}x_I + \alpha_{FK}x_K + \alpha_{FL}x_L + \alpha_{Ft}t + \alpha_{F.AGE}AGE + \alpha_{Fq1}q_1 + \alpha_{F.q2}q_2 = \frac{w_F}{p} - \theta\frac{g(\cdot)}{x_F}\beta_F \quad (2.24)$$

$$\alpha_L + \alpha_{FL}x_F + \alpha_{IL}x_I + \alpha_{KL}x_K + \alpha_{LL}x_L + \alpha_{Lt}t + \alpha_{L.AGE}AGE + \alpha_{L.q1}q_1 + \alpha_{L.q2}q_2 = \frac{w_L}{p} - \theta\frac{g(\cdot)}{x_L}\beta_L \quad (2.25)$$

A stochastic error term is appended in the above equations to accommodate optimization error, which is often labeled as allocative inefficiency. A linear quadratic AR aversion function $AR = \delta_1\mu_\pi + .5_2\mu_\pi^2$ is chosen. This gives an explicit parametric expression for the producers' downside risk aversion (DR). This particular form of AR allows us to test for several alternative forms of risk preferences. Despite the choice of a linear AR function, the risk preference function is highly non-linear. Equations 2.23–2.25 can be estimated jointly with or without any distributional assumptions on the error terms. Just and Pope (1979) considered estimation of only the production function in Eqn 2.23. Such a procedure can be justified based on the assumption that the realization of the production disturbance is not known to the producers when input decisions are made. However, estimation of Eqn 2.23 does not give estimates of the risk preference function θ. The advantage of estimating Eqns 2.24 and 2.25 jointly is that the optimality condition in feed and labor use is explicitly taken into account. Since the choice of feed and labor input quantity are affected by the presence of risk, use of the system approach can be estimated with and without distributional assumptions on the error terms.

Estimation of the above model can be done in three stages:

(i) Ignore heteroskedasticity in u (i.e. assume $g(.) = 1$) and estimate the mean output function $f(.)$ and the FOC for labor use equation simultaneously using three-stage least squares (3SLS). This approach gives consistent estimates of the production function parameters.

(ii) Regress the log of the absolute value of the residuals of the production function from the first stage, $\ln|\hat{u}|$, against the log of the $g(.)$ function to obtain estimates of the parameters of the $g(.)$ function (see Just and Pope (1979, pp. 74–81) for details).

(iii) The estimated $g(.)$ function is then used to correct for heteroskedasticity in u. Finally, the heteroskedasticity-corrected mean production function $f(.)$ is estimated together with the FOC as a system using iterated 3SLS procedure. The 3SLS estimates are then used to obtain revised estimates of the parameters of the risk function $g(.)$. The procedure can be repeated until convergence, but the estimates are consistent even without iteration. Kumbhakar and Tvetarås (2003) provides an application of this using Norwegian salmon farming.

2.6 Model with Inefficiency

Here we follow Kumbhakar (2002b) and consider two model specifications with only production risk but include technical inefficiency. In the additive model, an inefficiency term is added in the Just and Pope model. In the multiplicative model, the technical efficiency term appears in the mean output function in a multiplicative form. In both cases, risk preferences are introduced into the model via the FOC of expected utility maximization.

2.6.1 The additive model

We start with the formulation used by Battese *et al.* (1997) and Griffiths and Anderson (1982) in which technical inefficiency is introduced in an additive form in the production function

$$y = f(x,z) + g(x,z)\{\varepsilon - u\} \quad (2.26)$$

where y is output, x is a vector of J variable inputs, z is a vector of Q quasi-fixed inputs, ε is the error term (assumed to be *i.i.d.*(0,1)) representing production uncertainty and $u > 0$ represents technical inefficiency. If $u = 0$, the producer is technically efficient. Technical efficiency (TE) is thus defined as $TE = E(y|_{x,z,u}) / E(y|_{x,z,u=0}) = 1 - u.g(x,z)/f(x,z) \leq 1$. Technical inefficiency (TI), on the other hand, is the ratio of potential output loss, $E(y|_{x,z,u=0}) - E(y|_{x,z,u} = g(x,z)u)$ to $E(y|_{x,z,u=0})$, i.e. $TI = ug(x,z)/f(x,z)$. Thus, $TE = 1 - TI$. We use expected output conditional on u in defining TE and TI so that production uncertainty (ε) does not affect the measure of efficiency. This is important because production uncertainty is not in the control of any producer, and therefore it should not affect measured efficiency. The other reason is to maintain the relationship $TE = 1 - TI$, which is used as an identity in the efficiency literature. Note that both TE and TI depend on u and input quantities (x and z) in the present formulation. In this sense the present formulation of technical inefficiency deviates from the Aigner *et al.* (1977) formulation.

The mean output and production risk functions are defined at the frontier ($u = 0$). Thus, $E(y|_{u=0}) = f(x,z)$ is the mean output function, and $Var(y|_{u=0}) = g^2(x,z)$ is the risk function. Since the output variance is the square of the $g(x,z)$ function, it is considered as the output risk function.

As before, we assume that producers maximize expected utility of profit ($E[U(\Pi)]$) where $U(.)$ is the utility function which is assumed to be continuous, and the differentiable function of anticipated profit (Π), is normalized by output price p, defined as $\Pi = y - w.x - C$ where w is the prices of variable inputs relative to output price, and C is existing wealth, fixed costs or income from other sources (Chambers, 1983). Uncertainty in profit comes from production uncertainty (ε) as well as technical inefficiency (u). The FOC of the above problem can be expressed as

$$f_j(x,z) = w_j - \theta.g_j(x,z) + \lambda.g_j(x,z) + \eta_j \quad (2.27)$$

where $f_j(x,z) = \frac{\partial f(x,z)}{\partial x_j}$, $g_j(x,z) = \frac{\partial g(x,z)}{\partial x_j}$, $\theta = \frac{E[U'(\Pi)\varepsilon]}{E[U'(\Pi)]}$ and $\lambda = \frac{E[U'(\Pi)u]}{E[U'(\Pi)]}$. Note that θ and λ are not constants. Finally, η_j is the error term appended to the FOC which represents allocative inefficiency associated with input *j*.

The $f_j(x,z)$ term can be interpreted as the change in mean output for a unit change in the variable input x_j (marginal product of x_j defined in terms of the mean output function). Thus, the sign of $f_j(x,z)$ should be positive for all x_j . To interpret $g_j(x,z)$ we consider the variance function. The marginal production risk, defined as $\frac{\partial Var(y|_{u=0})}{\partial x_j} = 2g(x,z)g_j(x,z)$, can be positive or negative depending on the sign of $g_j(x,z)$. The variable input x_j is risk-increasing (risk-decreasing) if $g_j(x,z)$ is positive (negative). On the other hand, if $g_j(x,z) = 0$, input j is neither risk-increasing nor risk-decreasing.

Similar to the discussions before, risk preferences of the producers are captured by the θ and λ terms in Eqn 2.27. If producers are risk-averse, $\theta < 0$ (Chambers, 1983). On the other hand, λ will be positive if producers are risk-averse, because the effect of an increase in u on profit is the opposite of an increase in ε (when positive). If producers are risk-neutral, both θ and λ will be zero. If producers are fully efficient ($u = 0$), risk preference function is given by θ (Kumbhakar and Tveterås, 2003). Thus, adding technical inefficiency into the model adds another component to risk preferences.

It can be seen from Eqn 2.27 that input allocation is affected by the presence of technical inefficiency as well as production risk, via θ and λ . Thus, neglecting technical inefficiency, if any, is likely to give misleading information on input allocation and the predicted values of the risk preference functions. Consequently, predicted values of absolute, relative and downside risk aversion might all be invalid. On the other hand, neglecting production risk (assuming $g(x,z) =$ a constant, or a function of the z variables alone) is likely to give wrong estimates of technical inefficiency. The present model can test the presence of either technical inefficiency or production risk or both while taking producers' attitude toward risk into account.

2.6.2 The multiplicative model

The additive model of technical inefficiency in the Section 2.6.1 departs from the traditional stochastic frontier model (Aigner *et al.*, 1977) in which technical inefficiency is introduced in the mean output function in the form $f(x,z)\exp(-u)$. In view of this we now introduce technical inefficiency in a multiplicative form, and write the production technology as

$$y = f(x,z)\exp(-u) + g(x,z)\varepsilon \quad (2.28)$$

where as before $u > 0$ represents technical inefficiency. Defining TE and TI as before, we get $TE = \exp(-u)$ and $TI = u$. Note that in the present model TI and TE are independent of input quantities. This is a standard feature of the stochastic frontier models. The $g(x,z)\varepsilon$ term in Eqn 2.28 is kept in additive form to preserve the properties discussed in Just and Pope (1979).

Assuming that producers maximize expected utility of anticipated profit, the FOC can be expressed as

$$f_j(x,z).(1 - \lambda) = w - \theta.g_j(x,z) + \eta_j \quad (2.29)$$

where λ and θ are the same as before. The random term η_j represents allocative inefficiency that is associated with optimization error. It can be seen from Eqn 2.29 that risk preferences associated with production uncertainty and technical inefficiency are not additive as in Eqn 2.27. Apart from this difference in the form in which θ and λ appear in the FOC, the general conclusion from Eqn 2.29 is that both technical inefficiency and production uncertainty affect input allocation through the risk preference functions θ and λ. Consequently, omitting either θ or λ or both θ and λ is likely to give wrong results so far as allocation of variable inputs is concerned. (See Kumbhakar (2002b) for further details and estimation of these models.)

2.7 Certainty Equivalent Approach

2.7.1 With exogenous inputs

Following Freund (1956), the certainty equivalent (CE) of an outcome (income or revenue) in a risky situation can be approximated by:

$$CE = Y_1 - \frac{1}{2} \times ARA \times Var(Y_1 \mid \boldsymbol{X}) \tag{2.30}$$

In the context of production, Y_1 is gross revenue from outputs and $\boldsymbol{X}$ are inputs.

Empirically, one can use a risk-adjusted parametric translog function (where the risk-adjusted gross revenue is output). Additionally, one can use non-parametric production functions, with risk-adjusted (or CE) gross revenue as output. To make the CE approach operational, we proceed as follows.

1. Estimate the conditional mean for output, $E(Y \mid \boldsymbol{X})$, for each observation by non-parametric regression.
2. Calculate the conditional variance of output for each observation using the formula $Var(Y|\boldsymbol{X}) = [Y - E(Y|\boldsymbol{X})]^2$.
3. Use the risk attitude variable from the survey ($\boldsymbol{Z}$) to construct the relative risk-aversion coefficient with respect to wealth, $(CRRA)$, for each farmer.
4. Convert the relative risk aversion coefficient to the absolute risk aversion coefficient, (ARA), using the farmers' calculated/approximated wealth (W) level and the formula $ARA = CRRA / W$.
5. Calculate the CE using Eqn 2.30 and use it as the output variable in the production function below.

With the risk-adjusted parametric translog (TL) production function the model can be specified as (in a cross-sectional setup):

$$CE_i = TL_{CE}(\boldsymbol{X}_i) + u_i^{CE} \tag{2.31}$$

where $TL(.)$ stands for translog, CE_i is the logarithm of the risk-adjusted output for farm i ($i(i = 1, \cdots, n)$), X_i is a vector of inputs and u_i^{CE} is the noise term that reflects the effect of unobserved productivity shocks.

Similar to the parametric functions, the non-parametric risk-adjusted production can be specified as:

$$CE_i = f_{CE}(\boldsymbol{X}_i) + v_i^{CE} \tag{2.32}$$

where $f_{CE}(\boldsymbol{X}_i)$ is the non-parametric production function and u_i^{CE} is the noise term. All other variables are defined before.

2.7.2 Endogenous inputs with expected utility maximization

An alternative to the expected utility maximization approach is to consider the certainty equivalent profit maximization approach. Certainty equivalent profit π^* is defined as $\pi^* = E(\pi^e) - RP$ s, where RP is risk premium, which is the implicit cost of private risk taking. It is defined as the amount of money that would make the producer indifferent between uncertain profit π^e and certain profit $E(\pi^e) - RP$. In terms of expected utility, RP can be obtained from $E\{U(\pi^e)\} = U(E(\pi^e) - RP)$. That is, $RP = E(\pi^e) - U^{-1}\{E[U(\pi^e)]\}$.

Taking a first-order approximation of $U(E(\pi^e) - RP) = U(\mu_\pi - RP)$ around $RP = 0$, and a second-order approximation of $U(\pi^e)$ around μ_π, it can be easily shown that (Antle, 1983; Chavas and Holt, 1996):

$$RP = \frac{1}{2} AR(\mu_\pi) Var(\pi^e) \tag{2.33}$$

where $Var(\pi^e) = \{pf(x)\beta\}^2 + \{pg(x)\sqrt{1 + \beta^2}\}^2 + \sum_j w_j^2 x_j^2 \gamma_j^2$. Thus, certainty equivalent profit π is

$$\pi^* = E(\pi^e) - RP = \mu_\pi - \frac{1}{2}AR(\mu_\pi)[\{pf(x)\beta\}^2 + \left\{pg(x)\sqrt{1+\beta^2}\right\}^2 + \sum_j w_j^2 x_j^2 \gamma_j^2] \quad (2.34)$$

Now π^* can be maximized with respect to x_j to derive the optimal input quantities. The FOC for this problem, after a few lines of algebraic manipulations, can be expressed as

$$f_j(x)[1+\frac{1}{2}(DR-AR^2)Var(\pi^e) - ARpf(x)\beta^2] = \frac{w_j}{p}[1+\frac{1}{2}(DR-AR^2)Var(\pi^e) + ARw_jx_j\gamma_j^2] + AR[pg(x)g_j(x)(1+\beta^2)] \quad (2.35)$$

where $DR = U''(\mu_\pi)/U'(\mu_\pi)$ is a measure of downside risk aversion. Note that producers are averse to downside risk if they '... generally avoid situations which offer the potential for substantial gains but which also leave them even slightly vulnerable to losses below some critical level' (Menezes *et al.*, 1980, p. 921).

Because of different orders of approximation of the utility function, the FOC in Eqn 2.19 are different from those in Eqn 2.35. They are identical when AR is a constant (producers are CARA individuals) and $\varepsilon : N(0,1)$.

Although the risk preference functions λ, θ and ψ_j are not present in Eqn 2.35, the system of equations defined in Eqn 2.35 has the same parameters as in Eqn 2.10. The present system can be estimated only after some parametric forms of $f(x), g(x)$ and the AR functions are specified. In this sense the present system requires no additional information for estimation and provides no additional information compared to the system defined in Eqn 2.19. Although no risk premium was calculated for the original model, RP can be easily calculated once β^2, γ_j^2 and the parameters of the AR function are estimated. Similarly, once all the parameters are estimated, the λ, θ and ψ_j functions can also be evaluated.

Using the estimated parameters, one can obtain estimates of AR, RR and DR. Although λ, θ and ψ_j do not appear directly in Eqn 2.35, estimates of these risk functions can be obtained using the estimates of AR, β^2 and γ_j^2. Finally, estimates of RP can be obtained from Eqn 2.33.

2.8 Conclusion

In this chapter we surveyed models with uncertainty associated with output and input prices and analyzed their impacts on input allocation and efficiency under an expected utility of anticipated profit maximization hypothesis. Risk preference functions associated with each of these uncertainties are introduced without assuming a parametric utility function and without distributional assumptions on the random terms representing uncertainty of each type. It is shown that each of these risk components plays a role in input allocation decisions. Thus missing one or more of the risk components is likely to make the model misspecified. One can identify the sources of risk from a general model and test whether one or more of the uncertainties are present in the data.

The survey starts with some simpler models in which price uncertainties are ignored and inputs are assumed to be exogenous (not affected by production risk). We also consider models in which production risk or risk perception is observed in terms of some variables in the data. These variables are then either used in the production function as covariates or assumed to affect the parameters of the input variables non-parametrically. We discuss cases when inputs are treated as either exogenous or endogenous.

For estimation of the models with price uncertainties that affect inputs, we assume parametric functional form of the AR aversion function, instead of the utility function. Although a functional form on the AR implies a particular form of the utility function, the main advantage of working with the AR function is that one can use any functional form on it without worrying about the algebraic form of the underlying utility function. No additional assumptions are needed to estimate the risk preference functions since these functions are dependent on AR. We show that the risk preference functions are uniquely related to the AR function, the parameters of which are estimated jointly with those of the production function. Either a system approach or a multi-step approach can be used for estimation. Details on the estimation procedure are not discussed here and the readers are advised to go to the original papers.

References

Aigner, D.J, Lovell, C.A.K. and Schmidt, P. (1977) Formulation and estimation of stochastic frontier models. *Journal of Econometrics* 6, 21–37.

Antle, J.M. (1983) Testing the stochastic structure of production: a flexible moment-based approach. *Journal of Business & Economic Statistics* 1, 192–201.

Antle, J.M. and Goodger, W. (1984) Measuring stochastic technology: the case of Tulare milk production. *American Journal of Agricultural Economics* 66, 342–350.

Arrow, K.J. and Debreu, G. (1954) Existence of an equilibrium for a competitive economy. *Econometrica* 22, 265–290.

Ballivian, M.A. and Sickles, R.C., (1994) Product diversification and attitudes toward risk in agricultural production. *Journal of Productivity Analysis* 5, 271–286.

Battese, G.E., Rambaldi, A.N. and Wan, G.H. (1997) A stochastic frontier production function with flexible risk properties. *Journal Productivity Analysis* 8, 269–280.

Blair, R.D. and Lusky, R. (1975) A note on the influence of uncertainty on estimation of production function models. *Journal of Econometrics* 3, 391–394.

Chambers, R.G. (1983) Scale and productivity measurement under risk. *American Economic Review* 73, 802–805.

Chambers, R.G. (2022) Production under uncertainty. In: Ray, S.C., Chambers, R.G. and Kumbhakar, S.C. (eds) *Handbook of Production Economics.* Springer Nature, Singapore, pp. 575–610.

Chambers, R.G. and Quiggin, J. (2000) *Uncertainty, Production, Choice, and Agency*. Cambridge University Press, Cambridge, UK.

Chavas, J-P. and Holt, M. (1996) Economic behavior under uncertainty: a joint analysis of risk preferences and technology. *Review of Economics and Statistics* 78, 329–335.

Di Falco, S., Chavas, J-P. and Smale, M. (2007) Farmer management of production risk on degraded lands: the role of wheat variety diversity in the Tigray region, Ethiopia. *Agricultural Economics* 36, 147–156.

Freund, R.J. (1956) The introduction of risk into a programming model. *Econometrica* 24, 253–263.

Griffiths, W.E. and Anderson, J.R. (1982) Using time-series and cross-section data to estimate a production function with positive and negative marginal risks. *Journal of the American Statistical Association* 77, 529–536.

Hardaker, J.B. and Lien, G. (2014) An exposition of the state-contingent approach to the analysis of production under risk. In: *Proceedings of Australian Agricultural and Resource Economics Society Conference*. Port Macquarie, New South Wales.

Just, R.E. and Pope, R.D. (1979) Production function estimation and related risk considerations. *American Journal of Agricultural Economics* 61, 276–284.

Knight, F.H. (1921) *Risk, Uncertainty and Profit.* Houghton Mifflin, Boston, Massachusetts.

Kumbhakar, S.C. (1993) Production risk, technical efficiency, and panel data. *Economics Letters* 41, 11–16.

Kumbhakar, S.C. (2001) Risk preferences under price uncertainties and production risk. *Communications in Statistics – Theory and Methods* 30, 1715–1735.

Kumbhakar, S.C. (2002a) Risk preferences and technology: a joint analysis. *Marine Resource Economics* 17, 73–85.

Kumbhakar, S.C. (2002b) Specification and estimation of risk preferences, production risk and technical efficiency. *American Journal of Agricultural Economics* 84, 8–22.

Kumbhakar, S.C. and Tveterås, R. (2003) Production risk, risk preferences and firm-heterogeneity. *Scandinavian Journal of Economics* 105, 275–293.

Li, Q. and Racine, J.S. (2010) Smooth varying-coefficient estimation and inference for qualitative and quantitative data. *Econometric Theory* 26, 1607–1637.

Li, Q., Huang, C.J., Li, D. and Fu, T.T. (2002) Semiparametric smooth coefficient models. *Journal of Business & Economic Statistics* 20, 412–422.

Lien, G., Kumbhakar, S.C. and Hardaker, J.B. (2017) Accounting for risk in productivity analysis: an application to Norwegian dairy farming. *Journal of Productivity Analysis* 47, 247–257.

Love, H.A. and Buccola, S.T. (1991) Joint risk preference-technology estimation with a primal system. *American Journal of Agricultural Economics* 73, 765–774.

Menezes, C., Geiss, C. and Tressler, J. (1980) Increasing downside risk. *American Economic Review* 70, 921–932.

Mishra, A.K., Rezitis, A.N. and Tsionas, M.G. (2019) Estimating technical efficiency and production risk under contract farming: a Bayesian estimation and stochastic dominance methodology. *Journal of Agricultural Economics* 70, 353–371.

Nauges, C., O'Donnell, C.J. and Quiggin, J. (2011) Uncertainty and technical efficiency in Finnish agriculture: a state-contingent approach. *European Review of Agricultural Economics* 38, 449–467.
O'Donnell, C.J., Chambers, R.G. and Quiggin, J. (2010) Efficiency analysis in the presence of uncertainty. *Journal of Productivity Analysis* 33, 1–17.
Pratt, J.W. (1964) Risk aversion in the small and in the large. *Econometrica* 32, 122–136.
Regev, U., Gotsch, N. and Rieder, P. (1997) Are fungicides, nitrogen and plant growth regulators risk-reducing? Empirical evidence from Swiss wheat production. *Journal of Agricultural Economics* 48, 167–178.
Robinson, P.M. (1988) Root-*N*-consistent semiparametric regression. *Econometrica* 56, 931–954.
Robison, L.J. and Barry, P.J. (1987) *The Competitive Firm's Response to Risk*. Macmillan, New York.
Saha, A., Shumway, C.R. and Talpaz, H. (1994) Joint estimation of risk preference structure and technology using expo-power utility. *American Journal of Agricultural Economics* 76, 173–184.
Sandmo, A. (1971) On the theory of the competitive firm under price uncertainty. *American Economic Review* 61, 65–73.
Serra, T. and Lansink, A.O. (2014) Measuring the impacts of production risk on technical efficiency: a state-contingent conditional order-m approach. *European Journal of Operational Research* 239, 237–242.
Serra, T., Zilberman, D., Gil, J.M. and Featherstone, A. (2009) The effects of decoupling on land allocation. *Applied Economics* 41, 2323–2333.
Serra, T., Chambers, R.G. and Lansink, A.O. (2014) Measuring technical and environmental efficiency in a state-contingent technology. *European Journal of Operational Research* 236, 706–717.
Shankar, B., Bennett, R. and Morse, S. (2008) Production risk, pesticide use and GM crop technology in South Africa. *Applied Economics* 40, 2489–2500.
Traxler, G., Falck-Zepeda, J., Ortiz-Monasterio R.J. and Sayre, K. (1995) Production risk and the evolution of varietal technology. *American Journal of Agricultural Economics* 77, 1–7.
Tveterås, R. (1997) Econometric modelling of production technology under risk: the case of Norwegian salmon aquaculture industry. PhD dissertation. Norwegian School Economics and Business Administration, Bergen, Norway.
Tveterås, R., Flaten, O. and Lien, G. (2011) Production risk in multi-output industries: estimates from Norwegian dairy farms. *Applied Economics* 43, 4403–4414.
USDA (2023) Risk in Agriculture. US Department of Agriculture, Washington, DC. Available at: https://www.ers.usda.gov/topics/farm-practices-management/risk-management/risk-in-agriculture (accessed May 12, 2023).
Wan, G.H. and Anderson, J.R. (1990) Estimating risk effects in Chinese foodgrain production. *Journal of Agricultural Economics* 41, 85–93.
Zellner, A., Kmenta, J. and Drèze, J. (1966) Specification and estimation of Cobb-Douglas production function models. *Econometrica* 34, 784–795.

3 Subjective Probabilities and Farmers' Decision-Making in Developing Countries

Simone Cerroni[1,2]* and Ruggiero Rippo[1]
[1]*Department of Economics and Management, University of Trento, Trento, Italy;*
[2]*C3A, University of Trento, Italy*

Abstract
This chapter reviews studies exploring the role of subjective probabilities in explaining farmers' decision-making under risk and uncertainty in developing countries. These probabilities relate to a wide set of random variables, including climatic events, yield, price and income fluctuations. The chapter also provides a mapping of the most widely used methods to elicit subjective probabilities and a new framework to investigate the internal and external validity of the elicited probabilities.

3.1 Introduction

Farmers operating in developing countries are exposed to a wide variety of interrelated risks threatening farming activities and often the livelihood of entire households. These risks often are related to the following factors: (i) production and yield variability due to changing climatic conditions, plant or animal pests, and diseases and natural disasters; (ii) fluctuations of agricultural commodity prices and farm income related to macroeconomic factors and geopolitical instability; and (iii) variations of the operating cash flow that are often related to the use of credit (Komarek *et al.*, 2020). Most of the consequences related to these risks are characterized by a high degree of uncertainty, meaning that farmers have little or no knowledge of objective probabilities distributions (e.g. Manski, 2004; Trautmann and van de Kuilen, 2015).

Farmers can adopt many tools and strategies to cope with these risks. For example, they can diversify production (e.g., Chavas and Di Falco, 2012; Bozzola and Smale, 2020), adopt innovative technologies (e.g. Feder *et al.*, 1985; Janssen and Swinnen, 2019), enter contract farming (e.g. Bellemare and Bloem, 2018; Meemken and Bellemare, 2020) and buy insurance products (e.g. Miranda and Farrin, 2012; Mobarak and Rosenzweig, 2013). However, the benefits of these risk management strategies are often highly uncertain. This is particularly true in developing countries where farmers' access to information about the objective probability distribution of these benefits is more limited than in developed countries. Despite the profound changes in agricultural extension services over the years, the information delivery system is still inefficient, and farmers still have little access to information in many parts of the world (e.g. Aker, 2011; Norton and Alwang, 2020).

*E-mail: simone.cerroni@unitn.it

 Managing Risk in Agriculture: A Development Perspective (eds A.K. Mishra, S.C. Kumbhakar and G. Lien).
DOI: 10.1079/9781800622289.0003

When facing uncertain outcomes, farmers, like any other economic agent, rely on their beliefs or expectations to make decisions and maximize their utility (Savage, 1954). When expressed probabilistically, these beliefs or expectations are called subjective probabilities (e.g. Nelson and Bessler, 1989; Norris and Kramer, 1990; Hardaker and Lien, 2010; Cerroni, 2020). Other terms have also been used in the literature: probabilistic expectations (Manski, 2004), subjective expectations (Attanasio, 2009), subjective beliefs (Delavande *et al.*, 2011), beliefs (Schlag *et al.*, 2015; Charness *et al.*, 2021) and probabilistic beliefs (Marinacci, 2002; Cerroni *et al.*, 2019).

This chapter provides a critical review of the literature focusing on farmers' subjective probabilities and investigates their role in shaping farmers' decisions in developing countries. It also identifies gaps that future research could address.

Although there is a widespread agreement that studying subjective probabilities is vital to improve economists' understanding and predictions of economic agents' behavior, the number of studies eliciting farmers' subjective probabilities and investigating their role in shaping farmers' decisions is limited, especially in developing countries (see Delavande *et al.* (2011); Cerroni (2020) for a review). The paucity of studies eliciting and examining farmers' subjective probabilities in developing countries can be traced back to the concern that poor and illiterate individuals cannot understand the concept of probability (Delavande *et al.*, 2011; Delavande, 2014). However, recent empirical evidence suggests otherwise. Even in a context characterized by low literacy and low numeracy skills, farmers can demonstrate a sophisticated understanding of probabilities (e.g. Grisley and Kellogg, 1983; Delavande, 2014).

Since the early 1990s, a proliferation of methods to elicit subjective probabilities has been observed (Manski, 2004). The vast majority of studies eliciting farmers' subjective probabilities in both developed and developing countries used hypothetical direct questions paired with visual aids, which may be inadequate to capture truthful beliefs (see Delavande *et al.*, 2011; Cerroni, 2020). However, a number of methods to elicit subjective probabilities were developed within the decision analysis and experimental economic literature that were based on the use of incentivized and incentive-compatible mechanisms. In theory, these methods can elicit truthful beliefs (see Trautmann and van de Kuilen (2015); Schlag *et al.* (2015); Charness *et al.* (2021) for a review of these methods) but have never been used to elicit farmers' subjective probabilities in developing countries. In this chapter, we also propose a coherent set of criteria to explore the internal and external validity of farmers' subjective probabilities elicited in developing countries. This framework could facilitate the comparison of the quality of subjective probabilities retrieved using different elicitation mechanisms. These comparisons are rare in the current literature and are not based on a coherent and shared set of criteria.

The chapter is laid out as follows. Section 3.2 reviews the two main schools of thought on subjective probabilities. Section 3.3 provides a brief overview of the methods available to elicit subjective probabilities, and Section 3.4 proposes a framework to assess the internal and external validity of subjective probabilities. Section 3.5 presents a classification of the studies eliciting subjective probabilities in developing countries. This classification is based on the empirical domain under scrutiny. Section 3.6 provides the conclusions and areas for future research.

3.2 The Notion of Probability: Two Major Schools of Thought

Two main schools of thought exist regarding the definition and meaning of probability in economics. The *frequentist school* clearly distinguishes between the notions of risk and uncertainty. Under conditions of risk, definite numerical probabilities are known and can be objectively measured as the frequency of an event occurring in a number of homogeneous trials (Knight, 1921). Under conditions of uncertainty, the *frequentist school* speculates that economic agents could come up with probability judgments that are impossible to study and examine as they are unconscious and, therefore, unobservable (Feduzi *et al.*, 2014). In other words, there is no room for subjective probabilities, according to Knight and the *frequentist school.*

In contrast, the *subjectivist school*, led by de Finetti and Ramsey, argues that probabilities capture the personal degree of beliefs that an uncertain event will occur (Ramsey, 1931; de Finetti, 1931a, 1931b, 1933, 1937) and are by definition subjective. In his work, de Finetti, who holds a more radical perspective than Ramsey, speculates that objective probabilities do not exist, ruling out the distinction between risk and uncertainty proposed by Knight. Through the concepts of exchangeable events and coherence, de Finetti's notion of subjective probabilities is consistent with the basic axioms and rules of probability theory: (i) the probability for a given event is a number between 0 and 1; (ii) the sum of all probabilities of all the possible realizations is equal to 1; and (iii) probabilities for two mutually exclusive events must satisfy the additivity axiom. Furthermore, de Finetti's notion of subjective probabilities is intrinsically Bayesian (Galavotti, 1991, 2001). The idea of subjective probabilities, as formulated by de Finetti, is at the core of the subjective expected utility theory (SEUT) developed by Savage (1954). SEUT is considered the standard theory of decision-making under uncertainty for mainstream economists, and it is the natural development of the expected utility theory (EUT; von Neumann and Morgenstern, 1947) that explains economic agents' decision-making under risk.

3.3 Methods to Elicit Subjective Probabilities

Until recently, there was a widespread agreement that poor and illiterate individuals would struggle to understand and process the concept of probability (Delavande *et al.*, 2011). Hence, the vast majority of studies conducted in developing countries retrieved farmers' qualitative probability judgments instead of subjective probabilities. These judgments generally are elicited using Likert-scale questions asking farmers to state the likelihood of occurrence of given events on a scale from one (very unlikely) to five (very likely). In some cases, the upper bound of these scales could be seven or even nine. These methods are known in the literature under various appellatives: non-probabilistic methods (Delavande *et al.*, 2011), verbal encoding (Spetzler and Staël Von Holstein, 1975), or verbal questions (Manski, 2004, 2018). Eliciting qualitative judgments instead of subjective probabilities presents several limitations.

First, the use of Likert scales makes it difficult, if not impossible, to perform inter- and intrapersonal comparisons of elicited expectations. Empirical evidence suggests that different people interpret verbal probability statements differently and that probability judgments are context-dependent (Manski, 2004, 2018; McFadden *et al.*, 2005). In contrast, subjective probabilities are based on an absolute numerical scale that is comparable between subjects (Manski, 2004). In addition, the internal and external validity (using Cerroni *et al.*'s terminology, 2012) of subjective probabilities can be tested empirically (a more in-depth discussion is provided in Section 3.4). Second, while probability judgments cannot be incorporated in models of decision-making under risk and uncertainty, subjective probabilities can. If paired with observed choices (revealed preferences), subjective probabilities can improve our ability to understand and predict decision-making under risk and uncertainty (e.g. Manski, 2004; Hardaker and Lien, 2010; Fezzi *et al.*, 2021).

Recent empirical research has shown that valid subjective probabilities can be retrieved even in countries with low literacy and numeracy levels. Several studies demonstrated that farmers in developing countries understand and provide consistent answers to probabilistic questions (e.g. Grisley and Kellogg, 1983; Delavande *et al.*, 2011). The wide set of methods to elicit subjective probabilities can be classified into two main groups: direct and indirect methods (Spetzler and Staël Von Holstein, 1975). Direct methods, also known as introspection methods in the literature (Trautmann and van de Kuilen, 2015; Charness *et al.*, 2021), generally are implemented via surveys. These methods entail asking the respondents to directly state the probability that an event or a series of events (i.e. realizations of a random variable) will occur. These events could be binary or multiple (more than two) and could be represented as point estimates or intervals.

If the probabilities refer to mutually exclusive events, it is possible to infer a subjective probability or cumulative density distribution

(PDF or CDF) for the participants (Norris and Kramer, 1990). Direct methods are widely used to elicit farmers' subjective probabilities in developing countries and often are paired with the use of visual aids (e.g. Morgan *et al.*, 1990). For example, farmers are asked to allocate a number of stones, beans, or other familiar objects (out of n = total number of objects) in different buckets that represent different events (i.e. realizations of a random variable; e.g. Luseno *et al.*, 2003; Barrett and Santos, 2014). A limitation of direct approaches is the restricted amount of information that can be acquired unless participants state a subjective probability estimate for each of the possible realizations of the random variable (i.e. states of the world) under scrutiny. However, asking too many probability estimates can be cognitively demanding for respondents and trigger boredom or fatigue (Cerroni *et al.*, 2012). Another limitation is that respondents may not be able or willing to think in probabilistic terms and, therefore, may find it challenging to directly associate probabilities with different realizations of the random variable (see discussion in Weinstein and Diefenbach (1997)). Finally, direct methods have been criticized because of their hypothetical nature, which, in theory, undermines their incentive compatibility. In other words, participants have no incentive to report their 'true' subjective probabilities (Cerroni, 2020; Charness *et al.*, 2021).

Indirect methods can overcome some of these limitations. By asking respondents to make choices using lotteries and gambles, subjective probabilities are estimated indirectly by observing participants' behavior. Indirect methods are usually implemented within economic experiments in which participants receive a monetary reward that depends on the choices made during the tasks. Indirect methods are based on incentive-compatible belief elicitation mechanisms. Although there is a wide array of options, to date, only a limited number of studies have adopted indirect methods to elicit farmers' subjective probabilities (see Cerroni, 2020). Only two of these studies were conducted in developing countries (Grisley and Kellogg, 1983; Smith and Mandac, 1995). We briefly describe the most suitable approaches for agricultural research in Section 3.4, starting from methods characterized by a low degree of complexity and moving to very complex ones according to a scale developed by Charness *et al.* (2021).

The frequency method requires asking participants to report the occurrence frequency of different realizations of a random variable (i.e. states of the world). If the reported frequency matches exactly the empirical frequency (known a priori), participants are paid a fixed prize F. The method has been used extensively for random variables with two possible realizations (i.e. states of the world; e.g. Hurley and Shogren, 2005); however, it has the potential to be applied to problems in which the random variables have multiple possible realizations, as argued by Schlag and Tremewan (2021). A variation is the interval method. These two methods are equivalents except that, in the interval method, participants are paid a fixed prize F if the reported frequency approximately matches the empirical frequency (e.g. Charness and Dufwenberg, 2006). Both methods have a low complexity as the payment rule is quite simple to explain and peoples' ability to process probability is higher when these are expressed as frequencies (e.g. Gigerenzer and Hoffrage, 1995). These methods were never applied to elicit farmers' subjective probabilities in developing countries.

Another relatively low complexity method is the exchangeability method, which asks respondents to bet a given amount of money F on the occurrence of exchangeable events (*a la* de Finetti). A bi-sectioning process is used to create exchangeable events from the state space of the random variable under scrutiny. At the end of the experiment, one bet is randomly drawn, and respondents receive the prize F if the realization of the random variable falls within the exchangeable events they have bet on (e.g. Abdellaoui *et al.*, 2011; Cerroni *et al.*, 2012).

More complex methods are outcome matching and probability matching. In the former, participants are asked to make a series of pairwise choices (i.e. multiple price list format) between a certain payment and a lottery that pays a fixed prize F if and only if the realization of the random variable of interest is E. The certain payment increases from 0 to F (e.g. Heinemann *et al.*, 2009). One of these pairwise choices is randomly drawn and paid out at the end of the experiment. The subjective probability of the occurrence of event E is calculated indirectly by observing respondents' choice behaviors. This method is incentive-compatible if and only if participants are risk-neutral. However, *ex-post* correction mechanisms exist that allow the estimation of unbiased subjective probabilities for risk-averse (and risk-loving) subjects (see Trautmann and van de Kuilen, 2015).

In probability matching, respondents are asked to make a series of pairwise choices between two lotteries. One pays a fixed prize F if and only if the realization of the random variable of interest is E. The other pays the fixed price F with a probability p. The probability p increases from 0 to 1 (e.g. Baillon and Bleichrodt, 2015). The same approach can be implemented using the Becker–DeGroot–Marschak method (e.g. Holt and Smith, 2009) and second-price Vickrey auctions (Karni, 2009). We are not aware of studies using these methods to elicit farmers' subjective probabilities in either developed or developing countries.

Despite their very high level of complexity, proper scoring rules were widely used to retrieve subjective probabilities (Brier, 1950). In practice, respondents are asked to choose one binary lottery from a list of binary lotteries. These lotteries pay a prize F if the realization of the random variable of interest is E, otherwise, they pay a prize G. Each lottery is associated with a different subjective probability that the event E will occur, therefore, respondents' subjective probabilities can be inferred from their choices. Different proper scoring rules determine the prizes F and G. The most used proper scoring rules are the quadratic and the logarithmic (Charness *et al.*, 2021). Proper scoring rules are incentive-compatible only under risk neutrality; however, correction methods to retrieve unbiased estimates (e.g. Offerman *et al.*, 2009) and new elicitation mechanisms that mitigate the problem (e.g. binarized scoring rules, Hossain and Okui, 2013) have been developed over the years. Proper scoring rules can be extended to random variables with multiple realizations (see e.g. Harrison *et al.*, 2015, 2017). Only a small number of studies have implemented scoring rules to elicit farmers' subjective probabilities – a noticeable example is Cerroni (2020), who used quadratic scoring rules in a developed country. In the 1980s, two studies attempted to estimate farmers' subjective probabilities in developing countries using a linear scoring rule, which unfortunately is not incentive-compatible (Grisley and Kellogg, 1983; Smith and Mandac, 1995).

The methods presented so far share a common feature: participants are rewarded depending on how accurate their predictions are as compared with some empirical distributions. This implies that the random variable under study must sooner or later be verifiable. Unfortunately, in many instances, the random variable is not verifiable in the short term, for example, when we aim at eliciting subjective probabilities related to next year's crop prices, farm incomes, or temperature and rainfall levels (see discussion in Cerroni *et al.*, 2012). Two approaches can be used to resolve these types of uncertainty. On the one hand, it is possible to wait for the realization of the random variable and asks respondents to wait x months before they will receive the payment. However, this practice introduces another bias: respondents' choice behavior during the experiment can be influenced by the heterogeneity of time preferences in the sample. The other approach is to use results from simulation studies (e.g. Fiore *et al.*, 2009; Cerroni *et al.*, 2012) or experts' opinions (e.g. Cerroni, 2020) to resolve the uncertainty right away at the end of the experiment and pay respondents immediately. Using experts' opinions may be problematic as the researcher is *de facto* eliciting second-order subjective probabilities rather than first-order probabilities.

In other instances, the random variable under study is not verifiable at all (see discussion in Charness *et al.* (2021)), and hence the methods described above cannot be implemented. An example is the income loss due to exceptional events like volcano eruptions or earthquakes. Very complex methods such as Prelec's Bayesian Truth Serum (Prelec, 2004) and the choice-matching approach (Cvitanić *et al.*, 2019) can be used in these circumstances. A detailed description of these methods is beyond the scope of this chapter as their implementation in developing countries may be very problematic. We refer interested readers to the original works and other papers that have implemented these elicitation mechanisms to elicit preferences (not beliefs) in agricultural economics (e.g. Menapace and Raffaelli, 2020; Cerroni *et al.*, 2023).

In conclusion, indirect methods can elicit truthful subjective probabilities, but some come with a burdensome complexity. Respondents may fail to understand the logic behind the tasks to perform, and boredom and fatigue effects may arise (Norris and Kramer, 1990; Charness *et al.*, 2021). In addition, a recent paper indicates that complex belief elicitation mechanisms can generate biases in respondents' behavior that can undermine the incentive compatibility of elicited probabilities from a behavioral perspective (Danz *et al.*, 2022).

3.4 Evaluation of the Elicited Subjective Probabilities

The choice of the elicitation mechanism is not an easy task. A good starting point in this decision process is to consider the level of internal and external validity that each mechanism can guarantee. A few criteria have been developed in recent years (see Attanasio, 2009; Delavande *et al.*, 2011; Schlag *et al.*, 2015; Trautmann and van de Kuilen, 2015; Charness *et al.*, 2021) but, to date, a coherent and exhaustive framework has not been developed. In this chapter, we try to develop a coherent set of criteria that may be useful in assessing the quality of farmers' subjective probabilities.

Internal validity refers to the internal consistency (or coherence) of elicited subjective probabilities, and external validity refers to the generalizability of elicited subjective probabilities. We argue that two criteria can inform about the internal validity of elicited subjective probabilities:

A. Subjective probabilities' consistency with probability theory. For example, subjective probabilities should respect the additivity assumptions (i.e. sum up to one).
B. Subjective probabilities' ability to predict other choices that respondents make within the experimental context. In experimental settings, farmers may perform other tasks, for example, they can be asked whether or not they would purchase an insurance product. We expect that those who associate higher subjective probabilities with the occurrence of catastrophic weather events will be more likely to purchase the insurance product.

Regarding external validity, we envisage the possibility of using two criteria:

A. Subjective probabilities' ability to predict respondents' 'real life' choices. If revealed preference data on farmers' choice behavior is available, it would be possible to test whether elicited subjective probabilities are able to explain farmers' 'real life' decisions.
B. Subjective probabilities' consistency with previous empirical findings. It could be possible to test whether specific socio-demographics are correlated with elicited subjective probabilities in an expected way. For example, the literature suggests that men are more optimistic than women in several domains (e.g. Jacobsen *et al.*, 2014), therefore it is plausible to expect that subjective probabilities regarding the occurrence of a catastrophic weather event would be lower for men than for women.

Another criterion (C) that is commonly used to test external validity is the accuracy or reality check criterion: elicited subjective probability distributions are accurate when they are similar to historical evidence. However, under subjectivism, as formulated by de Finetti, there is no reason why subjective probabilities should correspond to objective probabilities (see also Manski's (2004) criticism of this approach).

Only a limited number of experimental studies compared the validity (internal and/or external) of subjective probabilities elicited via different mechanisms, and they were usually conducted in the lab with samples comprising students. Predictions refer to other participants' behavior in interactive games. We are not aware of any study comparing the validity of farmers' subjective probabilities elicited via different mechanisms in either developed or developing countries. We now report findings from studies comparing the performances of direct and indirect methods.

Regarding internal validity, Trautmann and van de Kuilen's study (2015) finds that subjective probabilities elicited using incentivized and incentive-compatible elicitation mechanisms (probability matching, outcome matching and quadratic scoring rules [QSR]) predict respondents' experimental choice behavior better than subjective probabilities elicited via direct methods (criterion B). Similarly, Wang (2011) and Massoni *et al.* (2014) find that quadratic scoring rules and probability matching outperform direct questions on the same ground, respectively. Cerroni *et al.* (2012) suggest that monetary incentives increase the internal validity of subjective probabilities elicited via the exchangeability method. This is the only study using a non-student sample and eliciting probabilities for a real-life outcome (i.e. the presence of pesticides in apples). However, empirical evidence is mixed, and it is impossible to establish whether indirect methods outperform direct methods in terms of internal validity. Some studies indicate that experimental methods are not superior to direct approaches (see Charness *et al.* (2021) for a review). For example, the very same study by Trautmann and van de Kuilen (2015) found that internal consistency measured by testing subjective probabilities additivity (i.e. validity or coherence) was not affected by the provision of monetary incentives and the use of incentive-compatible elicitation mechanisms (criterion A).

Pure experimental studies are generally not interested in testing the external validity of subjective probabilities using the criteria outlined in this chapter. Hence, the literature lacks studies directly comparing the performance of direct and indirect methods on this ground. However, a decent amount of works suggests that farmers' subjective probabilities are correlated with agricultural decisions in developing countries (criterion A) and are consistent with historical evidence (criterion C; see Delavande *et al.* (2011) for a review). More details on the topic are presented in Section 3.5.

In conclusion, as argued by Charness *et al.* (2021), we have to face and accept the fact that there is not enough reliable evidence to establish whether indirect methods outperform direct methods in terms of internal and/or external validity. Both approaches have limitations that preclude the elicitation of truthful subjective probabilities, and further research is surely needed, especially concerning farmers' beliefs in developing countries.

3.5 Eliciting Subjective Probabilities in Developing Countries

We review the studies that have elicited farmers' subjective probabilities in developing countries and categorize these studies into four groups depending on the nature of the random variable under scrutiny: climatic variables; innovation adoption; yield, price and income; and herd growth. The literature is not vast but is recent, denoting a relatively new interest in measuring subjective probabilities. For each study, we describe the method used, the sample size, the country where it was conducted, and the assessment of internal and external validity, if any (see Table 3.1 for a summary). Studies eliciting qualitative probability judgements are not considered here. It is interesting to note that the majority of studies used direct approaches to elicit subjective probabilities. Only two studies used incentivized linear scoring rules. To date, there are no studies on eliciting subjective probabilities of farmers using incentivized and incentive-compatible mechanisms in developing countries.

3.5.1 Climate and weather shocks

Subjective probability distributions associated with climatic variables may have a substantial influence on many farmers' decisions regarding planting, harvesting and timing of pesticide application. Luseno *et al.* (2003) and Lybbert *et al.* (2007) asked pastoralists in south Ethiopia and Kenya, respectively, to provide rain forecasts using a direct method. Participants were asked to allocate 12 stones into three piles, each representing a different level of precipitation of the coming rainy season (above normal, normal, or below normal). These studies also investigated belief updating by providing farmers with climate forecasts from computer-based climate models and found that only a limited group of respondents integrated and updated their beliefs accordingly with the information. Specifically, only below-normal rainfall expectations (i.e. pessimistic) were updated but not above-normal predictions (i.e. optimistic), suggesting an asymmetric belief updating process that is consistent with the so-called good news–bad news effect (i.e. people are more willing to accept good news than bad news; e.g. Moutsiana *et al.* (2013); Cerroni *et al.* (2019)).

Giné *et al.* (2009) elicited farmers' subjective probabilities on the onset of the monsoon in India. Using a direct method approach, they asked farmers to allocate ten stones in *n* different boxes that represented *n* different onset periods. Elicited subjective probabilities appeared to have

Table 3.1. Summary of studies used in the study. (Source: Authors' own work.)

Authors	Random variable	Elicitation method	Incentives	Sample size	Country	Internal validity	External validity
Luseno *et al.*, 2003	Climatic-rainfall	Direct methods paired with visual aids	No	323	South Ethiopia and north Kenya	–	–
Lybbert *et al.*, 2007	Climatic-rainfall	Direct methods paired with visual aids	No	245	Kenya	–	–
Giné *et al.*, 2009	Climatic-monsoon onset	Direct methods paired with visual aids	No	1054	India	–	A, C
Hill, 2009	Innovation adoption-price	Direct methods paired with visual aids	No	400	Uganda	–	A
Maertens, 2017	Innovation adoption-yield	Direct methods paired with visual aids	No	130	India	–	A, B, C
Tjernström *et al.*, 2021	Innovation adoption-yield	Direct methods paired with visual aids	No	200	Kenya	–	–
Grisley and Kellogg, 1983	Yield, price and income	Linear scoring rules	Yes	39	North Thailand	–	A, C
Smith and Mandac, 1995	Yield	Linear scoring rules	Yes	21	Philippines	–	A, C
Dillon, 2017	Yield and price	Direct methods paired with visual aids	No	300	Tanzania	–	C
Attanasio and Augsburg, 2016	Income	Direct methods paired with visual aids	No	1041	India	–	C
Barrett and Santos, 2014	Livestock herd growth	Direct methods paired with visual aids	No	116	Ethiopia	–	C

a high degree of external validity; in fact, these are highly correlated with farmers' decisions about the timing of crop planting activities (external validity criterion A). External validity was also tested by comparing subjective probability distributions with historical distributions (criterion C). The level of external validity using this criterion was high.

3.5.2 Innovation and insurance adoption

Adoption of innovation in developing countries remains low despite the availability of new and highly beneficial farming technologies. Farmers' decisions to innovate may be affected by expectations about future prices, yields and income associated with innovation. Hill (2009) explored whether farmers' subjective probabilities related to coffee prices influenced their decisions to buy price insurance in Uganda. Insurance products can be considered an innovation in Uganda, where the agricultural insurance market was virtually non-existent before the launch of the Uganda Agriculture Insurance Scheme (UAIS) in July 2016 (World Bank, 2019). Subjective probabilities were elicited using direct methods and visual aids. Farmers and traders were asked to allocate 20 beans among three squares representing different price categories (less than

US$0.10 [200 shillings], between US$0.10 and $0.20 [between 200 and 400 shillings] and more than US$0.20 [400 shillings]). Farmers' subjective probabilities were strongly anchored to the latest market information provided and were very important determinants of farmers' decisions to buy the insurance product. The latter result signals a potentially high external validity of elicited beliefs (criterion A).

Maertens (2017) investigated the role of social networks and learning on farmers' decisions to adopt an innovative type of genetically engineered cotton (i.e. Bt cotton) in India. The study also considered the impact that farmers' expectations about the yields of the new crop had on their decision to grow genetically modified (GM) cotton. Subjective probabilities related to this random variable were elicited using direct methods paired with visual aids. Following Lybbert and Just (2007), farmers were first asked to provide a state space for the random variable under investigation (a minimum and a maximum expected yield). Then, the state space was split into five intervals, and respondents were asked to allocate 20 stones among these intervals. Elicited subjective probabilities were characterized by a high degree of external validity based on three criteria: (i) elicited beliefs explain farmers' decisions to adopt the new crop (criterion A); (ii) elicited beliefs are correlated (in the expected direction) with socio-demographic characteristics (i.e. education) and farming strategies (i.e. irrigation; criterion B); and (iii) elicited beliefs are correlated with actual realizations of the random variable (criterion C).

Tjernström *et al.* (2021) investigated the influence of playing a farming-related simulation game on Kenyan farmers' subjective beliefs regarding the maize harvest obtained from adopting alternative improved inputs (i.e. fertilizers) as well as on farmers' real purchases of these inputs. The simulation game was played via an app delivering information on the maize harvest under different portfolios of input usage and different climatic conditions considering a virtual farm that resembled the holding of each participant. While the purchasing behavior was studied using an incentivized and incentive-compatible preference task, the subjective probabilities about expected yields were elicited using a direct method that is very similar to the one used by Lybbert and Just (2007) and Maertens (2017). Both subjective probabilities and purchasing behavior were investigated before and after farmers were exposed to the app. Results show that farmers updated their beliefs upward and revised their purchasing behavior after interacting with the application.

3.5.3 Yield, price and income

Beliefs about future yields, prices and income can have a substantial influence on farmers' decisions in developing countries, for example, on investments, crop diversification and input allocation choices (Delavande *et al.*, 2011; Delavande, 2014).

An early example is the study by Grisley and Kellogg (1983), who elicited farmers' subjective probabilities of yields, prices and net income of selected crops (i.e. glutinous rice tobacco, soybeans) in north Thailand using a linear scoring rule. Farmers were asked to provide a lower and upper bound for all random variables at harvest time. These state spaces were split into five intervals, and farmers were asked to distribute 25 coins in an allocation task. Farmers received a payment that depended on the number of coins allocated to the interval that would have realized in the future. Results indicated that elicited subjective probabilities were correlated with crop production behavior and were, in some cases, closely related to actual realizations of the random variables, suggesting that the study had a relatively high external validity (under criteria A and C). Smith and Mandac (1995) used an equivalent procedure in the Philippines to elicit farmers' subjective probabilities about the yields they thought they would have realized in the following season under three different levels of nitrogen use. External validity was tested by comparing subjective and actual probability distributions (criteria C). Although distributional means were similar, the variances were different. In both studies, farmers received pay that depended on their predictions and the realization of the random variable. Despite the provision of real incentives, linear scoring rules are not incentive-compatible (Winkler *et al.*, 1996).

Dillon (2017) explored whether the introduction of a subjective probability distribution into a stochastic model of production improved the understanding of farmers' behavior in Tanzania. Subjective probability distributions were elicited over time for two random variables: end-season yields and prices of cotton. Direct methods were used to elicit subjective probabilities. Specifically, farmers were asked to allocate 20 counters in ten different intervals for expected cotton yield and 14 counters in seven intervals for expected cotton price. External validity was tested using criterion C. Farmers' expectations about yields matched realized yields, while elicited price distributions showed a greater variance with a slight overestimation by the farmers.

The only study focusing on farmers' income expectations is by Attanasio and Augsburg (2016). They combined elicited subjective probability distributions about the future income of farmers and agricultural workers in rural India with data on actual income to characterize and estimate a model of household income dynamics. Subjective probabilities about next year's income variation were elicited by asking respondents to provide the state space of the random variable under study (i.e. lower and upper bound). The state space was then divided into four intervals characterized by three different cut-off points or thresholds. Respondents were asked about the probability that their income would be higher than the threshold in the coming year. The expected income levels were then compared with actual income realizations to test external validity (criterion C). The authors found that expected income varies in a way similar to actual income, showing a good degree of external validity of the elicited beliefs.

3.5.4 Livestock herd growth

A relatively recent study by Barrett and Santos (2014) used elicited subjective probabilities related to livestock herd growth conditional on rainfall states to simulate herd dynamics under different rainfall patterns in Ethiopia. The study shed light on the consequences that rainfall variability could have on the wealth and well-being of pastoralist households. The elicitation mechanism used is rather complex and consisted of two steps of direct questions. The first elicited subjective probabilities related to rainfall states in the next year implementing the same approach used by Luseno *et al.* (2003) and Lybbert *et al.* (2007). In the second stage, farmers were randomly assigned to different herd size ranges predefined by the researchers and asked to express their subjective probabilities regarding herd growth for the next year, assuming that the next year would be a bad year (similar to a major drought like 1999) or a good year. Again, the state space for this new random variable was retrieved from the respondents, and they were then asked to allocate 20 stones in interval herd sizes. In terms of external validity (criterion C), results demonstrate that simulations of herd dynamics based on elicited subjective probabilities were consistent with observed historical patterns of herd dynamics.

3.6 Conclusion

Subjective beliefs can influence on economic agents' decision-making. In particular, subjective probabilities (or probabilistic beliefs and expectations) are strong determinants of decision-making under risk and uncertainty (e.g. Spetzler and Staël Von Holstein, 1975; Manski, 2004). Farming is an inherently risky activity influenced by many interrelated sources of uncertainty, and many farmers' decisions are affected by subjective probabilities (e.g. Hardaker and Lien, 2010; Cerroni, 2020). In most developing countries, where the resolution of these uncertainties via information is challenging, farmers' subjective probabilities could be even more influential (Delavande, 2014). This chapter provides an overview of the studies and empirical applications eliciting farmers' subjective probabilities in developing countries with the final aim of identifying potential limitations and gaps in the literature that could be overcome and filled by further research. In particular, we categorized previous research into different groups depending on the nature of the random variables under study. This allows identification of the sub-domains in which the literature is less developed and further advances are needed. In this chapter, we put particular emphasis on the description of the methods that have been and could be used to elicit subjective probabilities in developing countries. In addition, we propose a new framework to evaluate the internal and external validity of subjective probabilities. This framework builds on previous research on the topic (Attanasio, 2009; Delavande *et al.*, 2011; Schlag *et al.*, 2015; Charness *et al.*, 2021) and could be interpreted as a toolbox to identify the most suitable methods to elicit farmers' subjective probabilities in developing countries.

Our review of the literature indicates that the number of studies looking at farmers' subjective probabilities in developing countries is small but growing. The majority of studies focused on expectations regarding future climate variability as well as future yields and price fluctuations. Sometimes these fluctuations were related to innovation that farmers could adopt, such as new inputs, crops and insurance products. The bulk of the research was conducted in Africa. All studies, except two, elicited subjective probabilities using direct methods that, by definition, are not incentive-compatible and hence may not be able to provide truthful beliefs. The only two studies applying indirect methods were based on a linear scoring rule which is a non-incentive-compatible mechanism. None of the literature tested internal validity, while there was widespread use of accuracy or reality checks to determine the interval validity of elicited subjective probabilities. This is rather peculiar because those who embrace the idea of subjectivism (see de Finetti's line of work) should not care much about whether subjective probabilities correspond to objective probabilities or not (see also Manski's (2004) criticism of this approach).

Based on our review, we try to define lines for further research. First, we believe there is an opportunity to test whether indirect approaches could be used to elicit farmers' subjective probabilities in developing countries. Since these approaches could be rather expensive, time-consuming and difficult to explain to participants, it could be wise to start with the least complex ones, like the frequency or the exchangeability method. Most importantly, future research should focus on testing whether these indirect approaches perform better in terms of internal and external validity when compared to direct approaches. Previous experimental research shows that this is not necessarily the case but further research is needed (e.g. Trautmann and van de Kuilen, 2015; Charness *et al.*, 2021).

Second, future research also should focus on exploring aspects related to the formation of subjective probabilities, as already suggested by Delavande *et al.* in 2011. Unfortunately, there has been little progress on this ground in the last 10 years. A better understanding of the mechanisms that drive the formation of farmers' subjective probabilities is fundamental to design personalized communication strategies that can facilitate innovation adoption or induce farmers to make more rational choices (i.e. crop diversification, input allocation). A noticeable example is a recent study by Freudenreich and Kebede (2022), who explored how previous experience with a wide range of shocks influences the formation of farmers' subjective probabilities in Kenya. Unfortunately, the study focused on non-probabilistic beliefs.

Finally, further research should be devoted to incorporating subjective probabilities into formal models developed to predict farmers' decisions in developing countries. A noticeable attempt was produced by Barrett and Santos (2014), who incorporated elicited subjective probabilities of livestock herd growth into a model predicting herd dynamics under different climatic scenarios. The potential for further research is huge thanks to the advances in mathematical programming, agent-based modeling and artificial intelligence experienced in recent years (e.g. Colen *et al.*, 2016). A noticeable example is a recent study by Huber *et al.* (2022), who attempted the introduction of behavioral factors in a standard bio-economic modeling of agricultural production.

References

Abdellaoui, M., Baillon, A., Placedo, L. and Wakker, P.P. (2011) The rich domain of uncertainty: source functions and their experimental implementation. *American Economic Review* 101, 695–723.

Aker, J.C. (2011) Dial 'A' for agriculture: a review of information and communication technologies for agricultural extension in developing countries. *Agricultural Economics* 42, 631–647. DOI: 10.1111/j.1574-0862.2011.00545.x.

Attanasio, O.P. (2009) Expectations and perceptions in developing countries: their measurement and their use. *American Economic Review* 99, 87–92. DOI: 10.1257/aer.99.2.87.

Attanasio, O. and Augsburg, B. (2016) Subjective expectations and income processes in rural India. *Economica* 83, 416–442. DOI: 10.1111/ecca.12190.

Baillon, A. and Bleichrodt, H. (2015) Testing ambiguity models through the measurement of probabilities for gains and losses. *American Economic Journal: Microeconomics* 7, 77–100.

Barrett, C.B. and Santos, P. (2014) The impact of changing rainfall variability on resource-dependent wealth dynamics. *Ecological Economics* 105, 48–54.

Bellemare, M.F. and Bloem, J.R. (2018) Does contract farming improve welfare? A review. *World Development* 112, 259–271. DOI: 10.1016/j.worlddev.2018.08.018.

Bozzola, M. and Smale, M. (2020) The welfare effects of crop biodiversity as an adaptation to climate shocks in Kenya. *World Development* 135, 105065. DOI: 10.1016/j.worlddev.2020.105065.

Brier, G.W. (1950) Verification of forecasts expressed in terms of probability. *Monthly Weather Review* 78, 1–3. DOI: 10.1175/1520-0493(1950)078<0001:VOFEIT>2.0.CO;2.

Cerroni, S. (2020) Eliciting farmers' subjective probabilities, risk, and uncertainty preferences using contextualized field experiments. *Agricultural Economics* 51, 707–724. DOI: 10.1111/agec.12587.

Cerroni, S., Notaro, S. and Shaw, W.D. (2012) Eliciting and estimating valid subjective probabilities: an experimental investigation of the exchangeability method. *Journal of Economic Behavior & Organization* 84, 201–215.

Cerroni, S., Notaro, S. and Raffaelli, R. (2019) Beliefs and preferences for food-safety policies: a discrete choice model under uncertainty. *European Review of Agricultural Economics* 46, 769–799. DOI: 10.1093/erae/jby038.

Cerroni, S., Derbyshire, D., Hutchinson, W.G. and Nayga, R.M. (2023) A choice matching approach for discrete choice analysis: an experimental investigation in the lab. *Land Economics* 99, 80–102. DOI: 10.3368/le.012621-0009R1 (in press).

Charness, G. and Dufwenberg, M. (2006) Promises and partnership. *Econometrica* 74, 1579–1601. DOI: 10.1111/j.1468-0262.2006.00719.x.

Charness, G., Gneezy, U. and Rasocha, V. (2021) Experimental methods: eliciting beliefs. *Journal of Economic Behavior & Organization* 189, 234–256. DOI: 10.1016/j.jebo.2021.06.032.

Chavas, J-P. and Di Falco, S. (2012) On the role of risk versus economies of scope in farm diversification with an application to Ethiopian farms. *Journal of Agricultural Economics* 63, 25–55. DOI: 10.1111/j.1477-9552.2011.00319.x.

Colen, L., Gomez y Paloma, S., Latacz-Lohmann, U., Lefebvre, M., Préget, R. *et al.* (2016) Economic experiments as a tool for agricultural policy evaluation: insights from the European CAP. *Canadian Journal of Agricultural Economics* 64, 667–694. DOI: 10.1111/cjag.12107.

Cvitanić, J., Prelec, D., Riley, B. and Tereick, B. (2019) Honesty via choice-matching. *American Economic Review: Insights* 1, 179–192. DOI: 10.1257/aeri.20180227.

Danz, D., Vesterlund, L. and Wilson, A.J. (2022) Belief elicitation and behavioral incentive compatibility. *American Economic Review*, 12, 2851–2883.

de Finetti, B. (1931a) Sul significato soggettivo della probabilità. *Fundamenta mathematicae* 17, 298–329. Translated into English as: On the subjective meaning of probability, in: Monari, P. and Cocchi, D. (eds) *Probability and Induction*, Clueb, Bologna, Italy, pp. 291–321.

de Finetti, B. (1931b) Probabilismo. *Logos* 14, 163–219, reprinted in B. de Finetti (1989) *La logica dell'incerto*, Il Saggiatore, Milan, Italy, pp. 3–70. Translated into English as: Probabilism. A critical essay on the theory of probability and on the value of science, *Erkenntnis* 31, 169–223.

de Finetti, B. (1933) Sul concetto di probabilità. *Rivista italiana di statistica, economia e finanza* 5, 723–747.

de Finetti, B. (1937) La prévision: ses lois logiques, ses sources subjectives. *Annales de l'Institut Henri Poincaré* 7, 1–68. Translated into English as: Foresight: its logical laws, its subjective sources, in: Kyburg, H.E. and Smokler, H.E. (eds) *Studies in Subjective Probability*, John Wiley & Sons, New York, pp. 93–158.

Delavande, A. (2014) Probabilistic expectations in developing countries. *Annual Review of Economics* 6, 1–20. DOI: 10.1146/annurev-economics-072413-105148.

Delavande, A., Giné, X. and McKenzie, D. (2011) Measuring subjective expectations in developing countries: a critical review and new evidence. *Journal of Development Economics* 94, 151–163.

Dillon, B. (2017) Private Information and Dynamic Risk. Mimeo.

Feder, G., Just, R.E. and Zilberman, D. (1985) Adoption of agricultural innovations in developing countries: a survey. *Economic Development and Cultural Change* 33, 255–298.

Feduzi, A., Runde, J. and Zappia, C. (2014) De Finetti on uncertainty. *Cambridge Journal of Economics* 38, 1–21. DOI: 10.1093/cje/bet054.

Fezzi, C., Menapace, L. and Raffaelli, R. (2021) Estimating risk preferences integrating insurance choices with subjective beliefs. *European Economic Review* 135, 103717. DOI: 10.1016/j.euroecorev.2021.103717.

Fiore, S.M., Harrison, G.W., Hughes, C.E. and Rutström, E.E. (2009) Virtual experiments and environmental policy. *Journal of Environmental Economics and Management, Frontiers of Environmental and Resource Economics* 57, 65–86. DOI: 10.1016/j.jeem.2008.08.002.

Freudenreich, H. and Kebede, S.W. (2022). Experience of shocks, household wealth and expectation formation: evidence from smallholder farmers in Kenya. *Agricultural Economics* 53, 756–774. https://doi.org/10.1111/agec.12718.

Galavotti, M.C. (1991) The notion of subjective probability in the work of Ramsey and de Finetti. *Theoria* 57, 239–259. DOI: 10.1111/j.1755-2567.1991.tb00839.x.

Galavotti, M.C. (2001) Subjectivism, objectivism and objectivity in Bruno de Finetti's Bayesianism. In: Corfield, D. and Williamson, J. (eds) *Foundations of Bayesianism, Applied Logic Series*. Springer, Dordrecht, Netherlands, pp. 161–174. DOI: 10.1007/978-94-017-1586-7_7.

Gigerenzer, G. and Hoffrage, U. (1995) How to improve Bayesian reasoning without instruction: frequency formats. *Psychological Review* 102, 684–704. DOI: 10.1037/0033-295X.102.4.684.

Giné, X., Townsend, R.M. and Vickery, J. (2009) Forecasting when it matters: evidence from semi-arid India. Mimeo, World Bank, Washington, DC.

Grisley, W. and Kellogg, E.D. (1983) Farmers' subjective probabilities in northern Thailand: an elicitation analysis. *American Journal of Agricultural Economics* 65, 74–82. DOI: 10.2307/1240339.

Hardaker, J.B. and Lien, G. (2010) Probabilities for decision analysis in agriculture and rural resource economics: the need for a paradigm change. *Agricultural Systems* 103, 345–350. DOI: 10.1016/j.agsy.2010.01.001.

Harrison, G.W., Martínez-Correa, J., Swarthout, J.T. and Ulm, E.R. (2015) Eliciting subjective probability distributions with binary lotteries. *Economics Letters* 127, 68–71. DOI: 10.1016/j.econlet.2014.12.027.

Harrison, G.W., Martínez-Correa, J., Swarthout, J.T. and Ulm, E.R. (2017) Scoring rules for subjective probability distributions. *Journal of Economic Behavior & Organization* 134, 430–448. DOI: 10.1016/j.jebo.2016.12.001.

Heinemann, F., Nagel, R. and Ockenfels, P. (2009) Measuring strategic uncertainty in coordination games. *The Review of Economic Studies* 76, 181–221.

Hill, R.V. (2009) Using stated preferences and beliefs to identify the impact of risk on poor households. *The Journal of Development Studies* 45, 151–171. DOI: 10.1080/00220380802553065.

Holt, C.A. and Smith, A.M. (2009) An update on Bayesian updating. *Journal of Economic Behavior & Organization* 69, 125–134. DOI: 10.1016/j.jebo.2007.08.013.

Hossain, T. and Okui, R. (2013) The binarized scoring rule. *The Review of Economic Studies* 80, 984–1001.

Huber, R., Xiong, H., Keller, K. and Finger, R. (2022) Bridging behavioural factors and standard bio-economic modelling in an agent-based modelling framework. *Journal of Agricultural Economics* 73, 35–63. DOI: 10.1111/1477-9552.12447.

Hurley, T.M. and Shogren, J.F. (2005) An experimental comparison of induced and elicited beliefs. *Journal of Risk Uncertainty* 30, 169–188. DOI: 10.1007/s11166-005-6565-5.

Jacobsen, B., Lee, J.B., Marquering, W. and Zhang, C.Y. (2014) Gender differences in optimism and asset allocation. *Journal of Economic Behavior & Organization, Empirical Behavioral Finance* 107, 630–651. DOI: 10.1016/j.jebo.2014.03.007.

Janssen, E. and Swinnen, J. (2019) Technology adoption and value chains in developing countries: evidence from dairy in India. *Food Policy* 83, 327–336. DOI: 10.1016/j.foodpol.2017.08.005.

Karni, E. (2009) A mechanism for eliciting probabilities. *Econometrica* 77, 603–606.

Knight, F.H. (1921) *Risk, Uncertainty and Profit.* Houghton Mifflin, Boston, Massachusetts.

Komarek, A.M., De Pinto, A. and Smith, V.H. (2020) A review of types of risks in agriculture: what we know and what we need to know. *Agricultural Systems* 178, 102738. DOI: 10.1016/j.agsy.2019.102738.

Luseno, W.K., McPeak, J.G., Barrett, C.B., Little, P.D. and Gebru, G. (2003) Assessing the value of climate forecast information for pastoralists: evidence from southern Ethiopia and northern Kenya. *World Development* 31, 1477–1494. DOI: 10.1016/S0305-750X(03)00113-X.

Lybbert, T.J. and Just, D.R. (2007) Is risk aversion really correlated with wealth? How estimated probabilities introduce spurious correlation. *American Journal of Agricultural Economics* 89, 964–979. DOI: 10.1111/j.1467-8276.2007.01034.x.

Lybbert, T.J., Barrett, C.B., McPeak, J.G. and Luseno, W.K. (2007) Bayesian herders: updating of rainfall beliefs in response to external forecasts. *World Development* 35, 480–497. DOI: 10.1016/j.worlddev.2006.04.004.

Maertens, A. (2017) Who cares what others think (or do)? Social learning and social pressures in cotton farming in India. *American Journal of Agricultural Economics* 99, 988–1007. DOI: 10.1093/ajae/aaw098.

Manski, C.F. (2004) Measuring expectations. *Econometrica* 72, 1329–1376. DOI: 10.1111/j.1468-0262.2004.00537.x.

Manski, C.F. (2018) Survey measurement of probabilistic macroeconomic expectations: progress and promise. *NBER Macroeconomics Annual* 32, 411–471.

Marinacci, M. (2002) Probabilistic sophistication and multiple priors. *Econometrica* 70, 755–764. DOI: 10.1111/1468-0262.00303.

Massoni, S., Gajdos, T. and Vergnaud, J-C. (2014) Confidence measurement in the light of signal detection theory. *Frontiers in Psychology* 5. DOI: 10.3389/fpsyg.2014.01455.

McFadden, D.L., Bemmaor, A.C., Caro, F.G., Dominitz, J., Jun, B-H., Lewbel, A. *et al.* (2005) Statistical analysis of choice experiments and surveys. *Market Letters* 16, 183–196. DOI: 10.1007/s11002-005-5884-2.

Meemken, E-M. and Bellemare, M.F. (2020) Smallholder farmers and contract farming in developing countries. *Proceedings of the National Academy of Sciences* 117, 259–264. DOI: 10.1073/pnas.1909501116.

Menapace, L. and Raffaelli, R. (2020) Unraveling hypothetical bias in discrete choice experiments. *Journal of Economic Behavior & Organization* 176, 416–430. DOI: 10.1016/j.jebo.2020.04.020.

Miranda, M.J. and Farrin, K. (2012) Index insurance for developing countries. *Applied Economic Perspectives and Policy* 34, 391–427. DOI: 10.1093/aepp/pps031.

Mobarak, A.M. and Rosenzweig, M.R. (2013) Informal risk sharing, index insurance, and risk taking in developing countries. *The American Economic Review* 103, 375–380.

Morgan, M.G., Henrion, M. and Small, M. (1990) *Uncertainty: A Guide to Dealing with Uncertainty in Quantitative Risk and Policy Analysis*. Cambridge University Press, Cambridge, UK.

Moutsiana, C., Garrett, N., Clarke, R.C., Lotto, R.B., Blakemore, S-J. and Sharot, T. (2013) Human development of the ability to learn from bad news. *Proceedings of the National Academy of Sciences* 110, 16396–16401. DOI: 10.1073/pnas.1305631110.

Nelson, R.G. and Bessler, D.A. (1989) Subjective probabilities and scoring rules: experimental evidence. *American Journal of Agricultural Economics* 71, 363–369. DOI: 10.2307/1241594.

Norris, P.E. and Kramer, R.A. (1990) The elicitation of subjective probabilities with applications in agricultural economics. *Review of Marketing and Agricultural Economics* 58, 127–147. DOI: 10.22004/ag.econ.12253.

Norton, G.W. and Alwang, J. (2020) Changes in agricultural extension and implications for farmer adoption of new practices. *Applied Economic Perspectives and Policy* 42, 8–20. DOI: 10.1002/aepp.13008.

Offerman, T., Sonnemans, J., Van De Kuilen, G. and Wakker, P.P. (2009) A truth serum for non-Bayesians: correcting proper scoring rules for risk attitudes. *Review of Economic Studies* 76, 1461–1489. DOI: 10.1111/j.1467-937X.2009.00557.x.

Prelec, D. (2004) A Bayesian truth serum for subjective data. *Science* 306, 462–466. DOI: 10.1126/science.1102081.

Ramsey, F.P. (1931) Foundations of mathematics and other logical essays. Routledge, London. DOI: 10.4324/9781315887814.

Savage, L.J. (1954) The foundations of statistics. *Naval Research Logistics Quarterly* 1, 236–236. DOI: 10.1002/nav.3800010316.

Schlag, K.H., Tremewan, J. and van der Weele, J.J. (2015) A penny for your thoughts: a survey of methods for eliciting beliefs. *Experimental Economics* 18, 457–490. DOI: 10.1007/s10683-014-9416-x.

Schlag, K. and Tremewan, J. (2021) Simple belief elicitation: an experimental evaluation. *Journal of Risk and Uncertainty* 62, 137–155.

Smith, J. and Mandac, A.M. (1995) Subjective versus objective yield distributions as measures of production risk. *American Journal of Agricultural Economics* 77, 152–161. DOI: 10.2307/1243897.

Spetzler, C.S. and Staël Von Holstein, C-A.S. (1975) Exceptional paper – Probability encoding in decision analysis. *Management Science* 22, 340–358. DOI: 10.1287/mnsc.22.3.340.

Tjernström, E., Lybbert, T.J., Hernández, R.F. and Correa, J.S. (2021) Learning by (virtually) doing: experimentation and belief updating in smallholder agriculture. *Journal of Economic Behavior & Organization* 189, 28–50. DOI: 10.1016/j.jebo.2021.03.001.

Trautmann, S.T. and van de Kuilen, G. (2015) Belief elicitation: a horse race among truth serums. *The Economic Journal* 125, 2116–2135. DOI: 10.1111/ecoj.12160.

von Neumann, J. and Morgenstern, O. (1947) *Theory of Games and Economic Behavior*, 2nd edn. Princeton University Press, Princeton, New Jersey.

Wang, S.W. (2011) Incentive effects: the case of belief elicitation from individuals in groups. *Economics Letters* 111, 30–33. DOI: 10.1016/j.econlet.2010.11.045.

Weinstein, N.D. and Diefenbach, M.A. (1997) Percentage and verbal category measures of risk likelihood. *Health Education Research* 12, 139–141.

Winkler, R.L., Muñoz, J., Cervera, J.L., Bernardo, J.M., Blattenberger, G. *et al.* (1996) Scoring rules and the evaluation of probabilities. *Test* 5, 1–60. DOI: 10.1007/BF02562681.

World Bank (2019) Unlocking Agriculture Finance and Insurance in Uganda: The Financial Sector's Role in Agricultural Transformation (Policy Note). World Bank, Washington, DC. DOI: 10.1596/32332.

4 Strategic Risk and Collective Action in Agriculture

Ariel Singerman[1]* and Sergio H. Lence[2]
[1]*University of Florida, Gainesville, USA;* [2]*Iowa State University, Ames, USA*

Abstract

Risk is strategic when agents' decisions depend on their beliefs about the action(s) of other individual(s). We discuss key issues arising from strategic risk among agricultural producers by looking at the case of voluntary pest control, which can be viewed as a collective-action problem analogous to that of contributing to the provision of a public good (i.e. the regional level of pest control). We focus on collective action because the agricultural economics literature does not seem to have sufficiently emphasized the uncertainty underlying the aggregate success of collective action efforts or the associated individual opportunity costs to collective action. From a policy standpoint, a better understanding of these issues is critical because strategic risk's underlying probabilities depend on individuals' beliefs about other individuals' decisions; therefore, the likelihood of success of collective-action efforts could in principle be managed by policy interventions designed to modify the beliefs of collective-action participants.

4.1 Introduction

Collective action has often been promoted as a way to solve problems usually faced by agricultural producers, including gaining access to new markets, spreading the costs of investing in capital equipment and bargaining with firms upstream and downstream along the agricultural supply chain (Bouamra-Mechemache and Zago, 2015). In recent years, interest also has been increasing in collective agri-environmental strategies to effectively provide public goods such as biodiversity, water availability and quality, resilience to flooding and fire, carbon storage to reduce greenhouse gas emissions and agricultural landscape (Vanni, 2014). Further, collective action has been proposed as a remedy to deal with issues associated with externalities and public goods among farmers and between farmers and other economic agents (Ayer, 1997). But the non-excludable defining characteristic of public goods (along with non-rivalry) generates the possibility of opportunistic behavior: public-good users have no incentive to voluntarily contribute to providing such goods because their payoff can be higher if they are able to free-ride on the contribution of others. The fundamental underlying issue – and the source of risk involved – is that individuals' payoffs depend not only on their own actions but also on the actions of others. Thus, each agent chooses an action that is the best response to their belief regarding

*E-mail: singerman@ufl.edu

 Managing Risk in Agriculture: A Development Perspective (eds A.K. Mishra, S.C. Kumbhakar and G. Lien).
DOI: 10.1079/9781800622289.0004

the proportion of other agents who will choose each action (Morris and Shin, 2006). Each agent's dilemma is whether to opt for the Pareto-optimal equilibrium or, given the strategic risk involved, opt for a Pareto-inferior equilibrium instead (Carlsson and van Damme, 1991). An equilibrium is Pareto-optimal if it is not possible to increase an agent's well-being without reducing the well-being of any other agent, and is Pareto-inferior otherwise (Lockwood, 2008).

Risk is strategic when agents' courses of action depend on their beliefs about the action(s) of other individual(s). In contrast, under non-strategic risk, agents' decisions do not depend on the actions of others. In this chapter, we discuss key issues arising from strategic risk among agricultural producers by looking at the case of voluntary pest control, which can be viewed as a collective-action problem analogous to that of contributing to providing a public good (i.e. the regional level of pest control). Based on the classification of collective-action problems postulated by Anomaly (2020), the strategic uncertainty we discuss in this chapter is the one associated with the assurance problem, which arises when individuals lack the assurance (hence, are uncertain) that other participants will in fact take the action required for the collective effort to succeed. We focus on collective action because the agricultural economics literature does not seem to have placed sufficient emphasis on the uncertainty underlying the aggregate success of collective action efforts or on the associated individual opportunity costs (i.e. payoffs from the alternatives to collective action). We bring attention to both issues by comparing the choices of individuals facing strategic risk versus their choices under non-strategic risk. Importantly, given that strategic risk's underlying probabilities depend on individuals' beliefs about other individuals' decisions, it is conceptually possible to design policy interventions to modify the beliefs of collective-action participants and, therefore, affect the likelihood of success of collective-action efforts.

Since the 1950s, synthetic pesticides have allowed farmers to control many pests within their farms cheaply, effectively and without regard for their neighbors' opinions and actions (Smith *et al.*, 1976; Hendrichs *et al.*, 2007). Thus, individual chemical pest management has become the most widely used strategy for pest control (Klassen, 2008). But despite the heavy use of chemicals, an estimated 10–35% of the US crop production is lost to pest damage (Pimentel *et al.*, 1993). Plant pests and diseases increasingly threaten plant health worldwide (Giovani *et al.*, 2020). For example, the bacterium *Xylella fastidiosa*, the causing agent of the olive quick decline syndrome, has decimated the Italian region of Puglia; the disease's potential economic impact is estimated at 5.2 billion euros (Schneider *et al.*, 2020). *Xylella fastidiosa* also causes Pierce disease, which is estimated to cost US$104.4 million per year in California alone (Alston *et al.*, 2013). Caused by the strain Tropical Race 4, Panama disease is threatening the production of Cavendish bananas worldwide – such a variety represents more than 40% of global production and virtually all exports (Dale *et al.*, 2017; Staver *et al.*, 2020). Citrus greening or Huanglongbing (HLB) disease is regarded as the most devastating citrus disease worldwide (FAO, 2015); in the state of Florida, it has caused production to decrease by approximately 80% and the industry to downsize significantly at all levels (Singerman and Rogers, 2020). In addition, invasive insects alone are estimated to cost US$70 billion annually worldwide (Bradshaw *et al.*, 2016).

A key issue arising from individual chemical pest management is that its effectiveness is compromised by the mobility of pests (Hendrichs *et al.*, 2005). For instance, Bagavathiannan *et al.* (2019) note that despite current best management practices and the efforts to promote them, weed species continue to spread and their management costs continue to rise as herbicide resistance evolves. Due to their mobility, pests (and the diseases they might vector) can be viewed as common property. Therefore, conducting only site-specific pest management control ignores the collective nature of the problem (see e.g. Perrings, 2001; Perrings *et al.*, 2002; Vreysen *et al.*, 2007). Due to reinfestation from neighboring farms, individual farmers' actions have little effect on the density of the pest in future periods (Lazarus and Dixon, 1984). Thus, from an economic perspective, individual pest management creates an externality that leads to a disparity between private and social optima (Regev *et al.*, 1976); more specifically, it results in an under-provision of pest control from a societal perspective (Miranowski and Carlson, 1986;

Yu and Leung, 2006). Effectively controlling mobile pests depends critically on the actions of neighboring farmers. This attribute characterizes the problem as a collective-action dilemma, in which trade-offs occur between individual and collective interests. By 'internalizing' the externalities associated with individual pest management, collective pest control (e.g. by means of area-wide pest management [AWPM] programs) has the potential to achieve the socially optimal outcome.

Collective-action dilemmas are typically dealt with by means of top-down regulation or a bottom-up approach. The goal of a government regulation for the case of pest control should be to induce producers to adopt levels of pest control that are closer to the social optimum. In principle, first-best policy would impose a tax on the use of pest control chemicals to capture their social cost (Waterfield and Zilberman, 2012). A government regulation could, alternatively, set a quota on the total use of a pest control treatment equivalent to the socially optimal level and allow farmers to trade their rights to the treatment among themselves. But taxes and quotas are generally not feasible to deal with pesticide externalities because of the heterogeneity of pesticide effects on the environment and human health, which vary by farm and region. Moreover, the external cost of pesticide pollution varies with the spatial and temporal characteristics of pesticide applications. Another hurdle to implementing an effective policy is the nonpoint source nature of pesticide pollution; monitoring costs are prohibitive, and precise determinations of the origins of pollution are seldom made (Sexton *et al.*, 2007). Thus, the obstacles to the design of market-based policies render the use of command-and-control, by which government decides whether a particular treatment can be used and under what conditions (Zilberman and Millock, 1997), an appealing alternative.

All chemical pest control contributes to generating the pest resistance externality because the stock of efficacy of a treatment is a common-pool resource (Regev *et al.*, 1976; Clark and Carlson, 1990) for all farmers within a region. Such common access of producers to the treatment gives rise to the externality because each producer's usage of the treatment impacts other farmers and has no compensation. In the absence of regulation, overuse and therefore resistance are likely. The variety of potential externalities suggests that many pest management decisions may be better made at the regional level rather than at the individual farm level (Dasgupta *et al.*, 2007). Growers coordinating pest control may internalize externalities and increase the productivity of pest-control inputs. In addition, area-wide pest management may contribute to reducing the risk of developing pesticide resistance (Hendrichs *et al.*, 2007). Therefore, a bottom-up approach may be more attractive given the financial constraints of governments, the resistance that top-down regulation typically encounters and the high cost and difficulties of enforcing a top-down regulation in agricultural settings (Ayer, 1997; Olmstead and Rhode, 2004; Ervin and Frisvold, 2016).

Although a community-based approach to deal with collective-action problems can be more effective and result in lower transaction costs than command-and-control or payment-based approaches (Ostrom, 1990, 2009, 2010), cooperative solutions may not be adopted without the appropriate institutions (Loehman and Dinar, 1994) or incentives to induce farmers to participate. The problem with building institutional capacity is that it takes time (Ervin and Frisvold, 2016; Frisvold, 2019a). But the risk of pest invasions depends on the human response to the threat (Perrings *et al.*, 2002). Thus, the time required to build institutions and trust among stakeholders for a centralized collective-action response contrasts starkly with the rapid reaction needed to address the immediate threat posed by a plant pathogen (Almeida, 2016). Similar issues arise with incentives, which may be positive (e.g. subsidies to participants, social prestige) as well as negative (e.g. taxes, penalties for non-participants), as their appropriate design and implementation could be time-consuming.

Under some circumstances, it may be desirable to first implement a quicker, albeit temporary, decentralized collective action approach (e.g. a voluntary area-wide pest management program) to deal with mobile pests/invasive species until the appropriate community-based institutions are put in place. This is particularly so for the case of pests vectoring plant diseases because, unlike animal diseases (which, in some cases, can be transmitted to humans), they are more likely to be dealt with by a slower policy response. However, voluntary coordination faces challenges similar to those of contributing to providing a public good, namely, stakeholder participation. The challenge that stakeholder participation poses to the success of the collective action of, for example, an area-wide pest management program has been widely recognized in the literature (see e.g. Klassen, 2000; Mumford, 2000; Hendrichs *et al.*, 2007). In their pioneering article, Rook and Carlson (1985) argue that farmers should participate in area-wide pest management if the benefits of doing so outweigh the costs. However, Hurley and Frisvold (2016) point out that economic factors are not farmers' only incentive, because gains from cooperation do not always result in a cooperative outcome (Loehman and Dinar, 1994). Behavioral and socio-economic factors may challenge rational economic behavior (Miranowski, 2016). Because agriculture is inherently a socio-ecological system, models and incentive schemes that integrate economics and biology would be uniquely positioned to provide insights on the interactions between farmer behavior and mobile pests/invasive species, enabling a better understanding of the rationale (and the consequences) of individual growers' choices and behavior, which are key for the design and implementation of effective public policies to deal with the externality. As an example, despite the evidence that adopting a voluntary area-wide pest management program to deal with citrus greening in Florida was profitable, strategic uncertainty hindered farmers' cooperation (Singerman *et al.*, 2017).

Contributing to providing a public good, such as the regional level of pest control provided by an area-wide pest management program, is risky because the successful provision requires a critical mass of players cooperating; if the critical mass is not met, the public good is not provided and those who contributed to it attain lower payoffs. Thus, opting out is a safe strategy for an individual player but results in a lower individual payoff and a sub-optimal social outcome. Hofstadter (1985) defined strategic uncertainty by using the expression 'reverberant doubt,' meaning that players' initial small doubts regarding the ability to collectively achieve the (better) cooperative outcome reverberate and eventually become large enough to undermine the individual player's original intention to stick to the cooperative strategy, causing the individual to opt out instead. Such a definition is illustrated by the finding of Singerman *et al.* (2017) that most citrus growers perceived (correctly or not) that other growers were reluctant to coordinate sprays, so they decided not to cooperate themselves. Moreover, Singerman and Useche (2019a) found that for citrus growers in Florida, key deterrents for successful coordination under strategic uncertainty are the coordination threshold (i.e. the minimum percentage of participants in the group that must coordinate to guarantee success) and the beliefs about others not coordinating.

A practical tool to assess the likelihood of success of a voluntary coordination program is the coordination frontier (CF) proposed by Lence and Singerman (2023). The CF quantifies the impact of two variables that are at the heart of the strategic uncertainty that area-wide pest management involves: first, the magnitude of the aggregate change in the probability of successful coordination due to a change in threshold, and second, the magnitude of the change in the probability of successful coordination as the (opportunity) cost of coordinating increases. The CF could be used not only to understand the conditions under which different levels of voluntary coordination can succeed, but also to establish the magnitude of the economic incentives needed to make it work. Importantly, the CF can help reduce the strategic uncertainty involved in a voluntary program. Providing public information can help align growers' beliefs, reduce strategic uncertainty and enhance coordination. Singerman and Useche (2019a) find that a substantial percentage of the growers choose to coordinate more after being provided public information on choices under strategic uncertainty. Morris and Shin (2006) illustrate that public information conveys more information than its face value, even if merely stating the obvious; it can convey important strategic information on the likely beliefs of others and thus contribute to coordinating outcomes.

Even though externalities associated with pest damage for annual crops may be considered to be mostly spatial, the negative spatial externalities from the lack of collective action in pest control can be more severe and difficult to deal with for the case of perennial crops. This is so because cultural practices effective at reducing disease inoculum in annual crops, such as crop rotation, delayed fall planting and tillage, are not feasible in perennial crops. In addition, the spatial externality also acquires a time (i.e. dynamic) dimension because the plants do not start anew every year (unless effectively treated, each successive year they start at a lower level of plant health), thus making farmers more dependent on neighboring farmers for successful management. In some cases, the plant is infected before the canopy shows any symptoms. For the case of HLB, for example, the latency period can take 6–12 months (Bové, 2006). Therefore, the spread and impact of the disease occur over multiple years, which prevents growers from removing all sources of bacteria from their groves.

Collective action among farmers is critical to address not only the externalities derived from decisions of whether to control disease by neighboring farmers, but also those related to pesticide resistance. Losses due to pesticide resistance are estimated at about US$9 billion annually in the United States alone (Palumbi, 2001). Moreover, the ever-increasing reliance on pesticides worldwide compounds the impact of resistance. Meanwhile, the use of pesticides continues as if resistance were a temporary issue that will be addressed by new chemicals. However, for example, no herbicides with new modes of action have been commercialized over the past 30 years (Gould *et al.*, 2018). Ervin and Frisvold (2016) find that when managing resistance of mobile weeds, farmers believe such a management depends on neighbors' actions, which discourages them from proactively managing resistance. Since pesticide resistance further promotes an increase in pesticide use to offset the pest's lower susceptibility, the development of resistance may exacerbate the negative externalities pesticides generate (Dover and Croft, 1986). Pesticide resistance also can create additional negative externalities: Van Deynze *et al.* (2022) find that the spread of glyphosate-resistant weeds has caused significant reductions in the use of conservation tillage in soybean production, resulting in water quality and climate damages via fuel emissions valued at nearly US$245 million.

4.2 Risk Preferences and Technology Adoption: Strategic Uncertainty versus Playing against Nature

Technology innovations are adopted because of their perceived benefits (Feder *et al.*, 1985). Since the benefits and the full range of consequences of a new technology are not completely understood before commercially implementing them in the field, Marra *et al.* (2003) emphasize the importance of uncertainty, information and learning in adoption decisions. As part of the learning process, adopters need to learn how a new technology works, how to adapt it to local conditions, what its return and risk profile are, and how such a profile compares to the alternatives (Rogers, 1962; Feder *et al.*, 1985; Conley and Udry, 2010). Such learning may be individual (e.g. if revealed to farmers through their own experience with a new technology) or social (e.g. as gained by observing their neighbors' choices [Conley and Udry, 2010]). This can explain previous research findings regarding the factors influencing the rate and speed of adoption among farmers, which characterize technology adoption as a gradual process based on the presence of heterogeneity among potential adopters (Griliches, 1957; Rogers, 1962; Feder *et al.*, 1985; Marra *et al.*, 2003; Foster and Rosenzweig, 2010).

Under the expected utility paradigm, which is generally used to model technology adoption, growers make their choices based on the highest expected utility conditional on their attitudes toward risk (Feder *et al.*, 1985; Isik and Khanna, 2003; Marra *et al.*, 2003; Foster and Rosenzweig, 2010). But, for example, ambiguity aversion also can play a role in farmers' decisions to adopt technological innovations (Chavas and Nauges, 2020). Barham *et al.* (2014) documented a case of farmers in the United States in which ambiguity aversion accelerated the pace of adoption; the authors hypothesized that the insect-resistance trait of the technology reduced the ambiguity related to pest damage, which is a trait valued by ambiguity-averse farmers.

4.2.1 The source of uncertainty influences technology adoption: Evidence from HLB in Florida

In Florida, as part of the citrus industry's strategic plan to address HLB, a voluntary area-wide pest management program to control the insect that transmits the disease was established in 2010. The scientific principles behind such a program and its implementation have been discussed in numerous studies, including Hendrichs *et al.* (2005), Klassen (2005), Jessup *et al.* (2007), Vreysen *et al.* (2007), Faust (2008) and Lloyd *et al.* (2010). The underlying premise of area-wide pest management is that the efforts to coordinate pest and disease control can be more effective than uncoordinated individual farmers' actions because they provide a larger and more lasting effect (Klassen, 2000). An area-wide pest management program is also aimed at reducing the risk of developing pesticide resistance (Vreysen *et al.*, 2007). Thus, by coordinating pest control, groups may internalize externalities and increase the productivity of pest-control inputs. In fact, there is evidence that such was the case in Florida. Coy *et al.* (2016) reported that resistance was on the rise for some classes of insecticides across the state in 2009, but fell back to susceptible levels in 2013 and 2014; they suggested the implementation of coordinated sprays contributed to the psyllids' heightened susceptibility to insecticides. In addition, Singerman *et al.* (2017) found evidence regarding the effectiveness of area-wide pest management to limit the impact of HLB. However, the authors also found that the combination of the high level of cooperation required and its voluntary character presented growers with (strategic) uncertainty derived from the actions and beliefs of others, which imposed too high of a cost for many growers and made it problematic to achieve widespread grower participation.

In March 2016, the Florida Commissioner of Agriculture declared a crisis that allowed growers to use foliar sprays of antibiotics to enhance the health of trees infected with HLB (FDACS, 2016). However, the expected enhancement of spraying those compounds on HLB-infected trees had not been scientifically proven. Zhang *et al.* (2014) had found some antibiotics to be effective against HLB but only in a lab setting, whereas Hu *et al.* (2018) had found that antibiotics provided control against HLB when using tree-trunk injections. Despite the lack of evidence about the effectiveness of antibiotic foliar sprays, and in contrast to findings that regard technology adoption as a gradual process, in Florida the adoption of antibiotics to combat HLB was immediate and widespread. Within the first year, survey data indicated that a vast majority of citrus growers adopted their use, while spraying coordination quickly deteriorated (Singerman, 2017). This evidence suggests that adoption follows different dynamics when decision makers face a pernicious disease. Moreover, the underlying source of uncertainty of a technology likely influences the dynamics of its adoption. Understanding such a link can be critical in the presence of externalities, so as to avoid worsening them.

Similar to insecticides, the social cost of using antibiotics in agriculture includes potential ecological harm. But, importantly, the use of antibiotics in agriculture can also contribute to the development of antibiotic resistance, with its consequent impact on public health (Laxminarayan, 2003). A conservative estimate of the mortality rate due to antibiotic resistance is 153,000 annual human deaths per year in the United States alone (Burnham *et al.*, 2019). As Laxminarayan and Brown (2001, p. 201) put it, '[A]t the heart of the problem of antibiotic resistance is the issue of the externality imposed by each beneficial use of antibiotics on its future effectiveness.' So, for example, although HLB is not transmitted to humans and therefore poses no direct risk for human health, the approval of the use of antibiotics to treat plant (and animal) diseases has potential consequences for human health by contributing to antibiotic resistance. The two antibiotics approved for field use against HLB, oxytetracycline and streptomycin, also are used to treat bacteria affecting humans and are regarded by the World Health Organization (WHO) as 'highly important' and 'critically important' for human medicine, respectively (WHO, 2016). Although growers' pest control decisions have wide-ranging impacts beyond their own farms, given that social costs are not reflected in the market price, growers have no incentive to include such costs as part of their adoption decisions.

Antibiotic foliar sprays were eventually found to be ineffective in enhancing the health of HLB-infected trees in both greenhouses and the field (Citrus Industry, 2018; Li *et al.*, 2019). Paradoxically enough, their use not only hampered the chances of dealing with HLB in Florida due to the compounds' ineffectiveness, but also had a debilitating impact on the coordination efforts for area-wide pest management (Singerman, 2017; Singerman *et al.*, 2017). Why did growers switch away from coordinated area-wide pest management sprays despite the lack of scientific evidence about the efficacy of antibiotic sprays?

The pattern of substitution of coordinated sprays for antibiotics observed in Florida implies that growers were choosing to avoid the strategic uncertainty that coordination posed, opting instead for the ambiguous outcome that the self-managed strategy of antibiotics presented them with, which they evidently perceived as having a more preferable payoff. The aversion to the ambiguity of the outcome may have been offset by the potential impact growers attributed to antibiotics to reduce the ambiguity related to the continued damage from the disease, an effect similar to that found by Barham *et al.* (2014).

Singerman and Lence (2023) used data from a framed field economic experiment to examine the aggregate decisions of citrus growers to adopt a technology based on its underlying source of uncertainty. After presenting growers with a scenario with non-strategic risk, also referred to as a 'game against nature,' and a game involving strategic interaction, aimed at emulating the type of uncertainty growers faced when adopting antibiotics and area-wide pest management, respectively, the authors found evidence that, on average, growers preferred the game against nature. While the adoption of antibiotics would have been better represented by an ambiguous lottery, Singerman and Lence's (2023) results allowed them to set a bound on the aggregate expected probability of success growers assigned to such an alternative.

Moreover, they found that growers preferred such a game even for ranges over which the game involving strategic interaction had a higher likelihood of success. The results obtained were similar when using data from another source in which the subjects were students instead of growers. The findings by Singerman and Lence (2023) have implications that are relevant for technology adoption, because farmers' preferences toward the source of uncertainty underlying a new technology can play a key role in their adoption decisions.

4.3 Analyzing Strategic Risk: The Case of Area-Wide Pest Management

To better understand some critical elements that differentiate the analysis of attitudes toward strategic uncertainty as opposed to risk involving non-strategic uncertainty (i.e. games against nature), we focus on the case of area-wide pest management. Such management provides an excellent example of strategic risk in agriculture because participating farmers' payoffs depend not only on their own actions, but also on those of neighboring farmers. Thus, individual farmers make decisions motivated by their beliefs about what other farmers will do. Such strategic uncertainty is a key consideration in growers' pest-control decision-making (Singerman and Useche, 2019a). As a result, each farmer chooses whether to coordinate as the best response to their beliefs regarding the behavior of other farmers.

To restrict attention to the key issues relevant to the strategic uncertainty underlying area-wide pest management, we follow Lence and Singerman (2023) by assuming that the program setup consists of a group of N identical risk-neutral farmers who individually decide whether to coordinate sprays to combat a pest, and that a minimum share or threshold $T \in [0, 1]$ of the farmers must coordinate for the collective effort to succeed. A farmer's profit equals (i) Π^{SC} if he or she coordinates and the area-wide management succeeds (i.e. successful coordination), (ii) $\Pi^{FC} < \Pi^{SC}$ if he or she coordinates but the area-wide management fails (i.e. failed coordination) and (iii) $\Pi^{FC} < \Pi^{NC} < \Pi^{SC}$ if he or she does not coordinate (i.e. no coordination). The problem faced by the ith farmer who must decide whether to coordinate his or her pesticide applications with that of others in the program ($W_i = 1$) or not ($W_i = 0$) can be summarized as

$$W_i^* = \begin{cases} 1 \; if \; L_i(1 + \sum_{n \neq i} W_n, T, N) \geq R\Pi = \dfrac{\Pi^{NC} - \Pi^{FC}}{\Pi^{SC} - \Pi^{FC}}, \\ 0 \text{ otherwise.} \end{cases} \tag{4.1}$$

In the above expression, $L_i(1 + \Sigma_{n \neq i} W_n, T, N)$ is farmer i's subjective probability of successful coordination (conditional on his or her decision to coordinate, i.e. $W_i = 1$), and $R\Pi$ is the 'relative profit,' computed as the amount by which the profit from not coordinating exceeds the profit from failed coordination, expressed as a share of the profit difference between successful and failed coordination.

A farmer benefits only if the threshold is met (i.e. $(\sum_{n=1}^{N} W_n)/N \geq T$). If the critical mass requirement is not met, coordination fails and the profit of the farmers who coordinated is lower than the profit of those who did not ($\Pi^{NC} > \Pi^{FC}$). In contrast, if the minimum threshold is met, the profit of those coordinating will be higher ($\Pi^{SC} > \Pi^{NC}$). Thus, not coordinating sprays provides a safer option for an individual, but the social outcome may be sub-optimal. In short, each farmer has to decide whether to coordinate sprays, but such a decision hinges upon his or her beliefs regarding the likelihood that a sufficiently large number of fellow farmers will coordinate as well.

Clearly, the solution to an individual farmer's decision problem depends on the relative profit ($R\Pi$), the number of farmers in the area-wide pest management program (N) and the minimum threshold for success (T). For any given equilibrium solution, it is conceptually possible to evaluate the aggregate probability of successful coordination among farmers as the function $\lambda_s(R\Pi, T, N) \equiv Prob_{F(W \mid R\Pi, T, N)}[(\sum_{n=1}^{N} W_n)/N \geq T]$, where $W \equiv [W_1, ..., W_N]$ is the vector of farmers' equilibrium decisions (based on their subjective beliefs), $F(W \mid R\Pi, T, N)$ is the corresponding joint distribution and the subscript s in λ_s is used to emphasize that the probability involves strategic uncertainty.

Another parameter of interest for the present purposes is the individual coordination probability (ICP) function $P_s(R\Pi, T, N) \equiv E_{F(W \mid R\Pi, T, N)}[(\sum_{n=1}^{N} W_n)/N]$, where $E_F(\cdot)$ denotes the expectation operator under probability measure $F(\cdot)$. For a given relative profit value $R\Pi$, function $P_s(R\Pi, T, N)$ gives the probability that a farmer selected at random will coordinate when his or her relative profit equals $R\Pi$ and he or she belongs to an area-wide pest management program comprising N farmers requiring a minimum share T of coordinating farmers to succeed in the collective effort. Alternatively, $P_s(R\Pi, T, N)$ is the share of individuals who prefer to coordinate at any given $\{R\Pi, T, N\}$ combination, that is, the proportion of individuals who favor the outcome involving strategic uncertainty over the risk-free alternative.

Figure 4.1 illustrates the ICP function $P_s(R\Pi, T, N)$ estimated using data from a laboratory experiment involving citrus growers, conducted by Singerman and Useche (2019a). The experiment required growers to decide whether to coordinate based on scenarios framed to mimic an area-wide pest management setup. Details about the experiment and the procedure used to estimate the P_s and P_n curves depicted in Fig. 4.1 are discussed at length in Lence and Singerman (2023).

It can be observed that, for a given threshold (T), the percentage of growers who chose to coordinate (P_s) falls as the relative profit of not coordinating ($R\Pi$) rises. In addition, holding constant the relative profit of not coordinating, it is clearly the case that the percentage of individuals who chose the risky alternative is negatively associated with the threshold for successful coordination (T).

The ICP function $P_s(R\Pi, T, N)$ shows the percentage of individuals who prefer the strategic risky outcome (i.e. to coordinate) over the risk-free alternative (i.e. to opt out). Figure 4.1 also graphs the function $P_n(R\Pi, \lambda_n)$ – estimated using citrus growers' data from the same experiment – that depicts the percentage of individuals preferring the risky outcome for an analogous decision-making problem involving non-strategic risk. For this game against nature the growers' decision was simply to choose between: (i) a lottery with a risky outcome consisting of profit Π^S if successful and profit $\Pi^F < \Pi^S$ in case of failure; and (ii) a risk-free alternative with profit $\Pi^N \in [\Pi^F, \Pi^S]$. Thus, the relative profit is computed as $R\Pi = (\Pi^N - \Pi^F)/(\Pi^S - \Pi^F)$. In this instance, the underlying uncertainty does not involve the decisions of other individuals, and the probability of success $\lambda_n \in (0, 1)$ was known to growers at the time of decision-making. Singerman and Lence (2023) exemplify this scenario with the use of antibiotics to combat citrus greening: the application of antibiotics did not rely on area-wide pest management, but its outcome was uncertain because the ability of foliar antibiotic applications to effectively combat citrus greening was unknown at the time they were approved for their use against the disease.

Similar to the situation with strategic uncertainty, schedule $P_n(R\Pi, \lambda_n)$ in Fig. 4.1 shows that the share of growers who preferred the risky outcome over the certain one in the case of non-strategic uncertainty decreases as the relative profit increases. Importantly, however, it is not appropriate to directly compare the curves depicting risky outcome preferences for the strategic game $P_s(R\Pi, T, N)$ with the one for the lottery curve $P_n(R\Pi, \lambda_n)$, because they typically involve different probabilities of success. More

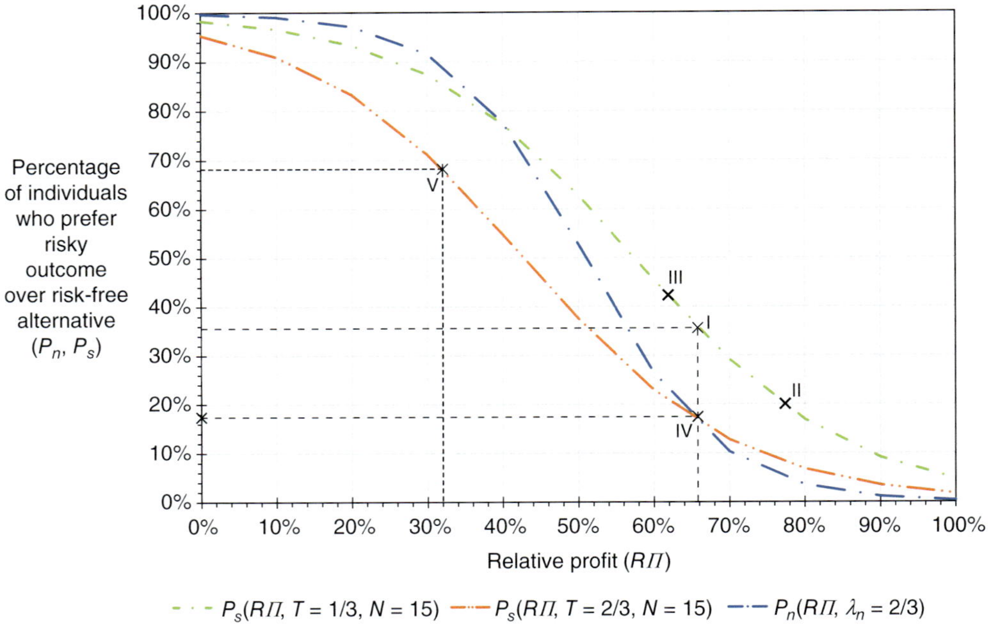

Fig. 4.1. Estimated average percentage of Florida citrus growers who prefer a risky outcome in a game against nature and in strategic games, for different levels of relative profits ($R\Pi$). (Source: Singerman, A. and Lence, S.H. (2023) The source of uncertainty influences technology adoption. *European Review of Agricultural Economics* 50, 591–623. DOI: 10.1093/erae/jbac036 by permission of Oxford University Press.)

concretely, all points along the lottery schedule $P_n(R\Pi, \lambda_n)$ are characterized by the same probability of attaining the risky payoff ($\lambda_n = 2/3$ in this case). In contrast, points along a particular strategic uncertainty curve $P_s(R\Pi, T, N)$ may have different probabilities of successful coordination (λ_s), i.e. a different probability of receiving the risky payoff.

To see why the probability of successful coordination may change as one moves along a given $P_s(R\Pi, T, N)$ curve, note that if the threshold for success is T, then at least $K \equiv \lceil T\,N \rceil$ of the N individuals in the group must coordinate, where $K \in \{1, 2, \ldots, N\}$ and $\lceil x \rceil$ denotes the closest integer greater than or equal to x. Then, the corresponding probability of successful coordination conditional on each individual coordinating with probability θ, $\lambda_s(T, N, \theta)$ is given by the complement of the binomial distribution function

$$\lambda_s(T, N, \theta) \equiv 1 - CumBinom(\lceil T\,N \rceil - 1, N, \theta) \quad (4.2)$$

$$= CumBinom(N - \lceil T\,N \rceil, N, 1-\theta) \quad (4.3)$$

In the above expressions, $CumBinom(k, n, q)$ represents the cumulative binomial distribution function of k successes in n independent trials with replacement, where q is the probability of success in each trial. According to Eqn 4.3, the probability of successful coordination $\lambda_s(T, N, \theta)$ is equal to the probability that at most $(N - \lceil T\,N \rceil) = (N - K)$ out of N individuals do not coordinate, where the probability of an individual not coordinating is $(1 - \theta)$.

Since the probability that a randomly chosen individual prefers to coordinate (θ) in the strategic game is given by the share of individuals who prefer the risky alternative over the certain one (P_s), Eqn 4.3 makes it straightforward to calculate the probability of successful coordination λ_s along the strategic uncertainty curves drawn in Fig. 4.1. For example, point I on the P_s curve for $T = 1/3$ is characterized by $R\Pi = 65.8\%$ and $P_s = \theta = 35.6\%$. Hence, for a group of $N = 15$ individuals, the probability of successful coordination at point I is given by

$$\lambda_s(T = 1/3, N = 15, \theta = 0.356) = CumBinom(N - \lceil T\,N \rceil, N, 1-\theta) \quad (4.4)$$

$$= 2/3$$

By repeating this exercise for other points along the P_s curve corresponding to $T = 1/3$, it becomes clear the probability of successful coordination decreases as the alternative certain payoff increases. For example, at point II (III) to the right (left) of point I, the probability of successful coordination is $\lambda_s(T=1/3, N=15, \theta=0.20) = 0.164 < 2/3$ ($\lambda_s(T=1/3, N=15, \theta=0.423) = 0.832 > 2/3$).

Similarly, the probability of successful coordination along the P_s curve corresponding to $T = 2/3$ falls monotonically as one moves from the top left to the bottom right of the graph. Further, for a given relative profit ($R\Pi$), the probability of successful coordination falls monotonically as the threshold (T) increases, because the ICP ($\theta = P_s$) decreases as the threshold rises while maintaining the relative profit constant. As an example, point IV on the P_s curve for T = 2/3 has the same relative profit $R\Pi = 65.8\%$ as point I on the P_s curve for T = 1/3, but its probability of successful coordination is only $\lambda_s(T=2/3, N=15, \theta=0.174) = 0.001 < 2/3 = \lambda_s(T=1/3, N=15, \theta=0.356)$. Notably, point IV demonstrates that a small number of growers may be considered coordination 'enthusiasts,' because they chose to coordinate even at relative profit values for which the probability of successful coordination is very small. In the case of point IV, $P_s = \theta = 17.4\%$ of individuals chose to coordinate, despite having a success probability of only $\lambda_s(T=2/3, N=15, \theta=0.174) = 0.001$.

Point V on the P_s curve for $T = 2/3$, which is characterized by $R\Pi = 32.0\%$ and $P_s = \theta = 68.2\%$, has the same probability of coordination as point I on the $T = 1/3$ curve, i.e. $\lambda_s(T=2/3, N=15, \theta=0.682) = 2/3 = \lambda_s(T=1/3, N=15, \theta=0.356)$. Given the same probability of success $\lambda_s = 2/3$ for points I and V, the higher threshold for point V than for point I requires that the ICP be higher for the former ($P_s = \theta = 68.2\%$) than for the latter ($P_s = \theta = 35.6\%$).

Figure 4.2 helps further clarify the distinction between the share of individuals choosing risky outcomes and the associated probabilities of success. The $P_s(R\Pi, T=2/3, N = 15)$ and $P_n(R\Pi, \lambda_n=2/3)$ participation curves are the same as the curves with such labels in Fig. 4.1. But Fig. 4.2 also depicts the $\lambda_s(R\Pi, N=15, T=2/3)$ and the '$\lambda_n=2/3$' curves, which represent the probabilities of success of the strategic game with $T = 2/3$ (and $N = 15$) and the lottery, respectively. For clarity, the success probabilities should be read from the right-hand side axis. The schedule $\lambda_s(R\Pi, N=15, T=2/3)$, denoted by the dashed black line, changes as the relative profit ($R\Pi$) does because, as explained above, participation in the strategic game ($\theta = P_s$) changes with $R\Pi$. In contrast, λ_n, portrayed as the dotted line, is a constant equal to 2/3, because that was the probability of success assigned to the lottery in the experiment. Points IV and V in Fig. 4.1 also are depicted in Fig. 4.2 to provide additional reference about the link between the fixed-T participation schedules graphed in Fig. 4.1 and the resulting strategic probabilities of success λ_s.

Interestingly, Fig. 4.2 reveals that choices are inconsistent over a large range of relative profit values, represented by the shaded regions. For $R\Pi < 32\%$, choices are inconsistent because the share of individuals who prefer the risky outcome over the risk-free alternative is greater for the game against nature than for the strategic game ($P_n > P_s$), even though the probability of success is smaller for the former than for the latter ($\lambda_n = 2/3 < \lambda_s$). Choices are also inconsistent for $R\Pi > 65.8\%$ for similar reasons. Over this relative profit range preferences are reversed, as the share of individuals who prefer the risky choice over the riskless alternative is greater for the strategic game than for the game against nature ($P_s > P_n$). However, the relative size of the probability of success is also reversed, as it is much smaller for the strategic game than for the lottery ($\lambda_n = 2/3 > \lambda_s \cong 0$).

The insights from Fig. 4.2 can be alternatively interpreted as evidence that growers under- and overestimated the probability of successful coordination for low and high values of $R\Pi$, denoted respectively by the grey and blue shaded areas in the graph. Only for relative profit values of $R\Pi \in (32\%, 65.8\%$; i.e. the non-shaded area) did a higher percentage of growers show a preference for the game against nature that was consistent with its probability of success being larger than that of the strategic game.

Referring back to Fig. 4.1, points I and V are of special interest because they allow one to compare participation in strategic games with the same probability of success $\lambda_s = 2/3$, albeit with different thresholds (i.e. $T = 1/3$ vs. $T = 2/3$). That is, a line drawn through points I and V – omitted from the graph to avoid cluttering –

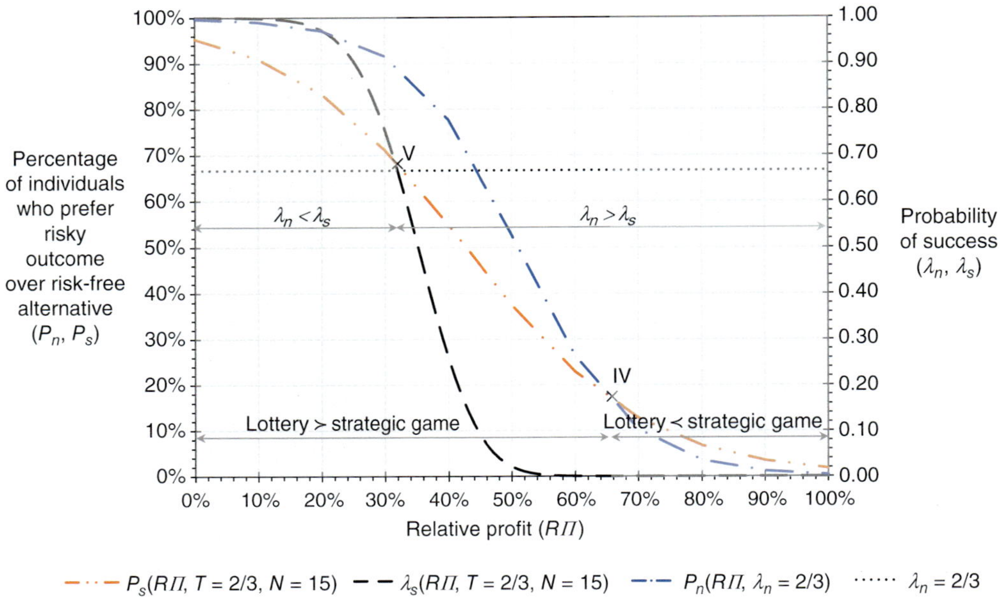

Fig. 4.2. Estimated average percentage of Florida citrus growers who prefer a risky outcome in a game against nature ($P_n(R\Pi, \lambda_n{=}2/3)$) and in a strategic game ($P_s(R\Pi, T{=}2/3, N{=}15)$), and their respective success probabilities (λ_n and $\lambda_s(R\Pi, T{=}2/3, N{=}15)$) for different levels of an alternative relative profit ($R\Pi$). (Source: Singerman, A. and Lence, S.H. (2023) The source of uncertainty influences technology adoption. *European Review of Agricultural Economics* 50, 591–623. DOI: 10.1093/erae/jbac036 by permission of Oxford University Press.)

would show the participation in a hypothetical strategic game with probability of success λ_s = 2/3. Importantly, it would be appropriate to compare such a line with the participation curve for a lottery with λ_n = 2/3, because both would represent the share of individuals opting for the risky alternative when the probability of the favorable outcome equals $\lambda_s = \lambda_n = 2/3$. For simplicity and to emphasize that participation curves with the same success probabilities in the lottery and the strategic game represent analogous concepts, hereafter we will refer to the one associated with the strategic game as the 'iso-λ_s' (strategic uncertainty) participation curve. Most notably, we propose that for a proper comparison of the share of individuals who prefer the risky alternative under strategic uncertainty (P_s) with the share of individuals who choose the game against nature (P_n), the probability of obtaining the risky alternative's high payoff should be the same for both kinds of uncertainty. That is, one should compare the iso-λ_s participation curve $P_s(R\Pi, \lambda_s = \lambda)$ versus $P_n(R\Pi, \lambda_n = \lambda)$ for $\lambda \in (0, 1)$.

For illustrative purposes, the estimated iso-λ_s participation curve for λ_s = 2/3 corresponding to the experimental data shown in Fig. 4.1 is drawn in Fig. 4.3, together with the participation line for the lottery with λ_n = 2/3. The construction and the properties of the iso-λ_s participation lines ($P_s(R\Pi, \lambda_s = \underline{\lambda})$) are explained in detail in Singerman and Lence (2023). Points I, III and V in Fig. 4.3 are the same as points I, III and V in Fig. 4.1 and are shown to help understand the relationship between the iso-λ_s participation curve $P_s(R\Pi, \lambda_s = \underline{\lambda})$ and the fixed-T participation schedules $P_s(R\Pi, T, N)$. Figure 4.3 shows that the iso-λ_s participation line is similar to its counterpart for the lottery, in that the share of individuals who prefer the risky alternative falls as the alternative certain payoff increases. In addition, the graph reveals a general pattern found in the experimental data: For a given probability of success, the share of individuals preferring the risky alternative is greater for the lottery than for strategic games at low relative profit values ($R\Pi$ < 49.8% in the graph), whereas the opposite is true at high relative profit values. This is so

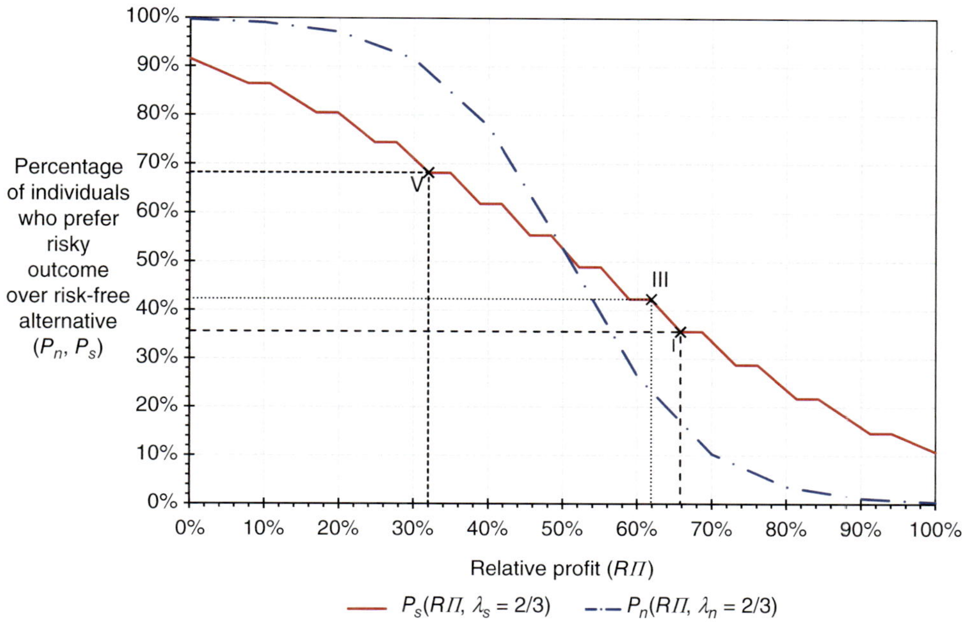

Fig. 4.3. Lottery and iso-λ_s participation lines for a success probability of $\lambda_s = \lambda_n = \lambda = 2/3$, estimated using experimental data from Florida citrus growers. (Source: Singerman, A. and Lence, S.H. (2023) The source of uncertainty influences technology adoption. *European Review of Agricultural Economics* 50, 591–623. DOI: 10.1093/erae/jbac036 by permission of Oxford University Press.)

because participation in the lottery falls relatively sharply as $R\Pi$ values get closer to the expected payoff of the lottery.

Figure 4.3 points to a potential reason why collective action may fail. If 'successful' collective action requires a substantial share of farmers to coordinate (e.g. $T \geq 60\%$), the graph shows that the relative payoff from the risk-free alternative must be sufficiently small (e.g. $R\Pi \leq 43.8\%$ for $P_s \geq 60\%$) to make the combination of risk and return of coordination attractive. However, when the relative profit value from the risk-free alternative is small, the percentage of individuals who prefer the risky outcome over the risk-free alternative is noticeably smaller when the risk involved is strategic compared to an otherwise identical situation involving non-strategic risk.

4.3.1 Sharing information about the probability of successful coordination

In addition to the differences in farmers' attitudes toward strategic risk versus non-strategic risk discussed in section 4.2, another fundamental difference arises from the fact that the probability of successful coordination depends on farmers' beliefs about the actions of their fellow farmers, instead of being set by nature. As such, in principle there is a potential for policy interventions aimed at changing farmers' beliefs and therefore the ensuing probability of a successful collective action.

Arguably, the most obvious intervention consists of improving farmers' awareness of the economic benefits from the collective action and sharing information with them regarding its aggregate probabilities of success based on previous observations, or inferred from experimental data. This type of intervention includes role-playing games designed to stimulate social learning and collective action. Salvinia *et al.* (2016) found that role-playing games tended to enhance engagement in collective action towards climate-smart agriculture. Self-help groups, defined by Desai and Joshi (2014) as 'membership-based organizations that aim to promote social cohesion through a mixture of education, access to finance, and linkages to wider development programs,' are another example of this type of intervention. In this regard, Singerman and Useche (2019a) present evidence that, in the same experimental setting from which the data to construct the present Figs 4.1 through 4.3 were obtained, a significant number of growers (approximately 30%) changed their decision from opting out to being willing to coordinate after being provided information about the benefits that accrued to growers in a well-performing real-world area-wide pest management program.

The authors interpreted this result as suggesting that either the information made those growers change their beliefs (about others) regarding the chances of achieving successful coordination, or they found the difference between the collective-action payoff and its alternative worth taking the risk. The findings by Singerman and Useche (2019a) support the notion that providing public information about positive outcomes could help increase the expected payoff and decrease the uncertainty about others' behavior. Thus, social learning, in which agents learn from the decisions of others, can be key to better manage externalities. Such social learning could take place through extension activities because they can play a key role in promoting collective action and therefore could lead to higher social welfare (Singerman and Useche, 2019a). But social learning could also take place though informal talks among producers; Conley and Udry (2010) studied the adoption of technology in Ghana and found that farmers learn about it from their neighbors.

Although it is conceivable that some growers would free-ride, thinking there is a high success probability when, say, $\lambda_s = 90\%$, two arguments can be made regarding the beneficial consequences of sharing information about the aggregate probabilities of success of collective action. First, in settings where coordination is important, public signals play a role in coordinating outcomes that exceeds the information content of those announcements (Morris and Shin, 2006). Second, Singerman *et al.* (2017) found that the main reason for area-wide pest management citrus growers not to coordinate was their beliefs about others not coordinating. Therefore, making the aggregate outcome publicly available should help reduce such (strategic) uncertainty. The other case that may discourage participation is when $\lambda_s < 50\%$, because growers might think the probability of success is low, so they may prefer to opt out instead of participating in the collective action. In such a case, before sharing the results with growers, policymakers and/or extension agents will have to come up with an incentive mechanism to foster cooperation among farmers so as to increase the probability of success.

The results from the study by Singerman and Useche (2019a) are consistent with the theoretical insights from the global games literature, according to which public information can convey strategic information on the individuals' beliefs about the beliefs of others (Morris and Shin, 2006). However, more empirical research is needed to support stronger policy recommendations regarding the impact of sharing information about coordination outcomes.

4.4 Policy Implications and Future Research

In this chapter, we focused on key issues arising when agricultural producers face strategic risk. We did so by illustrating the role of strategic uncertainty in the success (or lack thereof) of voluntary collective action to control pests, employing the outbreak of citrus greening disease in Florida as a case study. Although the negative spatial externalities from lack of collective action in pest control can be more severe and difficult to deal with for the case of perennial crops, the risks and decision-making behavior arising from strategic interaction to combat pests and diseases affecting other crops, or for dealing with the increasing pesticide resistance of weeds, are the same.

The key issues we discussed in connection with collective action included the uncertainty underlying the aggregate success of collective efforts, and the role that payoffs from the alternatives to collective action play in the decision whether to participate in the collective enterprise. We also brought attention to the critical fact that, unlike non-strategic risk, strategic risk involves probabilities that depend on individuals' beliefs about the decisions of other individuals. Therefore, the likelihood of success of collective-action efforts could in principle be managed by policy interventions designed to modify the beliefs of collective-action participants.

The multiple externalities that characterize pest management problems make a decentralized approach ineffective (Regev, 1984). The ability of individual farmers to control mobile pests depends critically on the actions of neighboring farmers; following only site-specific, individual farm-level practices ignores the collective nature of the problem. Interestingly, to date, in Florida there have been no public policies to coordinate efforts at the federal, state, or county level to deal with the disease. Therefore, there seems to be a need to increase policymakers' awareness of the availability of public research findings. Those findings can provide a scientific basis for policies to effectively deal with the externalities that arise due to the lack of coordination of individual actions (Singerman and Useche, 2019b), and the critical role that governments can play to address them (Frisvold, 2019b). Developing institutions and policies for cooperation in pest management would be valuable even if the long-term solution to HLB is genetic modification to convey host plant resistance to the disease. This is so because pest monitoring and integrated pest management practices still would need to be put into place. As Hurley (2016) showed, there is ample evidence of the development of resistance to genetically engineered crops, which, as he suggests, reflects the need to integrate resilient pest management strategies with genetically engineered crop production.

References

Almeida, R.P.P. (2016) Can Apulia's olive trees be saved? *Science* 353, 346–348.

Alston J.M., Fuller, K.B., Kaplan, J.D. and Tumber, K.P. (2013) The economic consequences of Pierce's disease and related policy in the California winegrape industry. *Journal Agricultural and Resource Economics* 38, 269–297.

Anomaly, J. (2020) Antibiotics and animal agriculture: the need for global collective action. In: Euzebiusz Jamrozik, E. and Selgelid, M. (eds) *Ethics and Drug Resistance: Collective Responsibility for Global Public Health. Ch. 18.* Springer, Cham, Switzerland, pp. 297–308.

Ayer, H.W. (1997) Grass roots collective action: agricultural opportunities. *Journal of Agricultural and Resource Economics* 22, 1–11.

Bagavathiannan, M.V., Graham, S., Ma, Z., Barney, J.N., Coutts, S.R. *et al.* (2019) Considering weed management as a social dilemma bridges individual and collective interests. *Nature Plants* 5, 343–351. DOI: 10.1038/s41477-019-0395-y.

Barham, B.L., Chavas, J.P., Fitz, D., Ríos Salas V. and Schechter. L. (2014) The roles of risk and ambiguity in technology adoption. *Journal of Economic Behavior & Organization* 97, 204–218.

Bouamra-Mechemache, Z. and Zago, A. (2015) Introduction: collective action in agriculture. *European Review of Agricultural Economics* 42, 707–711. DOI: 10.1093/erae/jbv027.

Bové, J.M. (2006) Huanglongbing: a destructive, newly-emerging, century-old disease of citrus. *Journal of Plant Pathology* 88, 7–37.

Bradshaw, C.J., Leroy, B., Bellard, C., Roiz, D., Albert, C. *et al.* (2016) Massive yet grossly underestimated global costs of invasive insects. *Nature Communications* 7, 12986.

Burnham, J.P., Olsen, M.A. and Kollef. M.H. (2019) Re-estimating annual deaths due to multidrug-resistant organism infections. *Infection Control and Hospital Epidemiology* 40, 112–113.

Carlsson, H. and van Damme, E. (1991) Equilibrium selection in stag hunt games. *Center Discussion Paper* Vol. 1991–70. Tilburg University, Netherlands.

Chavas, C.J. and Nauges, P. (2020) Uncertainty, learning, and technology adoption in agriculture. *Applied Economics Perspectives and Policy* 42, 42–53.

Citrus Industry (2018) 'No positive response' from bactericides. Available at: https://citrusindustry.net/2018/08/13/no-positive-response-from-bactericides/ (accessed May 7, 2023).

Clark, J.S. and Carlson, G.A. (1990) Testing for common versus private property: the case of pesticide resistance. *Journal of Environmental Economics and Management* 19, 45–60.

Conley, T. and Udry, C. (2010) Learning about a new technology: pineapple in Ghana. *American Economic Review* 100, 35–69.

Coy, M.R., Bin, L. and Stelinski, L. (2016) Reversal of insecticide resistance in Florida populations of *Diaphorina citri*. *Florida Entomologist* 99, 26–32.

Dale, J., James, A., Paul, J., Khanna, H., Smith, M. *et al.* (2017) Transgenic Cavendish bananas with resistance to fusarium wilt tropical race 4. *Nature Communications* 8, 1496.
Dasgupta, S., Meisner, C. and Wheeler, D. (2007) Is environmentally friendly agriculture less profitable for farmers? Evidence on integrated pest management in Bangladesh. *Applied Economic Perspectives and Policy* 29, 103–118.
Desai, R.M. and Joshi, S. (2014) Collective action and community development: evidence from self-help groups in rural India. *The World Bank Economic Review* 28, 492–524.
Dover, M.J. and Croft, B.A. (1986) Pesticide resistance and public policy. *BioScience* 36, 78–85.
Ervin, D.E. and Frisvold, G.B. (2016) Community-based approaches to herbicide-resistant weed management: lessons from science and practice. *Weed Science* 64, 609–626.
Faust, R.M. (2008) General introduction to areawide pest management. In: Koul, O., Cuperus, G.W. and Elliott, N. (eds) *Areawide Pest Management Theory and Implementation*. CAB International, Cambridge, Massachusetts, pp. 1–14.
Feder, G., Just, R.E. and Zilberman, D. (1985) Adoption of agricultural innovations in developing countries: a survey. *Economic Development and Cultural Change* 33, 255–298.
Florida Department of Agriculture and Consumer Services (FDACS) (2016) Citrus crisis declaration letter. Available at: http://files.ctctcdn.com/168a5f59201/e7bd14cf-0a3a-4f5c-9d3d-66ac12574bbc.pdf (accessed May 7, 2023).
Food and Agriculture Organization (FAO) (2015) Regional Management of Huanglongbing (HLB) in Latin America and the Caribbean. Available at: http://www.fao.org/americas/perspectivas/hlb/en/ (accessed May 7, 2023).
Foster, A.D. and Rosenzweig, M.R. (2010) Microeconomics of technology adoption. *Annual Review of Economics* 2, 395–424.
Frisvold, G.B. (2019a) How low can you go? Estimating impacts of reduced pesticide use. *Pest Management Science* 75, 1223–1233.
Frisvold, G.B. (2019b) Comment on the role of strategic uncertainty in area-wide pest management decisions of Florida citrus growers. *American Journal of Agricultural Economics* 101, 1012–1016.
Giovani, B., Blümel, S., Lopian, R., Teulon, D., Bloem, S. *et al.* (2020) Science diplomacy for plant health. *Nature Plants* 6, 902–905. DOI: 10.1038/s41477-020-0744-x.
Gould, F., Brown, Z.S. and Kuzma, J. (2018) Wicked evolution: can we address the sociobiological dilemma of pesticide resistance? *Science* 360, 728–732.
Griliches, Z. (1957) Hybrid corn: an exploration in the economics of technology change. *Econometrica* 25, 501–522.
Hendrichs, J., Vreysen, J.B., Enkerlin, W.R. and Cayol, J.P. (2005) Strategic options in using sterile insects for area-wide integrated pest management. In: Dyck, V.A., Hendrichs, J. and Robinson, A.S. (eds) *Sterile Insect Technique: Principles and Practice in Area-Wide Integrated Pest Management*. IAEA, Springer, Dordrecht, Netherlands, pp. 563–602.
Hendrichs, J., Kenmore, P., Robinson, A.S. and Vreysen, M.J.B. (2007) Area-wide integrated pest management (AW-IPM): principles, practice and prospects. In: Vreysen, M.J.B., Robinson, A.S. and Hendrichs, J. (eds) *Area-Wide Control of Insect Pests: From Research to Field Implementation*. Springer, Dordrecht, Netherlands, pp. 3–31.
Hofstadter, D.R. (1985) *Metamagical Themas: Questing for the Essence of Mind and Pattern*. Basic Books, New York.
Hu, J., Jiang, J. and Wang, N. (2018) Control of citrus Huanglongbing via trunk injection of plant defense activators and antibiotics. *Phytopathology* 108, 186–195.
Hurley, T.M. (2016) Shock and awe pest management: time for change. *Choices, Quarter* 4. Available at: http://www.choicesmagazine.org/choices-magazine/theme-articles/herbicide/shock-and-awe-pest-management-time-for-change (accessed May 7, 2023).
Hurley, T.M. and Frisvold, G. (2016) Economic barriers to herbicide-resistance management. *Weed Science* 64(sp1), 585–594.
Isik, M. and Khanna, M. (2003) Stochastic technology, risk preferences, and adoption of site-specific technologies. *American Journal of Agricultural Economics* 85, 305–317.
Jessup, A.J., Dominiak, B., Woods, B., De, C.P.F. *et al.* (2007) Area-wide management of fruit flies in Australia. In: Vreysen, M.J.B., Robinson, A.S. and Hendrichs, J. (eds) *Area-Wide Control of Insect Pests: From Research to Field Implementation*. Springer, Dordrecht, Netherlands, pp. 685–697.
Klassen, W. (2000) Area-wide approaches to insect pest management: history and lessons. In: Tan, K.H. (ed.) *Proceedings: Area-Wide Control of Fruit Flies and Other Insect Pests. International Conference*

on Area-Wide Control of Insect Pests, and the 5th International Symposium on Fruit Flies of Economic Importance, 28 May–5 June 1998, Penang, Malaysia. Penerbit Universiti Sains Malaysia, Pulau Pinang, Malaysia, pp. 21–38.

Klassen, W. (2005) Area-wide integrated pest management and sterile insect technique. In: Dyck, V.A., Hendrichs, J. and Robinson, A.S. (eds) *Sterile Insect Technique: Principles and Practice in Area-Wide Integrated Pest Management*. IAEA, Springer, Dordrecht, Netherlands, pp. 39–68.

Klassen, W. (2008) Area-wide insect pest management. In: Capinera, J.L. (ed.) *Encyclopaedia of Entomology*, Vol. 2. Springer, Dordrecht, Netherlands, pp. 266–282.

Laxminarayan, R. (2003) *Battling Resistance to Antibiotics and Pesticides: An Economic Approach*. Resources for the Future, Washington, DC.

Laxminarayan, R. and Brown, G.M. (2001) Economics of antibiotic resistance: a theory of optimal use. *Journal of Environmental Economics and Management* 42, 183–206.

Lazarus, W.F. and Dixon, B.L. (1984) Agricultural pests as common property: control of the corn rootworm. *American Journal of Agricultural Economics* 66, 456–465.

Lence, S.H. and Singerman, A. (2023) When does voluntary coordination work? Evidence from area-wide pest management. *American Journal of Agricultural Economics* 105, 243–264. DOI: 10.1111/ajae.12308.

Li, J., Pang, Z., Duan, S., Lee, D., Kolbasov, V.G. *et al.* (2019) The in planta effective concentration of oxytetracycline against '*Candidatus* Liberibacter asiaticus' for suppression of citrus Huanglongbing. *Phytopathology* 109, 2046–2054.

Lloyd, A.C., Hamacek, E.L., Kopittke, R.A., Peek, T., Wyatt, P.M., *et al.* (2010) Area-wide management of fruit flies (Diptera: Tephritidae) in the central Burnett district of Queensland, Australia. *Crop Protection* 29, 462–469.

Lockwood, B. (2008) Pareto efficiency. In: Vernengo, M., Perez Caldentey, E. and Rosser, B.J. (eds) *The New Palgrave Dictionary of Economics*. Palgrave Macmillan, London. DOI: 10.1057/978-1-349-95121-5_1823-2.

Loehman, E. and Dinar, A. (1994) Cooperative solution of local externality problems: a case of mechanism design applied to irrigation. *Journal of Environmental Economics and Management* 26, 235–256.

Marra, M., Pannell, D.J. and Ghadim, A.A. (2003) The economics of risk, uncertainty and learning in the adoption of new agricultural technologies: where are we on the learning curve? *Agricultural Systems* 75, 215–234.

Miranowski, J.A. (2016) Intervention to manage pest resistance: community-based or government regulation. *Choices* 31:1–8.

Miranowski, J.A. and Carlson, G. (1986) Economic issues in public and private approaches to preserving pest susceptibility. In: *Committee on Strategies for the Management of Pesticide Resistance Pest Populations, Pesticide Resistance: Strategies and Tactics for Management*. National Academy Press, Washington, DC.

Morris, S. and Shin, H.S. (2006) Global games: theory and applications. In: Dewatripont, M., Hansen, L.P. and Turnovsky, S.J. (eds) *Advances in Economics and Econometrics*. Cambridge University Press, Cambridge, Massachusetts.

Mumford, J. (2000) Economics of area-wide pest control. In: Tan K.H. (ed.) *Proceedings: Area-Wide Control of Fruit Flies and Other Insect Pests. International Conference on Area-Wide Control of Insect Pests, and the 5th International Symposium on Fruit Flies of Economic Importance, 28 May–5 June 1998, Penang, Malaysia*. Penerbit Universiti Sains Malaysia, Pulau Pinang, Malaysia, pp. 39–47.

Olmstead, A.L. and Rhode, P.W. (2004) An impossible undertaking: the eradication of bovine tuberculosis in the United States. *The Journal of Economic History* 64, 734–772.

Ostrom, E. (1990) *Governing the Commons: The Evolution of Institutions for Collective Action*. Cambridge University Press, Cambridge, UK.

Ostrom, E. (2009) A general framework for analyzing sustainability of social-ecological systems. *Science* 325, 419–422.

Ostrom, E. (2010) Analyzing collective action. *Agricultural Economics* 41, 155–166.

Palumbi, S.R. (2001) Humans as the world's greatest evolutionary force. *Science* 293, 1786–1790.

Perrings, C. (2001) The economics of biological invasions. *Land Use NS Water Resources Research* 1, 1–9.

Perrings, C., Williamson, M., Barbier, E.B., Delfino, D., Dalmazzone, S. *et al.* (2002) Biological invasion risks and the public good: an economic perspective. *Conservation Ecology* 6, 1.

Pimentel, D., Acquay, H., Biltonen, M., Rice, P., Silva, M. *et al.* (1993) Assessment of environmental and economic impacts of pesticide use. In: Pimentel, D. and Lehman, H. (eds) *The Pesticide Question: Environment, Economics and Ethics*. Chapman & Hall, New York, pp. 47–84.

Regev, U. (1984) An economics analysis of man's addiction to pesticides. In: Conway, G.R. (ed) *Pest and Pathogen Control*. Wiley, New York, pp. 441–453.

Regev, U., Gutierrez, A.P. and Feder, G. (1976) Pests as a common property resource: a case study of alfalfa weevil control. *American Journal of Agricultural Economics* 58, 186–197.

Rogers, E. (1962) *Diffusion of Innovations*. Free Press of Glencoe, New York.

Rook, S.P. and Carlson, G.A. (1985) Participation in pest management groups. *American Journal of Agricultural Economics* 67, 563–566.

Salvinia, G., van Paassen, A., Ligtenberg, A., Carrero, G.C. and Bregt, A.K. (2016) A role-playing game as a tool to facilitate social learning and collective action towards climate smart agriculture: lessons learned from Apuí, Brazil. *Environmental Science & Policy* 63, 113–121. DOI: 10.1016/j.envsci.2016.05.016.

Schneider, K., van der Werf, W., Cendoya, M., Mourits, M., Navas-Cortés, J.A. *et al.* (2020) Impact of *Xylella fastidiosa* subspecies *pauca* in European Olives. *Proceedings of the National Academy of Sciences* 117, 9250–9259. DOI: 10.1073/pnas.1912206117.

Sexton, S.E., Lei, Z. and Zilberman, D. (2007) The economics of pesticides and pest control. *International Review of Environmental and Resource Economics* 1, 271–326.

Singerman, A. (2017) Results in after bactericides put to the test. *Florida Grower Magazine*. Available at: https://www.growingproduce.com/citrus/insect-disease-update/results-in-after-citrus-bactericides-put-to-the-test/ (accessed May 7, 2023).

Singerman, A. and Lence, S.H. (2023) The source of uncertainty influences technology adoption. *European Review of Agricultural Economics* 50, 591–623. DOI: 10.1093/erae/jbac036.

Singerman, A. and Rogers, M. (2020) The economics challenges of dealing with citrus greening: the case of Florida. *Journal of Integrated Pest Management* 11, 1–7.

Singerman, A. and Useche, P. (2019a) The role of strategic uncertainty in area-wide pest management decisions of Florida citrus growers. *American Journal of Agricultural Economics* 101, 991–1011.

Singerman, A. and Useche, P. (2019b) Response to comment on the role of strategic uncertainty in area-wide pest management decisions of Florida citrus growers. *American Journal of Agricultural Economics* 101, 1012–1016.

Singerman, A., Lence, S.H. and Useche, P. (2017) Is area-wide pest management useful? The case of citrus greening. *Applied Economic Perspectives and Policy* 39, 609–634.

Smith, R.F., Apple, J.L. and Bottrell, D.G. (1976) The origins of integrated pest management concepts for agricultural crops. In: Apple, J.L. and Smith, R.F. (eds) *Integrated Pest Management*. Plenum Press, New York, pp. 1–16.

Staver, C., Pemsl, D.E., Scheerer, L., Perez, V.L. and Dita, M. (2020) Ex ante assessment of returns on research investments to address the impact of fusarium wilt tropical race 4 on global banana production. *Frontiers in Plant Science* 11, 884. DOI: 10.3389/fpls.2020.00844.

Van Deynze, B., Swinton, S.M. and Hennessy, D.A. (2022) Are glyphosate-resistant weeds a threat to conservation agriculture? Evidence from tillage practices in soybeans. *American Journal of Agricultural Economics* 104, 645–672.

Vanni, F. (2014) *Agriculture and Public Goods: The Role of Collective Action*. Springer, New York.

Vreysen, M.J.B., Robinson, A.S., Hendrichs, J. and Kenmore, P. (2007) Area-wide integrated pest management (AW-IPM): principles, practice and prospects. In: Vreysen, M.J.B., Robinson, A.S. and Hendrichs, J. (eds) *Area-Wide Control of Insect Pests: From Research to Field Implementation*. Springer, Dordrecht, Netherlands, pp. 3–33.

Waterfield, G. and Zilberman, D. (2012) Pest management in food systems: an economic perspective. *Annual Review of Environmental Resources* 37, 223–245.

World Health Organization (WHO) (2016) *Critically Important Antimicrobials for Human Medicine*. Available at: https://apps.who.int/iris/bitstream/handle/10665/255027/9789241512220-eng.pdf;jsessionid=4DE94E4630761B784D41A9BD4469CCD7?sequence=1 (accessed May 7, 2023).

Yu, R. and Leung, P. (2006) Optimal pest management: a reproductive pollutant perspective. *International Journal of Pest Management* 523, 155–166.

Zhang, M., Guo, Y., Powell, C.A., Doud, M.S., Yang, C. *et al.* (2014) Effective antibiotics against '*Candidatus* Liberibacter asiaticus' in HLB-affected citrus plants identified via the graft-based evaluation. *PLoS ONE* 9, e111032. DOI: 10.1371/journal.pone.0111032.

Zilberman, D. and Millock, K. (1997) Pesticide use and regulation: making economic sense out of an externality and regulation nightmare. *Journal of Agricultural and Resource Economics* 22(2), 321–332.

5 Assessing the Impact of Risk-Reducing Strategies on Food Security: Evidence from an Emerging Economy

Aditya R. Khanal[1]*, Ashok K. Mishra[2] and Gudbrand Lien[3,4]
[1]*Tennessee State University, Nashville, USA;* [2]*Arizona State University, Tempe, USA;* [3]*Inland Norway University of Applied Sciences, Lillehammer, Norway;* [4]*Norwegian Institute of Bioeconomy Research (NIBIO), Ås, Norway*

Abstract
This chapter analyzes the effects of farmers' risk-reducing strategies on productivity, which is a measure of food security. The production function is generalized to augment risk-reducing strategies variables. We used a semi-parametric smooth-coefficient (SPSC) framework in the estimation. The novelty of the SPSC approach is that it can explain the direct and indirect channels through which applying risk-reducing strategies to inputs can influence productivity. We examine four broad types of risk-reducing strategies, namely, farm diversification, the use of risk-reducing market-based instruments, the adoption of good agricultural practices (GAPs), and access to off-farm income sources. Using 2016 farm-level data from smallholder onion growers in India, we find that farm diversification and the use of GAPs as risk-reducing strategies for high-value onion crops significantly enhance productivity. Farm diversification enhances productivity by expanding the maximum revenue potential of a portfolio of combined agricultural enterprises. Finally, we find that smallholder onion farmers are operating at increasing returns to scale, which suggests that the growers could further increase productivity by increasing farm size or reducing production costs.

5.1 Introduction

In most developing and emerging economies (DEEs), smallholder households dominate the farming sector. Farmers are risk-averse in general (Binswanger, 1980). Consequently, smallholder households do not tolerate risks passively. Instead, they use a number of risk-reducing measures, as documented by the empirical literature (Bhattamishra and Barrett, 2010; Jodha *et al.*, 2012; Hardaker *et al.*, 2015). These risk-reducing measures include traditional farm management practices, such as cultivating stress-tolerant crops and seeds, diversification of farm enterprises, conserving soil and water, and the use of supplementary irrigation. Another strategy farmers use is the adoption of market-based instruments, such as contracts and crop and livestock insurance. Several studies have evaluated the benefits of adopting crop insurance and other risk-mitigating measures (e.g. Dalton *et al.*, 2004; Barham *et al.*, 2011; Di Falco and Veronesi, 2014; Vigani and Kathage, 2019). For instance, Dalton *et al.* (2004) found that crop insurance was less effective than supplemental irrigation in reducing farmers' exposure to downside risks

*E-mail: akhanal1@tnstate.edu

DOI: 10.1079/9781800622289.0005

in humid environments. Barham *et al.* (2011) reported that crop insurance was an effective means of mitigating risk at lower irrigation levels. In a study of Ethiopian farmers, Di Falco and Veronesi (2014) showed that crop diversification was a substitute for crop insurance. More recently, Vigani and Kathage (2019) concluded that adopting crop insurance provided fewer benefits than other adaptation measures (i.e. crop varieties, on-farm diversification and contracts). In summary, the literature has indicated that farmers in DEEs adopt multiple risk-mitigating strategies, involving both on- and off-farm risk management.

Take the case of on-farm risk management strategies. In the early 21st century, smallholders have often made trade-offs between crop choices and income because of market imperfections, resulting in the absence of insurance and risk-sharing instruments. In other words, smallholders trade vital portions of their incomes as a form of insurance premium (Dercon and Christiaensen, 2011). Thus, smallholders engage in intercropping, increase the number of their crops and livestock enterprises, and use variations in geographical terrain to diversify their crops. Dercon and Christiaensen (2011) noted that risk management is inherently connected to technology adoption and crop portfolio choices. In other words, smallholders habitually use risk management strategies to reduce risk and uncertainty in food production that result from weather and biological hazards. For instance, Dercon (1996) found that farmers in Shinyanga, despite facing a 20% reduction in income, planted sweet potatoes for greater food security. Similarly, Kurosaki and Fafchamps (2002) found that basmati rice farmers in Punjab, Pakistan, modified their crop portfolio to reduce risk.

Farmers have used other risk management strategies to manage their exposure to farming risk. These include adopting improved production and marketing management and off-farm work (off-farm income) strategies. Kim and Chavas (2003) and Chavas (2008) discussed the adoption of technological and managerial innovations in hybrid corn to manage production risks. In the same vein, we argue that the adoption of good agricultural practices (GAPs) is an important management decision for the safe production of vegetables. In other words, GAPs provide smallholders with the implementation of best practices that reduce microbial contamination of vegetables and disease infestation in crops after harvest and during storage. Actors along the food supply chain (growers, packers, family members, transporters, importers, exporters and retailers) should minimize food safety hazards by adopting strategies outlined in GAPs. Additionally, in a recent study, Bairagi *et al.* (2019) concluded that adopting GAPs increased farm income and reduced fertilizer usage (cost reduction) of smallholders in Nepal. Regarding off-farm work, studies have found that the smallholders in DEEs acquire a large part of their income from off-farm sources (Reardon, 1997; Ruben *et al.*, 2001). Off-farm income reduces the variance in overall income risk, serves as collateral for credit, and improves food security (Reardon, 1997). Several studies in DEEs have discussed the roles of income diversification through farm enterprises (crops and livestock) and off-farm activities (Benin *et al.*, 2004; Kijima *et al.*, 2006; Matsumoto *et al.*, 2006; Di Falco and Chavas, 2009; Di Falco *et al.*, 2010; Bezabih and Di Falco, 2012; Ullah and Shivakoti, 2014; Birthal and Hazrana, 2019).

Although several studies in the literature have investigated the impact of contract farming and adoption of GAPs on farm profitability, employment, and output, none has simultaneously estimated the impact of several risk-reducing strategies on farm profitability (gross revenue). Therefore, this chapter investigates the impact of four risk-reducing strategies – namely, farm enterprise diversification, use of risk-reducing market instruments (crop and livestock insurance, contract farming), use of GAPs, and off-farm income – on the productivity of smallholder onion farmers in India.

This chapter contributes to the literature on several fronts. First, it examines the direct and indirect effects of the four risk-reducing strategies simultaneously on farm productivity. Second, by including the adoption of GAPs and off-farm income in risk-reducing and risk-coping strategies to enhance farm productivity, it fills a gap in the literature, as these issues have been relatively neglected. Third, we use a semi-parametric smooth-coefficient (SPSC; see, Li and Racine, 2010) production function to estimate our empirical model, which makes a methodological contribution to the literature. Finally, we directly connect the use of risk management strategies to food security. We link farm productivity, farm returns, and profitability potential – measured by gross farm revenue in crops – with the livelihoods of growers and the performance of the agriculture sector at large.

5.2 Smallholder Risk Aversion, GAPs and Off-Farm Income

In DEEs, individual families decide on what to produce and derive their livelihoods to protect their income and subsistence. Farmers consider the relative riskiness of GAPs when making their production decisions. Such practices are becoming increasingly important in DEEs, not only because of national mandates but also because of regulations imposed by international companies that are engaged in value chains and export-oriented production. For instance, under GAPs, if smallholders are engaged in onion or ginger farming, they need to be aware of clean storage, temperature and humidity control, hygiene procedure, worker health when handling produce, and ensuring that packages are properly labeled and coded for unique identification that can be traced back. However, the literature has shown that risk-averse farmers view new technologies, such as those involved in GAPs, as risky and thus are less likely to adopt them (Knight *et al.*, 2003; Vargas Hill, 2009; Hurley, 2010; Liu, 2013). Similarly, Mohan (2020) and Ayuya *et al.* (2015) argued that risk-averse farmers regard organic certification as a risky proposition. Conversely, the literature on contract farming in DEEs shows that smallholders regard contracts as reducing risk, such that relatively risk-averse smallholders are more likely to participate in contract farming than in GAPs (Mishra *et al.*, 2018).

Thus, in this study we posit that smallholders view the adoption of GAPs as involving more risk and higher returns compared with their counterparts. Like Mohan (2020), we argue that smallholders' risk aversion does not vary over time. Finally, we note that smallholders are free to choose whether to adopt GAPs. The use of such practices as a risk management strategy has not received much attention in the literature. One exception is Mohan (2020), who recently found a positive and significant relationship between risk preferences and the adoption of organic production practices. Specifically, the author investigated the impact of the risk attitudes of Nepalese smallholder tea growers on their decisions regarding whether to obtain organic certification. Mohan (2020) found that more risk-averse tea farmers in Nepal had a higher propensity to obtain certification compared with less risk-averse tea farmers. Thus, we posit that risk-averse onion farmers in our study are more likely to adopt a higher number of GAPs compared with their less risk-conscious counterparts. Consequently, we consider the adoption of these practices as a risk-reducing strategy.

Off-farm work is another risk management strategy used by smallholders around the globe to mitigate the effects of agricultural risks. Smallholders face variability in farm income, and to smooth income and consumption behavior, farm families diversify and supplement their household income by working off the farm. Off-farm income is often a reliable source of income. Following Markowitz (1952), risk-averse households are likely to have portfolios or engage in activities whose individual returns are uncorrelated or negatively correlated. Additionally, off-farm income could be used to invest in the farm, relax credit constraints, and adopt new farming technologies. In a study of Ghanaian farmers, Asravor (2018) found that small farmers and farmers who face production risks are more likely to prioritize off-farm work as a risk management strategy as compared to the counterpart farms not affected by production risks.

5.3 Modeling Framework

5.3.1 The conceptual model

In measuring farm-level productivity, the production function is typically (one of) the main 'work horses' for econometric analysis. The standard production function is a technical relationship between inputs and output. It can be written as $Y = f(X)$, where Y is the output and X is typically a vector for inputs. Specification of $f(\cdot)$ is frequently done with a Cobb–Douglas function, or a more flexible translog function (Coelli *et al.*, 2005). However, in reality, the amount of output produced given a certain level of inputs often depends not only on the inputs but also on other environmental factors, Z. The standard production function model can be extended to include these Z variables as control variables in the production function, $Y = f(X,Z)$. Both the standard production function and the slightly extended production function can be estimated as linear parametric functions using the ordinary least squares method. In a further extension of $Y = f(X,Z)$, risk indices (parametric functions of Z variables) can enter the production function as inputs ($Y = f(X,g(Z))$), as in Lien *et al.* (2017), which can then be estimated using non-linear regression. However, a problem with both of these extensions is that Z is different from X because, unlike X, Z is not a standard input.

In our model, the Z variables denote risk-reducing strategies. Thus, we use a framework in which the Z variables change the production technology. That is, $Y = f|_Z(X)$, where $f|_Z(.)$ indicates that the production function changes (shifts) with Z; thus, there is no single production function for all levels of Z. This specification is different from $Y = f(X,Z)$, in which the technology is the same for all producers. In our specification $Y = f|_Z(X)$, the parameters are specific to the Z variables. For example, in a Cobb–Douglas specification of $Y = f|_Z(X)$, i.e. $Y_i = X_i'\beta$, the β parameters (the input elasticities) that define the technology are functions of Z. That is, $Y_i = X_i'\beta(Z)$. To make our specification more flexible, we assume that $\beta(Z)$ is non-parametric. This modeling framework is referred to as the SPSC regression model, in which the parameters (both the intercept and the slope parameters) are non-parametric functions of Z (Li *et al.*, 2002; Li and Racine, 2010).

5.3.2 Model specification

The SPSC Cobb–Douglas regression model can be written as:

$$Y_i = \alpha(Z_i) + X_{1i}\beta_1(Z_i) + X_{2i}\beta_2(Z_i) + X_{3i}\beta_3(Z_i) + u_i, i = 1,\cdots,n \quad (5.1)$$

where Y_i is the logarithm (log) of output (measured using gross revenue) for farm *i*, X_{1i} is the log of land input, X_{2i} is the log of labor inputs, X_{3i} is the log of materials (Indian rupees), and Z_i is a vector of risk-related variables. We include four risk-reducing strategy variables in our empirical model: farm diversification (Z_1); use of risk-reducing instruments (Z_2); GAPs (Z_3); and access to off-farm income (Z_4). Finally, the β parameters are smooth non-parametric functions to be estimated, and u_i is a random disturbance term.

Note that the model has a parametric structure in terms of the input variables. The SPSC model is very flexible because it allows the parameters to change with Z and thus the parameters are observation-specific if the Z variables are observation-specific. The overall marginal effect of the Z variables on output can be decomposed into direct and indirect marginal effects. If all the β parameters are constants (i.e. $\beta_j(Z_i) = \beta_j$, $j = 1, \cdots 3$), the model collapses to the partially linear model of Robinson (1988), where only the intercept $\alpha(Z_i)$ is the non-parametric part of the model. If the β_j parameters as well as the intercept are linear functions of the Z variables, the model reduces to a restricted translog function (without the square and cross-products of the inputs and the Z variables). In matrix form, Eqn 5.1 can be written as:

$$Y_i = X_i'\delta(Z_i) + u_i, \quad (5.2)$$

where $X_i' = (1, X_{1i}, X_{2i}, X_{3i})$ and $\delta(Z_i) = (\alpha(Z_i), \beta_1(Z_i), \beta_2(Z_i), \beta_3(Z_i))'$. This model can be estimated by employing the kernel method (Li *et al.*, 2002). Pre-multiplying Eqn 5.2 by X_i and taking the conditional expectation $E(\cdot|Z_i)$ yields the following:

$$E(X_iY_i|Z_i) = E(X_iX_i'|Z_i)\delta(Z_i) \quad (5.3)$$

because $E(X_iu_i|Z_i) = 0$. Rearranging Eqn 5.3, we obtain:

$$\delta(Z_i) = \left[E(X_iX_i'|Z_i)\right]^{-1} E(X_iY_i|Z_i) \quad (5.4)$$

Then, the parameter function $\delta(Z_i)$ can be estimated by:

$$\hat{\delta}(Z_i) = \left[\sum_{i=1}^{n} X_i X_i' K(Z_i, z)\right]^{-1} \sum_{i=1}^{n} X_i Y_i K(Z_i, z) \quad (5.5)$$

where $K(Z_i, z)$ is the kernel function. As discussed in the literature (for example, Henderson and Parmeter (2015)), there are several ways to specify the kernel function (for example, Gaussian, uniform, or Epanechnikov) and several ways to specify the bandwidth (for instance, least-squares cross-validation or likelihood cross-validation). For this study, we chose Gaussian kernel functions and least-squares cross-validation to select the bandwidths.

5.3.3 Constrained estimation

According to neoclassical production theory, marginal products should be non-negative. However, in some cases in empirical work, and especially when applying the flexible SPSC model, marginal products (elasticities) can be negative. Because this is counterintuitive, it is desirable to impose constraints to ensure that they are non-negative. One way to mitigate this potential problem is to consider constrained estimation. There is extensive literature on constrained semi-parametric estimation (see Henderson and Parmeter (2015), Chapter 12 for an overview). To impose the required constraints, we used the approach described by Bhaumik *et al.* (2018) and rewrite Eqn 5.5 as:

$$\hat{\delta}(Z_i) = \sum_{i=1}^{n} A(X_i, Z_i, z) Y_i \tag{5.6}$$

where

$$A(X_i, Z_i, z) = \left[\sum_{i=1}^{n} X_i X_i' K(Z_i, z)\right]^{-1} \sum_{i=1}^{n} X_i Y_i K(Z_i, z).$$

This equation can be written as:

$$\hat{\delta}(Z_i) = n.\sum_{i=1}^{n} A(Xi, Zi, z).p_u.Y_i \tag{5.7}$$

where $p_u = n^{-1}$ denotes uniform weights. Then, we introduced a re-weighting (from the uniform weight) scheme that changes the dependent variable as little as possible, but at the same time avoids negative input elasticities. We can specify the constrained SPSC estimator as:

$$\hat{\delta}*(Z_i) = n.\sum_{i=1}^{n} A(Xi, Zi, z).p_i.Y_i \tag{5.8}$$

where p_i denotes the observation-specific weights and $\sum_i p_i = 1$. To select the optimal p_i, we minimize $\sum_i (p_i - p_u)^2$ subject to $\beta(Z_i) \geq 0$ using quadratic programming (the quadprog package in R).

5.3.4 Marginal effects of the Z variables

The change in the expected value of Y_i (which is in log form) with respect to a change in a particular element of X_i, say $X_{ji} = X_j$ ($j = 1, 2, \cdots$) (which are also in log form), is $\beta_j(Z_i)$, based on Eqn 5.1. That is, the input elasticities $\beta_j(Z_i)$ are defined as:

$$\frac{\partial \hat{Y}}{\partial \hat{X}_j} = \beta_j(Z_i) \tag{5.9}$$

Note that as long as the Z variables are observation-specific and these input elasticities are functions of Z_i, we can obtain observation-specific input elasticities.

Based on Eqn 5.1, the change in the expected value of Y_i with respect to a change in a particular element of Z_i, say Z_{ki} ($k = 1, 2, \ldots.$), is defined as:

$$\frac{\partial \hat{Y}}{\partial Z_{ki}} = \frac{\partial \alpha(Z_i)}{\partial Z_{ki}} + \sum_{j=1}^{j} X_j \frac{\partial \beta_j(Z_i)}{\partial Z_{ki}} \tag{5.10}$$

Then, the marginal impacts of the risk-related Z variables depend on each of the Z and X variables included in the analysis. The first component of Eqn 5.10 is the direct marginal impact of Z_{ki}, and the second component is the indirect marginal impact of Z_{ki}, which can be further decomposed into *J* elements of X.

5.4 Data

Our study is based on a primary survey conducted during March and April 2016 in the Nashik and Jalgaon districts of the Indian state of Maharashtra. Farmers in the Jalgaon district had formal contracts for the production and supply of white onions with Jain Farm Fresh Foods Limited, a subsidiary owned by Jain Irrigation System Limited. The company is based in the Jalgaon district and processes white onions for export. The onion growers who had contracts with Jain Farm Fresh Foods Limited were from the villages of Shirsoli, Vadali, Pasardi, Dhanora, Panchak, Nashirabad and Mhaswad in Jalgaon district. Non-contract smallholders and independent farmers were selected from the Nashik district, next to the Jalgaon district. The Nashik district is the largest onion-producing district in Maharashtra, accounting for more than 25% of state onion production. Both districts,

Jalgaon and Nashik, are situated in the same agro-climatic zone, the western Maharashtra agro-climatic zone.

We commenced by identifying independent onion growers using the Nashik District's information (Nashik District, 2022) and data on agriculture and onion blocks in collaboration with district and local offices. We randomly chose independent onion growers from 15 villages in three blocks in Nashik, namely Nifad, Sinnur and Baglan. The blocks were classified into three categories, representing low, medium, and high onion production areas, and we randomly selected one block from each category. The sample of independent onion growers was distributed almost equally among the three blocks, Nifad (172), Sinnur (169) and Baglan (188). Five sample villages were chosen randomly from each of the three blocks, and a sample of about 35 households was randomly chosen from each sample village. The data collected included farm and farmer characteristics, cropping patterns, economics of cultivation, marketing channels, information on GAPs, assets, income, expenditure, social networks, perception of weather and pest risks, and smallholders' risk attitudes. In addition, the survey collected data on gross output and inputs, including land, labor and other materials.

The smallholder onion growers were queried on the number of enterprises in which they were engaged on their farm, that is, whether they grew a single crop (indicating specialized farms) or multiple crops (rice, wheat and millet) and whether they kept livestock, poultry and other animals (oxen). The survey indicated that smallholders could diversify their farms into up to eight enterprises. In addition, the smallholders engaged in other market-based risk management strategies, including buying crop insurance, buying livestock insurance and contract farming. Thus, farmers could participate in all three possible market-based risk management strategies. Finally, the adoption of risk management strategies such as GAPs was also linked to the smallholder's risk preferences, and the survey also collected specific information on types of GAPs from the onion growers.

The survey included 45 questions related to GAPs. Smallholders were asked if they used such practices or if they considered such practices not applicable to them. The questions on GAPs covered the use of organic fertilizer, pesticides, water and micro-nutrient use, soil testing, prevention of water flow from sewers and from undesirable sources like municipal landfills, hospitals and industrial waste dump areas. GAP questions also covered whether farmers followed soil treatment suggested by the National Agricultural Research Council, their maturity standards for crops before they were harvested and their awareness of the maximum residue level permitted in their target market/country. They were also asked about their hygiene procedures while handling harvested produce and produce packed and handled directly in the field, orchard, or greenhouse, including whether farm workers received basic instructions on hygiene before handling produce and whether they implemented the hygiene instructions on handling produce. Other questions included whether their packages were properly labeled and coded for unique identification and traceback, whether there were controls for temperature and humidity (where applicable) where produce was packed and/or stored, whether they used curing and ripening agents approved/recommended by relevant government agencies, and whether product store walls, floors, and horizontal surfaces of any storage, holding, or reception facilities were cleaned and, where appropriate, treated with insecticide before use.

The rural economy in India has undergone a slow but steady shift toward the non-farm sector, which has provided smallholders and their families with opportunities to acquire non-farm jobs and incomes. Such diversification into non-farm employment has been attractive to smallholders for several reasons. Income earned from non-farm jobs helps overcome fragmented land and other land constraints, increases household income, absorbs shocks to farm income, and provides funds for purchasing more and better-quality farming inputs and technologies (Collier *et al.*, 1986; Reardon and Taylor, 1996). Birthal *et al.* (2014) noted that about 46% of rural farmers in India engaged in off-farm work to earn off-farm wages and salaries and found that 20% of farmers' income originated from off-farm wages and salaried employment. Similarly, the authors note that about 64% of sub-marginal farmers (0.002–0.5 ha) reported off-farm work, leading to more than 36% of income from off-farm wages and salaried employment. Due to lack of data on off-farm work (hours allocated to off-farm work by operator, spouse and adults in the family), in this study we use the share of off-farm income to total household income to measure the effect of off-farm work on farm productivity.

After excluding surveys that were incomplete or missing information on the dependent and independent variables, our sample consisted of 505 households. Table 5.1 presents the descriptive statistics of the variables used in the analysis. Drawing on the survey data, we specified the production function using gross revenue as output (Y), and three inputs: land (X_1), measured in acres; labor (X_2), measured in person-year equivalents; and materials (X_3), measured in Indian rupees. On average, our sample onion growers had gross revenue of approximately 122,000 rupees and held about 4 acres ($\approx$ 1.6 hectares) of land. However, there was significant variability in both gross revenue and land holdings among the sampled growers, with gross revenue ranging from 1200 rupees to 960,000 rupees, and land holdings ranging from 0.5 acre to 60 acres. As noted earlier, the risk-reducing strategies are included as Z variables within the production function. Our Z variables comprise the following: Z_1, representing farm diversification, based on the number of farm enterprises; Z_2, which denotes the use of contract farming, crop insurance, and livestock insurance as risk-reducing instruments, with possible values ranging from 0 to 3; Z_3, which denotes the number of GAPs adopted by farmers (from a total of 40 possible practices); and Z_4, which denotes off-farm income, measured as the share of off-farm income in total income. The use of farm diversification, contracts, and insurance was discussed at length by Hardaker *et al.* (2015) and Moschini and Hennessy (2001). However, the use of GAPs as a risk management strategy has received little attention in the literature. One recent exception, as noted earlier, is Mohan (2020), who found a positive and significant relationship between risk preferences and the adoption of organic production practices.

5.5 Empirical Results

First, we present our findings on how productivity was influenced by risk-reducing strategies. Then, we turn to our findings on input elasticities and returns to scale.

5.5.1 Direct, indirect and overall marginal impacts of the Z variables

Figure 5.1 shows the direct, indirect and overall marginal impacts of the four categories of risk-reducing strategies, namely farm diversification (*DF*), use of risk-reducing instruments (*RI*), good agricultural practices (*GAP*) and off-farm income (*OF*). *DF* as a risk-reducing strategy had a negative direct impact on productivity at the median but positive indirect and overall impacts, indicating that, on average, farm diversification involving the adoption of a greater number/type of enterprises (e.g. crops and livestock) increased overall productivity. Table 5.2 complements the figures by presenting the overall positive and negative estimates in percentage

Table 5.1. Descriptive statistics (N = 505). (Source: Authors' own work.)

Variables	Label	Mean	Std. dev.	Min	Max
Output					
Gross revenue (Rupees)	Y_1	121,956	116,759	1,200	960,000
Inputs					
Land (acre)	X_1	4.32	6.24	0.50	60.00
Labor (person-year equivalents)	X_2	0.34	0.32	0.05	3.08
Materials (Rupees)	X_3	82,975	57,773	8,600	549,900
Risk-reducing strategies					
Farm diversification (*DF*) (0–8)	Z_1	2.94	1.63	0.00	8.00
Use of risk-reducing instruments (*RI*) (0–3)	Z_2	0.26	0.55	0.00	2.00
Good agricultural practices (*GAP*) (0–40)	Z_3	13.81	6.64	0.00	40.00
Off-farm income (*OF*) (share, %)	Z_4	0.06	0.16	0.00	0.93

Table 5.2. Summary statistics of mean and median overall impacts of the risk-reducing strategies, and the extent of positive and negative impacts, in percentage terms. (Source: Authors' own work.)

			Positive estimates	Statistically significant (5%)		
				Negative	Not significant	Positive
Overall impacts	Mean	Median	%	%	%	%
Farm diversification (*DF*)	0.002	0.002	80	0	81	19
Use of risk-reducing instruments (*RI*)	−0.001	0.00002	50	12	77	11
Good agricultural practices (*GAP*)	0.007	0.005	70	2	76	22
Off-farm income (*OF*)	−0.00001	−0.00001	45	2	94	4

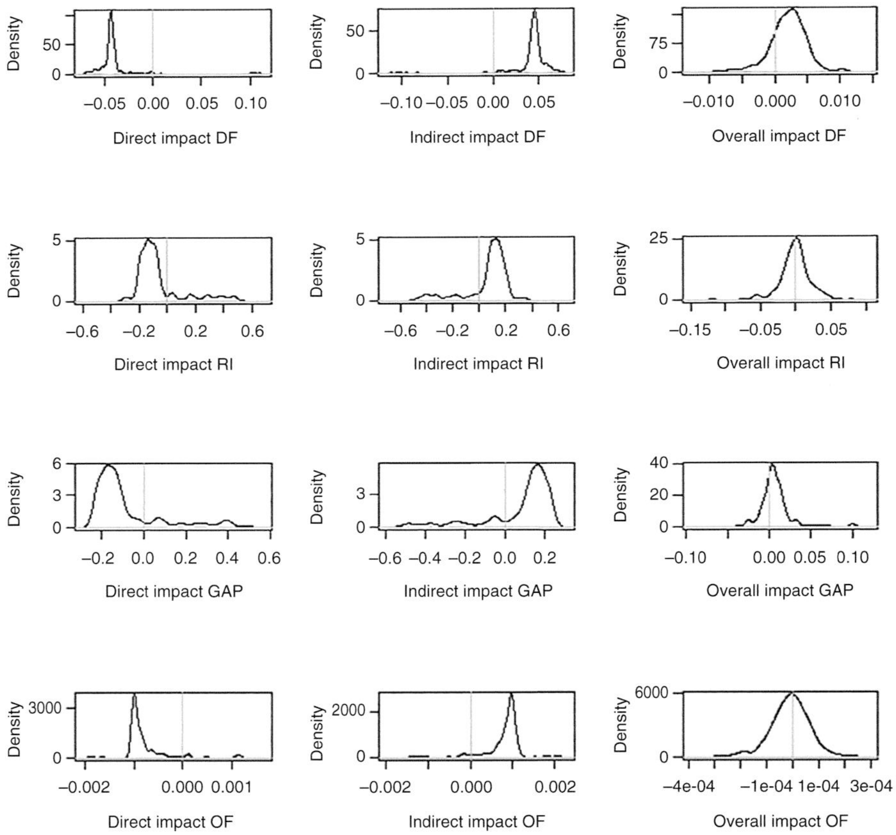

Fig. 5.1. Density plots of direct, indirect and overall marginal impacts of risk-reducing strategies of farm diversification (*DF*), use of risk-reducing instruments (*RI*), good agricultural practices (*GAP*) and off-farm income (*OF*).

terms. As the table shows, the positive estimate of 80% for *DF* suggests that farm diversification was an important risk-reducing strategy that positively affected productivity.

Our finding is consistent with that of Mishra *et al.* (2004), who identified enterprise diversification as a self-insuring risk management strategy to effectively manage farming risk, and with Kim *et al.* (2012), who found positive but small productivity gains from farm diversification among Korean rice farmers. Similarly, Joshi *et al.* (2006) found that smallholders in Uttar Pradesh,

India, who diversified their farms into high-value crops, such as vegetable and fruit production, reported higher income than did smallholders who did not diversify production. Joshi *et al.* (2006) argued that the additional enterprises on the farm utilized family labor, used complementary resources, and increased efficiency.

Our results indicated that the GAP variables were important for risk reduction, as shown in Fig. 5.1 and Table 5.2. Figure 5.1 illustrates that GAPs had negative direct impacts, but these were offset by positive indirect impacts, resulting in a positive impact overall of these practices on productivity. Table 5.2 shows a positive estimate of 70% for GAPs on productivity.

The results for the remaining Z variables in Fig. 5.1 and Table 5.2 do not provide a clear indication of the direction of the overall effect on productivity; that is, it is not clear whether it was positive or negative. For example, we found that the use of risk-reducing market instruments (Z_2) had negative direct impacts and positive indirect impacts, resulting in neutral overall impacts. Similarly, off-farm income (Z_3) had negative direct and positive indirect impacts, which resulted in a neutral and insignificant overall impact.

Thus, our results suggest that, of the four categories of risk-reducing strategies, farm diversification and the adoption of GAPs were the most effective strategies for reducing risk and, hence, they had positive effects on productivity and, in turn, on food security. It is likely that GAPs enhance input utilization and enable value standardization for growers. Indeed, Mariano *et al.* (2012) concluded that farmers are likely to adopt these practices because they are profit-oriented, suggesting that farmers perceive these practices as increasing profits. Our results are consistent with those of Bairagi *et al.* (2019), who found that smallholder Nepalese farmers who engaged in farming both high-value crops (ginger and tomatoes) and low-value crops (lentils and hybrid rice seeds) had higher farm incomes than their counterpart non-adopters if they adopted GAPs. Similarly, Kumar *et al.* (2011) found that dairy farmers in India who adopted food safety practices had higher farm income than other farmers. Handschuch *et al.* (2013) found that the adoption and implementation of GAPs by Chilean raspberry growers had positive effects on raspberry quality and on net farm income. Finally, Krause *et al.* (2016) found that mango farmers in Thailand who adopted such practices earned significantly higher incomes than those who did not.

To explore and visualize the marginal impacts of the Z variables further, we used the 45° plot method, described by Henderson *et al.* (2012). The results are presented in Fig. 5.2. Because the SPSC model provides observation-specific estimates (e.g. for marginal impacts or other measures), we generated observation-specific confidence intervals for the estimates by using the residual-based wild bootstrap method (Li and Racine, 2010). For each observation-specific estimate, we plotted the estimate (the gradient in this case) against itself, resulting in a 45° line (represented with black triangles (△)). Next, we plotted upper and lower 95% confidence intervals, represented by red △ and green △ triangles, respectively, around each of the estimated points on the 45° line. When the upper and lower confidence bounds are both in the upper right (lower left) quadrant, then the single-point estimate for that observation is positive (negative) and statistically significant from zero. When the upper and lower confidence bounds straddle the horizontal line, the single-point estimate is not statistically different from zero. As reported in Fig. 5.2, the impacts of the risk-reducing strategy variables on productivity were mixed and not always statistically significant from zero. Moreover, the variability in the estimates was indicative of the heterogeneous impact of the risk-reducing strategies for farmers. However, the robustness of our findings regarding farm diversification and GAPs is confirmed by the fact that for both variables, the observations in the overall impact plots are weighted toward the upper right quadrant in Fig. 5.2.

5.5.2 The estimated elasticities and returns to scale

The SPSC model does not give 'standard' parameter estimates because the 'parameters' are nonparametric functions. As a result, we cannot adopt the standard approach of presenting parametric estimates in a table with standard errors (Henderson and Parmeter, 2015). The 'parameters' are observation-specific or, more precisely, they are the gradient estimates which, in the Cobb–Douglas case, represent the elasticities of output with respect to the inputs.

In Fig. 5.3, we plotted the kernel density of the intercept, the input elasticities, and the

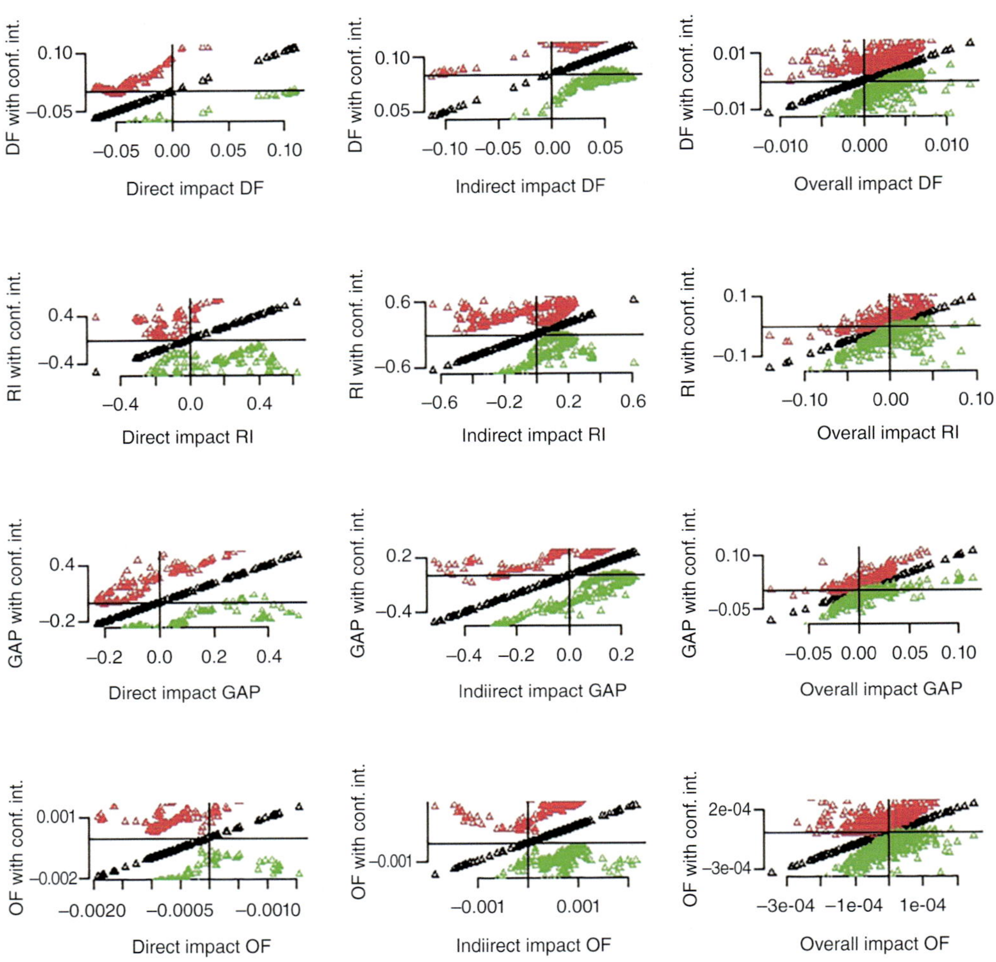

Fig. 5.2. 45° plots of estimated gradients of the four categories of risk-reducing strategies, farm diversification (*DF*), use of risk-reducing instruments (*RI*), good agricultural practices (*GAP*) and off-farm income (*OF*) and confidence interval (conf. int.) of the estimates. The estimate (the gradient) against itself, resulting in a 45° line, is represented by black triangles; the upper and lower 95% confidence intervals are represented by red and green triangles respectively.

returns to scale (sum of the input elasticities). Figure 5.3 shows that the elasticity with respect to materials is the highest (mean 0.44, median 0.43) of the input elasticities. Labor (mean 0.34, median 0.38) and land (mean 0.32, median 0.30) have very similar distributions. Returns to scale range from 0.98 to 1.15, with a mean of 1.11 and a median of 1.12, implying increasing returns to scale. This result suggests that onion growers in India could reduce their costs of production by increasing the scale of their farming operations.

Figure 5.4 shows the distribution of the estimates for the intercept and three inputs (single point estimates for the intercept and the elasticities for the land, labor, and material inputs), and the returns to scale (the black triangles (△)). For the three inputs, the elasticities for the whole sample are positive and statistically significantly different from zero. Indeed, these estimates are consistent with theory and agree with the results of Mishra *et al.* (2020). Figure 5.4 also shows statistically significant increasing returns to scale for all farms in the sample. Some onion farmers had returns to scale close to 1.10, although for the majority of farms, returns to scale ranged between 1.12 and 1.16. Our finding contrasts with Mishra *et al.* (2019),

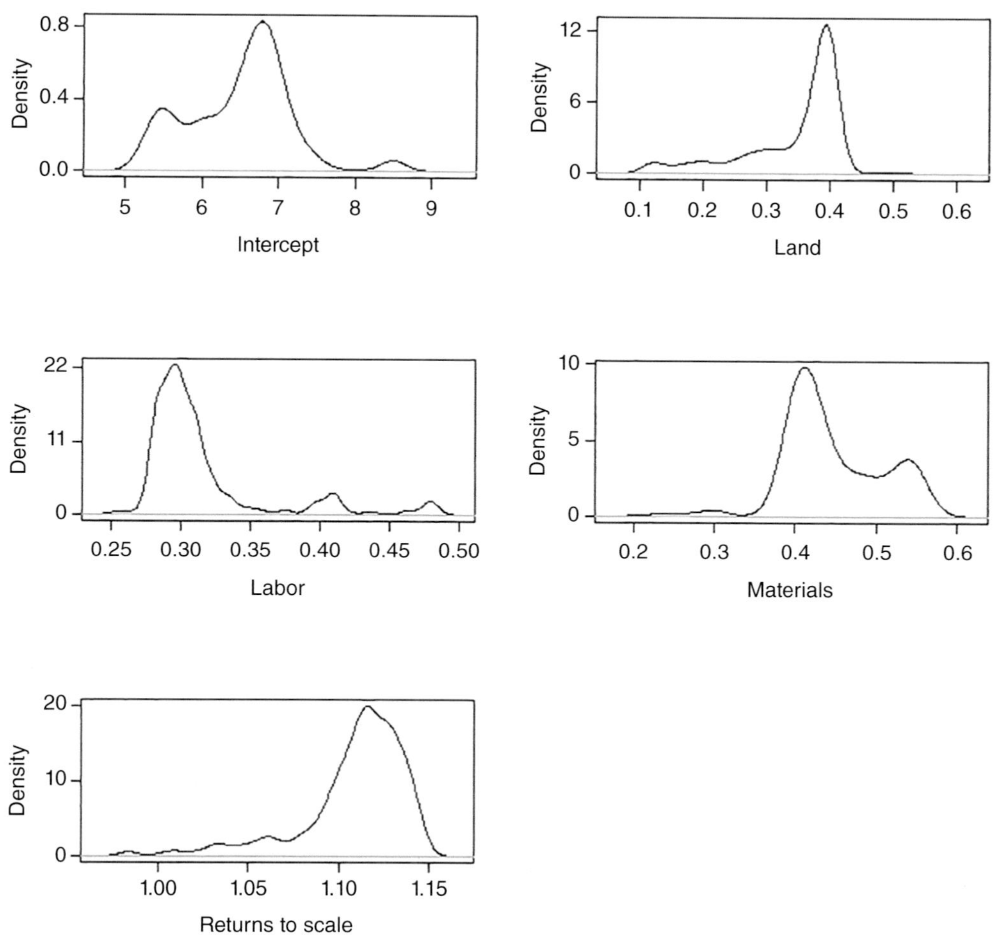

Fig. 5.3. Kernel density plots of the intercept, elasticities and the returns to scale.

who found decreasing returns to scale for high-value crop growers in Nepal.

5.6 Concluding Remarks and Implications

We examined the risk management strategies adopted by onion farmers in India and estimated the effect of four risk-reducing strategies on farm productivity. Farm productivity (measured by gross revenue) is inherently linked to the food security of the smallholder onion growers, who derive their household income and livelihood from the revenues generated from agriculture. We considered four risk-reducing strategies, namely farm diversification, market-based risk instruments (insurance and contracts), the adoption of GAPs, and access to off-farm income. In addition to our methodological contribution of using the SPSC approach to analyze high-value crop productivity, our study makes several other contributions to the literature. Specifically, we examined the direct and indirect effects of four risk-reducing strategies simultaneously on farm productivity. Two of these strategies, the adoption of GAPs and access to off-farm income, are not commonly included as risk-reducing and risk-coping strategies in the productivity literature.

In a semi-parametric SPSC framework, we specifically tested for the effects of the above four risk-reducing strategies on productivity. We found that farm diversification and the adoption of GAPs were the most important risk-reducing strategies in terms of positive effects on the prod-

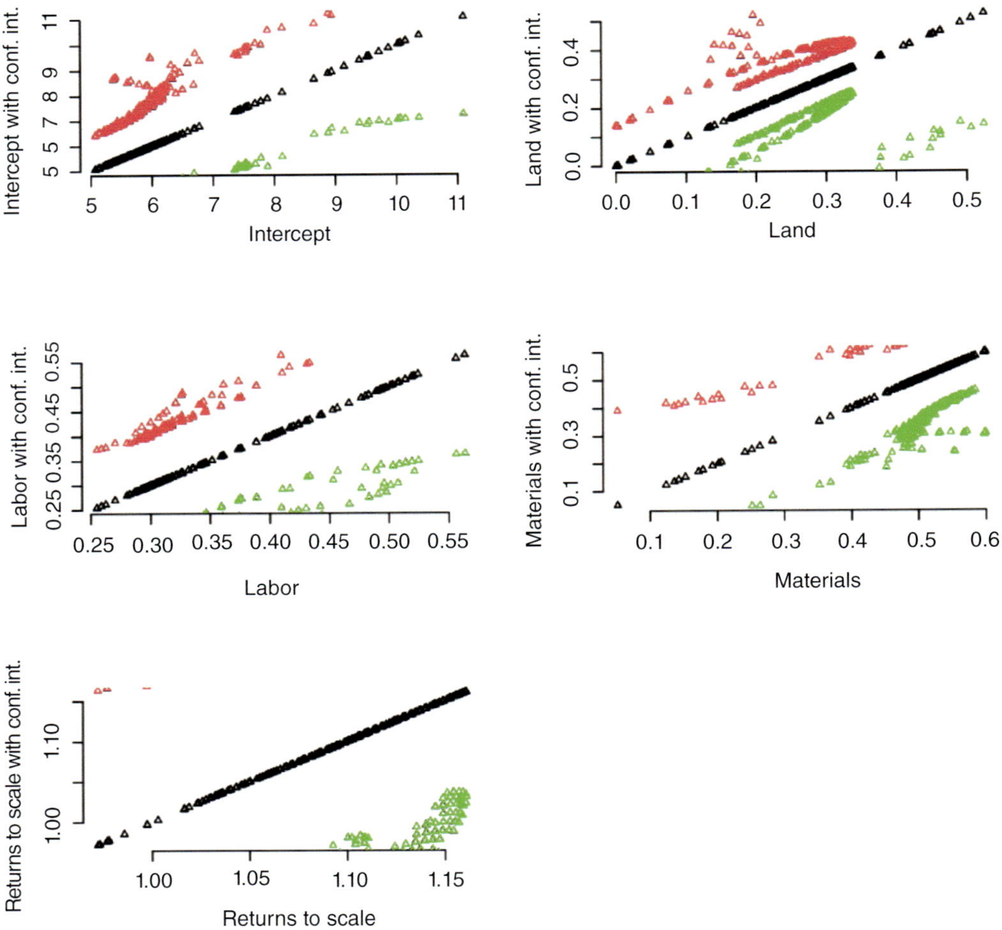

Source: Authors' own work.

Fig. 5.4. 45° plots of estimated gradients and their confidence intervals (conf. int.) for the intercept, three coefficients (land, labor and materials) and returns to scale. The estimate (the gradient) against itself, resulting in a 45° line, is represented by black triangles; the upper and lower 95% confidence intervals are represented by red and green triangles respectively.

uctivity of onion growers. Smallholders specializing in high-value crops such as onions can enhance their productivity by embracing enterprise diversification (effectively using family labor and complementary inputs), which would enable them to maximize their revenue from a combined portfolio of enterprises. In addition, farmers can increase their revenue and productivity by adopting GAPs that augment input utilization and value standardization. These practices are especially important if farmers wish to maximize their profits and serve the needs of their consumers, especially when their commodities are export-oriented. Our results on returns to scale indicated that the smallholder onion growers could further increase productivity by increasing farm size or reducing production costs.

Although we do not imagine that many Indian onion farmers will read this chapter, we think our findings have implications for advice given by government organizations, non-governmental organizations (NGOs), extension workers, and policies developed by policymakers. Diversification, or 'don't put all your eggs in one basket,' is one of our main findings about risk-reducing strategies for increased productivity and food security. Indeed, our finding is a well-known strategy in the literature. However, it does not necessarily mean that farmers should diversify as much as possible, as it can have diminishing benefits (Hardaker *et al.*, 2015). Perhaps three forms of production capture most of the gain, while management of multiple activities gets tricky, and benefits may be lost.

Extension workers should focus intensely on the guidelines for GAPs in their efforts and outreach work. Additionally, policymakers should increase the focus on incentives that can stimulate the uptake of such practices. Besides, if smallholders are convinced and observe empirical evidence that the adoption of these practices may increase farmers' productivity, the farmers could be motivated to hear and adopt the practices. Finally, adopting GAPs may reduce the probability of overuse of farming inputs (e.g. pesticides, herbicides and irrigation), which result in soil degradation and environmental damage. Similarly, adopting GAPs reduces the risk of a lower-standard product and increases the export opportunities for their products, all with the possibility of increasing farmers' income.

Acknowledgment

We would like to thank Emeritus Professor J. Brian Hardaker for useful comments in improving this chapter.

References

Asravor, R. (2018) Smallholder farmers' risk perceptions and risk management responses: evidence from the semi-arid region of Ghana. *African Journal of Economic and Management Studies* 9, 367–387.

Ayuya, O.I., Gido, E.O., Bett, H.K., Lagat, J.K., Kahi, A.K. *et al.* (2015) Effect of certified organic production systems on poverty among smallholder farmers: empirical evidence from Kenya. *World Development* 67, 27–37.

Bairagi, S., Mishra, A.K. and Giri, A. (2019) Good agricultural practices, farm performance, and input usage by smallholders: empirical evidence from Nepal. *Agribusiness* 35, 471–491.

Barham, E.H.B., Robinson, J.R., Richardson, J.W. and Rister, M.E. (2011) Mitigating cotton revenue risk through irrigation, insurance, and hedging. *Journal of Agricultural and Applied Economics* 43, 529–540.

Benin, S., Smale, M., Pender, J., Gebremedhin, B. and Ehui, S. (2004) The economic determinants of cereal crop diversity on farms in the Ethiopian highlands. *Agricultural Economics* 31, 197–208.

Bezabih, M. and Di Falco, S. (2012) Rainfall variability and food crop portfolio choice: evidence from Ethiopia. *Food Security* 4, 557–567.

Bhattamishra, R. and Barrett, C.B. (2010) Community-based risk management arrangements: a review. *World Development* 38, 923–932.

Bhaumik, S.K., Dimova, R., Kumbhakar, S.C. and Sun, K. (2018) Is tinkering with institutional quality a panacea for firm performance? Insights from a semiparametric approach to modeling firm performance. *Review of Development Economics* 22, 1–22.

Binswanger, H.P. (1980) Attitudes toward risk: experimental measurement in rural India. *American Journal of Agricultural Economics* 62, 395–407.

Birthal, P.S. and Hazrana, J. (2019) Crop diversification and resilience of agriculture to climatic shocks: evidence from India. *Agricultural Systems* 173, 345–354.

Birthal, P.S., Negi, D.S., Jha, A.K. and Singh, D. (2014) Income sources of farm households in India: Determinants, distributional consequences and policy implications. *Agricultural Economics Research Review* 27, 37–48.

Chavas, J.P. (2008) A cost approach to economic analysis under state-contingent production uncertainty. *American Journal of Agricultural Economics* 90, 435–446.

Coelli, T.J., Rao, D.S.P., O'Donnell, C.J. and Battese, G.E. (2005) *An Introduction to Efficiency and Productivity Analysis*. Springer Science & Business Media Inc., New York.

Collier, P., Radwan, S.M., Wangwe, S. and Wagner, A. (1986) *Labor and Poverty in Rural Tanzania; Ujamaa and Rural Development in the United Republic of Tanzania*. Clarendon Press, Oxford, UK.

Dalton, T.J., Porter, G.A. and Winslow, N.G. (2004) Risk management strategies in humid production regions: a comparison of supplemental irrigation and crop insurance. *Agricultural and Resource Economics Review* 33, 220–232.

Dercon, S. (1996) Risk, crop choice, and savings: evidence from Tanzania. *Economic Development and Cultural Change* 44, 485–513.
Dercon, S. and Christiaensen, L. (2011) Consumption risk, technology adoption and poverty traps: evidence from Ethiopia. *Journal of Development Economics* 96, 159–173.
Di Falco, S. and Chavas, J.P. (2009) On crop biodiversity, risk exposure, and food security in the highlands of Ethiopia. *American Journal of Agricultural Economics* 91, 599–611.
Di Falco, S. and Veronesi, M. (2014) Managing environmental risk in presence of climate change: the role of adaptation in the Nile Basin of Ethiopia. *Environmental and Resource Economics* 57, 553–577.
Di Falco, S., Bezabih, M. and Yesuf, M. (2010) Seeds for livelihood: crop biodiversity and food production in Ethiopia. *Ecological Economics* 69, 1695–1702.
Handschuch, C., Wollni, M. and Villalobos, P. (2013) Adoption of food safety and quality standards among Chilean raspberry producers – do smallholders benefit? *Food Policy* 40, 64–73.
Hardaker, J.B., Lien, G., Anderson, J.R. and Huirne, R.B. (2015) *Coping with Risk in Agriculture: Applied Decision Analysis*, 3rd edn. CAB International, Wallingford, UK.
Henderson, D.J. and Parmeter, C.F. (2015) *Applied Nonparametric Econometrics.* Cambridge University Press, New York.
Henderson, D.J., Kumbhakar, S.C. and Parmeter, C.F. (2012) A simple method to visualize results in nonlinear regression models. *Economics Letters* 117, 578–581.
Hurley, T.M. (2010) A review of agricultural production risk in the developing world. Working Paper 188476, HarvestChoice. DOI: 10.22004/ag.econ.188476.
Jodha, N.S., Singh, N.P. and Bantilan, M.C.S. (2012) Enhancing farmers' adaptation to climate change in arid and semi-arid agriculture of India: evidences from indigenous practices: developing international public goods from development-oriented projects. *Working Paper Series no. 32*. International Crops Research Institute for Semi-Arid Tropics, Andhra Pradesh, India.
Joshi, P., Joshi, L. and Birthal, P.S. (2006) Diversification and its impact on smallholders: evidence from a study on vegetable production. *Agricultural Economics Research Review* 19, 219–236.
Kijima, Y., Matsumoto, T. and Yamano, T. (2006) Nonfarm employment, agricultural shocks, and poverty dynamics: evidence from rural Uganda. *Agricultural Economics* 35, 459–467.
Kim, K. and Chavas, J.P. (2003) Technological change and risk management: an application to the economics of corn production. *Agricultural Economics* 29, 125–142.
Kim, K., Chavas, J.P., Barham, B. and Foltz, J. (2012) Specialization, diversification, and productivity: a panel data analysis of rice farms in Korea. *Agricultural Economics* 43, 687–700.
Knight, J., Weir, S. and Woldehanna, T. (2003) The role of education in facilitating risk-taking and innovation in agriculture. *The Journal of Development Studies* 39, 1–22.
Krause, H., Lippe, R.S. and Grote, U. (2016) Adoption and income effects of public gap standards: evidence from the horticultural sector in Thailand. *Horticulturae* 2, 18.
Kumar, A., Wright, I.A. and Singh, D.K. (2011) Adoption of food safety practices in milk production: implications for dairy farmers in India. *Journal of International Food & Agribusiness Marketing* 23, 330–344.
Kurosaki, T. and Fafchamps, M. (2002) Insurance market efficiency and crop choices in Pakistan. *Journal of Development Economics* 67, 419–453.
Li, Q. and Racine, J.S. (2010) Smooth varying-coefficient estimation and inference for qualitative and quantitative data. *Econometric Theory* 26, 1607–1637.
Li, Q., Huang, C.J., Li, D. and Fu, T.T. (2002) Semiparametric smooth coefficient models. *Journal of Business & Economic Statistics* 20, 412–422.
Lien, G., Kumbhakar, S.C. and Hardaker, J.B. (2017) Accounting for risk in productivity analysis: an application to Norwegian dairy farming. *Journal of Productivity Analysis* 47, 247–257.
Liu, E.M. (2013) Time to change what to sow: risk preferences and technology adoption decisions of cotton farmers in China. *Review of Economics and Statistics* 95, 1386–1403.
Mariano, M.J., Villano, R. and Fleming, E. (2012) Factors influencing farmers' adoption of modern rice technologies and good management practices in the Philippines. *Agricultural Systems* 110, 41–53.
Markowitz, H. (1952) Portfolio selection. *The Journal of Finance* 7, 77–91.
Matsumoto, T., Kijima, Y. and Yamano, T. (2006) The role of local nonfarm activities and migration in reducing poverty: evidence from Ethiopia, Kenya, and Uganda. *Agricultural Economics* 35, 449–458.
Mishra, A.K., El-Osta, H.S. and Sandretto, C.L. (2004) Factors affecting farm enterprise diversification. *Agricultural Finance Review* 64, 151–166.

Mishra, A.K., Kumar, A., Joshi, P.K., D'Souza, A. and Tripathi, G. (2018) How can organic rice be a boon to smallholders? Evidence from contract farming in India. *Food Policy* 75, 147–157.

Mishra, A.K., Rezitis, A.N. and Tsionas, M.G. (2019) Estimating technical efficiency and production risk under contract farming: a Bayesian estimation and stochastic dominance methodology. *Journal of Agricultural Economics* 70, 353–371.

Mishra, A.K., Rezitis, A.N. and Tsionas, M.G. (2020) Production under input endogeneity and farm-specific risk aversion: evidence from contract farming and Bayesian method. *European Review of Agricultural Economics* 47, 591–618.

Mohan, S. (2020) Risk aversion and certification: evidence from the Nepali tea fields. *World Development* 129, 104903.

Moschini, G. and Hennessy, D.A. (2001) Uncertainty, risk aversion, and risk management for agricultural producers. In: Gardner, B.L. and Rausser, G.C. (eds) *Handbook of Agricultural Economics*, Vol. 1. Elsevier Science B.V., Amsterdam, pp. 87–153.

Nashik District (2022) Nashik District, Government of Maharastra. Available at: https://nashik.gov.in/ (accessed January 2022).

Reardon, T. (1997) Using evidence of household income diversification to inform study of the rural nonfarm labor market in Africa. *World Development* 25, 735–747.

Reardon, T. and Taylor, J.E. (1996) Agroclimatic shock, income inequality, and poverty: evidence from Burkina Faso. *World Development* 24, 901–914.

Robinson, P.M. (1988) Root-*N*-consistent semiparametric regression. *Econometrica* 56, 931–954.

Ruben, R. *et al.* (2001) Nonfarm employment and poverty alleviation of rural farm households in Honduras. *World Development* 29, 549–560.

Ullah, R. and Shivakoti, G.P. (2014) Adoption of on-farm and off-farm diversification to manage agricultural risks: are these decisions correlated? *Outlook on Agriculture* 43, 265–271.

Vargas Hill, R. (2009) Using stated preferences and beliefs to identify the impact of risk on poor households. *The Journal of Development Studies* 45, 151–171.

Vigani, M. and Kathage, J. (2019) To risk or not to risk? Risk management and farm productivity. *American Journal of Agricultural Economics* 101, 1432–1454.

Section II

Using Experimental Economics to Measure Risk, Preferences and Risk Management Strategies

Farming is an inherently risky business. Among other factors, risk and uncertainty preferences are important drivers of farmers' decision-making. Risk refers to situations where definite numerical probabilities are known and can be measured objectively. Uncertainty refers to situations in which definite numerical probabilities are unobservable. Literature suggests numerous risk elicitation methods suitable for different research endeavors and target groups. Numerous studies have been carried out to understand and better manage decision-making under risk. However, studies conducted in emerging and developing countries examining the determinants of agricultural risk management and coping decisions either did not consider risk attitudes or relied on self-reported measures.

With a better understanding of the risk and uncertainties in agricultural production, the second section of this book adds to the discussion of risk and uncertainty preferences in farming in developing and emerging economies. Many approaches have been developed to elicit risk and uncertainty preferences over the years. Elicitation of reliable risk and uncertainty preferences is still an open field. The viewpoints gained from chapters and case studies in point are critical to understanding how farmers perceive risk, evaluate risk and plan for coping strategies. Because products, farmers and markets have unique attributes that influence risk perception and risk management strategies, this section explores the missing pieces in the measurement of risk attitudes, risk preferences and risk management strategies via an experimental economics method and survey data.

Chapter 6 examines an elicitation of uncertainty preferences that relies on the use of elicited subjective probability distributions. The approach is based on implementing incentivized and incentive-compatible mechanisms for eliciting beliefs and preferences. The authors conduct their experiment in a pilot study conducted in Zimbabwe with a sample of horticultural producers.

Chapter 7 elicits farmers' risk preferences via an experimental method. Using survey data from India and Nepal, it determines whether risk preferences explain the decision to adopt practices and strategies often reported to be driven by risk and risk aversion. Chapter 8 provides a practical orientation for researchers aiming to measure risk attitudes of the rural poor by briefly presenting two frontline theories for decision under risk, nine common risk elicitation methods and several approaches to verify the explanatory power of their suitability to generate quality data in the context of developing countries.

Chapter 9 develops a model of agricultural risk at the farm level, including both price risk and production risk. The chapter examines the role of risk preferences with a focus on aversion to risk and downside risk. The author evaluates the cost of risk under general conditions and methods available to assess risk exposure. Chapter 10, using experimental economics, elicits the risk preferences of smallholder rice farmers in West Africa and investigates how risk preferences impact access to technology in African rice.

Crop insurance has been a popular tool used in most countries to mitigate economic risks from farming. Chapter 11, using choice experiment and contingent valuation methods, assesses China's crop insurance and smallholders' preference for alternative types of insurance. Chapter 12 uses a lottery choice experiment to elicit the risk attitudes of rice farming households in Odisha, India. The author uses the Holt–Laury lottery method to construct risk attitudes and then applies it to studying farmers' responses to crop insurance in natural disasters like floods and in drought-prone areas of Odisha.

Finally, understanding farmers' decision-making under risk will improve the ability to adopt innovations in the farm sectors of developing and emerging economies. In recent years, drought-tolerant crops and weather index insurance have received attention in minimizing risk in farming. Chapter 13 presents two case studies in farmers' willingness to pay for risk-reducing innovations, including drought-tolerant maize in Nigeria and weather-based index insurance in Togo.

6 Eliciting Farmers' Risk and Uncertainty Preferences Using Contextualized Field Experiments in Developing Countries: A Pilot Study

Simone Cerroni,[1,2]* Martina Bozzola[3] and Ruggiero Rippo[1]
[1]*Department of Economics and Management, University of Trento, Trento, Italy;* [2]*C3A, University of Trento, Italy;* [3]*Queen's University, Belfast, Belfast, UK*

Abstract
In this chapter, we use a novel approach to elicit risk and uncertainty preferences using contextualized field experiments in developing countries. The elicitation of uncertainty preferences relies on the use of elicited subjective probability distributions. The approach presented here is based on implementing incentivized and incentive-compatible mechanisms for eliciting beliefs and preferences. We present results from a pilot study conducted in Zimbabwe with a sample of horticultural producers. Our contextualization relates to price risk mitigation and contract farming.

6.1 Introduction

Farming is an inherently risky business (Moschini and Hennessy, 2001). This is particularly true in developing countries where farmers' activity is increasingly threatened by global warming, climatic extreme events, and local and global geopolitical instability (e.g. Barbier and Hochard, 2018; Cottrell *et al.*, 2019). Understanding factors that determine farmers' coping strategies in the face of these risks is very important in developing actions that help farmers escape poverty traps and mitigate food insecurity. Among other factors, risk and uncertainty preferences are important drivers of farmers' decision-making (Moschini and Hennessy, 2001). Here, we refer to the terms risk and uncertainty in a 'Knightian' fashion (1921). Risk refers to situations where definite numerical probabilities are known and can be objectively measured. Uncertainty refers to situations where definite numerical probabilities are unobservable. In developing countries, farmers generally own few assets, have low incomes, and face numerous and threatening sources of risk. In this setting, it is of great importance to understand how farmers' risk and uncertainty preferences (alongside other factors) may contribute to their inability to undertake activities and investments that could generate high returns and reduce chronic poverty (e.g. Rosenzweig and Binswanger, 1993; Barnett *et al.*, 2008; Visser *et al.*, 2020). Risk and uncertainty preferences have been shown to influence many farmers' decisions that could mitigate the

*E-mail: simone.cerroni@unitn.it

 Managing Risk in Agriculture: A Development Perspective (eds A.K. Mishra, S.C. Kumbhakar and G. Lien).
DOI: 10.1079/9781800622289.0006

risk of falling into poverty traps: innovation adoption (e.g. Liu, 2013; Ward and Singh, 2015), crop diversification (e.g. Warnick *et al.*, 2011; Kijima, 2019), insurance decisions (e.g. Belissa *et al.*, 2020; Ali *et al.*, 2021), credit access (e.g. Bauer *et al.*, 2012; Visser *et al.*, 2020), contract farming (e.g. Mishra *et al.*, 2018; Bellemare *et al.*, 2020), adaptation to climate change (e.g. Alpizar *et al.*, 2011; Jianjun *et al.*, 2015) and participation in agri-environmental schemes (e.g. Marenya *et al.*, 2014; Mohan, 2020).

Many approaches have been developed to elicit risk and uncertainty preferences over the years (e.g. Charness *et al.*, 2013; Trautmann and van de Kuilen, 2015; Iyer *et al.*, 2020). The reliability of elicited risk and uncertainty preferences has been debated, because different elicitation methods lead to different results (e.g. Schildberg-Hörisch, 2018). A number of studies have shown that farmers' risk preferences depend on the method used to elicit them in both developed and developing countries (e.g. Reynaud and Couture, 2012; Charness and Viceisza, 2016; Iyer *et al.*, 2020).

The race to the elicitation of reliable risk and uncertainty preferences is still open. Recently, a few studies have attempted to elicit farmers' risk preferences using contextualized field experiments (e.g. Sanou *et al.*, 2018; Cerroni, 2020). Instead of asking farmers to make choices among standard monetary lotteries (an example is the well-known Holt and Laury's (2002) multi-price list), these experiments add context to the lottery, so that the farmers' decision problem resembles a real-life situation (for example, the adoption of a new crop variety). Adding context can improve farmers' understanding of the experimental tasks, thus reducing inconsistent choice behavior, which is common in non-contextualized experiments (i.e. internal validity). Contextualization can also improve the ability of elicited preferences to explain farmers' decisions on their farms, which is the main purpose of eliciting risk and uncertainty preferences (i.e. external validity). A study by Sanou *et al.* (2018) shows that risk preferences elicited via contextualized experiments better explain farmers' on-farm decisions in rural Niger than risk preferences from non-contextualized experiments. However, the use of context can generate issues. A general critique of contextualization is that farmers could use heuristics that muddle elicited preferences, undermining their generalizability (e.g. availability bias, representativeness bias).

Although a few studies elicited risk preferences using contextualized tasks in developing countries (e.g. Elabed and Carter, 2015; Visser *et al.*, 2020), to the best of the authors' knowledge, no studies have employed this technique to elicit uncertainty preferences. This chapter fills this gap in the literature by using and testing a novel approach to elicit uncertainty preferences using contextualized field experiments that builds on the work conducted by Cerroni (2020) in a developed country (i.e. Scotland). Here, we present results from a pilot study conducted with 14 horticultural farmers in Zimbabwe. Although Cerroni (2020) conducted only a contextualized experiment, in this within-subject pilot experiment we elicited risk and uncertainty preferences using both contextualized and non-contextualized experiments. This improved experimental design allows for a direct comparison between risk and uncertainty preferences elicited via contextualized and non-contextualized experiments. When eliciting uncertainty preferences using contextualized experiments, one needs to control for farmers' subjective probability distributions of the random variable used to contextualize the lottery tasks (see Cerroni, 2020). Therefore, in our pilot experiment, farmers were exposed to three tasks: (i) the subjective probability distribution task; (ii) the uncertainty preference task; and (iii) the risk preference elicitation task. The non-contextualized experiment has the same structure, so farmers are exposed to six tasks in total.

Building upon the rural context where this study was conducted, the contextualization is about entering contract farming for sugar beans as a price risk mitigation strategy. Contract farming is still a hot topic in the development policy debate. In theory, contract farming can promote smallholder farmers' participation in the market, thus improving farm income, reducing the risk of poverty traps and facilitating rural development. However, evidence of these beneficial impacts of contract farming is limited and mixed (e.g. Meemken and Bellemare, 2020). Similarly, the literature exploring factors mediating participation in contract farming is vast, but results are contrasting, and there are still gaps to be filled. Although it is assumed that behavioral factors such as subjective beliefs and risk and uncertainty preferences can certainly play a role in farmers' decisions to enter contract farming, the number of studies exploring this topic is small, and evidence is very limited (e.g. Bellemare and Bloem, 2018).

The rest of the chapter is laid out as follows. Section 6.2 offers an overview of the literature eliciting risk and uncertainty preferences in developing countries. Section 6.3 describes the context where our pilot study was conducted, and Section 6.4 presents the experimental design and the methods used to elicit risk and uncertainty preferences. In Section 6.5 we discuss the main results, and Section 6.6 provides some conclusions and a discussion of possible future research.

6.2 Eliciting Risk and Uncertainty Preferences in Developing Countries

Although the number of studies that elicited risk preferences in developing countries is large, only a few studies retrieved uncertainty preferences. Table 6.1 lists all studies that elicited quantitative measures of risk and/or uncertainty preferences in developing countries using incentivized experiments and the incentive-compatible elicitation method. Studies eliciting qualitative risk and/or uncertainty preferences are not surveyed here. Qualitative measures of risk and uncertainty preferences can be gathered using survey methods and Likert scales (Dohmen *et al.*, 2011; Charness *et al.*, 2013). Many studies conducted in developing countries have asked farmers to directly state how much they think they are risk (or uncertainty) averse (or loving) on scale from one to five (e.g. Giné *et al.*, 2009; Sanou *et al.*, 2018; Crentsil *et al.*, 2020). A preference elicitation method is considered incentive-compatible when, under a proper financial reward (or incentive) scheme, it induces participants to reveal their true preferences. For each study presented in Table 6.1 we report the main characteristics of these studies, namely: the main focus of the study, the country where it was conducted, whether the study elicited risk and/or uncertainty preferences, the methods employed to elicit risk and/or uncertainty preferences, and whether contextualization was used.

A number of incentive-compatible approaches can be used to elicit risk and uncertainty preferences. Most of these consist of asking farmers to choose between pairwise lotteries (see Charness *et al.* (2013); Trautmann and van de Kuilen (2015); Cerroni (2020) for comprehensive reviews). Based on these decisions between pairwise lotteries, it is possible to estimate parameters of risk and uncertainty aversion that can subsequently be incorporated in models of decision-making under risk and uncertainty to explain and/or predict farmers' behavior. By reviewing the related literature, it appears that the most-used methods to elicit farmers' risk preferences in developing countries are: (i) Holt and Laury's multiple price list (MPL) (2002) and Tanaka *et al.*'s development (2010) and (ii) the certainty equivalent (CE) MPL (Eckel and Grossman, 2008; Abdellaoui *et al.*, 2011; Barham *et al.*, 2014). Farmers' uncertainty preferences in developing countries, on the other hand, are usually elicited through techniques that are variations of the classic Ellsberg experiments (1961). The most popular is probably the CE-MPL method (e.g. Abdellaoui *et al.*, 2011; Barham *et al.*, 2014), which elicits certainty equivalents for a lottery characterized by unknown probabilities. Only a few studies do not provide monetary incentives to farmers depending on their choices in the lottery tasks and the realization of events presented in these lotteries (e.g. Ross *et al.*, 2012; Bryan, 2019), while the vast majority do.

Recently, a small but growing number of studies attempted to frame these monetary lotteries as a decision problem that farmers commonly face on the farm – for example, the decision to purchase a crop insurance product or the decision to stipulate a future contract with a buyer. Table 6.1 highlights that the bulk of the literature uses standard monetary lotteries, while a small but increasing number of studies use contextualized lotteries. Contextualized experiments have been used to elicit farmers' risk preferences in developing countries (e.g. Yesuf and Bluffstone, 2009; Sanou *et al.*, 2018; Visser *et al.*, 2020), but we are not aware of any study using this type of experiment to elicit uncertainty preferences. A noticeable exception is Cerroni (2020), which elicited farmers' risk and uncertainty preferences in Europe.

Adding an agricultural context to the experimental lotteries may increase the realism of the experimental task and hence increase farmers' familiarity with the decision problem. This can improve farmers' understanding of the experimental tasks and hence raise the internal validity of the elicited preferences. Adding context could also improve the ability of the elicited preferences to explain farmers' behavior in

Table 6.1 Summary of the studies eliciting risk and uncertainty preferences in developing countries using incentivized and incentive-compatible mechanisms. (Source: Authors' own work.)

Authors	Focus	Country	Risk/ uncertainty	Method risk	Method uncertainty	Context risk	Context uncertainty
Harrison *et al*., 2010	Preference elicitation study	Ethiopia, India and Uganda	R	MPL	–	No	–
Tanaka *et al*., 2010	Preference elicitation study	Vietnam	R	MPL (Tanaka *et al*., 2010)	–	No	–
Akay *et al*., 2012	Preference elicitation study	Ethiopia	R & U	CE-MPL (Barham *et al*., 2014)	CE-MPL (Barham *et al*., 2014)	No	No
Liebenehm and Waibel, 2014	Preference elicitation study	Mali and Burkina Faso	R	MPL (Tanaka *et al*., 2010)	–	No	–
Charness and Viceisza, 2016	Preference elicitation study	Senegal	R	MPL (Holt and Laury, 2002); CE-MPL (based on Gneezy and Potters, 1997)	–	No	–
He *et al*., 2018	Preference elicitation study	Rural China	R	MPL (Holt and Laury, 2002) and Andreoni and Harbaugh's method (2010)	–	No	–
Hill and Viceisza, 2012	Weather index insurance product adoption	Ethiopia	R	Binswanger's method (1980)	–	No	–
Elabed and Carter, 2015	Cotton production microinsurance	Mali	R	Eckel and Grossman's method (2008)	–	Yes	No
Freudenreich and Mußhoff, 2018	Insurance and higher yielding hybrid seed	Mexico	R	CE-MPL	–	No	–
Belissa *et al*., 2020	Innovation adoption – insurance	Ethiopia	R & U	CE-MPL (Barham *et al*., 2014)	MPL Ellsberg's type experiment (1961)	No	No
Kemeze *et al*., 2020	Index insurance and production technology adoption	North Ghana	R	MPL (Holt and Laury, 2002)	–	Yes	–
Visser *et al*., 2020	Uptake of credit and insurance	South Africa	R	MPL (Tanaka *et al*., 2010)	–	Yes	–

Ali *et al*., 2021	Cocoa farmers' participation in an insurance program	Ghana	R & U	CE-MPL	CE-MPL	No	No
Warnick *et al*., 2011	Crop diversification	Peru	R & U	Eckel and Grossman's method (2008)	Ellsberg's type experiment (1961)	No	No
Liu, 2013	Adoption of agricultural biotechnology for cotton	China	R	MPL (Tanaka *et al*., 2010)	–	No	–
Ward and Singh, 2015	Technology adoption – drought-tolerant rice	India	R & U	MPL (Tanaka *et al*., 2010)	CE-MPL (Barham *et al*., 2014)	No	No
Sanou *et al*., 2018	Technology adoption – fertilizer micro-dosing	Niger	R & U	CE-MPL (based on Gneezy and Potters, 1997) CE-MPL (Holt and Laury, 2002)	CE-MPL (Holt and Laury, 2002)	Yes	No
Vieider *et al*., 2019	Innovation adoption	Ethiopia	R	CE-MPL	–	No	–
Gars and Ward, 2019	Hybrid rice adoption	India	R & U	CE-MPL (Barham *et al*., 2014)	CE-MPL (Barham *et al*., 2014)	No	No
Kijima, 2019	Crop diversification	Uganda	R	CE-MPL (Barham *et al*., 2014)	–	No	–
Crentsil *et al*., 2020	Technology adoption in aquaculture	Ghana	R & U	MPL (Tanaka *et al*., 2010)	Ellsberg's type experiment (1961)	No	No
Channa *et al*., 2021	Food safety technology adoption in grain	Kenya	R	Choice task based on Eckel and Grossman's method (2008)	–	No	–
Binswanger, 1980	Income and credit	India	R	Binswanger (1980)	–	No	–
Yesuf and Bluffstone, 2009	Poverty traps	Ethiopia	R	Choice task based on Binswanger (1980)	–	Yes	–
Bauer *et al*., 2012	Microcredit	India	R	Choice task based on Binswanger (1980)	–	No	–
Jianjun *et al*., 2015	Climate change adaptation	China	R	MPL (Tanaka *et al*., 2010)	–	No	–

Continued

Table 6.1 Continued.

Authors	Focus	Country	Risk/ uncertainty	Method risk	Method uncertainty	Context risk	Context uncertainty
Alpizar *et al.*, 2011	Climate change adaptation	Costa Rica	R & U	CE-MPL (similar to Barham *et al.*, 2014)	CE-MPL (similar to Barham *et al.*, 2014)	Yes	No[a]
Brick *et al.*, 2012	Compliance with fisheries regulations	South Africa	R	MPL (Tanaka *et al.*, 2010)	–	No	–
Ruhinduka *et al.*, 2020	Rice farmers' storage and processing decisions	Tanzania	R & U	CE-MPL (Barham *et al.*, 2014)	CE-MPL (Barham *et al.*, 2014)	No	No
Ihli *et al.*, 2022	Participation into an agri-environmental scheme	Uganda	R	MPL (Holt and Laury, 2002)	–	No	–
Mohan, 2022	Organic certification in tea fields	Nepal	R & U	Eckel and Grossman's method (2008); CE-MPL (Barham *et al.*, 2014)	CE-MPL (Barham *et al.*, 2014)	No	No

Note: [a]Alpizar *et al.* (2011) elicited uncertainty preferences using contextualized field experiments. However, their design did not allow the estimation of quantitative measurement of uncertainty preferences.

'real-life' situations (i.e. external validity; Alekseev *et al.*, 2017). However, these benefits come with a cost. Contextualized experiments may generate confounding factors i.e. greater complexity of the experimental task (Smith, 1976) and are generally more difficult to replicate because they are more context-dependent (Gneezy and Imas, 2017) compared to non-contextualized experiments. A few studies explore the impact of contextualization on elicited risk preferences in developed countries (i.e. Europe) (Rommel *et al.*, 2017, 2019; Meraner *et al.*, 2018). Only Sanou *et al.* (2018) investigate the same topic in a developing country, finding that risk preferences elicited via contextualized experiments are better predictors of farmers' adoption of fertilizer micro-dosing in rural Niger (than risk preferences elicited via non-contextualized experiments).

6.3 Empirical Application

Zimbabwe is a landlocked southern African country with a total land area of more than 39 million hectares, with 33.3 million hectares used for agricultural purposes (FAO, 2022). Agriculture is a crucial sector of the country's economy and contributes 15–20% of the national gross domestic product. Between 60–70% of the population relies on income from agriculture. This sector accounts for the 40% of total exports (ZAS and La Gazzette, 2019).

Maize is the country's most-produced crop, accounting for about 25% of total production, and it is the main staple crop in the country. Farmers grow maize mainly for their household's subsistence. However, maize also represents a source of income as it is sold as a cash crop, mainly in local markets. Agricultural production, including maize and sugar beans, has generally declined since the introduction in 2001 of the agricultural reform known as the Fast Track Land Reform Programme (FTLRP). This reform promoted the redistribution of agricultural land from large-scale farms (i.e. high input, high output) and prompted the creation of a number of small and medium-size farms. To overcome food insecurity caused by the decline of agricultural production, the government through the years launched some fiscal intervention to support and revamp the agricultural sector. One of these actions was the introduction of a state-led contract farming system in 2016 coined the Targeted Command Agricultural Programme (Mazwi *et al.*, 2019; Makuwerere Dube, 2021).

We conducted our pilot study in the Murehwa district in the province of Mashonaland East in Zimbabwe. This province accounts for about 16% of total sugar beans production (ZAS and La Gazzette, 2019) and is the third-largest maize-producing province after Manicaland and Mashonaland Central.

Zimbabwe is divided into five main agro-ecological regions, known as natural regions, on the basis of the rainfall regime, soil quality and vegetation, among other factors (FAO, 2006). The Murewa district falls within the IIb Natural Farming Region, which is suitable for semi-intensive farming of maize, sugar beans, tobacco and cotton and livestock production.

The vast majority of agricultural production (approximately 75%) in Murehwa district comes from communal farmers. Communal farmers are characterized by individual households generally owning about 2–4 ha of arable land and communal grazing. A larger survey complementing this pilot study shows that these farmers produce mainly for their own consumption and sell a little bit of excess. Other typologies are A1 farms and small-scale farms. The A1 farms were generated under the FTLRP's resettlement scheme and have an average size of about 37 ha (ZAS and La Gazzette, 2019). Approximately 10% of the production is generated by A1 farms and 15% by small-scale farms. More specifically, these farms are developed using a 'village' model; that is, each farmer owns an individual residential plot of about 0.5 ha within a defined village block and ten arable hectares in the drier regions. The village has communal grazing land (25–60 ha, depending on the village size and natural region of the settlement) and a set of communal infrastructures such as water points, access roads, woodlots and other social services.

6.4 Methods

6.4.1 Sample and data collection

The pilot study was conducted in Zimbabwe in the first week of September 2019 and had a dual purpose: (i) to test farmers' understanding of the experimental instructions and tasks and (ii) to train the enumerators who were recruited to collect data for a larger study in the following months. The pilot's sample consisted of 14 farmers operating in the Murewha region. Farmers were recruited with the support of the local agricultural extension services. Participants were paid a flat show-up fee corresponding to US$3. They could receive an additional payment ranging from an equivalent of US$0 to US$6, depending on their decisions in the experimental tasks. More details are provided in the following sections.

6.4.2 Experimental design

The experiment consisted of two parts. In the first part, farmers were asked to make choices in three non-contextualized tasks. In the second part, they were asked to make choices in three contextualized tasks. The contextualized and non-contextualized tasks were identical apart from the framing. Half of the sample was exposed first to the non-contextualized tasks, while the other half was exposed to the contextualized tasks.

Regardless of contextualization, farmers were exposed to three tasks in the following order. The first task elicited farmers' uncertainty preferences (task 1), the second task elicited farmers' subjective probability distributions (task 2), and the third task elicited their risk preferences (task 3). The order of the tasks was fixed to avoid learning effects. Farmers' additional earning depended on their decisions in both the contextualized and the non-contextualized parts of the experiment. Specifically, one contextualized task and one non-contextualized task were randomly selected at the end of the experiment. These tasks could generate an additional earning that ranged from US$0 to US$3. However, famers were told that, due to budget constraints, these earnings would be divided by ten, hence they could earn up to US$3 from each task, up to a maximum of US$6 in total. More details are provided in the following sections.

6.4.3 Eliciting subjective probability distribution

In the non-contextualized experiment, farmers were exposed to an induced belief experiment. This type of experiment is generally used to test the internal validity (i.e. accuracy) of elicited beliefs (Schlag *et al.*, 2015) and is very similar to induced value experiments (e.g. Smith, 1976; Cerroni *et al.*, 2019). Farmers were first informed that a bag contained 100 colored balls: 20 purple balls, 35 green balls, 25 blue balls, 15 yellow balls and 5 orange balls (i.e. beliefs were induced). They were then asked to predict the probability that a purple, green, blue, yellow, or orange ball would be drawn randomly at the end of the study using a quadratic scoring rule (QSR) (see Harrison *et al.* (2015, 2017) for earlier examples).

The QSR is an incentive-compatible technique that, under a proper incentive scheme and the risk neutrality assumption, elicits truthful beliefs. We acknowledge that many other possible ways to elicit subjective probabilities exist in the literature (see Chapter 3 for a review). In our study, each farmer n was asked to allocate a number of stones r_i (out of 100) in different colored buckets i = (purple, green, blue, yellow and orange). The QSR calculates the payoff S that farmers receive depending on the accuracy of their reports r_i. Specifically, it rewards participants when their reports r_i are accurate, but punishes them when their reports r_i are inaccurate. The algorithm that generates farmers' payoff is derived from Matheson and Winkler (1976):

$$S = \alpha + \beta\left[\left(2 \times r_k\right) - \sum_{i=1\ldots K}\left(r_i\right)^2\right] \tag{6.1}$$

where k is the color of the randomly drawn ball, and r_k is the report associated to the color k. In our study, $\alpha = \beta = 15$, so that farmers can earn up to US$3 from this task. Fixing α and β equal to certain values allows the researcher to control respondents' earnings from the QSR task. In this particular study, after consultation with local stakeholders, we agreed that an additional incentive of US$3 would be enough to properly incentivize farmers to reveal their true preferences.

In the contextualized experiment, farmers were exposed to standard belief elicitation experiments (i.e. beliefs were not induced). Farmers were asked to express their subjective probability distributions regarding the price of a packet of sugar beans (20 kg) in the nearest market in June 2020 using a QSR identical to the one used in the non-contextualized experiment.

This random variable presented five states of the world (i): (i) between US$12.00 and US$15.99, more precisely US$14.00; (ii) between US$16.00 and US$19.99, more precisely US$18.00; (iii) between US$20.00 and US$23.99, more precisely US$22.00; (iv) between US$24.00 and US$27.99, more precisely US$26.00; and (v) between US$28.00 and US$31.99, more precisely US$30.00.

Farmers were asked to allocate 100 stones across the five different states of the world i. The potential additional earning from the contextualized experiment depended on farmers' predictions. The payoff was calculated based on a simulated realization of the random variables. These simulations were performed using predictions made by ten farmers and five extension services in a previous stage of the study. Farmers could earn up to US$3 from this task. This procedure was disclosed to farmers via our experimental instructions, and the payoff was disclosed at the end of the experiment.

6.4.4 Eliciting risk and uncertainty preferences

In the non-contextualized experiment, uncertainty and risk preferences were obtained eliciting the certainty equivalents (CEs) for a lottery with five states of the world: US$14.00, US$18.00, US$22.00, US$26.00 and US$30.00. The tool for eliciting the CE was a classic multi-price list (MPL) with CE ranging from US$18.00 to US$26.00 (e.g. Abdellaoui *et al.*, 2011; Cerroni, 2020).

CEs for the uncertain lottery $CE(UL)$ were elicited using a multi-price list in which the probabilities associated with the states of the world presented in the lottery are unknown to farmers. CEs for the risky lottery $CE(RL)$ were elicited using a multi-price list in which these probabilities are known to subjects and are equivalent to those presented in the task that elicited subjective probabilities distributions. At the end of the experiment, a row number of the multi-price list is randomly drawn, and the payoff is calculated depending on the farmer's choice (i.e. receiving the sure payment versus playing the lottery) and the realization of the state of the world.

The elicitation of the CEs allows the calculation of the relative risk premium and uncertainty premium (*RRP* and *RUP*, respectively) using the following algorithms (e.g. Bruhin *et al.*, 2010; Robinson and Botzen, 2020):

$$RRP = \left[EV(RL) - CE(RL)\right] / \left|EV(RL)\right| \quad (6.2)$$

$$RUP = \left[EV(UL) - CE(UL)\right] / \left|EV(UL)\right| \quad (6.3)$$

where $EV(RL)$ and $EV(UL)$ are the expected values of the risky and uncertain lotteries, respectively. *EV*s are calculated using the induced beliefs used in the non-contextualized belief elicitation task. A negative risk (or uncertainty) premium indicates risk-loving (or uncertainty-loving) behavior, and a positive premium indicates risk-averse (or uncertainty-averse) behavior.

In the contextualized experiment, uncertainty and risk preferences are obtained using an equivalent approach. However, the MPLs resemble a farmer's decision to enter contract farming or not. Farmers face an MPL in which the CE would represent the sure price guaranteed by a forward contract and agreed to before planting the crop. This ranged from US$18.00 to US$26.00 for sugar beans. These ranges were identified thanks to three farmers, personnel from the extension services, and historical data on crop prices. The lottery represents the scenario in which the farmer decides not to enter contract farming. In this scenario, the price that he or she will be paid is the realization of the price for that crop at the nearest market in June 2020. The lottery presents five states of the world, which are those presented in the description of the subjective probability distributions for the contextualized task.

In the MPL related to the *UL*, probabilities associated with these states of the world are unknown. The *RUP* can be calculated using the subjective probability distributions elicited in task 1. In the MPL related to the *RL*, probabilities associated with these states of the world are those simulated thanks to other farmers' and extension services' predictions. The *RRP* is calculated using these simulations. It is interesting to note that the MPLs presented in the non-contextualized experiment and in the contextualized experiment for sugar beans are perfectly comparable in terms of CEs, lottery outcomes, and probabilities. This allows a direct comparison between risk and uncertainty preferences elicited using contextualized and non-contextualized experiments.

At the end of the experiment, if task 2 or task 3 is selected at random, a row number of the MPL is randomly drawn. The payoff is calculated depending on the farmer's choice (i.e. receiving the sure payment versus playing the lottery) and the realization of the state of the world, which is based on the simulation procedure explained in the previous section.

6.5 Results

6.5.1 Subjective probability distribution

Subjective probability distributions elicited in the contextualized and non-contextualized experiments are reported in Fig. 6.1. The non-contextualized experiment is an induced belief experiment that allows us to test the accuracy of elicited subjective probability distributions. Figure 6.1(a) suggests that the average subjective probability distribution is very similar to the induced subjective probability distribution, meaning that elicited beliefs are accurate. Fisher's exact and likelihood ratio tests indicate that elicited and induced probability distributions are not statistically different from each other (p-values = 0.767 and 0.777, respectively). This result is supported by the fact that the *EV*s of the elicited and induced subjective probability distributions are not statistically different according to Wilcoxon matched-pair tests and binomial tests (p-value = 0.890 and 0.774, respectively). The contextualized experiments elicit subjective probabilities for real agricultural outcomes. Figure 6.1(b) indicates that farmers' subjective probability distributions regarding sugar bean prices are consistent with those of experts.

6.5.2 Risk and uncertainty preferences elicited via contextualized and non-contextualized experiments

RRP and *RUP* reported in Table 6.2 indicate these measures are rather consistent when elicited using contextualized and non-contextualized experiments. Farmers appear to be generally risk- and uncertainty-neutral; in fact, *RRP* and *RUP* are very close to zero. Although farmers in developing countries were historically depicted as very risk- and uncertainty-averse (see Haushofer and Fehr (2014) for a recent literature review), recent evidence debunks this myth. A few studies indicate that farmers in developing countries are more risk-tolerant than in Western societies, and in some instances they are risk-neutral (e.g. Vieider *et al.*, 2019; Bellemare *et al.*, 2020). Similarly, another branch of the literature shows that ambiguity aversion is not

Table 6.2. *RRP* and *RUP* elicited using contextualized and non-contextualized experiments. (Source: Authors' own work.)

Variable	Obs	Mean	Std. Dev.	Min	Max
RRP_color	14	0.00	0.17	–0.30	0.10
RUP_color	14	–0.04	0.24	–0.54	0.18
RRP_beans	14	–0.02	0.16	–0.30	0.10
RUP_beans	14	0.02	0.20	–0.48	0.21

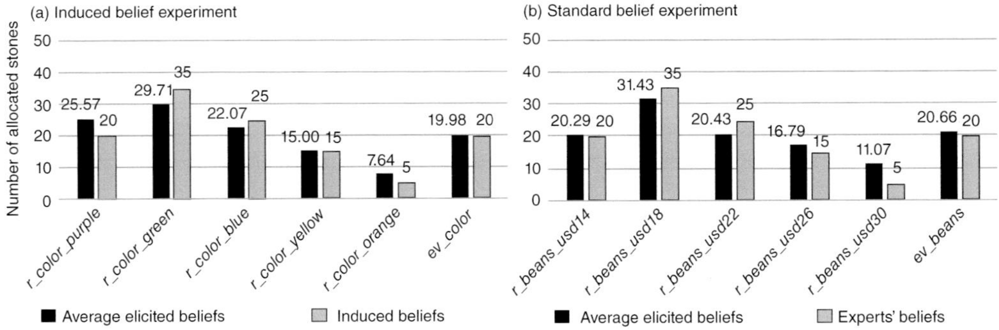

Fig. 6.1. Subjective probability distributions elicited via contextualized and non-contextualized experiments. (Source: Authors' own work) Notes: r: reported belief; ev: expected value of the subjective probability distribution.

universal but is highly dependent on context (e.g. Kocher *et al.*, 2018). Recent studies found that farmers can be quite tolerant of uncertainty (e.g. Ward and Singh, 2015).

A set of Wilcoxon matched-pair tests and binomial tests were also performed to statistically check whether contextualized and non-contextualized experiments produced different *RRP* and *RUP* (see Table 6.3). These are never statistically different from each other. Another set of Wilcoxon matched-pair tests and binomial tests indicated that farmers' preferences for risky and uncertain outcomes were the equivalent. In fact, *RRP* are never statistically different from *RUP* (within random variables – crop or color). This confirms that farmers display neutrality toward risk and uncertainty.

6.6 Conclusion

This chapter uses and tests a novel approach to elicit farmers' uncertainty preferences using contextualized and non-contextualized field experiments developed by Cerroni (2020). Specifically, we test whether uncertainty and risk preferences elicited using contextualized field experiments are consistent with those elicited via non-contextualized experiments. We used data from a pilot study with a sample of 14 horticultural farmers from Zimbabwe who were asked to make decisions in some contextualized and non-contextualized tasks.

In the contextualized experiment, uncertainty preferences are elicited by asking farmers to report their certainty equivalents for a contextualized lottery. The lottery is characterized by states of the world that depict different realizations of an agricultural random variable. Probabilities associated with these states of the world are unknown. In our study, the random variable is the future market price of a packet (20 kg) of maize sugar beans. The certainty equivalent is framed as the minimum fixed price that farmers are willing to accept for a packet of sugar beans to sign a forward contract at planting time. This experiment resembles a farmer's decision to enter contract farming. Uncertainty preferences can be identified if and only if farmers' subjective probability distributions of this random variable are known. Therefore, the method implies a task in which farmers are asked to report their expectations regarding the probability of realization of each state of the world (future prices) depicted in the contextualized lottery. Uncertainty preferences are elicited using a certainty equivalent multi-price list (CE-MPL; e.g. Abdellaoui *et al.*, 2011; Barham *et al.*, 2014) and subjective probability distributions using a QSR (e.g. Harrison *et al.*, 2015, 2017). Our experiment also elicited risk preferences using a contextualized CE-MPL task in which subjective probability distributions of future market prices were disclosed to farmers (e.g. Barham *et al.*, 2014; Cerroni, 2020). In the non-contextualized experiments, farmers faced the same typology of tasks. The only difference is that these tasks were unframed.

Results from the pilot study suggest that risk and uncertainty preferences elicited using contextualized and non-contextualized experiments are equivalent. In addition, we found that farmers are both risk- and uncertainty-neutral. This result may appear to be surprising at first glance, but it is consistent with recent findings suggesting that risk and uncertainty preferences may not be universal. Other studies found that farmers in developing countries are rather tolerant toward risk and uncertainty. Recent studies found that farmers can be quite tolerant of uncertainty (e.g. Ward and Singh, 2015; Kocher *et al.*, 2018).

Based on our findings and our review of the literature, we can suggest additional lines of research that may be worth pursuing. First, given the paucity of studies exploring farmers' uncertainty preferences in developing countries, future research could focus on the elicitation

Table 6.3. Comparison of *RRP* and *RUP* (within random variable). (Source: Authors' own work.)

Null hypothesis	Obs	Binomial test p-value	Wilcoxon matched-pair test p-value
RRP_beans = RRP_color	14	0.25	0.08
RUP_beans = RUP_color	14	0.42	0.16

of such preferences and the investigation of whether these explain and predict farmers' decisions more parsimoniously than risk preferences. After all, many agricultural outcomes (i.e. future production, income, market prices) are often surrounded by a great deal of uncertainty in developing countries. Second, assuming that practitioners aim to use uncertainty preferences to explain and predict farmers' decisions, such preferences must be accurate and reliable. The large majority of studies elicited qualitative measurements of these preferences via Likert scales. Unfortunately, these cannot be incorporated in standard models of decision-making under risk and uncertainty that explain farmers' behavior. Future research could focus instead on the use of incentivized and incentive-compatible approaches to elicit uncertainty preferences quantitatively. Recent studies conducted in developing countries found that qualitative and quantitative measurements of risk and uncertainty preferences are inconsistent (e.g. He *et al.*, 2018; Sanou *et al.*, 2018). Finally, our results may indicate that using contextualized tasks is not worth the hassle, as risk and uncertainty preferences are equivalent when using framed and unframed tasks. However, we must consider that the sample of our pilot study is very small (i.e. 14 farmers) and that the only previous study comparing risk preferences elicited via contextualized and non-contextualized experiments found opposite results (Sanou *et al.*, 2018). The use of agricultural framings to elicit risk and uncertainty preference can be rather appealing in developing countries, considering that exposing farmers to lotteries that resemble real-life situations could facilitate their understanding of the experimental task. One of the main concerns regarding the elicitation of subjective probabilities and risk and uncertainty preferences is that farmers in developing countries are often illiterate and may not be able to process numerical probabilities (e.g. Delavande *et al.*, 2011). Further research investigating whether contextualized field experiments can improve the internal and external validity of elicited risk and uncertainty preference is definitively needed in developing countries.

Acknowledgments

We would like to thank Lloyd Elliott Nyasha Nyemba, Angela Msipha, Gibson Mandikuwaza and Anesu Jambwa, who helped us collect the data for this project; without their support, it wouldn't have been possible. We are also grateful to the Alliances for Action team at the International Trade Centre, in particular to Hernan Manson, Benjamin Morrison, Nicholas Mudungwe and Grace Musarurwa for many useful discussions. The authors gratefully acknowledge funding from the Department for the Economy Global Challenge Research Fund (GCRF).

References

Abdellaoui, M., Baillon, A., Placedo, L. and Wakker, P.P. (2011) The rich domain of uncertainty: source functions and their experimental implementation. *American Economic Review* 101, 695–723.

Akay, A., Martinsson, P., Medhin, H. and Trautmann, S.T. (2012) Attitudes toward uncertainty among the poor: an experiment in rural Ethiopia. *Theory Decision* 73, 453–464. DOI: 10.1007/s11238-011-9250-y.

Alekseev, A., Charness, G. and Gneezy, U. (2017) Experimental methods: when and why contextual instructions are important. *Journal of Economic Behavior & Organization* 134, 48–59. DOI: 10.1016/j.jebo.2016.12.005.

Ali, W., Abdulai, A., Goetz, R. and Owusu, V. (2021) Risk, ambiguity and willingness to participate in crop insurance programs: evidence from a field experiment. *Australian Journal of Agricultural and Resource Economics* 65, 679–703. DOI: 10.1111/1467-8489.12434.

Alpizar, F., Carlsson, F. and Naranjo, M.A. (2011) The effect of ambiguous risk, and coordination on farmers' adaptation to climate change – a framed field experiment. *Ecological Economics* 70, 2317–2326. DOI: 10.1016/j.ecolecon.2011.07.004.

Andreoni, J. and Harbaugh, W.T. (2010) Unexpected utility: experimental tests of five key questions about preferences over risk. University of Oregon Economics Department Working Papers 2010–14, University of Oregon Economics Department, Eugene, Oregon.

Barbier, E.B. and Hochard, J.P. (2018) The impacts of climate change on the poor in disadvantaged regions. *Review of Environmental Economics and Policy* 12, 26–47. DOI: 10.1093/reep/rex023.

Barham, B.L., Chavas, J-P., Fitz, D., Salas, V.R. and Schechter, L. (2014) The roles of risk and ambiguity in technology adoption. *Journal of Economic Behavior & Organization* 97, 204–218. DOI: 10.1016/j.jebo.2013.06.014.

Barnett, B.J., Barrett, C.B. and Skees, J.R. (2008) Poverty traps and index-based risk transfer products. *World Development*, 36, 1766–1785. DOI: 10.1016/j.worlddev.2007.10.016.

Bauer, M., Chytilová, J. and Morduch, J. (2012) Behavioral foundations of microcredit: experimental and survey evidence from rural India. *American Economic Review* 102, 1118–1139. DOI: 10.1257/aer.102.2.1118.

Belissa, T.K., Lensink, R. and van Asseldonk, M. (2020) Risk and ambiguity aversion behavior in index-based insurance uptake decisions: experimental evidence from Ethiopia. *Journal of Economic Behavior & Organization* 180, 718–730. DOI: 10.1016/j.jebo.2019.07.018.

Bellemare, M.F. and Bloem, J.R. (2018) Does contract farming improve welfare? A review. *World Development* 112, 259–271. DOI: 10.1016/j.worlddev.2018.08.018.

Bellemare, M.F., Lee, Y.N. and Just, D.R. (2020) Producer attitudes toward output price risk: experimental evidence from the lab and from the field. *American Journal of Agricultural Economics* 102, 806–825. DOI: 10.1002/ajae.12004.

Binswanger, H.P. (1980) Attitudes toward risk: experimental measurement in rural India. *American Journal of Agricultural Economics* 62, 395–407. DOI: 10.2307/1240194.

Brick, K., Visser, M. and Burns, J. (2012) Risk aversion: experimental evidence from South African fishing communities. *American Journal of Agricultural Economics* 94, 133–152.

Bruhin, A., Fehr-Duda, H. and Epper, T. (2010) Risk and rationality: uncovering heterogeneity in probability distortion. *Econometrica* 78, 1375–1412. DOI: 10.3982/ECTA7139.

Bryan, G. (2019) Ambiguity aversion decreases the impact of partial insurance: evidence from African farmers. *Journal of the European Economic Association* 17, 1428–1469. DOI: 10.1093/jeea/jvy056.

Cerroni, S. (2020) Eliciting farmers' subjective probabilities, risk, and uncertainty preferences using contextualized field experiments. *Agricultural Economics* 51, 707–724. DOI: 10.1111/agec.12587.

Cerroni, S., Notaro, S. and Raffaelli, R. (2019) Beliefs and preferences for food-safety policies: a discrete choice model under uncertainty. *European Review of Agricultural Economics* 46, 769–799. DOI: 10.1093/erae/jby038.

Channa, H., Ricker-Gilbert, J., De Groote, H. and Bauchet, J. (2021) Willingness to pay for a new farm technology given risk preferences: evidence from an experimental auction in Kenya. *Agricultural Economics* 52, 733–748. DOI: 10.1111/agec.12651.

Charness, G. and Viceisza, A. (2016) Three risk-elicitation methods in the field: evidence from rural Senegal. *Review of Behavioral Economics* 3, 145–171. DOI: 10.1561/105.00000046.

Charness, G., Gneezy, U. and Imas, A. (2013) Experimental methods: eliciting risk preferences. *Journal of Economic Behavior & Organization* 87, 43–51. DOI: 10.1016/j.jebo.2012.12.023.

Cottrell, R.S., Nash, K.L., Halpern, B.S., Remenyi, T.A., Corney, S.P., Fleming, A. *et al.* (2019) Food production shocks across land and sea. *Nature Sustainability* 2, 130–137. DOI: 10.1038/s41893-018-0210-1.

Crentsil, C., Gschwandtner, A. and Wahhaj, Z. (2020) The effects of risk and ambiguity aversion on technology adoption: evidence from aquaculture in Ghana. *Journal of Economic Behavior & Organization* 179, 46–68. DOI: 10.1016/j.jebo.2020.07.035.

Delavande, A., Giné, X. and McKenzie, D. (2011) Measuring subjective expectations in developing countries: a critical review and new evidence. *Journal of Development Economics* 94, 151–163.

Dohmen, T., Falk, A., Huffman, D., Sunde, U., Schupp, J. and Wagner, G.G. (2011) Individual risk attitudes: measurement, determinants, and consequences. *Journal of the European Economic Association* 9, 522–550. DOI: 10.1111/j.1542-4774.2011.01015.x.

Eckel, C.C. and Grossman, P.J. (2008) Forecasting risk attitudes: an experimental study using actual and forecast gamble choices. *Journal of Economic Behavior & Organization* 68, 1–17. DOI: 10.1016/j.jebo.2008.04.006.

Elabed, G. and Carter, M.R. (2015) Compound-risk aversion, ambiguity and the willingness to pay for microinsurance. *Journal of Economic Behavior & Organization* 118, 150–166. DOI: 10.1016/j.jebo.2015.03.002.

Ellsberg, D. (1961) Risk, ambiguity, and the Savage axioms. *The Quarterly Journal of Economics* 75, 643–669.

FAO (2006) *Fertilizer Use by Crop in Zimbabwe*. FAO, Rome.

FAO (2022) Zimbabwe at a Glance. FAO, Rome. Available at: https://www.fao.org/zimbabwe/fao-in-zimbabwe/zimbabwe-at-a-glance/ar/ (accessed October 2022).

Freudenreich, H. and Mußhoff, O. (2018) Insurance for technology adoption: an experimental evaluation of schemes and subsidies with maize farmers in Mexico. *Journal of Agricultural Economics* 69, 96–120. DOI: 10.1111/1477-9552.12226.

Gars, J. and Ward, P.S. (2019). Can differences in individual learning explain patterns of technology adoption? Evidence on heterogeneous learning patterns and hybrid rice adoption in Bihar, India. *World Development* 115, 178–189. DOI: 10.1016/j.worlddev.2018.11.014.

Giné, X., Townsend, R.M. and Vickery, J. (2009) Forecasting when it matters: evidence from semi-arid India. Mimeo. World Bank, Washington, DC.

Gneezy, U. and Imas, A. (2017) Chapter 10 – Lab in the field: measuring preferences in the wild. In: Banerjee, A.V. and Duflo, E. (eds) *Handbook of Economic Field Experiments*. North-Holland, Amsterdam, pp. 439–464. DOI: 10.1016/bs.hefe.2016.08.003.

Gneezy, U. and Potters, J. (1997) An experiment on risk taking and evaluation periods. *The Quarterly Journal of Economics* 112, 631–645.

Harrison, G.W., Humphrey, S.J. and Verschoor, A. (2010) Choice under uncertainty: evidence from Ethiopia, India and Uganda. *The Economic Journal* 120, 80–104. DOI: 10.1111/j.1468-0297.2009.02303.x.

Harrison, G.W., Martínez-Correa, J., Swarthout, J.T. and Ulm, E.R. (2015) Eliciting subjective probability distributions with binary lotteries. *Economics Letters* 127, 68–71. DOI: 10.1016/j.econlet.2014.12.027.

Harrison, G.W., Martínez-Correa, J., Swarthout, J.T. and Ulm, E.R. (2017) Scoring rules for subjective probability distributions. *Journal of Economic Behavior & Organization* 134, 430–448. DOI: 10.1016/j.jebo.2016.12.001.

Haushofer, J. and Fehr, E. (2014) On the psychology of poverty. *Science* 344, 862–867. DOI: 10.1126/science.1232491.

He, J., Rejesus, R., Zheng, X. and Yorobe Jr, J. (2018) Advantageous selection in crop insurance: theory and evidence. *Journal of Agricultural Economics* 69, 646–668. DOI: 10.1111/1477-9552.12267.

Hill, R.V. and Viceisza, A. (2012) A field experiment on the impact of weather shocks and insurance on risky investment. *Experimental Economics* 15, 341–371. DOI: 10.1007/s10683-011-9303-7.

Holt, C.A. and Laury, S.K. (2002) Risk aversion and incentive effects. *American Economic Review* 92, 1644–1655.

Ihli J.H., Chiputwa, B., Winter, E. and Gassner, A. (2022) Risk and time preferences for participating in forest landscape restoration: the case of coffee farmers in Uganda. *World Development* 150, 105713. DOI: 10.1016/j.worlddev.2021.105713.

Iyer, P., Bozzola, M., Hirsch, S., Meraner, M. and Finger, R. (2020) Measuring farmer risk preferences in Europe: a systematic review. *Journal of Agricultural Economics* 71, 3–26. DOI: 10.1111/1477-9552.12325.

Jianjun, J., Yiwei, G., Xiaomin, W. and Nam, P.K. (2015) Farmers' risk preferences and their climate change adaptation strategies in the Yongqiao District, China. *Land Use Policy* 47, 365–372. DOI: 10.1016/j.landusepol.2015.04.028.

Kemeze, F.H., Miranda, M.J., Kuwornu, J.K.M. and Anim-Somuah, H. (2020) Smallholder farmer risk preferences in Northern Ghana: evidence from a controlled field experiment. *The Journal of Development Studies* 56, 1894–1908. DOI: 10.1080/00220388.2020.1715945.

Kijima, Y. (2019) Farmers' risk preferences and rice production: experimental and panel data evidence from Uganda. *Plos One* 14, e0219202. DOI: 10.1371/journal.pone.0219202.

Knight, F.H. (1921) *Risk, Uncertainty and Profit.* Houghton Mifflin, Boston, Massachusetts.

Kocher, M.G., Lahno, A.M. and Trautmann, S.T. (2018) Ambiguity aversion is not universal. *European Economic Review* 101, 268–283. DOI: 10.1016/j.euroecorev.2017.09.016.

Liebenehm, S. and Waibel, H. (2014) Simultaneous estimation of risk and time preferences among small-scale cattle farmers in West Africa. *American Journal of Agricultural Economics* 96, 1420–1438. DOI: 10.1093/ajae/aau056.

Liu, E.M. (2013) Time to change what to sow: risk preferences and technology adoption decisions of cotton farmers in China. *The Review of Economics and Statistics* 95, 1386–1403. DOI: 10.1162/REST_a_00295.

Makuwerere Dube, L. (2021) Command agriculture and food security: an interrogation of state intervention in the post-fast track land redistribution era in Zimbabwe. *Journal of Asian and African Studies* 56, 1626–1643. DOI: 10.1177/0021909620979330.

Marenya, P., Smith, V.H. and Nkonya, E. (2014) Relative preferences for soil conservation incentives among smallholder farmers: evidence from Malawi. *American Journal of Agricultural Economics* 96, 690–710. DOI: 10.1093/ajae/aat117.

Matheson, J.E. and Winkler, R.L. (1976) Scoring rules for continuous probability distributions. *Management Science* 22, 1087–1096. DOI: 10.1287/mnsc.22.10.1087.

Mazwi, F., Chemura, A., Mudimu, G.T. and Chambati, W. (2019) Political economy of command agriculture in Zimbabwe: a state-led contract farming model. *Agrarian South: Journal of Political Economy* 8, 232–257. DOI: 10.1177/2277976019856742.

Meemken, E-M. and Bellemare, M.F. (2020) Smallholder farmers and contract farming in developing countries. *Proceedings of the National Academy of Sciences* 117, 259–264. DOI: 10.1073/pnas.1909501116.

Meraner, M., Mußhoff, O. and Finger, R. (2018) Using involvement to reduce inconsistencies in risk preference elicitation. *Journal of Behavioral and Experimental Economics* 73, 22–33. DOI: 10.1016/j.socec.2018.01.001.

Mishra, A.K., Kumar, A., Joshi, P.K. and D'Souza, A. (2018) Production risks, risk preference and contract farming: impact on food security in India. *Applied Economic Perspectives and Policy* 40, 353–378. DOI: 10.1093/aepp/ppy017.

Mohan, S. (2020) Risk aversion and certification: evidence from the Nepali tea fields. *World Development* 129, 104903. DOI: 10.1016/j.worlddev.2020.104903.

Moschini, G. and Hennessy, D.A. (2001) Uncertainty, risk aversion, and risk management for agricultural producers. In: Gardner, B.L. and Rausser, G.C. (eds) *Handbook of Agricultural Economics*. North-Holland, Amsterdam, pp. 87–153. DOI: 10.1016/S1574-0072(01)10005-8.

Reynaud, A. and Couture, S. (2012) Stability of risk preference measures: results from a field experiment on French farmers. *Theory Decision* 73, 203–221. DOI: 10.1007/s11238-012-9296-5.

Robinson, P.J. and Botzen, W.J.W. (2020) Flood insurance demand and probability weighting: the influences of regret, worry, locus of control and the threshold of concern heuristic. *Water Resources and Economics* 30, 1–19. DOI: 10.1016/j.wre.2019.100144.

Rommel, J., Hermann, D., Müller, M. and Mußhoff, O. (2017) Comprehension in risk elicitation experiments. *Applied Economics Letters* 24, 627–634. DOI: 10.1080/13504851.2016.1217302.

Rommel, J., Hermann, D., Müller, M. and Mußhoff, O. (2019) Contextual framing and monetary incentives in field experiments on risk preferences: evidence from German farmers. *Journal of Agricultural Economics* 70, 408–425. DOI: 10.1111/1477-9552.12298.

Rosenzweig, M.R. and Binswanger, H.P. (1993) Wealth, weather risk and the composition and profitability of agricultural investments. *The Economic Journal* 103, 56–78. DOI: 10.2307/2234337.

Ross, N., Santos, P. and Capon, T. (2012) Risk, ambiguity and the adoption of new technologies: experimental evidence from a developing economy. In: *Paper prepared for the International Association of Agricultural Economists*, August 18–24, 2012, Foz do Iguaçu, Brazil.

Ruhinduka, R.D., Alem, Y., Eggert, H. and Lybbert, T. (2020) Smallholder rice farmers' post-harvest decisions: preferences and structural factors. *European Review of Agricultural Economics* 47, 1587–1620. DOI: 10.1093/erae/jbz052.

Sanou, A., Liverpool-Tasie, L.S.O. and Shupp, R. (2018) Eliciting risk attitudes in the field: surveys or experimental methods? An empirical comparison in rural Niger. *The Journal of Development Studies* 54, 1450–1470. DOI: 10.1080/00220388.2017.1404034.

Schildberg-Hörisch, H. (2018) Are risk preferences stable? *Journal of Economic Perspectives* 32, 135–154. DOI: 10.1257/jep.32.2.135.

Schlag, K.H., Tremewan, J. and van der Weele, J.J. (2015) A penny for your thoughts: a survey of methods for eliciting beliefs. *Experimental Economics* 18, 457–490. DOI: 10.1007/s10683-014-9416-x.

Smith, V.L. (1976) Experimental economics: induced value theory. *The American Economic Review* 66, 274–279.

Tanaka, T., Camerer, C.F. and Nguyen, Q. (2010) Risk and time preferences: linking experimental and household survey data from Vietnam. *American Economic Review* 100, 557–571. DOI: 10.1257/aer.100.1.557.

Trautmann, S.T. and van de Kuilen, G. (2015) Belief elicitation: a horse race among truth serums. *The Economic Journal* 125, 2116–2135. DOI: 10.1111/ecoj.12160.

Vieider, F.M., Martinsson, P., Nam, P.K. and Truong, N. (2019) Risk preferences and development revisited. *Theory Decision* 86, 1–21. DOI: 10.1007/s11238-018-9674-8.

Visser, M., Jumare, H. and Brick, K. (2020) Risk preferences and poverty traps in the uptake of credit and insurance amongst small-scale farmers in South Africa. *Journal of Economic Behavior & Organization* 180, 826–836. DOI: 10.1016/j.jebo.2019.05.007.

Ward, P.S. and Singh, V. (2015) Using field experiments to elicit risk and ambiguity preferences: behavioural factors and the adoption of new agricultural technologies in rural India. *The Journal of Development Studies* 51, 707–724. DOI: 10.1080/00220388.2014.989996.

Warnick, J.C.E., Escobal, J. and Laszlo, S.C. (2011) Ambiguity aversion and portfolio choice in small-scale Peruvian farming. *The Behavioral Economics Journal of Economic Analysis & Policy* 11, 1–56. DOI: 10.2202/1935-1682.2331.

Yesuf, M. and Bluffstone, R.A. (2009) Poverty, risk aversion, and path dependence in low-income countries: experimental evidence from Ethiopia. *American Journal of Agricultural Economics* 91, 1022–1037. DOI: 10.1111/j.1467-8276.2009.01307.x.

ZAS and La Gazzette (2019) The State of Zimbabwe's Agricultural Sector. Inaugural Zimbabwe Agricultural Survey 2018–2019. Available at: https://www.aedsafrica.com/wp-content/uploads/2021/06/Inaugural-Zimbabwe-Agricultural-Sector-Survey-2018-2019.pdf (accessed August 8, 2022).

7 Examining the Role of Risk Attitudes in Agricultural Management Decisions in Emerging and Developing Countries: Experimental Evidence

Toritseju Begho[1]* and Kehinde Odeniyi[2]
[1]*Scotland's Rural College, Edinburgh, UK;* [2]*Osun State University, Osogbo, Nigeria*

Abstract

Farmers are constantly faced with numerous agricultural risks and have to make decisions in an uncertain and changing environment. Numerous studies have been carried out to understand and better manage decision-making under risk. However, many of these studies conducted in emerging and developing countries examining the determinants of agricultural risk management and coping decisions either did not consider risk attitudes or relied on self-reported measures. In addition, some of the experimental research conducted in developed countries or research that used non-farmer convenience samples lacks external validity due to variations in socio-economic, institutional, environmental, contextual and circumstantial factors. Thus, this chapter's objective was to elicit farmers' risk preferences via an experimental method and determine if risk preferences explain the decision to adopt practices and strategies often reported to be driven by risk and risk aversion. This chapter analyzes survey data from India (5732) and Nepal (570). The results show that farmers in the sample mostly avoided taking risks when faced with a decision with financial outcomes. Those averse to risk are more likely to stick with traditional crops than to adopt hybrid or improved crop varieties and practice crop diversification. Contrary to most findings, farmers more averse to risk are less likely than their risk-loving counterparts to use synthetic fertilizer and less likely to have off-farm sources of income. However, risk preferences do not influence the likelihood that farmers will take up sharecropping and crop insurance. The recommendations are that policy should focus on reducing the risk farmers face and should target empowering farmers to manage risk on their own through increased access to a range of instruments and strategies. Direct technical, financial and institutional support will help shift many farmers from *ex-post* crisis management to *ex-ante* risk management and help ensure that farmers' livelihoods are resilient.

7.1 Introduction

In agriculture, many strategic decisions involve considerably high levels of uncertainty and risk. The risks farmers face are spread across production, price or market, financial, institutional and human or personal risks (Komarek *et al.*, 2020). These risks mainly result in yield and price volatility. Such volatility disproportionately triggers or exacerbates food insecurity and livelihood vulnerability in developing countries where the majority of the world's extreme poor are located

*E-mail: toritseju.begho@sruc.ac.uk

 Managing Risk in Agriculture: A Development Perspective (eds A.K. Mishra, S.C. Kumbhakar and G. Lien).
DOI: 10.1079/9781800622289.0007

and where a significant share of the population work in agriculture (World Bank, 2013, 2019). Thus, farmers, policymakers, communities and stakeholders can gain value from understanding the drivers of farmers' decision-making, being able to predict economic behavior and helping farmers make better decisions.

Studies on decision-making in recent years have shown that risk perceptions and attitudes are important determinants of farm production, investment and management. Specifically, empirical evidence suggests that risk preferences and attitudes elicited via experiments can reasonably be used to predict important farm decisions. Also, a considerable amount of experimental research has been carried out on the techniques for analyzing risky decisions, modeling the processes and predicting the consequences of the decisions farmers make, as the benefit of having this understanding is well known. However, many of these studies are conducted in developed countries or have relied on convenience samples.

Notably, recent evidence calls into question the extent to which such studies contribute to our understanding of the behavior of farmers in developing countries. There is evidence of the lack of external validity of convenience samples and the demerits of generalizing such findings. In addition, some studies conducted in emerging and developing countries examining the determinants of agricultural risk management and coping decisions did not consider risk attitudes or relied on self-reported measures (Sarker *et al.*, 2013; Bahinipati and Venkatachalam, 2015; Ali, 2017; Bishu *et al.*, 2018). Hence, there is a need to measure risk attitudes and the effect on agricultural decision problems using samples from the population of interest, i.e. farmers in emerging and developing countries. Thus, this study builds upon past experimental evidence by eliciting risk preferences and identifying whether risk preferences explain farmers' decisions, which theoretically are linked with risk attitudes.

Section 7.2 summarizes the relevant findings on the association between farm management strategies and risk preferences/attitudes, provides the foundation to support the risk preference elicitation method used in this study and specifies the hypotheses. Section 7.3 presents the data. Section 7.4 details the experimental procedure and estimation methods. Section 7.5 presents the results and discussion and Section 7.6 concludes the chapter.

7.2 Synthesis of the Literature

7.2.1 Risk management and coping strategies

Risk management predominantly refers to *ex-ante* strategies, while risk-coping strategies are largely synonymous with *ex-post* strategies. Thus, the reference to *ex-post* strategies in the literature is to those strategies put in place after a risk event occurs (Collins-Sowah and Henning, 2019). However, it is worth noting that the distinction between *ex-ante* and *ex-post* risk-coping strategies is not entirely unambiguous. For example, adopting certain measures after a shock can transition to a permanent measure that will reduce exposure to future shocks. In contrast, the availability of certain *ex-post* strategies depends on their implementation *ex-ante* (Lekprichakul, 2009). Notably, these management strategies are also not independent of one another, as previous studies found that adopting one risk management strategy results in the use of or reluctance to adopt another (Akhtar *et al.*, 2019; Knapp *et al.*, 2021). We highlight this aspect in the discussion that follows. Past studies also have distinguished between risk-seeking farmers who are more likely to apply *ex-ante* risk management strategies and risk-averse farmers who prefer to cope with the consequences of a risky outcome (van Winsen *et al.*, 2016).

Risk management and coping strategies can be broadly categorized as diversifying (e.g. income, production), optimizing (e.g. via technical or scale enlargement), external (e.g. insurance or contract agreements) and buffering strategies (van Winsen *et al.*, 2016). The synthesis of the literature thus focuses on practices and strategies often reported to be heavily driven by risk and risk aversion.

Technology adoption

The role of risk attitude in low rates of technology adoption in emerging and developing countries is being increasingly recognized (Janssen and Swinnen, 2019; Curry *et al.*, 2021). It is expected a priori that risk-averse farmers would be more inclined to adopt a technology if it could reduce variability in economic outcomes. However, the uncertainty surrounding new technologies can drive risk-averse farmers to be extra cautious (Liu, 2013). This is due to the possibility that the aims of adopting a technology may not be met at the early stages due to higher exigencies of knowledge in its use.

Adopting new crop varieties is often advanced as a risk-reducing strategy. For example, new crop varieties are often promoted as having the capacity to reduce climate change's negative impacts on crop productivity, reduce the risk of total crop failure and build resilience and adaptive capacity. Hence, Bannor *et al.* (2020) found that risk aversion positively influenced the decision to adopt modern rice varieties. Bannor *et al.* (2020) agree with Ward and Singh's (2015) findings that risk-averse farmers are more likely to adopt new seed varieties with risk-reducing capabilities. Risk aversion is also known to determine the specific new crop varieties that farmers adopt depending on salient attributes in reducing downside risk. The definition of downside risks aligns with Chavas and Nauges (2020). It refers to the risk associated with adverse shocks at the lower tail of the payoff distribution. For example, Holden and Quiggin (2017) found a higher chance of risk-averse households adopting drought-tolerant maize but a lower likelihood of them adopting other improved maize varieties. Similarly, results from Mehar *et al.* (2017) suggest that risk aversion increases the chances that a farmer will choose a stress-resistant crop variety over varieties with market-oriented attributes. As with most technologies, more risk-averse farmers have been observed to place a greater value on the attributes of new crop varieties that reduce variances in yield (Paik *et al.*, 2020).

Insurance

Crop insurance is an important market-based risk management instrument that provides economic support to farmers and reduces loss or variability in income in the event of crop failure (Swain, 2014; Panda, 2021). There are reasons to contend that risk attitudes influence insurance decisions. Purchasing crop insurance is like a gamble, which depends on the uncertainty of events, with payment made only when the event occurs. Findings in the literature mostly suggest that high-risk aversion increases the willingness to purchase agricultural insurance (Lyu and Barré, 2017; Zhang *et al.*, 2021). However, due to moral hazard problems, insurance may have a negative impact on the use of risk-reducing inputs (Vigani and Kathage, 2019). Notably, farmers who take up insurance may have less incentive for on-farm diversification, considering that the insurance will deal with risks. However, this relationship is found to vary by type of diversification activities (Wuepper *et al.*, 2018; Knapp *et al.*, 2021).

On-farm and off-farm diversification

It is not always the case that risk-averse farmers choose not to diversify but to focus on activities that, by nature, have a lower risk. Risk-averse farmers may spread income sources to protect against risks and shocks. This may be achieved through an on-farm reallocation of agricultural production resources. For example, previous studies reported that risk-averse farmers mostly undertook crop diversification (Sarwosri and Mußhoff, 2020; Adnan *et al.*, 2021), especially in zones of high climate risk (Saqib *et al.*, 2021).

Also, diversifying income may be achieved through engaging in off-farm income-earning jobs. Off-farm income is one of the important strategies for managing production risk and ensuring income stability (Akhtar *et al.*, 2019). Most findings suggest that among farmers in emerging and developing countries in particular, risk aversion is a factor that increases the likelihood of engaging in income-earning activities off-farm as a risk management strategy (for example, see Iqbal *et al.*, 2016; Lu *et al.*, 2017; Akhtar *et al.*, 2019, 2021; Khanal *et al.*, 2021). However, there are also contradictory findings with respect to the relationship between off-farm work and risk aversion. For example, for high-value crops or among farmers in high-income economies, it has been shown that participation in off-farm work increases with higher risk-taking behavior or that risk attitude is not an important determinant of off-farm decisions (see Hellerstein *et al.*, 2013; Ullah and Shivakoti, 2014; Lawin and Tamini, 2017). Such findings challenge the proposition that farmers engage in off-farm work as a risk management tool, justifying the need for further investigation.

Input use

There is empirical evidence of the effect of risk attitudes on input use decisions. One such important input to crop production is synthetic fertilizer. In Asia and regions notable for problems with the quality of synthetic fertilizer (Cook *et al.*, 2016; Praveen, 2021), the effect of risk attitudes on decisions about the use of synthetic fertilizer is likely to be more salient. Assefa *et al.* (2021) reported that applying fertilizer at scientifically recommended rates is likely to result in insufficient returns to the risk-averse farmer. Results from Khor *et al.* (2018) show that risk aversion impacts the use of synthetic fertilizer but is mediated by farmers' wealth levels such that among low-wealth farmers, the intensity of fertilizer use was lower when their risk aversion was higher. From a different perspective, Qiao and Huang's (2021) results indicate that, depending on the elasticity of fertilizer use and the probability of achieving the desired effects from fertilizer use, the fertilizer application rate may be higher or lower for risk-averse farmers than for risk-seeking farmers. However, the findings are not unanimous. For example, Le Cotty *et al.* (2018) did not observe any statistically significant effect of risk aversion on fertilizer use.

There are explanations for why farmers may apply synthetic fertilizer over or under the recommended rates, depending on whether farmers perceive fertilizer as a risk-enhancing or risk-reducing input (Khor *et al.*, 2018; Begho *et al.*, 2022). For example, according to Moser and Mußhoff (2017), fertilizer usage reportedly has a risk-reducing effect. However, Chai *et al.* (2022) observe the opposite effect. Therefore, following the argument in the latter, the expectation is that there should be a negative relationship between risk aversion and fertilizer use when farmers perceive fertilizer as a risk-increasing technology.

Risk-sharing land contracts

Sharecropping is a strategy farmers use to distribute production risks (Dubois, 2000). Although sharecropping may have declined in many countries, farm surveys show that it still exists in emerging and developing countries in Asia (Ahmed and Billah, 2018; Mukhamedova and Pomfret, 2019). The role of risk attitudes in sharecropping contracts becomes particularly evident in the absence of insurance markets and when the farmer is unable or reluctant to be solely responsible for the risk of agricultural production (Das *et al.*, 2019). The risk-based models of sharecropping suggest that the risk aversion of both landowner and tenant drives sharecropping. The farmers are more likely to choose a sharecropping land contract because they have a greater risk aversion than the landlords (Federico, 2006). Risk aversion also interacts with moral hazards and financial constraints to influence sharecropping decisions (Singh, 2000).

7.2.2 Hypotheses

Based on the summarized literature, the expectation is that farmers who adopt risk management and coping strategies can be distinguished from those who do not by differences in risk aversion. Accordingly, the hypotheses formulated were:

H_1. Risk aversion has predictive power on farmers' decisions regarding the use of synthetic fertilizer.

H_2. Risk aversion has predictive power on engaging in off-farm income-earning jobs.

H_3. Risk aversion has predictive power on sharecropping contracts.

H_4. Risk aversion has predictive power on the uptake of crop insurance.

H_5. Risk aversion has predictive power on the adoption of improved crop varieties.

H_6. Risk aversion has predictive power on decisions about on-farm diversification.

7.2.3 Risk preference elicitation methods

The risk preference elicitation methods commonly featured in experimental elicitation tasks are Gneezy and Potters (1997), Holt and Laury (2002), Eckel and Grossman (2002, 2008), Tanaka *et al.* (2010) and Crosetto and Filippin (2013), among others. Most of these tasks feature multiple price list formats in which participants must choose between one 'safe' and one 'risky' option, with the former having a small difference between the unfavorable and the favorable payoff and the latter having a larger difference between the unfavorable and the favorable payoff. For some elicitation methods, the payoffs remain constant while the probabilities change. In contrast with other methods, the probability remains constant (readers can refer to Harrison and Rutström, 2008; Crosetto and Filippin, 2016, for details of these tasks).

Previous studies have compared risk preferences elicited via the different lottery tasks. The findings, however, are mixed. Some studies find that participants exhibit different levels of risk aversion by task (for example, Dave *et al.*, 2010; Reynaud and Couture, 2012), while other studies report fairly stable risk preferences across the tasks under comparison (e.g. Dasgupta *et al.*, 2016; Sauter *et al.*, 2018).

Similarly, there are comparisons between risk preferences elicited through a lottery task and self-reported measures. These studies also reveal mixed risk preferences. Findings such as Dohmen *et al.* (2011) suggest that self-reported measures align with the lottery task, but several other studies report contradictory findings (Sauter *et al.*, 2018).

Several important points from the literature are that, first, there is a consensus that the more cognitively demanding tasks do not predict behavior as well as simpler tasks do. This is especially the case for subjects with low levels of education, quantitative skills and comprehension (Dasgupta *et al.*, 2016). Second, there is some agreement that risk preference evaluations are not stable across elicitation techniques. Third, although many elicitation tasks have the same theoretical foundation, the different methods yield different results, making a direct comparison of findings difficult.

7.3 Data

This study relies on data from two sources. The 2014–2015 International Rice Research Institutes' Rice Monitoring Survey: South Asia (RMS-SA) data was collected to monitor rice systems that capture varietal turnovers over time. The surveys elicited information from about 9000 rice-farming households across South Asia, with respondents in India totaling 5868. The respondents were obtained by randomly selecting 600 villages in India and then randomly selecting households in these villages. We also used the 2016 International Food Policy Research Institute (IFPRI) and the Seed Entrepreneur Association Nepal (SEAN) survey of 600 vegetable-farming households in Nepal based on a multi-stage stratified random sampling from 20 districts. The data was collected as part of a wider study on the demand for and supply of vegetable seeds in Nepal. After excluding incomplete responses, we analyzed the data from 5732 households in India and 570 households in Nepal. The details of how the sample was obtained can be found in Malabayabas *et al.* (2015) and IFPRI-SEAN (2017). Notably, small farmers mainly constitute the samples in this study. Restricting the sample based on this criterion prevents findings and conclusions driven by potentially confounding variations in socio-economic, contextual and circumstantial factors. The decision to use datasets from different countries was made in order to examine whether risk attitudes play a role in agricultural decisions in emerging and developing countries. However, a comparison between the countries is beyond the scope of this chapter.

7.4 Methodology

Following the synthesis of the literature, we focus on selected determinants of risk management and coping strategies informed by the review of global literature and on those commonly adopted by farmers in Asia (Sulewski and Kłoczko-Gajewska, 2014; van Winsen *et al.*, 2016; Duong *et al.*, 2019; Adnan *et al.*, 2020). The measurement of these strategies is summarized in the first part of Table 7.1. The study also accounts for other factors, besides risk preferences, that influence risk management and coping strategies, as reported in previous studies.

7.4.1 Experimental procedure

The study elicits risk preferences using an adaptation of the Eckel and Grossman task. In line with the recommendations of Dave *et al.* (2010) and Hossack and An (2015), considering the numerical skills of respondents that constitute the sample, the choice of the Eckel and Grossman

Table 7.1. Description of variables. (Source: Authors' computation of the data from Yamano, 2014–2015, and International Food Policy Research Institute (IFPRI) Seed Entrepreneur Association Nepal (SEAN), 2017.)

Variable	Description
Dependent variables	
Uses synthetic fertilizer	1 if respondent uses synthetic fertilizer, 0 otherwise
Has off-farm job	1 if respondent has an off-farm job which provides income, 0 otherwise
In a sharecropping contract	1 if respondent's agreed tenure of any of the plot is sharecropping, 0 otherwise
Has crop insurance	1 if respondent has any crop insurance, 0 otherwise
Cultivates hybrid varieties	1 if respondent cultivates at least one hybrid or inbred variety of crop, 0 otherwise
Practices crop diversification	1 if respondent cultivates more than one crop per season, 0 otherwise
Independent variables	
Risk preference – CRRA	Mean of the implied constant relative risk aversion ranges estimated from the lottery tasks
Earthquake	1 if respondent's household experienced earthquake that affected their income and assets in the past, 0 otherwise
Fire	1 if respondent experienced fire on their farm that affected their income and assets in the past, 0 otherwise
Drought	1 if respondent experienced drought on their farm that affected their income and assets in the past, 0 otherwise
Flooding	1 if respondent experienced flooding on their farm that affected their income and assets in the past, 0 otherwise
Salinity	1 if respondent experienced soil salinity on their farm that affected their income and assets in the past, 0 otherwise
Pests and diseases	1 if respondent experienced pest and disease infestation on their farm that affected their income and assets in the past, 0 otherwise
Number of plots	Number of plots owned
Experience growing crop	Number of years of growing the crop
Subsistence oriented	1 if respondent is producing mainly for household consumption, 0 otherwise
Mechanized	1 if respondent uses machines in crop production
Uses organic fertilizer	1 if respondent uses organic fertilizer in crop production
Intent to adopt new varieties	Willingness to adopt new varieties on a scale of 1–4
Knowledge of local climate	Number of years residing in the location
Extension visits	Number of extension visits per year
Location	Location from northernmost to southernmost (1–4)
Gender	1 if respondent is male, 0 otherwise
Age	Respondent's age in number of years
Household size	Number of adults and children in the household
Income	Income from the crop cultivated per year

task is appropriate as it is less demanding cognitively, simpler and relatively easy to implement. As part of the wider survey and before the task was presented, participants read the study information to acquaint them with the study procedure, confidentiality and the right to withdraw. Participants were told they would be offered several payment options from which to choose. They also were told that although the questions were hypothetical, they should assume they had been given actual cash as their choices had real consequences because an actual payment in mobile phone credit would be made. Hence, they should think carefully about the choice they were making. Participants were also informed that the payout would be determined by throwing a die. For the Nepalese data, there was no information to suggest that it was not purely hypothetical or that their choices had real consequences. As presented in Table 7.2, participants had to make a single choice from the five lotteries over small stakes. In addition to a sure outcome, the other four risky outcomes had a 50% probability of obtaining a higher payoff, thus permitting the categorization of participants into four risk categories from the linearly increasing expected payoffs and risk. In summary, participants had to choose between receiving a sure payoff (i.e. Rs. 40 in the case of India or Rs. 1250 for participants from Nepal) and any options with varying payoffs.

7.4.2 Estimation methods

We estimate a constant relative risk aversion (CRRA) parameter using a widely applied parametric family of utility functions given as: $u(c|\sigma)=\frac{c^{1-\sigma}}{1-\sigma}$, where $\sigma \geq 0$ is the CRRA coefficient as implied by the bounds of the Eckel and Grossman task and based on expected utility maximization. Readers can refer to De Palma *et al.* (2008) for a discussion of the decision-making theories. Finally, we follow Menapace *et al.* (2016) by estimating a probit regression model to determine the relationship between risk preferences from the implied CRRA ranges estimated from the lottery tasks and actual risk management and coping decisions while controlling for farmer-specific factors. The choice of the probit model is due to the binary nature of the dependent variables. The (dependent and independent) variables used in the probit model are described in Table 7.1.

7.5 Results and Discussion

7.5.1 Description of the sample

The sample was composed mostly of men with an average age of 48 years. The summary statistics in Table 7.2 show that most of the farmers either chose to avoid risk (61.1% among Nepal farmers) or mostly chose safer options in both experiments. Previous studies have observed that risk aversion and avoidance of situations that have a high probability of failure are greater among farmers in emerging and developing countries than farmers in developed countries. Specifically, the overall proportion of the risk aversion levels aligns with previous studies (Akter *et al.*, 2017; Patil and Veettil, 2018) that employed similar elicitation methods on farmers in South Asia and found that most farmers were risk-averse.

7.5.2 Regression results and discussion

The results of the probit model estimating the relationship between risk preferences and various

Table 7.2. Eckel and Grossman lottery adapted to local currency. (Source: Authors' computation of the data from Yamano, 2014–2015, and International Food Policy Research Institute (IFPRI) Seed Entrepreneur Association Nepal (SEAN), 2017.)

	Lottery choice			
Option	50% (potential payoff)	50% (losing payoff)	CRRA ranges	% of respondents
India (Indian rupees)				
1	Rs. 40	Rs. 40 (i.e. 40 for sure)	$2 < \sigma$	46.5
2	Rs. 60	Rs. 30	$0.67 < \sigma < 2$	10.1
3	Rs. 80	Rs. 20	$0.38 < \sigma < 0.67$	11.4
4	Rs. 100	Rs. 10	$0.20 < \sigma < 0.38$	21.8
5	Rs. 120	Rs. 0	$0 < \sigma < 0.20$	10.2
Nepal (Nepalese rupees)				
1	Rs. 1250	Rs. 1250 (i.e. 1250 for sure)	$3.25 < \sigma$	61.1
2	Rs. 2000	Rs. 1000	$1.10 < \sigma < 3.25$	22.2
3	Rs. 2750	Rs. 750	$0.68 < \sigma < 1.10$	11.2
4	Rs. 3500	Rs. 500	$0.33 < \sigma < 0.68$	1.9
5	Rs. 5000	Rs. 0	$0 < \sigma < 0.33$	3.7

other factors on the adoption of risk management and coping strategies are presented in Table 7.3. The test statistic based on likelihood ratio (LR) indicates that the estimated models are statistically significant at the 1% level.

The results show that farmers more averse to risk are less likely to use synthetic fertilizer than farmers more willing to take risks. With this finding, H_1 specifying that risk aversion has predictive power on farmers' decisions about fertilizer use is confirmed. These results support the argument (in Fufa and Hassan, 2003; Villano *et al.*, 2005) that fertilizer is a risk-increasing input. Also, the association between risk attitudes and the use of synthetic fertilizer points to the importance of country- and perhaps farm-type-specific studies, since generalizations using findings in developed countries, for example, could result in misleading predictions. Using synthetic fertilizer increases yield variability and profit more so in situations in which farmers must rely on high-priced fertilizer from informal sources due to the inadequate and untimely supply of fertilizers from formal sources. Considering that most farmers in the sample reported using organic fertilizer, the findings highlight that, unlike organic fertilizer, synthetic fertilizer potentially increases price risk and is likely to further impact overall risk. Thus, it explains why risk-averse farmers in developing countries may avoid using synthetic fertilizers.

Against expectation, risk preference was negatively associated with having income from an off-farm job. That is, farmers who are more averse to risk are less likely to have off-farm sources of income than their risk-loving counterparts. Hence, the hypothesis (H_2) that risk aversion has predictive power on engaging in off-farm income-earning jobs is confirmed. That risk-averse farmers mostly do not diversify income beyond the farm is not supported by most of the previous empirical analyses. Although there is a basis to contend that farmers may increase off-farm work, especially after a loss from recent shocks or adverse events, the findings presented here possibly corroborate Reardon *et al.* (1998), who found that compared to wealthier farmers, poorer farmers (who are arguably more risk-averse) lack the capacity to increase off-farm work and that off-farm income accounts for much less of their total income. An alternative point of view is that we observe the possibility that off-farm income reduces risk aversion (Picazo-Tadeo and Wall, 2011; Saqib *et al.*, 2016). Notably, investigating the effect of off-farm work on risk aversion is beyond the focus of this chapter.

Farmers who are more averse to risk are more likely than farmers who are willing to take risks to adopt hybrid or inbred crop varieties, and H_5 was confirmed. The result supports Ward and Singh (2015) and Bannor *et al.* (2020), who found that risk aversion positively influenced the decision to adopt improved varieties. Similarly, farmers more averse to risk are more likely than their risk-loving counterparts to practice crop diversification and support H_6. This result differs from Kijima (2019), who found no effect of risk aversion on diversification, but aligns with Sarwosri and Mußhoff (2020), who found that farmers who diversify their crop portfolio are more risk-averse than farmers who do not diversify. Notably, the marginal effects of the risk preference variables are small.

The results also indicate that the measure of risk attitude used in this study has no direct predictive power for sharecropping and crop insurance. Thus, H_3 and H_4 are not supported. However, these findings should be generalized with caution because of the particularities of vegetable farming, i.e. mainly done in the homestead, which is either owned or rented. Also, the insurance and credit markets, which are not so well developed and are targeted mostly at principal cash crops, may be the reason for the deviation from a priori expectations. In addition, in Asia, rice crops are mainly insured, and insurance purchase is encouraged by the government subsidizing the cost of the insurance premium (Yanuarti *et al.*, 2019).

Compared to rare shocks like earthquakes, shocks such as flooding, drought, salinity, agricultural fire and pests and diseases predicted the adoption of farm management and coping strategies. In addition, as found in previous studies, farm and farmer characteristics (gender, age, household size, income, farming experience and location) explain at least one of the farm management strategies examined.

Some aspects of this topic require further investigation. First, the study did not investigate whether adopting one or more of the management strategies covered in this study increased or decreased overall productivity. Second, the

Table 7.3. Probit regression and marginal effect of the determinants of selected management and coping strategies. (Source: Authors' own work.)

	Uses synthetic fertilizer		Has off-farm job		In a sharecropping contract		Has crop insurance		Cultivates hybrid or improved varieties		Practices crop diversification	
Variable	Coef.	ME	Coef.	ME	Coef.	ME	Coef.	ME	Coef.	ME	Coef.	ME
Risk preference – CRRA	−0.19** (−2.03)	−0.042	−0.314** (−2.18)	−0.002	−0.065 (−0.55)	−0.004	0.082 (0.98)	0.022	0.175*** (4.21)	0.046	0.139*** (3.21)	0.031
Earthquake	0.006 (0.71)	0.001	0.528 (0.96)	0.003	0.00 (0.00)		−0.005 (−0.72)	−0.001				
Fire	0.001 (0.11)	0.00	0.001 (0.12)	0.00	0.00 (0.00)		0.012* (1.84)	0.003				
Drought	0.009 (1.07)	0.002	0.003 (0.28)	0.00	−0.007 (−0.17)	0.00	−0.001 (−0.09)	0.00	−0.047 (−1.02)	−0.013	−0.15*** (−3.16)	−0.035
Flooding									0.122*** (2.93)	0.033	0.149*** (3.44)	0.034
Salinity									−0.887*** (−14.68)	−0.213	−0.851*** (−13.28)	−0.173
Pests and diseases	0.008 (1.13)	0.002	−0.001 (−0.1)	0.00	0.00		0.018 *** (3.32)	0.005				
Number of plots	0.22*** (2.75)	0.049	−0.059 (−0.79)	0.00	0.379 *** (4.39)	0.023	−0.068 (−1.14)	−0.018	−0.069*** (−5.11)	−0.018	−0.063*** (−4.34)	−0.014
Experience growing crop	−0.004 (−0.81)	−0.001	0.003 (0.51)	0.00	0.009 (1.21)	0.001	0.01** (2.35)	0.003				
Subsistence oriented	−0.007 (−0.63)	−0.001	−0.012 (−1.32)	0.00	0.021* (1.85)	0.001	−0.021 (−0.96)	−0.005				
Mechanized	0.694 *** (4.17)	0.151	0.163 (0.83)	0.001	−0.082 (−0.3)	−0.005	−0.028 (−0.88)	−0.007				
Uses organic fertilizer	−0.391 (−0.64)	−0.068	1.12** (2.01)	0.031	−0.525 (−0.53)	−0.052	0.847 (1.33)	0.145				
Intent to adopt new varieties	−0.226 (−1.53)	−0.049	−0.268 (−1.32)	−0.001	0.105 (0.47)	0.006	0.085 (0.61)	0.022				
Knowledge of local climate	0.003 ** (2.44)	0.001	−0.001 (−1.07)	0.00	0.002 (1.27)	0.00	0.001 (0.73)	0.00				
Extension visits									0.038* (1.93)	0.01	0.004 (0.33)	0.001

Continued

Table 7.3. Continued.

Variable	Uses synthetic fertilizer		Has off-farm job		In a sharecropping contract		Has crop insurance		Cultivates hybrid or improved varieties		Practices crop diversification	
	Coef.	ME	Coef.	ME	Coef.	ME	Coef.	ME	Coef.	ME	Coef.	ME
Location	−0.06 ***	−0.013	−0.002	0.00	−0.032	−0.002	−0.03**	−0.008	−0.029	−0.008	−0.046**	−0.01
	(−5.1)		(−0.11)		(−1.27)		(−2.43)		(−1.43)		(−2.19)	
Gender	−0.141	−0.031	−0.049	0.00	−0.79**	−0.049	0.169 (0.74)	0.045	−0.046	−0.012	−0.193*	−0.047
	(−0.64)		(−0.17)		(−2.4)				(−0.45)		(−1.87)	
Age	−0.01*	−0.002	−0.004	0.00	0.00	0.00	0.008 (1.51)	0.002	−0.003*	−0.001	0.002	0.00
	(−1.93)		(−0.57)		(0.04)				(−1.7)		(1.25)	
Household size	−0.018	−0.004	0.098*	0.001	−0.021	−0.001	−0.005	−0.001	0.011**	0.003	0.005	0.001
	(−0.48)		(1.82)		(−0.36)		(−0.14)		(2.41)		(0.98)	
Income	0.001	0.00	0.00	0.00	−0.003*	0.00	0.00	0.00	0.00	0.00	0.00	0.00
	(1.39)		(0.03)		(−1.81)		(1.1)		(−0.44)		(0.1)	
Constant	8.768***		1.721		2.902 (0.95)		0.259 (0.16)		−0.285*		−0.406***	
	(5.44)		(0.79)						(−1.93)		(−2.68)	
Number of observations	570		570		419[a]		570		5732		5732	
Akaike crit. (AIC)	528.567		294.778		179.324		5332.906		557.760		4821.405	
Bayesian crit. (BIC)	606.789		373.000		239.892		5412.751		635.981		4901.331	
Chi squared	88.808		20.442		45.577		804.027		51.461		637.620	
LR test statistic	160.055***		39.794***		69.535***		87.6186***		523.689***		422.795***	

Notes:
t-statistics in parentheses.
*p<0.10, **p<0.05, ***p<0.01.
ME = Marginal effect.
The reason that some of the variables in the first four models are missing from the last two models is due to the estimation of two completely different datasets.
[a]There was some incomplete information for a small part of our sample in this case, hence the total number of observations was below 570.

study did not examine the joint adoption of risk management practices. There is empirical evidence (see Ullah and Shivakoti, 2014; Knapp *et al.*, 2021) that farmers tend to jointly adopt bundles of risk management practices. Third, although the procedure used to elicit risk attitude was justified, and alternative methods were not captured in the data, there is a concern about the validity of the measurement instruments, which can be alleviated only by comparing methods. Fourth, the survey data did not permit a deeper investigation into the association between risk preferences and fertilizer application rates, a gap that future studies could aim to fill. Extending the current work to cover these aspects in future studies will improve our understanding of the drivers of farm management decision-making and strengthen policy recommendations.

7.6 Conclusion

Risk is embedded in many farming decisions. In recent years, climate-related risks and crop-destroying pests and diseases, in addition to other risks, have brought attention to the vulnerability of the livelihoods of farmers in emerging and developing countries. There is value to be gained from understanding the drivers of farmers' decisions and specifically identifying the role that risk attitudes play in farm risk management and coping strategies. The objective of the study was to elicit farmers' risk preferences and determine if risk preferences could explain the decision to adopt practices and strategies often reported to be driven by risk and risk aversion.

This study contributes to the literature that examines risk management and coping strategies in emerging and developing countries by analyzing data from South Asia (Nepal and India). The study shows that simple experimental economic methods can be used to elicit risk preferences, which can then be used to predict decisions. The study also highlights that risk attitudes are an important factor shaping farm management decisions. Finally, the findings indicate that not all risk management and coping strategies reported in the literature apply universally to farmers.

The recommendation from the findings in this study is that policy should focus on reducing the risk that farmers face, as this will be more feasible and effective than aiming to change farmers' risk attitudes. Where extreme risk aversion evidently acts as a barrier, e.g. to adopting risk-reducing technologies, intervention should aim to account for risk attitudes and incorporate methods for extra encouragement or nudges to such farmers to take some manageable risks. Further, farmers in emerging and developing countries have mostly been responsible for managing risks on their own. With justification for it to remain so, policies should be targeted at empowering farmers through increased access to a range of instruments and strategies. Besides, other direct technical, financial and institutional support will help shift many farmers from *ex-post* crisis management to *ex-ante* risk management and help ensure that farmers' livelihoods are resilient.

References

Adnan, K.M., Ying, L., Ayoub, Z., Sarker, S.A., Menhas, R., *et al.* (2020) Risk management strategies to cope catastrophic risks in agriculture: the case of contract farming, diversification and precautionary savings. *Agriculture* 10, 351.

Adnan, K.M., Ying, L., Sarker, S.A., Yu, M.M. and Tama, R.A.Z. (2021) Simultaneous adoption of diversification and agricultural credit to manage catastrophic risk for maize production in Bangladesh. *Environmental Science and Pollution Research* 28, 58258–58270.

Ahmed, M.F. and Billah, M.M. (2018) Impact of sharecropping on rice productivity in some areas of Khulna district. *Bangladesh Journal of Agricultural Research* 43, 417–430.

Akhtar, S., Li, G.C., Nazir, A., Razzaq, A., Ullah, R. *et al.* (2019) Maize production under risk: the simultaneous adoption of off-farm income diversification and agricultural credit to manage risk. *Journal of Integrative Agriculture* 18, 460–470.

Akhtar, S., Abbas, A., Iqbal, M.A., Rizwan, M., Samie, A. *et al.* (2021) What determines the uptake of multiple tools to mitigate agricultural risks among hybrid maize growers in Pakistan? Findings from field-level data. *Agriculture* 11, 578.

Akter, S., Krupnik, T.J. and Khanam, F. (2017) Climate change skepticism and index versus standard crop insurance demand in coastal Bangladesh. *Regional Environmental Change* 17, 2455–2466.

Ali, A. (2017) Coping with climate change and its impact on productivity, income, and poverty: evidence from the Himalayan region of Pakistan. *International Journal of Disaster Risk Reduction* 24, 515–525.

Assefa, B.T., Reidsma, P., Chamberlin, J. and van Ittersum, M.K. (2021) Farm- and community-level factors underlying the profitability of fertiliser usage for Ethiopian smallholder farmers. *Agrekon* 60, 460–479.

Bahinipati, C.S. and Venkatachalam, L. (2015) What drives farmers to adopt farm-level adaptation practices to climate extremes: empirical evidence from Odisha, India. *International Journal of Disaster Risk Reduction* 14, 347–356.

Bannor, R.K., Kumar, G.A.K., Oppong-Kyeremeh, H. and Wongnaa, C.A. (2020) Adoption and impact of modern rice varieties on poverty in eastern India. *Rice Science* 27(1), 56–66.

Begho, T., Eory, V. and Glenk, K. (2022) Demystifying risk attitudes and fertiliser use: a review focusing on the behavioural factors associated with agricultural nitrogen emissions in South Asia. *Frontiers in Sustainable Food Systems* 6, 991185. DOI: 10.3389/fsufs.2022.991185.

Bishu, K.G., O'Reilly, S., Lahiff, E. and Steiner, B. (2018) Cattle farmers' perceptions of risk and risk management strategies: evidence from northern Ethiopia. *Journal of Risk Research* 21, 579–598.

Chai, Y., Pannell, D.J. and Pardey, P.G. (2022) Reducing water pollution from nitrogen fertilizer: revisiting insights from production economics. Staff Papers 320519, University of Minnesota, Department of Applied Economics. DOI: 10.22004/ag.econ.320519.

Chavas, J.P. and Nauges, C. (2020) Uncertainty, learning, and technology adoption in agriculture. *Applied Economic Perspectives and Policy* 42, 42–53.

Collins-Sowah, P.A. and Henning, C.H. (2019) Risk management and its implications on household incomes. Working Papers of Agricultural Policy WP2019-05, University of Kiel, Department of Agricultural Economics, Chair of Agricultural Policy.

Cook, S., Henderson, C., Kharel, M. and Begum, A. (2016) Collaborative action on soil fertility in South Asia: experiences from Bangladesh and Nepal. IIED Working Paper.

Crosetto, P. and Filippin, A. (2013) The 'bomb' risk elicitation task. *Journal of Risk and Uncertainty* 47, 31–65.

Crosetto, P. and Filippin, A. (2016) A theoretical and experimental appraisal of four risk elicitation methods. *Experimental Economics* 19, 613–641.

Curry, G.N., Nake, S., Koczberski, G., Oswald, M., Rafflegeau, S. *et al.* (2021) Disruptive innovation in agriculture: socio-cultural factors in technology adoption in the developing world. *Journal of Rural Studies* 88, 422–431.

Das, N., de Janvry, A. and Sadoulet, E. (2019) Credit and land contracting: a test of the theory of sharecropping. *American Journal of Agricultural Economics* 101, 1098–1114.

Dasgupta, U., Mani, S., Sharma, S. and Singhal, S. (2016) Eliciting risk preferences: firefighting in the field. IZA Discussion Paper No. 9765. DOI: 10.2139/ssrn.2742553.

Dave, C., Eckel, C.C., Johnson, C.A. and Rojas, C. (2010) Eliciting risk preferences: when is simple better? *Journal of Risk and Uncertainty* 41, 219–243.

De Palma, A., Ben-Akiva, M., Brownstone, D., Holt, C., Magnac, T., *et al.* (2008) Risk, uncertainty and discrete choice models. *Marketing Letters* 19, 269–285.

Dohmen, T., Falk, A., Huffman, D., Sunde, U., Schupp, J. *et al.* (2011) Individual risk attitudes: measurement, determinants and behavioral consequences. *Journal of the European Economic Association* 9, 522–550.

Dubois, P. (2000) Consumption insurance with heterogeneous preferences. Can sharecropping help complete markets? Econometric Society World Congress 2000 Contributed Papers 1597.

Duong, T.T., Brewer, T., Luck, J. and Zander, K. (2019) A global review of farmers' perceptions of agricultural risks and risk management strategies. *Agriculture* 9, 10.

Eckel, C.C. and Grossman, P.J. (2002) Sex differences and statistical stereotyping in attitudes toward financial risk. *Evolution and Human Behavior* 23, 281–295.

Eckel, C.C. and Grossman, P.J. (2008) Forecasting risk attitudes: an experimental study using actual and forecast gamble choices. *Journal of Economic Behavior & Organization* 68, 1–17.

Federico, G. (2006) The 'real' puzzle of sharecropping: why is it disappearing? *Continuity and Change* 21, 261–285.

Fufa, B. and Hassan, R.M. (2003) Stochastic maize production technology and production risk analysis in Dadar district, east Ethiopia. *Agrekon* 42, 116–128.

Gneezy, U. and Potters, J. (1997) An experiment on risk taking and evaluation periods. *The Quarterly Journal of Economics* 112, 631–645.

Harrison, G.W. and Rutström, E.E. (2008) Risk aversion in the laboratory. *Research in Experimental Economics* 12, 41–196.

Hellerstein, D., Higgins, N. and Horowitz, J. (2013) The predictive power of risk preference measures for farming decisions. *European Review of Agricultural Economics* 40, 807–833.

Holden, S.T. and Quiggin, J. (2017) Climate risk and state-contingent technology adoption: shocks, drought tolerance and preferences. *European Review of Agricultural Economics* 44, 285–308.

Holt, C.A. and Laury, S.K. (2002) Risk aversion and incentive effects. *American Economic Review* 92, 1644–1655.

Hossack, F. and An, H. (2015) Does payment type affect willingness-to-pay? Valuing new seed varieties in India. *Environment and Development Economics* 20, 407–423.

International Food Policy Research Institute (IFPRI); Seed Entrepreneur Association Nepal (SEAN) (2017) *Nepal Vegetable Seed Study*. IFPRI, Washington, DC. DOI: 10.7910/DVN/9BRU7N.

Iqbal, M.A., Ping, Q., Abid, M., Ahmed, U.I., Nazir, A. *et al.* (2016) Adoption of off-farm diversification income sources in managing agricultural risks among cotton farmers in Punjab Pakistan. *Journal of Applied Environmental and Biological Sciences* 6, 47–53.

Janssen, E. and Swinnen, J. (2019) Technology adoption and value chains in developing countries: evidence from dairy in India. *Food Policy* 83, 327–336.

Khanal, A.R., Mishra, A.K. and Lien, G. (2021) Effects of risk attitude and risk perceptions on risk management decisions: evidence from farmers in an emerging economy. *Journal of Agricultural and Resource Economics* 47, 495–512.

Khor, L.Y., Ufer, S., Nielsen, T. and Zeller, M. (2018) Impact of risk aversion on fertiliser use: evidence from Vietnam. *Oxford Development Studies* 46, 483–496.

Kijima, Y. (2019) Farmers' risk preferences and rice production: experimental and panel data evidence from Uganda. *PloS One* 14, e0219202.

Knapp, L., Wuepper, D., Dalhaus, T. and Finger, R. (2021) Revisiting the diversification and insurance relationship: differences between on- and off-farm strategies. *Climate Risk Management* 32, 100315.

Komarek, A.M., De Pinto, A. and Smith, V.H. (2020) A review of types of risks in agriculture: what we know and what we need to know. *Agricultural Systems* 178, 102738.

Lawin, K.G. and Tamini, L.D. (2017) Risk preferences and crop diversification amongst smallholder farmers in Burkina Faso (No. 1910-2017-1299). Canadian Agricultural Economics Society, Annual Meeting, 2017, June 18–21, Montreal, Canada 258058. DOI: 10.22004/ag.econ.258058.

Le Cotty, T., Maître d'Hôtel, E., Soubeyran, R. and Subervie, J. (2018) Linking risk aversion, time preference and fertiliser use in Burkina Faso. *The Journal of Development Studies* 54, 1991–2006.

Lekprichakul, T. (2009) *Ex Ante and Ex Post Risk Coping Strategies: How Do Subsistence Farmers in Southern and Eastern Province of Zambia Cope?* Research Institute for Humanity and Nature, Kyoto, Japan.

Liu, E.M. (2013) Time to change what to sow: risk preferences and technology adoption decisions of cotton farmers in China. *The Review of Economics and Statistics* 95:1386–1403.

Lu, W., Latif, A. and Ullah, R. (2017) Simultaneous adoption of contract farming and off-farm diversification for managing agricultural risks: the case of flue-cured Virginia tobacco in Pakistan. *Natural Hazards* 86, 1347–1361.

Lyu, K. and Barré, T.J. (2017) Risk aversion in crop insurance program purchase decisions: evidence from maize production areas in China. *China Agricultural Economic Review* 9, 62–80.

Malabayabas, M.L., Yamano, T., Gupta, I., Panda, A. and Bandyopadhyay, S. (2015) Assessing the use of stress-tolerant rice varieties in eastern India and Bangladesh. *STRASA Economic Brief* 5, 1–2.

Mehar, M., Yamano, T. and Panda, A. (2017) The role of gender, risk, and time preferences in farmers' rice variety selection in eastern India. *Asian Journal of Agriculture and Development* 14, 17–36.

Menapace, L., Colson, G. and Raffaelli, R. (2016) A comparison of hypothetical risk attitude elicitation instruments for explaining farmer crop insurance purchases. *European Review of Agricultural Economics* 43, 113–135.

Moser, S. and Mußhoff, O. (2017) Comparing the use of risk-influencing production inputs and experimentally measured risk attitude: do decisions of Indonesian small-scale rubber farmers match? *German Journal of Agricultural Economics* 66, 124–139.

Mukhamedova, N. and Pomfret, R. (2019) Why does sharecropping survive? Agrarian institutions and contract choice in Kazakhstan and Uzbekistan. *Comparative Economic Studies* 61, 576–597.

Paik, S., Le, D.T.P., Nhu, L.T. and Mills, B.F. (2020) Salt-tolerant rice variety adoption in the Mekong River Delta: farmer adaptation to sea-level rise. *PloS One* 15, e0229464.

Panda, A. (2021) *Climate Change and Agricultural Insurance in the Asia and Pacific Region*. The Asian Development Bank, Mandaluyong, Philippines.
Patil, V. and Veettil, P.C. (2018) Experimental evidence of risk attitude of farmers from risk-preference elicitation in India. International Association of Agricultural Economists Conference, July 28–August 2, 2018, Vancouver, British Columbia.
Picazo-Tadeo, A.J. and Wall, A. (2011) Production risk, risk aversion and the determination of risk attitudes among Spanish rice producers. *Agricultural Economics* 42, 451–464.
Praveen, K.V. (2021) Indian fertiliser policy: retrospect and prospect. In: Rakshit, A., Singh, S.K., Abhilash, P.C. and Biswas, A. (eds) *Soil Science: Fundamentals to Recent Advances*. Springer, Singapore, pp. 765–779.
Qiao, F.B. and Huang, J.K. (2021) Farmers' risk preference and fertilizer use. *Journal of Integrative Agriculture* 20, 1987–1995.
Reardon, T., Stamoulis, K., Balisacan, A., Cruz, M.E., Berdegué, J. *et al.* (1998) Rural non-farm income in developing countries. *The State of Food and Agriculture*, 1998, 283–356.
Reynaud, A. and Couture, S. (2012) Stability of risk preference measures: results from a field experiment on French farmers. *Theory and Decision* 73, 203–221.
Saqib, S.E., Ahmad, M.M., Panezai, S. and Rana, I.A. (2016) An empirical assessment of farmers' risk attitudes in flood-prone areas of Pakistan. *International Journal of Disaster Risk Reduction* 18, 107–114.
Saqib, S.E., Arifullah, A. and Yaseen, M. (2021) Managing farm-centric risks in agricultural production at the flood-prone locations of Khyber Pakhtunkhwa, Pakistan. *Natural Hazards* 107, 853–871.
Sarker, M.A.R., Alam, K. and Gow, J. (2013) Assessing the determinants of rice farmers' adaptation strategies to climate change in Bangladesh. *International Journal of Climate Change Strategies and Management* 5, 382–403.
Sarwosri, A.W. and Mußhoff, O. (2020) Are risk attitudes and time preferences crucial factors for crop diversification by smallholder farmers? *Journal of International Development* 32, 922–942.
Sauter, P.A., Hermann, D. and Mußhoff, O. (2018) Are foresters really risk-averse? A multi-method analysis and a cross-occupational comparison. *Forest Policy and Economics* 95, 37–45.
Singh, N. (2000) Theories of sharecropping. In: Bardham, P. and Udry, C. (eds) *Readings in Development Microeconomics*, Vol. 2. The MIT Press, Cambridge, Massachusetts, pp. 19–71.
Sulewski, P. and Kłoczko-Gajewska, A. (2014) Farmers' risk perception, risk aversion and strategies to cope with production risk: an empirical study from Poland. *Studies in Agricultural Economics* 116, 140–147.
Swain, M. (2014) Crop insurance for adaptation to climate change in India. Working Paper No. 61. Asia Research Centre (ARC), London School of Economics and Political Science, London, UK.
Tanaka, T., Camerer, C.F. and Nguyen, Q. (2010) Risk and time preferences: linking experimental and household survey data from Vietnam. *American Economic Review* 100, 557–571.
Ullah, R. and Shivakoti, G.P. (2014) Adoption of on-farm and off-farm diversification to manage agricultural risks: are these decisions correlated? *Outlook on Agriculture* 43, 265–271.
van Winsen, F., de Mey, Y., Lauwers, L., Van Passel, S., Vancauteren, M. and Wauters, E. (2016) Determinants of risk behaviour: effects of perceived risks and risk attitude on farmer's adoption of risk management strategies. *Journal of Risk Research* 19, 56–78.
Vigani, M. and Kathage, J. (2019) To risk or not to risk? Risk management and farm productivity. *American Journal of Agricultural Economics* 101, 1432–1454.
Villano, R.A., O'Donnell, C.J. and Battese, G.E. (2005) An investigation of production risk, risk preferences and technical efficiency: evidence from rainfed lowland rice farms in the Philippines (No. 1730-2016-140228). Working Papers 12953, University of New England, School of Economics, Armidale, Australia. DOI: 10.22004/ag.econ.12953.
Ward, P.S. and Singh, V. (2015) Using field experiments to elicit risk and ambiguity preferences: behavioural factors and the adoption of new agricultural technologies in rural India. *The Journal of Development Studies* 51, 707–724.
World Bank (2013) *Warmer World Threatens Livelihoods in South East Asia*. Press Release. Available at: https://www.worldbank.org/en/news/press-release/2013/06/19/warmer-world-threatens-livelihoods-in-south-east-asia (accessed September 24, 2022).
World Bank (2019) *Agricultural Innovation & Technology Hold Key to Poverty Reduction in Developing Countries, says World Bank Report*. Press Release. Available at: https://www.worldbank.org/en/news/press-release/2019/09/16/agricultural-innovation-technology-hold-key-to-poverty-reduction-in-developing-countries-says-world-bank-report (accessed September 24, 2022).

Wuepper, D., Yesigat Ayenew, H. and Sauer, J. (2018) Social capital, income diversification and climate change adaptation: panel data evidence from rural Ethiopia. *Journal of Agricultural Economics* 69, 458–475.

Yamano, T. (2014–2015) Rice Monitoring Survey: South Asia. Farm Household Survey Database, Panel Data. Available at: http://ricestat.irri.org/fhsd/php/panel.php (accessed May 12, 2023).

Yanuarti, R., Aji, J.M.M. and Rondhi, M. (2019) Risk aversion level influence on farmer's decision to participate in crop insurance: a review. *Agricultural Economics* 65, 481–489.

Zhang, R., Fan, D., Lai, G., Wu, J. and Li, J. (2021) Rank-dependent preferences, social network and crop insurance uptake: field experimental evidence from rural China. *Agricultural Finance Review* 85, 636–656.

8 How to Measure Risk Attitudes Among Poor Rural Households: An Overview of Popular Risk Elicitation Methods

Selina Bruns*
University of Göttingen, Göttingen, Germany

Abstract
Risk attitude can play a key role in the vicious cycle of poverty. Reliable instruments to elicit the risk attitudes of the rural poor in developing countries are therefore essential to contribute to designing instruments to combat poverty. However, literature suggests numerous risk elicitation methods, suitable for different research endeavors and target groups. This chapter aims to provide a practical orientation for researchers aiming to measure risk attitudes of the rural poor by briefly presenting: (i) two frontline theories for decision under risk; (ii) eight common risk elicitation methods; and (iii) several approaches to verify the explanatory power for their suitability to generate quality data in the context of developing countries. As a general rule of thumb for this specific study group, simpler methods – or at least methods with a very intuitive illustration – tend to outperform complex methods. Instruments to properly assess external validity are limited and remain a critical research gap.

8.1 Introduction

According to the most recent statistics, about 767 million people live in poverty. The vast majority of these global poor reside in rural areas, depending to some degree on agriculture as their main source of livelihood (World Bank, 2018). These individuals live under particularly risky conditions, e.g. due to unstable environmental, market and household conditions (World Bank, 2016, 2023). However, to interrupt the downward spiral of poverty, individuals need to take the risk of investing in human, physical, or social capital, which can be a difficult endeavor without any safety net. Individuals who are risk-averse to the extent that they refuse to invest in new technologies will be less able to cope with shocks than risk-seeking individuals and hence might end up in chronic poverty (Mosley and Verschoor, 2005). Thus, risk attitude analysis can serve as one crucial step in contributing to combating poverty, by supporting custom-fit policy interventions.

Critical to this analysis is having a reliable instrument to elicit risk attitudes. In recent years, various experimental methods, such as playing lotteries, investment games, or rather intuitive games, have been tested in the lab or the field to understand how individuals make decisions under risk (Charness *et al.*, 2013). However, the methods being used to elicit risk attitudes often involve complex and abstract tasks. For example, the relatively popular Holt–Laury task – a very precise yet complex binary

*E-mail: selina.bruns@uni-goettingen.de

 Managing Risk in Agriculture: A Development Perspective (eds A.K. Mishra, S.C. Kumbhakar and G. Lien).
DOI: 10.1079/9781800622289.0008

lottery (Charness *et al.*, 2013) – often appears to be among the top choices for measuring risk attitudes among low-literacy subjects (Verschoor *et al.*, 2016). However, literature suggests that complex tasks might not be well understood by the poor, who often lack basic schooling (Byerlee *et al.*, 2008; IFAD, 2011; McGuire, 2015; Rapsomanikis, 2015). Thus, such tasks can yield flawed results (Brick *et al.*, 2012).

The aim of this chapter is to give a practical orientation to researchers searching for a suitable measure to elicit risk attitudes among low-literacy subject groups. This chapter does not claim to cover all the tools that measure risk attitudes. Rather, it serves as a practical overview and starting point for researchers aiming to find a suitable method for their specific research endeavor and for further investigations. I will briefly touch on: (i) the two leading theories serving as a foundation for the most common risk elicitation methods; (ii) eight different risk elicitation methods, selected due to popularity, ranging from simple to complex, incentivized and non-incentivized tools; and (iii) approaches to identify the explanatory power of risk elicitation methods. These elements will be discussed in light of their suitability for a developing country's circumstances. The chapter concludes with final remarks.

8.2 Theoretical Background

The following briefly presents the theoretical background of choice behavior. Even though a number of theories aim to explain how people make decisions, this chapter focuses on the two 'front runners' for choice under uncertainty: expected utility theory (EUT) and prospect theory (PT; Harrison and Rutström, 2009).

8.2.1 Expected utility theory

When the axioms of rationality (Arrow, 1982; Levin, 2006; von Neumann and Morgenstern, 2007 [1944]) are fulfilled, the expected utility function can hold with a rational player aiming to maximize his or her utility. Figure 8.1 depicts the typical shapes of the function. The x-axis represents Z, an uncertain value, and the y-axis, the utility of Z. No negative domain is considered,

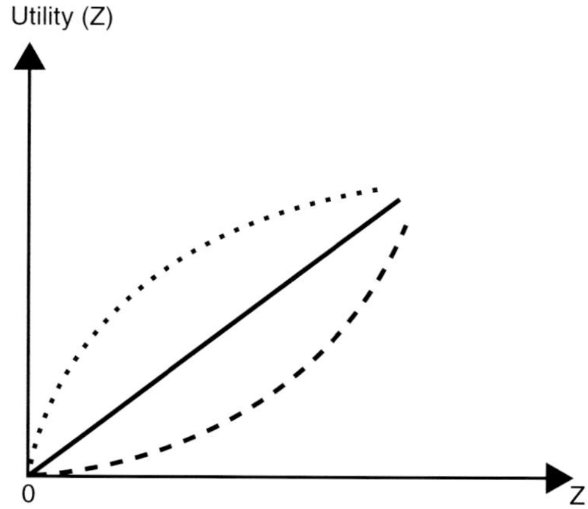

Fig. 8.1. Typical utility function. (Source: Author's own work)

as the function originates at 0. Regardless of the player's risk attitude, the utility function will be strictly increasing. As risk aversion is typically assumed, the standard utility function is concave in shape (dotted). The risk-neutral player will be represented by a linear course, meaning that U(Z) = Z, and the risk-seeking individual will have a convex utility function (dashed; Mußhoff and Hirschauer, 2010). See Varian (2010) for further details.

8.2.2 Non-expected utility theory

Several variants to EUT have been developed, e.g. subjective expected utility theory (Savage, 1954), generalized expected utility theory (Quiggin, 1994), weighted utility theory (Chew and Maccrimmon, 1979; Hong and Waller, 1986), rank dependent utility theory (Quiggin, 1982), theory of disappointment (Bell, 1985), regret theory (Loomes and Sugden, 1982) and salience theory (Bordalo *et al.*, 2012). See Starmer (2000) for an extensive review of alternative models. For the sake of relevance, this chapter will focus on prospect theory, as coined by Kahneman and Tversky (1979). Building on their empirical findings, indicating that individuals tend to violate axioms of rationality, Kahneman and Tversky developed prospect theory. Here, instead of a utility function, we look at a value function, as seen in Fig. 8.2, with which it is possible to elicit a more nuanced profile of an individual's risk preference.

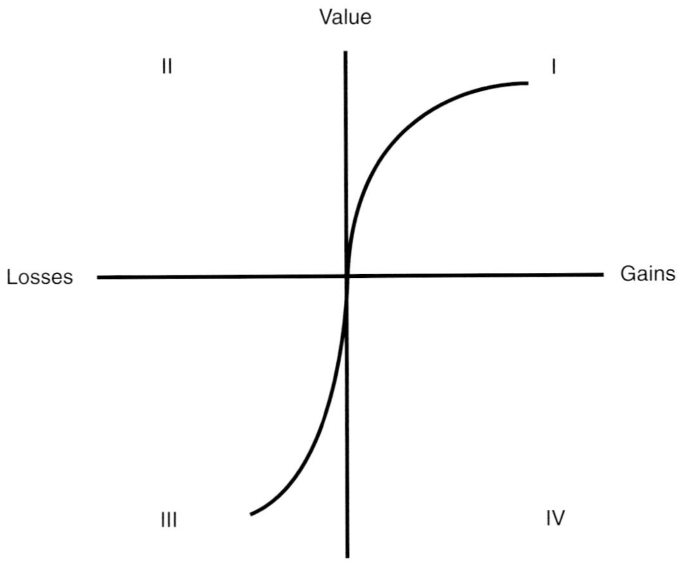

Fig. 8.2. Typical value function. (Source: Based on Kahneman and Tversky, 1979; p. 259.)

The value function (1) revolves around a reference point and thus accounts for gains and losses rather than for absolute states of wealth; (2) is concave for gains and convex for losses, and steeper for losses than for gains to account for loss aversion; (3) replaces probabilities with decision weights, which follow from a non-linear transformation of the probability scale. Small probabilities can be overweighted, and moderate and high probabilities can be underweighted (Tversky and Kahneman, 1992). In the gain sphere (Quadrant I), the concave shape of the value function implies that the actor is assumed to act in a rather risk-averse way, and in the loss sphere (Quadrant III), the convex shape implies a risk-seeking behavior for losses (Kahneman and Tversky, 1979).

8.3 Methods

This chapter now presents a number of previously used methods to elicit risk attitudes in the field. Robison *et al.* (1984) divides approaches to risk attitude measurement into three broad categories: (i) direct elicitation of utility functions (see e.g. Anderson *et al.* (1977); Hardaker *et al.* (2015) for details); (ii) experimental procedures (e.g. Binswanger, 1980); and (iii) empirical inference of economic behavior, based on a hypothesized relationship between the observed behavior of a decision maker and the behavior forecasted from empirically specified models (e.g. Moscardi and de Janvry, 1977; Brink and McCarl, 1978; Antle, 1987; Kumbhakar, 2002a, 2002b; Lien, 2002; Kumbhakar and Tveterås, 2003). The methods shown and discussed in this chapter mainly fall into the second category. For further reading on the other categories, see Robison *et al.* (1984). The chapter extends the categorization of risk elicitation methods put forward by Charness and Viceisza (2016) by dividing the simple/incentivized dimension into numeric and non-numeric instruments (see Table 8.1).

Incentivized approaches include some kind of monetary or in-kind stimulus, included in the game's design to incentivize players to play at their best – to yield true results. These approaches can be complex, for example, multiple price lists (MPL), from which the player has to choose the preferred option. Binswanger (1980) is a well-known pioneer in this context, using a lottery-choice approach among farmers in rural

Table 8.1. Taxonomy of methods. (Source: Based on Charness and Viceisza, 2016, p. 148.)

	Complex	Simple	
Incentivized	– Holt and Laury (2002) – Tanaka *et al.* (2010)	Numeric – Binswanger (1980) – Gneezy and Potters (1997) – Eckel and Grossman (2002)	Non-numeric – Lejuez *et al.* (2002) – Crosetto and Filippin (2013) – Slovic (1966)
Non-incentivized	– Weber *et al.* (2002)	– Dohmen *et al.* (2011)	

India (Charness and Viceisza, 2016). Incentivized methods can also be simple (numeric), for example, a simple investment game (Eckel and Grossman, 2002). Simple (non-numeric) methods are those in which the player is not confronted with numbers and probabilities but rather with more intuitive tasks, e.g. pumping a balloon (Lejuez *et al.*, 2002).

Non-incentivized approaches are mainly embedded in surveys. Most commonly, individuals are asked to state their general (Dohmen *et al.*, 2011) or domain-specific (Weber *et al.*, 2002) preferences (Charness and Viceisza, 2016).

The examples in Table 8.1 are explained further in the following paragraphs and have been chosen for their popularity in the field. The Holt–Laury (HL) method (Holt and Laury, 2002), the Binswanger (BW) method (Binswanger, 1980), the Eckel and Grossman (EG) method (Eckel and Grossman, 2002) and the Gneezy and Potters (GP) method (Gneezy and Potters, 1997) make up the frontline methods that have been used in the context of a developing country (Verschoor *et al.*, 2016). The Tanaka *et al.* (TCN) method (Tanaka *et al.*, 2010) depicts a relatively new approach that, from the methods chosen, is the only one to build on prospect theory (Charness and Viceisza, 2016).

8.3.1 Holt and Laury (2002)

Praised throughout literature (Anderson and Mellor, 2009) as the gold standard for eliciting risk preferences, the EUT-based lottery developed by Charles A. Holt and Susan K. Laury in 2002 seems to be the first-choice instrument for researchers interested in understanding risk attitudes. For measurement, subjects are presented with paired lottery choices to make over ten rounds. Throughout the ten rounds, the payoffs in Option A and Option B stay unchanged. As seen in Table 8.2 (Holt and Laury, 2002), the payoffs of Option A fall in a narrow range (US$2–$1.60), and those of Option B are more variable ($3.85–$0.10). What changes during the lottery are the probabilities with which the payoffs occur. Thus, within the ten rounds the player is expected to switch from Option A to Option B. The switching point is crucial as it determines the subject's risk attitude. A risk-neutral individual is expected to switch in round five, as the difference in expected payoffs from Option A to Option B turns negative (−$0.18). Even an extremely risk-averse person would show a switching point, namely in round ten, as the probability of receiving the high payoff of $3.85 is certain. A risk-loving person would switch before round five, with the most extreme starting with Option B in the first place and thus being the only exception of not switching at all (Holt and Laury, 2002).

Although the switching point allows for estimating individuals' risk attitudes, it may also serve as an indicator for accuracy to the researchers. The MPL method is designed in such a way that ideally the subject should not switch more than once. However, when the list is presented, individuals are allowed to switch freely between Option A and Option B as they play through the ten rounds. This might yield inconsistent results, as individuals might switch back and forth between the two options. In general, it is interpreted that the higher the inconsistency rate, the lower the level of accuracy due to participants' lack of understanding (Charness *et al.*, 2013).

8.3.2 Tanaka *et al.* (2010)

Tanaka *et al.* (2010) conducted research among Vietnamese villagers, pioneering in developing an MPL built on prospect theory and testing it in a rural setting. Thus, they were able to play lotteries involving not only gains, but also losses. To elicit risk attitudes, they played three series of binary lottery choices, as demonstrated in Tables 8.3 to 8.5. In rounds one and two, while the probabilities stay constant, so do almost all of the payouts. The only thing that changes is the higher payouts in Option B. These two rounds are played to determine the shape of the weighting functioning and the concavity of the value function. Round three then defines the degree of loss aversion. In the third round, again, the probabilities of payouts are fixed, but the payoffs themselves, except for the high payouts in Option B, change as one moves down the rows. In Option A, the negative payouts increase (from negative four to negative eight) while the positive payoffs decrease (from 25 to 1). In Option B, the negative payouts decrease (from −21 to −11). In the field, Tanaka *et al.* illustrated the probabilities by using balls numbered from 1 to 35. To ensure consistency, the researchers did not play the lottery row by row but specifically asked the participants for their switching point (Tanaka *et al.*, 2010).

Table 8.2. Holt and Laury lottery. (Source: Based on Holt and Laury, 2002)

Option A		Option B	
0.1 of $2	0.9 of $1.60	0.1 of $3.85	0.9 of $0.10
0.2 of $2	0.8 of $1.60	0.2 of $3.85	0.8 of $0.10
0.3 of $2	0.7 of $1.60	0.3 of $3.85	0.7 of $0.10
0.4 of $2	0.6 of $1.60	0.4 of $3.85	0.6 of $0.10
0.5 of $2	0.5 of $1.60	0.5 of $3.85	0.5 of $0.10
0.6 of $2	0.4 of $1.60	0.6 of $3.85	0.4 of $0.10
0.7 of $2	0.3 of $1.60	0.7 of $3.85	0.3 of $0.10
0.8 of $2	0.2 of $1.60	0.8 of $3.85	0.2 of $0.10
0.9 of $2	0.1 of $1.60	0.9 of $3.85	0.1 of $0.10
1 of $2	0 of $0.10	1 of $3.85	0 of $0.10

Table 8.3. Tanaka *et al.* lottery (in 1000 dong), round 1. (Source: Based on Tanaka *et al.*, 2010.)

Option A		Option B		Expected payoff difference
30%	**70%**	**10%**	**90%**	
40	10	68	5	7.7
40	10	75	5	7.0
40	10	83	5	6.0
40	10	93	5	5.2
40	10	106	5	3.9
40	10	125	5	2.0
40	10	150	5	−0.5
40	10	185	5	−4.0
40	10	220	5	−7.5
40	10	300	5	−15.5
40	10	400	5	−25.5
40	10	600	5	−45.5
40	10	1000	5	−85.5
40	10	1700	5	−155.5

Table 8.4. Tanaka *et al.* lottery (in 1000 dong), round 2. (Source: Based on Tanaka *et al.*, 2010.)

Option A		Option B		Expected payoff difference
90%	**10%**	**70%**	**30%**	
40	30	54	5	−0.3
40	30	56	5	−1.7
40	30	58	5	−3.1
40	30	60	5	−4.5
40	30	62	5	−5.9
40	30	65	5	−8.0
40	30	68	5	−10.1
40	30	72	5	−12.9
40	30	77	5	−16.4
40	30	83	5	−20.6
40	30	90	5	−25.5
40	30	100	5	−32.5
40	30	110	5	−39.5
40	30	130	5	−53.5

Table 8.5. Tanaka *et al.* lottery (in 1000 dong), round 3. (Source: Based on Tanaka *et al.*, 2010.)

Option A		Option B		Expected payoff difference
50%	**50%**	**50%**	**50%**	
25	−4	30	−21	6.0
4	−4	30	−21	−4.5
1	−4	30	−21	−6.0
1	−4	30	−16	−8.5
1	−8	30	−16	−10.5
1	−8	30	−14	−11.5
1	−8	30	−11	−13.0

8.3.3 Eckel and Grossman (2002)

Eckel and Grossman (2002) developed a simple and incentivized gamble that they conducted as a laboratory experiment among undergraduate students from two US universities.

For the game, individuals are confronted with five different gambles, from which they have to choose one. Each gamble has two possible outcomes: Option A, with a 50% chance of occurrence, and Option B, also with a 50% chance of occurrence (Table 8.6). Expected payoffs increase as one proceeds from Gamble 1 to Gamble 5, but so does the level of risk (measured in standard deviation of expected payoff). Under EUT, an extremely risk-averse person would prioritize safety over higher gains and go for the safe bet: Gamble 1. A moderately risk-averse individual would choose any gamble between 2 and 5. The risk-seeking person should also go for Gamble 5 (Eckel and Grossman, 2002).

Table 8.6. Eckel and Grossman gamble. (Source: Based on Eckel and Grossman, 2002.)

	Payoff		
Gamble	A (50%)	B (50%)	Expected payoff
1	16	16	16
2	24	12	18
3	32	8	20
4	40	4	22
5	48	0	24

8.3.4 Gneezy and Potters (1997)

Conducting a decision-making experiment among Dutch university students, Gneezy and Potters (1997) designed a simple investment game, with its basic mechanics shown in Fig. 8.3. An individual gets an initial endowment of money; for the sake of simplicity, one can assume to start with 200 cents. The individual has the choice to invest any amount of his or her initial endowment (X). This investment will yield 2.5 times X with a probability of 50%, or 0 with a probability of 50%. Whatever the individual does not invest (200 − X) is safe to keep. The amount X then can be used to estimate risk attitudes. Under EUT, the risk-seeker would invest the total endowment. An extremely risk-averse player would not invest at all and would keep the total endowment. The risk-neutral player would also invest the total endowment, as this yields the highest expected outcome (Gneezy and Potters, 1997).

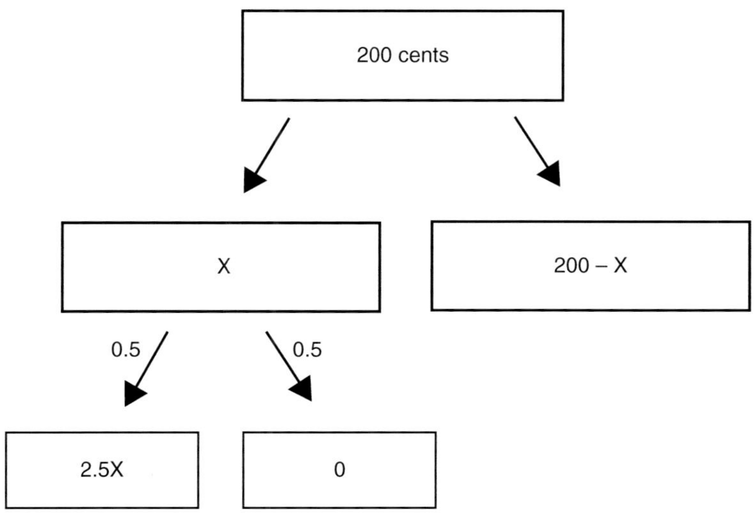

Source: Gneezy and Potters, 1997; pp. 633–634.

Fig. 8.3. Gneezy and Potters gamble.

8.3.5 Lejuez *et al.* (2002)

One of the earliest simple, non-numeric risk elicitation tasks was constructed by the psychologist Paul Slovic in 1966, who aimed to understand the risk behavior of children. For this, he set up a simple game: participants have the option to switch or pull knife switches. Out of ten switches, nine are 'safe switches,' for which the child gets a handful of M&Ms. One switch is the 'disaster switch,' which triggers an alarm. With the sound of the alarm, the experiment ends and all the earnings are lost (Slovic, 1966).

A more recent version is the Balloon Analogue Risk Task (BART; Lejuez *et al.*, 2002). In this game, a computerized balloon is presented to the participants. The task is to inflate the balloon. Every pump inflates the balloon by 3 cm and yields five cents to the participant. The participant needs to decide how often he or she wishes to pump as he or she goes, as the balloon has an explosion point after n pumps. If it pops, the player loses all his or her earnings. Therefore, in every new round, the subjects must decide if they wish to inflate an additional time or if they want to stop the game and take the earnings with them. Using the number of pumps decided by the participant, the researchers can determine the participant's risk attitude (Lejuez *et al.*, 2002; White *et al.*, 2008). If the participants are aware of n, and let's assume, for simplicity, that n is 100, then the game tests for uncertainty. In this case, a risk-neutral person would stop after $n = 50$, a risk-averse person would stop at any time $n < 50$, and a risk-seeking person would stop at any time $n > 50$ (Lejuez *et al.*, 2002).

8.3.6 Crosetto and Filippin (2013)

Another non-conventional risk elicitation method is put forward by Crosetto and Filippin (2013b), who designed the Bomb Risk Elicitation Task (BRET) method and tested it among German adults. This intuitive task required low numeracy skills, as participants are asked to select up to 100 boxes, one of which contains a 'bomb' (b). Within a 10 × 10 square, in which each cell represents a square, subjects can choose a number (N) between 0 and 100 that gives the number of boxes they wish to collect. After the participant has made his or her choice, the position of the bomb is set by drawing a number (b^*) between 1 and 100. Then, if $N \geq b^*$ the subject has collected the bomb. It explodes and he or she has lost all of his or her earnings. If $N < b^*$, the subject survived and receives some payout y for every box collected. A risk-neutral participant should choose $N = 50$, a risk-averse participant chooses $N < 50$, and the risk seeker $N > 50$ (Crosetto and Filippin, 2013b).

8.3.7 Dohmen *et al.* (2011)

Another approach to understanding risk attitudes is to ask directly. Dohmen *et al.* (2011) used this method, among others, in their analysis. One core source of data is from the German Socio-Economic Panel, a provider of large-scale panel data for Germany. To measure risk behavior, they ask participants the following question:

> How do you see yourself: are you generally a person who is fully prepared to take risks or do you try to avoid taking risks? Please tick a box on a scale, where the value 0 means: 'not at all willing to take risks' and the value 10 means: 'very willing to take risks.' (Dohmen *et al.*, 2011, p. 525)

8.3.8 Weber *et al.* (2002)

Asking a general question implies the assumption of a constant risk attitude across all domains of life. Building on the issue of domain as opposed to general predicting power of measured risk attitudes, Weber *et al.* (2002) developed a domain-specific risk-taking (DOSPERT) scale. Among US university students, they conducted a questionnaire aimed at covering the whole range of risk-taking situations that young adults encounter in Western cultures. They came up with a variety of items, ranging from 40 to 101, in five domains: financial, health/safety, recreational, ethics and social. Subjects then had to answer the questionnaire multiple times. First, they had to indicate their likelihood of taking part in a risk on a scale from 1 (extremely unlikely) to 5 (extremely likely). Second, they had to rate the level of riskiness involved in the activity from 1 (not at all risky) to 5 (extremely risky) as well as the benefits they assume to get out of the activity on a five-point scale (Weber *et al.*, 2002).

8.4 Explanatory Power

Understanding whether the results measured by risk elicitation methods are mirrored in real life behavior is essential for (i) finding a robust method and (ii) making results reliable to give proper political advice. However, testing for the predictive power of the different methods is a rather complex endeavor, and more research is needed (Hellerstein *et al.*, 2013). Nevertheless, some approaches have been developed and tested in the field.

One approach is to test the explanatory power with the help of accountancy. The bookkeeping data of an enterprise run by the subject, e.g. a farm, or regional financial monitoring data can give hints of risk behavior over a longer time span. Vollmer *et al.* (2017) follow that idea and analyze how elicited risk behavior (determined using the Holt–Laury task) is correlated with production risk. Building on official production data of German farmers, they use a production function for their analysis, assuming that greater variance in output mirrors risk-seeking behavior. For their work in a developed country, this approach appeared to be a valid indicator for risk behavior, as they found that a more risk-averse farmer has significantly less variance in output.

Verschoor *et al.* (2016) follow a similar approach but adapted it to subsistence smallholder farmers in Uganda. They look for agricultural investments made by farmers, which will increase the expected value as well as the variability of outputs and thus mirror rather riskier farming practices. Although the purchase of fertilizer serves as the classic example for this (Duflo *et al.*, 2008), they also developed a more specific indicator: growing cash crops and buying the needed inputs. As a third candidate, they looked at market participation. Their results indicate that fertilizer can be a candidate for revealing the predictive power of the Gneezy and Potters method, which they use to elicit risk attitudes, but the other two candidates cannot be included. They explain their finding by drawing a comparison between the Gneezy and Potters task and buying fertilizer: both are straight-forward, single choices, while the other two candidates presented more complex decision-making. Therefore, they conclude, so as to be more likely to have predictive power, methods should reflect a similar real-life domain (Verschoor *et al.*, 2016). Other examples studying risk preferences of farmers in developing countries (Vargas Hill, 2009; Gong *et al.*, 2016) find that farmers' production decisions seem to be a valid candidate for indicating predictive power.

Hellerstein *et al.* (2013) attempt to validate the predictive power of risk attitudes elicited by a Holt–Laury task through information gathered from American farmers about their farming decisions. To identify potential candidates for external validity, they look at farm management decisions that have been shown to depend on risk and risk aversion, narrowing them down to decisions about diversification, crop insurance, farming practices, farm ownership and farm income. Their approach, however, did not yield any significant results. Only when they relaxed their analysis were they able to find a relationship between risk aversion and insurance. Menapace *et al.* (2015) also use insurance uptake as a measure for external validity for risk attitudes of north Italian farmers. They too do not find a strong relationship.

Several more examples come from outside the agricultural realm. Lusk and Coble (2005) set out to find if the risk attitudes of US undergraduate students, measured through a Holt–Laury task, can explain their tendency to try genetically modified food, considered rather risky food. To come up with a measure for external validity, they use a two-step approach. First, they ask the subjects about their risk perception of genetically modified food. Second, the subjects are asked to state their willingness to (i) eat, (ii) purchase, (iii) accept and (iv) consume the food on a scale from one to nine. Their approach seems promising, finding a significant relationship between risk attitude and willingness to try risky food.

Anderson and Mellor (2008) check to what extent risk attitude revealed through a Holt–Laury task can explain the health-related behavior of US adults. For this, the subjects are asked to fill out a survey stating their frequency of engaging in certain risky behavior: smoking, drinking heavily, not using seat belts and speeding (in a car). They also check for the subjects' Body Mass Index to measure obesity. The authors find a significant negative relationship between risk aversion and smoking, as well as drinking heavily, being overweight, and not using seat belts.

A similar approach is used by Menkhoff and Sakha (2017) among a rural population in Thailand, comparing popular risk elicitation methods of a modified binary lottery, two modifications of the Eckel and Grossman method, a modification of the Gneezy and Potters method, and three non-incentivized survey items, two following Dohmen *et al.* (2011) asking for general and financial risk-taking attitudes, and the last being a hypothetical investment question used by Barsky *et al.* (1997). The authors tested for predictive power by relating the above-mentioned measures of risk attitude to areas of risk behavior that according to theory should be greatly affected by risk attitude, asking the subjects about playing the lottery, risky employment, financial behavior, risk avoidance and behavior toward health.

A number of other psychologically developed measures of personality traits related to risk can be used to understand a method's predictive power. One measure is the Sensation Seeking Scale (Zuckerman *et al.*, 1978), a 40-item forced-choice scale divided into four major areas: thrill- and adventure-seeking, disinhibition, experience-seeking and boredom susceptibility. Similar examples are the Risk-taking Propensity Measures (Maccrimmon and Wehrung, 1985) and the Zuckerman-Kuhlman Personality Questionnaire (Zuckerman *et al.*, 1993).

8.5 Discussion

8.5.1 Choice of method

The complex/incentivized approaches, especially the Holt–Laury method, stand out as the most popular despite their cognitive difficulty for participants, as detected by the high inconsistency rates of the Holt–Laury lottery. They are between 10% and 15% in developed-country settings (Charness and Viceisza, 2016), but they tend to be crucially higher in developing countries. Brick *et al.* (2012) came up with an inconsistency rate of 41% in South Africa, Doerr *et al.* (2011) found 39% in Ethiopia, Jacobson and Petrie (2009) found 50% in Rwanda, Galarza (2009) found that 50% of the farmers in Peru switch at least two times, and Bruns *et al.* (2022) found up to 70% of inconsistent behavior in Cambodia. Liu (2013) also found signs of a number of her subjects behaving inconsistently in the TCN method. Although the standard version of this method allows no switching, and therefore no inconsistencies, as in the Holt–Laury task, Liu observed that monotonous decision behavior (only Option A or only Option B throughout all rows) can be interpreted as a sign of inconsistent behavior. Holden (2014) observes that in the TCN method, which he used among poor households in Malawi, individuals find it difficult to understand the first two tasks.

Anderson and Mellor (2009) suggest that inconsistencies might indicate indifference toward the alternatives, but it is more likely that they indicate a lack of understanding of the task (Brick *et al.*, 2012). Dave *et al.* (2010) test this by conducting a simple math test before running the actual experiment. Later, they analyze subjects who made inconsistent choices in the Holt–Laury task by math skill level. Results show that the share of individuals making inconsistent choices is about four times larger in populations with low math ability than in populations with higher math ability. Bruns *et al.* (2022) further that investigation with a study among smallholder farmers in Cambodia and show that inconsistency levels in the Holt–Laury lottery can be explained by the level of cognitive skills. Brick *et al.* suggest that MPL methods 'although widely used in both student samples and in the field, might not be the most appropriate elicitation tool within a developing country context' (Brick *et al.*, 2012, p. 141). Many other researchers (Dave *et al.*, 2010; Nielsen *et al.*, 2013; Holden, 2014; Charness and Viceisza, 2016) share this opinion.

However, the Holt–Laury task has promising applications in the field. Verschoor *et al.* (2016) suggest that the Holt–Laury lottery is actually suitable for eliciting risk preferences among low-income households in developing countries. They point out that how the lottery is designed and implemented is crucial. Careful application of the Holt–Laury method has been done, e.g. by Ihli *et al.* (2018) among smallholder farmers in Uganda, yielding considerably lower inconsistency levels than comparable field research (Ihli *et al.*, 2013, 2018). A good starting point, therefore, to reduce noise in complex lotteries is to remodel lab designs for use in the specific field setting. Strategies include translating probabilities into easily comprehensible illustrations such as different colored balls for different probabilities (Harrison *et al.*, 2010; Tanaka *et al.* 2010; Liu, 2013; Ihli *et al.*, 2018); a coin flip (Brick *et al.*, 2012); relying on different colored counters, where the vertical arrangement signals the probabilities (Verschoor and D'Exelle, 2022); or easy demonstrations with cards and fingers (Holden, 2014). Also, researchers commonly apply 'salient probabilities' of 50/50 chances in gambles (Brick *et al.*, 2012) or ¼, ½, ¾ and 1 in a Holt–Laury task (Harrison *et al.*, 2010), rather than the more complex and more frequently changing probabilities of lab designs.

Incentivized simple non-numeric methods such as the balloon task stand out for their simplicity. They are highly intuitive to grasp, and no education level is needed to accomplish them. However, they are seldom applied in the field, which might be due to their low level of precision, e.g. the narrow range of variables generated (Lejuez *et al.*, 2002). Simple incentivized methods, such as the Eckel and Grossman gamble, have been shown to be more easily understood by individuals (Dave *et al.*, 2010; Charness and Viceisza, 2016; Verschoor *et al.*, 2016) than an MPL method. Another advantage is that, compared to most other tasks, an investment game can reflect a real-life decision and therefore might be superior, at least in predicting behavior in a financial domain (Menapace *et al.*, 2015; Verschoor *et al.*, 2016). However, in simple incentivized methods it is more difficult, if not impossible, to detect confusion (Charness and Viceisza, 2016) and to disentangle risk-neutral from risk-seeking behavior (Crosetto and Filippin, 2013a).

Non-incentivized methods, be it the DOSPERT or the DO approach, are praised in literature for being cheap and easy to conduct and easy for participants to understand (Menkhoff and Sakha, 2017), but the precision of the methods is seen as ambivalent. Menapace *et al.* (2015) are concerned that results might not capture true attitudes as the general DO question is not embedded in any real setting and therefore individuals cannot associate it with their past behavior. Also, hypothetical survey questions might lack precision as they are not incentive-compatible and suffer from framing effects and potentially biased answers (Dohmen *et al.*, 2011). Despite their carefully laid-out concerns regarding hypothetical survey questions, Dohmen *et al.* (2011) further compare and find that, nevertheless, the best comprehensive indicator of risk attitude appears to be the general question. Many researchers, such as Charness and Viceisza (2016), find it to be an accurate alternative to an MPL task.

Regarding the choice of a suitable method, a general rule of thumb found in literature is 'simpler is better' (Dave *et al.*, 2010; Crosetto and Filippin, 2013a; Charness and Viceisza, 2016). Simpler and more intuitive methods appear to trump more complex methods, as the ability of low-numeracy, low-literacy individuals to understand such methods yields more qualitative data. Using simpler designs makes researchers more likely to find a closer approximation of real risk attitudes (Dave *et al.*, 2010). One interesting insight comes from Menkhoff and Sakha (2017), who find that a multiple-item risk indicator, e.g. comprising results from differing risk elicitation methods, outperforms isolated results from any single risk elicitation method. Charness and Viceisza (2016) suggest that researchers aiming to identify suitable risk elicitation methods in developing countries follow a four-step procedure: (i) conduct a pilot; (ii) include multiple methods; (iii) consider framing, especially with complex methods so that subjects can understand; and (iv) conduct an *ex-ante* survey to understand the risk perceptions of specific subjects.

8.5.2 Check for explanatory power

To test for explanatory power is generally not a simple task, and it is further complicated in the setting of a developing country due to issues such as information constraints. The best case for such an indicator would be based on exogenous information, as it curtails bias. This is done by Vollmer *et al.* (2017) by using official national data for their production function. However, getting this kind of exogenous information might be difficult when operating in a developing country. As explained in previous sections, researchers instead use an indicator for risky behavior that can be either *revealed* or *stated*. *Revealed* could be observed, e.g. crop diversification, investment in cash crops, or market participation (Hellerstein *et al.*, 2013; Verschoor *et al.*, 2016), or asked about, e.g. engagement in risky behavior such as smoking, drinking, or speeding (Anderson and Mellor, 2008; Menkhoff and Sakha, 2017). *Stated* would be to simply ask subjects about their risk-taking behavior, as done in the DO method (Dohmen *et al.*, 2011).

This last point might represent the confusion involved. Some researchers use the DO task or other measures of stated preferences as a risk elicitation method (which needs to be validated on its own), but others use it to validate risk elicitation methods (Frey *et al.*, 2017). Drawing the line between the two might not be clear from literature, and more information is needed to rely on stable exogenous indicators to validate experimentally revealed risk preferences. Until then, researchers need to look at their specific study setting and follow Hellerstein *et al.* (2013) or Verschoor *et al.* (2016): search for site-specific candidates that reveal risky behavior. Regarding the aim of the research, it might be useful to further distinguish approaches that yield domain-specific risk attitudes, e.g. regarding financial concerns (Verschoor *et al.*, 2016), or general risk attitude (Dohmen *et al.*, 2011).

8.6 Conclusion

Individuals residing in rural areas of developing countries are among the most vulnerable and those who are most exposed to shocks. Risk attitude can play a crucial role in turning the vicious cycle of poverty into a virtuous cycle of sustainable growth. It is therefore crucial to select reliable tools for measuring risk attitudes, as the goal is to accurately capture – and then account for – risk attitudes when designing policy interventions. To do so, the researcher's main interest needs to be accurately measuring risk attitudes: not how risk attitudes should be, but how they actually are. Only when empirically understanding the heterogeneity in risk attitudes – even though some individuals' decisions might seem irrational – can we identify and approach especially risk-averse decision makers by offering additional safety nets and thus make policy interventions potentially more effective.

However, which method(s) should a researcher choose when aiming to elicit risk attitudes of low-literacy subjects in remote rural settings? A battery of methods is available in the literature, as discussed in this chapter. Researchers on the quest to identify suitable methods by comparing complex and simple (incentivized and non-incentivized) methods in the field and conducting a within-subject analysis (Charness *et al.*, 2013; Crosetto and Filippin, 2016; He *et al.*, 2018) typically find low consistency among the answers given in the different experimental tasks. However, when contrasting the within-subject results with real-life behavior, studies do find hints for a general rule of thumb. In synthesis, simple incentivized methods, or those with better illustration, have been shown to be easier than complex methods for low-literacy individuals to understand (Dave *et al.*, 2010; Charness *et al.*, 2013, 2020; Verschoor *et al.*, 2016). Also, simple survey measures (such as Dohmen *et al.*, 2011) appear to be a fair indicator of risk attitude (Dohmen *et al.*, 2011; Charness and Viceisza, 2016). However, in simple incentivized methods it is more difficult, if not impossible, to detect confusion (Charness and Viceisza, 2016) as well as to disentangle risk-neutral from risk-seeking behavior (Filippin and Crosetto, 2016). Thus, although simple methods might be better understood, they tend to lack the level of precision needed (Charness *et al.*, 2013). Researchers thus may want to carefully adjust a method suitable for their research endeavor to the specific field setting, after careful piloting.

This chapter also touched on how to account for a method's explanatory power. This issue, in general, requires more research, but it is highly difficult for researchers in a developing country setting to understand if the experimentally elicited risk attitudes can predict real-life behavior due to issues such as lack of transparency. With respect to the aim of their specific research, researchers might want to find unique, exogenous indicators mirroring risky or non-risky behavior and look for correlations and further distinguish if they aim to check for domain-specific or general risk attitudes.

References

Anderson, L.R. and Mellor, J.M. (2008) Predicting health behaviors with an experimental measure of risk preference. *Journal of Health Economics* 27, 1260–1274.

Anderson, L.R. and Mellor, J.M. (2009) Are risk preferences stable? Comparing an experimental measure with a validated survey-based measure. *Journal of Risk and Uncertainty* 39, 137–160.

Anderson, J.R., Dillon, J. and Hardaker, J.B. (1977) *Agricultural Decision Analysis*. Iowa State University Press, Ames, Iowa.

Antle, J.M. (1987) Economic estimation of producers' risk attitudes. *American Journal of Agricultural Economics* 69, 509–522.

Arrow, K.J. (1982) Risk perception in psychology and economics. *Economic Inquiry* 20, 1–9.

Barsky, R.B., Juster, F.T., Kimball, M.S. and Shapiro, M.D. (1997) Preference parameters and behavioral heterogeneity: an experimental approach in the health and retirement study. *The Quarterly Journal of Economics* 112, 537–579.

Bell, D.E. (1985) Disappointment in decision making under uncertainty. *Operations Research* 33, 1–27.

Binswanger, H. (1980) Attitudes toward risk: experimental measurement in rural India. *American Journal of Agricultural Economics* 62, 395–407.

Bordalo, P., Gennaioli, N. and Shleifer, A. (2012) Salience theory of choice under risk. *The Quarterly Journal of Economics* 127, 1243–1285.

Brick, K., Visser, M. and Burns, J. (2012) Risk aversion: experimental evidence from South African fishing communities. *American Journal of Agricultural Economics* 94, 133–152.

Brink, L. and McCarl, B. (1978) The tradeoff between expected return and risk among cornbelt farmers. *American Journal of Agricultural Economics* 60, 259–263.

Bruns, S., Hermann, D. and Mußhoff, O. (2022) Investigating inconsistencies in complex lotteries: the role of cognitive skills of low-numeracy subjects. *Journal of Behavioral and Experimental Economics* 97, 101840.

Byerlee, D., De Janvry, A., Sadoulet, E., Townsend, R. and Klytchnikova, I. (2008) World Development Report 2008: agriculture for development. *World Development Report*, 30. World Bank Group, Washington, DC.

Charness, G. and Viceisza, A. (2016) Three risk-elicitation methods in the field: evidence from rural Senegal. *Review of Behavioral Economics* 3, 145–171.

Charness, G., Gneezy, U. and Imas, A. (2013) Experimental methods: eliciting risk preferences. *Journal of Economic Behavior & Organization* 87, 43–51.

Charness, G., Garcia, T., Offerman, T. and Villeval, M.C. (2020). Do measures of risk attitude in the laboratory predict behavior under risk in and outside of the laboratory? *Journal of Risk and Uncertainty*, 60, 99–123.

Chew, S.H. and Maccrimmon, K.R. (1979) *Alpha-nu Choice Theory: A Generalization of Expected Utility Theory*. Working Paper No. 669, University of British Columbia, Vancouver, Canada.

Crosetto, P. and Filippin, A. (2013a) A theoretical and experimental appraisal of five risk elicitation methods. *Jena Economic Research Paper 9*, Jena University, Jena, Germany.

Crosetto, P. and Filippin, A. (2013b) The 'bomb' risk elicitation task. *Journal of Risk and Uncertainty* 47, 31–65.

Dave, C., Eckel, C.C., Johnson, C.A. and Rojas, C. (2010) Eliciting risk preferences: when is simple better? *Journal of Risk and Uncertainty* 41, 219–243.

Doerr, U., Toman, O.M. and Schmidt, U. (2011) Overconfidence and risk management of Ethiopian farmers. Working Paper, University of Kiel, Kiel, Germany.

Dohmen, T., Falk, A., Huffman, D., Sunde, U., Schupp, J. and Wagner, G.G. (2011) Individual risk attitudes: measurement, determinants, and behavioral consequences. *Journal of the European Economic Association* 9, 522–550.

Duflo, E., Kremer, M. and Robinson, J. (2008) How high are rates of return to fertilizer? Evidence from field experiments in Kenya. *The American Economic Review* 98, 482–488.

Eckel, C.C. and Grossman, P.J. (2002) Sex differences and statistical stereotyping in attitudes toward financial risk. *Evolution and Human Behavior* 23, 281–295.

Filippin, A. and Crosetto, P. (2016) A reconsideration of gender differences in risk attitudes. *Management Science* 62, 3138–3160.

Frey, R., Pedroni, A., Mata, R., Rieskamp, J. and Hertwig, R. (2017) Risk preference shares the psychometric structure of major psychological traits. *Science Advances* 3, e1701381.

Galarza, F. (2009) *Choices under Risk in Rural Peru*. MPRA Paper 17708, University Library of Munich, Germany.
Gneezy, U. and Potters, J. (1997) An experiment on risk taking and evaluation periods. *The Quarterly Journal of Economics* 112, 631–645.
Gong, Y., Baylis, K., Kozak, R. and Bull, G. (2016) Farmers' risk preferences and pesticide use decisions: evidence from field experiments in China. *Agricultural Economics* 47, 411–421.
Hardaker, J.B., Lien, G., Anderson, J.R. and Huirne, R.B.M. (2015) *Coping with Risk in Agriculture, Applied Decision Analysis*, 3rd edn. CAB International, Wallingford, UK.
Harrison, G.W. and Rutström, E.E. (2009) Expected utility theory and prospect theory: one wedding and a decent funeral. *Experimental Economics* 12, 133–158.
Harrison, G.W., Humphrey, S.J. and Verschoor, A. (2010) Choice under uncertainty: evidence from Ethiopia, India and Uganda. *The Economic Journal* 120, 80–104.
He, P., Veronesi, M. and Engel, S. (2018). Consistency of risk preference measures: an artefactual field experiment from rural China. *The Journal of Development Studies* 54, 1955–1973.
Hellerstein, D., Higgins, N. and Horowitz, J. (2013) The predictive power of risk preference measures for farming decisions. *European Review of Agricultural Economics* 40, 807–833.
Holden, S.T. (2014) Risky choices of poor people: comparing risk preference elicitation approaches in field experiments. CLTS Working Papers 10/14, Centre for Land Tenure Studies, Norwegian University of Life Sciences, Ås, Norway.
Holt, C.A. and Laury, S.K. (2002) Risk aversion and incentive effects. *The American Economic Review* 92, 1644–1655.
Hong, C.S. and Waller, W.S. (1986) Empirical tests of weighted utility theory. *Journal of Mathematical Psychology* 30, 55–72.
IFAD (2011) *Rural Poverty Report 2011: New Realities, New Challenges: New Opportunities for Tomorrow's Generation.* IFAD, Rome, 322 pp.
Ihli, H.J., Chiputwa, B., Baumeister, G.-F. and Mußhoff, O. (eds) (2013) Measuring risk attitudes of smallholder farmers in Uganda: how consistent are results of different methods. *The Second International Agricultural Risk, Finance and Insurance Conference (IARFIC)*, June 16–18, 2013, Vancouver, Canada.
Ihli, H. J., Gassner, A. and Mußhoff, O. (2018) Experimental insights on the investment behavior of small-scale coffee farmers in central Uganda under risk and uncertainty. *Journal of Behavioral and Experimental Economics* 75, 31–44.
Jacobson, S. and Petrie, R. (2009) Learning from mistakes: what do inconsistent choices over risk tell us? *Journal of Risk and Uncertainty* 38, 143–158.
Kahneman, D. and Tversky, A. (1979) Prospect theory: an analysis of decision under risk. *Econometrica: Journal of the Econometric Society* 47, 263–291.
Kumbhakar, S.C. (2002a) Risk preferences and technology: a joint analysis. *Marine Resource Economics* 17, 73–85.
Kumbhakar, S.C. (2002b) Specification and estimation of risk preferences, production risk and technical efficiency. *American Journal of Agricultural Economics* 84, 8–22.
Kumbhakar, S.C. and Tveterås, R. (2003) Production risk, risk preferences and firm-heterogeneity. *Scandinavian Journal of Economics* 105, 275–293.
Lejuez, C.W., Read, J.P., Kahler, C.W., Richards, J.B., Ramsey, S.E. *et al.* (2002) Evaluation of a behavioral measure of risk taking: the Balloon Analogue Risk Task (BART). *Journal of Experimental Psychology: Applied* 8, 75–84.
Levin, J. (2006) Choice under uncertainty [Online]. *Lecture Notes*. Stanford University, Stanford, California. Available at: https://web.stanford.edu/~jdlevin/Econ%20202/Uncertainty.pdf (accessed May 12, 2023).
Lien, G. (2002) Non-parametric estimation of decision makers' risk aversion. *Agricultural Economics* 27, 75–83.
Liu, E.M. (2013) Time to change what to sow: risk preferences and technology adoption decisions of cotton farmers in China. *Review of Economics and Statistics* 95, 1386–1403.
Loomes, G. and Sugden, R. (1982) Regret theory: an alternative theory of rational choice under uncertainty. *The Economic Journal* 92, 805.
Lusk, J.L. and Coble, K.H. (2005) Risk perceptions, risk preference, and acceptance of risky food. *American Journal of Agricultural Economics* 87, 393–405.
Maccrimmon, K.R. and Wehrung, D.A. (1985) A portfolio of risk measures. *Theory and Decision* 19, 1–29.
McGuire, S. (2015) FAO, IFAD and WFP. The state of food insecurity in the world 2015. Meeting the 2015 international hunger targets: taking stock of uneven progress. *Advances in Nutrition* 6, 623–624.

Menapace, L., Colson, G. and Raffaelli, R. (2015) A comparison of hypothetical risk attitude elicitation instruments for explaining farmer crop insurance purchases. *European Review of Agricultural Economics* 43, 113–135.
Menkhoff, L. and Sakha, S. (2017) Estimating risky behavior with multiple-item risk measures. *Journal of Economic Psychology* 59, 59–86.
Moscardi, E. and de Janvry, A. (1977) Attitudes toward risk among peasants: an econometric approach. *American Journal of Agricultural Economics* 59, 710–716.
Mosley, P. and Verschoor, A. (2005) Risk attitudes and the 'vicious circle of poverty.' *The European Journal of Development Research* 17, 59–88.
Mußhoff, O. and Hirschauer, N. (2010) *Modernes Agrarmanagement: Betriebswirtschaftliche Analyse- und Planungsverfahren.* Vahlen, Munich, Germany.
Nielsen, T., Keil, A. and Zeller, M. (2013) Assessing farmers' risk preferences and their determinants in a marginal upland area of Vietnam: a comparison of multiple elicitation techniques. *Agricultural Economics* 44, 255–273.
Quiggin, J. (1982) A theory of anticipated utility. *Journal of Economic Behavior & Organization* 3, 323–343.
Quiggin, J. (1994) Generalized expected utility theory: the rank-dependent model. *The Economic Journal* 104, 1490–1491. DOI: 10.2307/2235481.
Rapsomanikis, G. (2015) *The Economic Lives of Smallholder Farmers: An Analysis Based on Household Data from Nine Countries*. FAO, Rome, 40 pp.
Robison, L.J., Barry, P.J., Kliebenstein, J.B. and Patrick, G.F. (1984) Risk attitudes: concepts and measurements approaches. In: Barry, P.J. (ed.) *Risk Management in Agriculture*. Iowa State University Press, Ames, Iowa, pp. 11–30.
Savage, L.J. (1954) *The Foundations of Statistics.* Wiley, NY.
Slovic, P. (1966) Risk-taking in children: age and sex differences. *Child Development* 37, 169.
Starmer, C. (2000) Developments in non-expected utility theory: the hunt for a descriptive theory of choice under risk. *Journal of Economic Literature* 38, 332–382.
Tanaka, T., Camerer, C. and Nguyen, Q. (2010) Risk and time preferences: linking experimental and household survey data from Vietnam. *The American Economic Review* 100, 557–571.
Tversky, A. and Kahneman, D. (1992) Advances in prospect theory: cumulative representation of uncertainty. *Journal of Risk and Uncertainty* 5, 297–323.
Vargas Hill, R. (2009) Using stated preferences and beliefs to identify the impact of risk on poor households. *The Journal of Development Studies* 45, 151–171.
Varian, H.R. (2010) *Intermediate Microeconomics: A Modern Approach*. Norton, New York.
Verschoor, A. and D'Exelle, B. (2022) Probability weighting for losses and for gains among smallholder farmers in Uganda. *Theory and Decision*, 92, 223–258.
Verschoor, A., D'Exelle, B. and Perez-Viana, B. (2016) Lab and life: Does risky choice behaviour observed in experiments reflect that in the real world? *Journal of Economic Behavior & Organization* 128, 134–148.
Vollmer, E., Hermann, D. and Mußhoff, O. (2017) Is the risk attitude measured with the Holt and Laury task reflected in farmers' production risk? *European Review of Agricultural Economics* 44, 399–424.
von Neumann, J. and Morgenstern, O. (2007 [original work published in 1944]) *Theory of Games and Economic Behavior*, commemorative edn. Princeton University Press, Princeton, New Jersey.
Weber, E.U., Blais, A-R. and Betz, N.E. (2002) A domain-specific risk-attitude scale: measuring risk perceptions and risk behaviors. *Journal of Behavioral Decision Making* 15, 263–290.
White, T.L., Lejuez, C.W. and de Wit, H. (2008) Test-retest characteristics of the Balloon Analogue Risk Task (BART). *Experimental and Clinical Psychopharmacology* 16, 565–570.
World Bank (2016) Poverty: Overview. Available at: https://www.worldbank.org/en/topic/poverty/overview (accessed May 12, 2023).
World Bank (2018) *Poverty and Shared Prosperity 2018: Piecing Together the Poverty Puzzle*. World Bank, Washington, DC.
World Bank (2023) World Development Indicators Database. Available at: https://databank.worldbank.org/source/world-development-indicators (accessed May 12, 2023).
Zuckerman, M., Eysenck, S.B. and Eysenck, H.J. (1978) Sensation seeking in England and America: cross-cultural, age, and sex comparisons. *Journal of Consulting and Clinical Psychology* 46, 139–149.
Zuckerman, M., Kuhlman, D.M., Joireman, J., Teta, P. and Kraft, M. (1993) A comparison of three structural models for personality: the Big Three, the Big Five, and the Alternative Five. *Journal of Personality and Social Psychology* 65, 757–768.

9 Agricultural Risk Management and Technology

Jean-Paul Chavas*
University of Wisconsin, Madison, USA

Abstract
This chapter provides an overview of the economics of risk in agriculture. It develops a farm-level model of agricultural risk with a focus on price risk and production risk. The analysis examines the role of aversion to risk and downside risk, followed by an evaluation of the cost of risk under general conditions. It reviews methods used in assessing risk exposure, with a special emphasis on the role of downside risk (including the risk of crop failure). Implications for risk management and the role of technology are discussed.

9.1 Introduction

Risk is a fundamental characteristic of agricultural production (e.g. Harwood *et al.*, 1999; Just and Pope, 2002; Kimura and Le Thi, 2011). Weather shocks and unpredictable pest damages can have large adverse effects on farm production. Farmers also face significant price risk as agricultural markets exhibit high price volatility. Thus, uncertainty plays a role in almost every economic aspect of the agricultural sector. In addition, climate change and the increased frequency of extreme weather events have heightened the effects of risk on farm production. These concerns stress the need to better understand the management options available to farmers in dealing with agricultural risk. This chapter's objective is to present an overview of the economics of agricultural risk and its implications for farm management.

This chapter develops a model of agricultural risk at the farm level, including both price risk and production risk. It examines the role of risk preferences with a focus on aversion to risk and downside risk. The analysis evaluates the cost of risk under general conditions. Methods available to assess risk exposure are discussed. The linkages between risk, downside risk (including the risk of crop failure), and its management are examined in detail. This includes farm diversification (including crop rotation) and the use of risk-reducing inputs (such as irrigation). The analysis also discusses technology's role in reducing risk exposure.

9.2 Decisions under Risk

Consider a farmer making production decisions under risk. The production process involves n

*E-mail: jchavas@wisc.edu

 Managing Risk in Agriculture: A Development Perspective (eds A.K. Mishra, S.C. Kumbhakar and G. Lien).
DOI: 10.1079/9781800622289.0009

market inputs $x = (x_1, \ldots, x_n)$ along with k other goods $z = (z_1, \ldots, z_k)$ involved in the production of m marketed outputs $y = (y_1, \ldots, y_m)$. Let $p = (p_1, \ldots, p_m) > 0$ be the market prices for outputs y and $r = (r_1, \ldots, r_n) > 0$ be the market prices for inputs x. The farm net revenue is

$$\pi = \sum_{j=1}^{m} p_j y_j - \sum_{i=1}^{n} r_i x_i$$

There are two important sources of risk: production risk due to unanticipated weather shocks and pest damages, and price risk due to volatile agricultural markets. Denote the risk by the states $e = (e_p, e_y) \in \Omega$, where e_p are random variables affecting output prices p (reflecting unforeseen changes in market conditions), e_y are random variables affecting outputs y (e.g. weather shocks, diseases, pest damages), and Ω is the set of all possible states. The production technology is represented by the feasible set $T(z, t, e_y) \in \mathbb{R}^{n+m}$ where $(x, y) \in T(z, t, e_y)$ means that outputs y can be produced using inputs x conditional on z, technology t, and state e_y. Inputs x include market goods chosen at the beginning of the growing season (e.g. seeds, fertilizers). The goods z include factors other than x involved in the production process (e.g. labor, capital). Any change in the technology index t captures technological change, with technological progress being represented by a shift from t to t' satisfying $T(z, t, e_y) \subset T(z, t', e_y)$. And having $T(z, t, e_y) \subset T(z, t, e_y')$ means that states e_y' are more favorable than states e_y (as e_y' expands the possibility set compared to e_y).

Assume that the set T is closed. Letting $y = (y_1, y_a)$ where $y_a = (y_2, \ldots, y_m)$, consider the production function for output y_1

$$f(x, t, z, y_a, e_y) = Max_{y_1} \{y_1 : (x, y_1, y_a) \in T(z, t, e_y)\} \text{ if a maximum exists} \tag{9.1}$$
$$= -\infty \text{ otherwise.}$$

The production function $f(x, t, z, y_a, e_y)$ in Eqn 9.1 is conditional on (x, t, z, y_a, e_y) and provides a general representation of the production technology under production risk. Note that the production function $f(x, t, z, y_a, e_y)$ given in Eqn 9.1 is expressed in units of output y_1. In a multi-output context, a more general representation of the production technology under risk would be given by a directional distance function (Chambers *et al.*, 1996; Chavas *et al.*, 2019). Define technical efficiency as a situation where $y_1 = f(x, t, z, y_a, e_y)$, i.e. where output y_1 is on the upper bound of the set T. Assuming technical efficiency, net farm income can be written as

$$\pi(x, t, z, y_a, e) = p_1(e_p) f(x, t, z, y_a, e_y) + \sum_{j=2}^{m} p_j(e_p) y_j - \sum_{i=1}^{n} r_i x_i \tag{9.2}$$

where $p(e_p) = (p_1(e_p), \ldots, p_m(e_p))$ denotes the prices of outputs y under state e_p. Given $e = (e_p, e_y) \in \Omega$, Equation 9.2 reflects that farm net income $\pi(x, t, z, y_a, e)$ is affected by both price risk e_p and production risk e_y.

Due to production lags, assume that the states $e = (e_p, e_y) \in \Omega$ are not known at the time of the input decisions x. Let e be a random vector with a subjective probability distribution reflecting the information available to the farm manager. Here we rely on probabilities in risk assessment. Note that there are uncertain situations where probability assessments can be difficult (Knight, 1921), leading to investigations about the role of ambiguity (Halevy, 2007; Barham *et al.*, 2014b). More generally, a 'state-contingent' approach to risk analysis can be applied without probability assessment (Debreu, 1959, p. 98; Chambers and Quiggin, 2000; Chavas, 2008; Chavas and Briec, 2019). The choice of the variables in (x, t, z, y_a) depends on the information available to the manager at decision time. We assume that inputs x and technology t are chosen at planting time. Let $z = (z_a, z_b)$ where the variables z_a are also chosen at planting time (e.g. labor and land allocation at planting time), while z_b are variables chosen after planting time (e.g. labor allocation during the growing season). Below, we assume that the output variables y_a are chosen at harvest time when the states e are observed, implying that the choice of y_a can be written in terms of the decision rule $y_a(e)$. And we assume that the variables z_b are chosen during the growing season based on information that becomes available after planting time but before harvest time. Denote this information by the partition $P(\Omega)$ of Ω, where the farmer can distinguish between states that are in different cells of $P(\Omega)$ but cannot distinguish between states that are within a given cell of $P(\Omega)$ (e.g. Debreu, 1959, pp. 98–100; Chavas and Briec,

2019). In this context, the choice of the variables z_b can be written in terms of the decision rule $z_b(P(\Omega))$. For simplicity, in our discussion below, we will take the information partition $P(\Omega)$ as given.

Denote by $F(\eta\,|\,x, t, z_a, z_b(\cdot), y_a(\cdot)) = Prob_{e \in \Omega}[\pi(x, t, z_a, z_b(P(\Omega)), y_a(e), e) \leq \eta]$ the distribution function of net income π given in Eqn 9.2 conditional on $(x, t, z_a, z_b(\cdot), y_a(\cdot))$. In this context, consider the case where the decision maker has preferences represented by the preference functional $V(F(\cdot))$. Our focus on monetary payoff π simplifies the analysis by avoiding issues related to intertemporal preferences. Note that it has some limitations. For example, it neglects situations where the variables z are non-market goods that have non-monetary benefits. The presence of non-monetary benefits would require considering a utility functional more general than $V(F)$. We assume that the functional $V(F)$ is continuous and non-satiated in income. Non-satiation means that $V(F^b) > V(F^a)$ for any $F^a \neq F^b$ satisfying $F^a(\eta) \geq F^b(\eta)$ for all $\eta \in \mathbb{R}$, corresponding to first-order stochastic dominance under any rightward shift in the distribution function from F^a to F^b.

Using backward induction, optimal management decisions are represented by the following maximization problem:

$$Max_{x,t,z_a}\ Max_{z_b(\cdot)}\ Max_{y_a(\cdot)}\ \{V(F(\bullet)):$$

$$F(\eta) = Prob_{e\in\Omega}[\pi(x, t, z_a, z_b(\bullet), y_a(\bullet)) \leq \eta],\ \eta \in \mathbb{R}\}. \tag{9.3}$$

Equation 9.3 reflects that decisions are made at three different times: (x, z_a, t) are chosen *ex-ante* at planting time, $z_b(P(\Omega))$ are chosen during the growing season based on the information represented by the partition $P(\Omega)$, and $y_b(e)$ are chosen at harvest time after the states e are observed. Methods for choosing (x, t, z_a) in Eqn 9.3 are standard in the literature on static optimization. Methods for choosing the decision rules $(z_b(\cdot), y_a(\cdot))$ in Eqn 9.3 are more complex (e.g. see the literature on dynamic programming). In this context, uncertainty, information and risk preferences affect the farmer's decisions.

The preference functional $V(F(\cdot))$ has many models as special cases. These include the rank-dependent utility model where $V(F(\cdot)) = \int_{\eta\in\mathbb{R}} U(\eta)\, dG(F(\eta\,|\, x, t, z_a, z_b(\cdot), y_a(\cdot))$, $U(\eta)$ is a von Neumann–Morgenstern utility function reflecting risk preferences and $G(F)$: $[0, 1] \to [0, 1]$ is a strictly increasing function satisfying $G(0) = 0$ and $G(1) = 1$. When $G(F) \neq F$, the model corresponds to cumulative prospect theory (Tversky and Kahneman, 1992), which allows for non-linearity in probabilities (e.g. Kahneman and Tversky, 1979; Quiggin, 1993; Barberis, 2013). The model also includes as a special case the expected utility (EU) model when $V(F(\cdot)) = \int_{\eta\in\mathbb{R}} U(\eta)\, dF(\eta\,|\, x, y_a, z, t)$, i.e. when preferences are linear in probabilities and the objective function reduces to $EU(\pi)$ where E is the expectation operator.

To illustrate the role of risk preferences, consider the case where

$$V(F(\cdot)) = \int_{\eta\in\mathbb{R}} U(\eta)\, dG(F(\eta|\, x, t, z_a, z_b(\cdot), y_a(\cdot))) \tag{9.4a}$$

$$U(\pi) = \begin{cases} \dfrac{\pi^{1-\gamma_1}-1}{1-\gamma_1} - \dfrac{L^{1-\gamma_1}-1}{1-\gamma_1} \\ k\left[\dfrac{\pi^{1-\gamma_2}-1}{1-\gamma_2} - \dfrac{L^{1-\gamma_2}-1}{1-\gamma_2}\right] \end{cases} if \begin{Bmatrix} \pi \leq L \\ \pi > L \end{Bmatrix} \tag{9.4b}$$

with $k > 0$ and L being some income threshold, and

$$G(F) = exp\left[-(-log(F))^{\alpha}\right] \tag{9.4c}$$

where $\alpha \in (0, 1]$ is a probability weight (Prelec, 1998). When $k = 1$ and $\alpha = 1$, Equations 9.4a–9.4c reduce to the EU model. In addition, when $\gamma_1 = \gamma_2 = \gamma$, Equations 9.4a–9.4c become an EU model under constant relative risk aversion (CRRA) where γ is the Arrow-Pratt relative risk aversion coefficient (Pratt, 1964). As argued by Gollier (2001), most individuals exhibit risk aversion, with a coefficient of relative risk aversion γ typically being in the range from 0 (corresponding to risk neutrality) to 5 (corresponding to a high level of risk aversion).

When $k \neq 1$ and $\alpha \in (0, 1)$, Equations 9.4a–9.4c correspond to cumulative prospect theory (CPT; Kahneman and Tversky, 1979; Tversky and Kahneman, 1992). This theory departs from the standard EU model in two ways. First, having $k \neq 1$ in Eqn 9.4b introduces a kink in the function $U(\pi)$ at the threshold point L, allowing for 'loss aversion' as risk preferences can differ between favorable outcomes (when $\pi \geq L$)

and unfavorable outcomes (when $\pi < L$) (Kahneman and Tversky, 1979; Liu, 2013). Second, having $\alpha < 1$ in Eqn 9.4c introduces non-linearity in the probabilities: the function $G(F)$ then has an inverted S-shape overweighting the probability of rare events (see Tversky and Kahneman, 1992; Prelec, 1998; Barberis, 2013). Evidence that farmers tend to overweight such probabilities is presented in Liu (2013) and Bocquého *et al.* (2014). Compared to the EU model, the overweighting of small probabilities under cumulative prospect theory implies greater behavioral sensitivity to rare events.

9.3 The Cost of Risk

To assess risk management, we need to evaluate risk exposure and examine how risk affects welfare. Let $F_0(\pi_0)$ be a distribution function where all the probability mass of net income is located at the sure income π_0.

Under the preference functional $V(F(\cdot))$, define the *certainty equivalent* associated with a distribution function $F(\pi \mid x, t, z_a, z_b(\cdot), y_a(\cdot))$ as the sure income CE satisfying

$$V(F(\pi| x, t, z_a, z_b(\bullet), y_a(\bullet)) = V\left(F_0\left(CE\right)\right) \quad (9.5)$$

Denote the solution of Eqn 9.5 by $CE(x, t, z_a, z_b(\cdot), y_a(\cdot))$. The certainty equivalent $CE(x, t, z_a, z_b(\cdot), y_a(\cdot))$ is the sure monetary amount that makes the decision maker indifferent between facing the random payoff $\pi(x, t, z_a, z_b(P(\Omega)), y_a(e), e)$ and facing $CE(x, t, z_a, z_b(\cdot), y_a(\cdot))$ with certainty. The certainty equivalent depends on uncertainty, information and risk preferences. It provides a convenient welfare measure under risk. Indeed, under non-satiation, using Eqn 9.5 and substituting Eqns 9.4a–9.4b into Eqn 9.3, the maximization problem in Eqn 9.3 is equivalent to maximizing the certainty equivalent $CE(x, t, z_a, z_b(\cdot), y_a(\cdot))$. It means that the optimal management in Eqn 9.3 can be written alternatively as the solution to the optimization problem

$$Max_{x,t,z_a}\ Max_{z_b(\bullet)}\ Max_{y_a(\bullet)}\ \{CE(x,t,z_a,z_b(\bullet),\ y_a(\bullet))\} \quad (9.6)$$

Under the preference functional $V(F(\cdot))$ and for a given $(x, t, z_a, z_b(\cdot), y_a(\cdot))$, define the implicit cost of risk as the *risk premium* $R(x, y_a, z, t)$ satisfying

$$\begin{aligned} R(x,t,z_a,z_b(\bullet),y_a(\bullet)) &= M_1(x,t,z_a,z_b(\bullet),y_a(\bullet)) \\ &\quad - CE(x,t,z_a,z_b(\bullet),y_a(\bullet)) \end{aligned} \quad (9.7)$$

where $CE(x, t, z_a, z_b(\cdot), y_a(\cdot))$ is defined in Eqn 9.5, and

$$\begin{aligned} M_1\left(x, t, z_a, z_b(\bullet), y_a(\bullet)\right) &\equiv \int_{\pi\in\mathbb{R}} \pi dF(\pi \quad x, t, z_a, \\ z_b(\bullet), y_a(\bullet)) &= E[p_1(e_p) f(x, t, z_a, z_b(\bullet), \\ y_a(\bullet), e_y) &+ \textstyle\sum_{j=2}^{m} p_j(e_p) y_j(\bullet)] - rx \end{aligned} \quad (9.8)$$

is expected net farm income conditional on $(x, t, z_a, z_b(\cdot), y_a(\cdot))$.

Conditional on $(x, t, z_a, z_b(\cdot), y_a(\cdot))$, the risk premium $R(x, t, z_a, z_b(\cdot), y_a(\cdot))$ in Eqn 9.7 is the sure amount of money the decision maker is willing to pay to replace random income π by its mean M_1. In general, the risk premium depends on uncertainty, information and risk preferences. The sign of $R(\cdot)$ can be used to characterize the decision maker's risk preferences. In general, the decision maker is said to be $\left\{\begin{matrix}\text{risk averse}\\ \text{risk neutral}\\ \text{risk lover}\end{matrix}\right\}$ when $R(\cdot)\left\{\begin{matrix}>\\=\\<\end{matrix}\right\}0$. There is strong empirical evidence that most individuals and most farmers are risk-averse (e.g. Lin *et al.*, 1974; Dillon and Scandizzo, 1978; Binswanger, 1981; Antle, 1987; Holt and Laury, 2002; Barseghyan *et al.*, 2018).

Combining Eqns 9.6 and 9.7, it follows that optimal management in Eqn 9.3 or Eqn 9.6 can be written alternatively as the solution to the optimization problem

$$\begin{aligned} Max_{x,t,z_a}\ Max_{z_b(\bullet)}\ Max_{y_a(\bullet)}\ \{M_1(x, t, z_a, \\ z_b(\bullet), y_a(\bullet)) - R(x, t, z_a, z_b(\bullet), y_a(\bullet))\} \end{aligned} \quad (9.9)$$

Equation 9.9 identifies efficient decisions as the ones that maximize expected income $M_1(x, t, z_a, z_b(\cdot), y_a(\cdot))$ minus the cost of risk as measured by $R(x, t, z_a, z_b(\cdot), y_a(\cdot))$. In general, the cost for risk $R(x, t, z_a, z_b(\cdot), y_a(\cdot))$ depends on risk exposure, risk preferences and decisions $(x, t, z_a, z_b(\cdot), y_a(\cdot))$.

Evaluating the determinants of $R(x, t, z_a, z_b(\cdot), y_a(\cdot))$ is of significant interest. With $M_1(x, t, z_a, z_b(\cdot), y_a(\cdot))$ being the expected net income given in Eqn 9.8, let

$$\begin{aligned} M_i(x, t, z_a, z_b(\bullet), y_a(\bullet)) = E\,[\pi(x, t, z_a, z_b(\bullet), \\ y_a(\bullet), e) - M_1(x, t, z_a, z_b(\bullet), y_a(\bullet))]^i \end{aligned} \quad (9.10)$$

be the i-th central moment of net income, $i = 2, 3, \ldots$ In this context, M_2 is the variance of π and M_3 is a measure of skewness of π reflecting asymmetry in the distribution of risk exposure (with $M_3 < 0$ indicating exposure to downside risk).

The properties of the risk premium have been analyzed under the EU model (Pratt, 1964). Under the EU model, Equation 9.5 can be written as $EU(\pi) = U(E(\pi) - R)$. In this context, R in Eqn 9.7 reduces to the Arrow-Pratt definition of the risk premium, where risk aversion corresponds to a concave function $U(\pi)$ (e.g. Pratt, 1964). Under EU and assuming that $U(\pi)$ is differentiable, Pratt (1964) showed that the cost of risk can be approximated. Given $\frac{\partial U}{\partial \pi}(\pi) > 0$, such an approximation is

$$R\big(x, t, z_a, z_b(\bullet), y_a(\bullet)\big) \cong \frac{1}{2} R_2 M_2 (x, t, z_a, z_b(\bullet), y_a(\bullet)) + \frac{1}{6} R_3 M_3 (x, t, z_a, z_b(\bullet), y_a(\bullet)) \quad (9.11)$$

where $R_i = -\frac{\partial^i U}{\partial \pi^i}(M_1) / \frac{\partial U}{\partial \pi}(M_1)$ is the i-th risk aversion coefficient, $i = 2, 3$ (Pratt, 1964; Chavas, 2004, p. 49). Equation 9.11 states that the cost of risk R can be approximated as a linear function of the central moments of income, where R_2 is the coefficient of the variance M_2 and R_3 is the coefficient of skewness M_3. Note that R_2 is the Arrow-Pratt absolute risk aversion parameter showing how the cost of risk R varies with the variance M_2, where $\frac{\partial^2 U}{\partial \pi^2}(\pi) < 0$ and $R_2 > 0$ correspond to risk aversion (Pratt, 1964). And $\frac{\partial^3 U}{\partial \pi^3}(\pi) > 0$ and $R_3 < 0$ correspond to downside risk aversion, where R_3 can be interpreted as a downside risk aversion parameter indicating how R varies with skewness (Antle, 1987).

To illustrate, consider the case of CRRA where $U(\pi)$ is given by Eqn 9.4b, $k = 1$, $\gamma = \gamma_1 = \gamma_2$ is the Arrow-Pratt relative risk aversion parameter, and $M_1 > 0$. Then, $R_2 = \gamma / M_1$ and $R_3 = -\gamma(1+\gamma)/M_1^2$, and Eqn 9.11 becomes

$$Rx, t, z_a, z_b(\bullet), y_a(\bullet) \cong \frac{\gamma}{2} \frac{M_2\big(x, t, z_a, z_b(\bullet), y_a(\bullet)\big)}{M_1\big(x, t, z_a, z_b(\bullet), y_a(\bullet)\big)} - \frac{\gamma(1+\gamma)}{6} \frac{M_3\big(x, t, z_a, z_b(\bullet), y_a(\bullet)\big)}{M_1\big(x, t, z_a, z_b(\bullet), y_a(\bullet)\big)^2} \quad (9.12a)$$

or

$$\frac{R(x, t, z_a, z_b(\bullet), y_a(\bullet))}{M_1(x, t, z_a, z_b(\bullet), y_a(\bullet))} \cong \frac{\gamma}{2} \frac{M_2(x, t, z_a, z_b(\bullet), y_a(\bullet))}{M_1(x, t, z_a, z_b(\bullet), y_a(\bullet))^2} - \frac{\gamma(1+\gamma)}{6} \frac{M_3(x, t, z_a, z_b(\bullet), y_a(\bullet))}{M_1(x, t, z_a, z_b(\bullet), y_a(\bullet))^3} \quad (9.12b)$$

Under CRRA with $\gamma > 0$ and $M_1 > 0$, it follows from Eqn 9.12a that the risk premium R increases with the variance M_2, reflecting that the variance acts as a proxy measure of risk exposure; and R decreases with the skewness M_3, indicating that CRRA exhibits 'downside risk aversion' (as a decrease in skewness M_3 corresponds to an increased exposure to downside risk and a rise in the cost of risk R). To the extent that CRRA and $\gamma > 0$ provide a good approximation of risk preferences for many individuals, this has three important implications: (i) risk aversion and downside risk aversion are pervasive characteristics of risk behavior; (ii) any increase in risk exposure (as measured by the variance M_2) increases the cost of risk R, making the decision maker worse off; and (iii) any increase in exposure to downside risk (as measured by decline in skewness M_3) increases the cost of risk R, again making the decision maker worse off. These arguments reflect a common attribute of risk behavior: most individuals worry most about their exposure to risk and downside risk, including exposure to losses (reflecting 'loss aversion').

Under risk aversion and CRRA (when $R > 0$ and $\gamma > 0$), Equation 9.12a shows that, holding other central moments constant, any increase in $M_1 > 0$ tends to reduce the risk premium R. Thus, CRRA preferences exhibit income effects: the larger the expected payoff M_1, the smaller the cost of risk R. In other words, *ceteris paribus*, high-income individuals tend to be less risk-averse than low-income individuals. Pratt (1964) called this property 'decreasing absolute risk aversion' (DARA). There is strong evidence that most decision makers are risk-averse and exhibit DARA preferences (e.g. Binswanger, 1981; Chavas and Holt, 1996). This indicates that the adverse welfare impacts of exposure to a given risk decline with income or wealth.

Given $M_1 > 0$, Equation 9.12b also shows how $\frac{R}{M_1}$ can be approximated as a simple function of γ, M_1, M_2 and M_3. The ratio $\frac{R}{M_1}$ is the relative risk premium measuring the proportion of expected income M_1 the decision maker is willing to pay to eliminate risk exposure. Noting that $\frac{R}{M_1}$ is a unit-free measure of the relative cost of risk, it follows that the right-hand side in Eqn 9.12b is also 'unit-free,' reflecting that γ is a unit-free measure of the degree of risk aversion. This is a useful result when risk preferences vary across individuals: the relative risk aversion coefficient γ can be used in evaluating the heterogeneity of risk preferences across decision makers.

Although Eqns 9.7–9.9 hold under general conditions, note that the approximations given in Eqns 9.11 and 9.12 apply only under the EU model. Under the more general cumulative prospect theory model given in Eqns 9.4a–9.4c, estimating the moments of risky payoffs can still be used in the evaluation of risk exposure. But the approximations of R given in Eqns 9.11–9.12 would no longer apply when $k \neq 1$ (as the function $U(\pi)$ in Eqn 9.4b would no longer be differentiable). In addition, when $\alpha \neq 1$ and the $G(F)$ is an inverted-S shape in Eqn 9.4c, then under risk aversion and CRRA, Equations 9.11–9.12 would underestimate the cost of risk R in two ways: they would underestimate the variance effect (as the G distribution exhibits greater risk exposure than F in the tails); and they would underestimate the skewness effect (as G exhibits greater exposure to downside risk than F). In such situations, Equation 9.9 (as an implicit equation for R) would still provide a correct evaluation of the cost of risk R.

Finally, the above discussion makes it clear that the cost of risk R depends on both risk exposure and risk preferences. Evaluating the separate effects of risk exposure and preferences can be challenging. To see that, consider the EU model $EU(\pi(e)) = \sum_{s \in S} Prob(e_s)\ U(\pi(e_s))$, along with another EU model $E_0U_0(\pi(e)) = \sum_{s \in S} Prob_0(e_s)\ U_0(\pi(e_s))$. Letting $h_s \equiv \frac{Prob(e_s)}{Prob_0(e_s)} \neq 1$, consider the case where $U_0(\pi(e_s)) = h_s\ U(\pi(e_s))$. Then, the two models (representing different risk exposure and exhibiting different preferences) would generate identical behavior and welfare. This makes it clear that the separate identification of risk exposure and risk preferences must rely on different data sources. Experimental methods have been commonly used to identify and estimate the nature of individual risk preferences (e.g. Binswanger, 1981; Liu, 2013), and econometric methods are available to assess the nature of risk exposure.

Another challenge comes from the heterogeneity in both risk exposure and risk preferences among farmers. Risk exposure typically depends on local agro-climatic conditions (e.g. weather shocks). The spatial heterogeneity in risk exposure makes the assessment of farm risk empirically difficult. This issue is compounded by the presence of heterogeneity in risk preferences across individuals. Although there is strong evidence that most farmers are risk-averse and downside risk-averse, much heterogeneity remains in the degree of risk aversion across individuals (e.g. Halek and Eisenhauer, 2001; Dohmen *et al.*, 2011; Barseghyan *et al.*, 2018). As a result, different farmers can react to the same risk exposure in different ways. Again, it means that specific risk management recommendations that seem appropriate for some farmers may be inappropriate for others. These arguments stress the empirical challenges of dealing with heterogeneity in the evaluation of both risk preferences and risk exposure for farmers.

9.4 Risk Assessment

The usefulness of the above analysis relies on knowing the risk exposure facing each farmer. This section discusses alternative ways to evaluate the nature and extent of risk exposure, including both production risk and price risk.

9.4.1 Moment-based approach

Risk exposure can be evaluated in terms of the central moments of net income π, including its mean M_1 (a measure of central tendency), its variance M_2 (a measure of dispersion), and its skewness M_3 (a measure of asymmetry of the probability function around the mean). Estimating the moments of the probability distribution of payoff has been commonly used in applied risk analysis. Using variance as a measure of risk exposure, the simplest approach is to rely on a mean-variance analysis in the evaluation of risk management. This approach has been very popular because of its simplicity (e.g. Just and

Pope, 1979; Harwood *et al.*, 1999). In general, the variance M_2 takes the form

$$M_2(x, t, z_a, z_b(\bullet), y_a(\bullet)) = Var[p_1(e_p) f(x, t, z_a, z_b(\bullet), y_a(\bullet), e_y) + \sum_{j=2}^{m} p_j(e_p)\, y_j(\bullet)]$$
$$= Var[p_1(e_p) f(x, t, z_a, z_b(\bullet), y_a(\bullet), e_y)] + \sum_{j=2}^{m} Var[p_j(e_p) y_j(\bullet)] + \sum_{j=2}^{m} 2\, cov\, [p_1(e_p) f(x, t, z_a, z_b(\bullet), y_a(\bullet), e_y), p_j(e_p)\, y_j(\bullet)] \quad (9.13)$$

Equation 9.13 shows that both variances and covariances play a role: the variances of revenue for each output: $Var[p_1(e_p) f(x, t, z_a, z_b(\cdot), y_a(\cdot), e_y)]$ and $Var[p_j(e_p)\, y_j(\cdot)]$, $j = 2, \ldots, m$; and the covariances $Cov[p_1(e_p)\, f(x, t, z_a, z_b(\cdot), y_a(\cdot), e_y), p_j(e_p)\, y_j(\cdot)]$. In the special case where $p_1(e_p)$ and $f(x, t, z_a, z_b(\cdot), e_y)$ are independently distributed, note that $Var(p_1 f) = Var(p_1)\, Var(f) + [E(f)]^2\, Var(p_1) + [E(p_1)]^2\, Var(f)$. Several risk management options are available. From Eqn 9.13, risk exposure can be reduced by producing outputs with low price variance $Var(p_j)$, by choosing inputs x that reduce production risk (i.e. where $\frac{\partial Var(f)}{\partial x} < 0$), by producing outputs that satisfy $\frac{\partial Var(f)}{\partial y_j} < 0$, by using z that are risk-decreasing with $\frac{\partial Var(f)}{\partial z} < 0$, and by using a technology t that reduces $Var(f)$. In addition, it follows from Eqn 9.13 that negative covariances (or correlations) between $[p_j(e_p)\, y_j(\cdot)]$ and $[p_1(e_p) f(x, t, z_a, z_b(\cdot), y_a(\cdot), e_y)]$ would contribute to decreasing M_2. This is a scenario where output diversification would help reduce risk exposure (see below).

Although the mean-variance approach is very convenient in applied risk analysis, questions have been raised about its general validity. A mean-variance approach is valid when the probability function is 'normal' (because mean and variance are sufficient statistics for a normal distribution). A mean-variance approach is also valid when risk management options affect only the mean and the standard deviation of the random variable (Meyer, 1987). But beyond these two cases, a mean-variance analysis can be very restrictive. Indeed, the variance does not distinguish between 'upside risk' and 'downside risk' and is unable to evaluate exposure to specific downside risk. But introducing skewness in risk analysis can (as increasing skewness means reducing downside risk). This suggests a need to go beyond the first two moments (mean and variance) in risk analysis. Empirical estimates of risk exposure can be extended to include any number of moments of the probability function (e.g. Antle, 1987). Including at least three moments (mean, variance and skewness) is important when the focus is on managing downside risk. Other possibilities include the use of partial moments (Antle, 2010) and quantile moments (Kim *et al.*, 2014). To illustrate, under the EU model and using Eqn 9.11, Eqn 9.9 can be approximated as

$$Max_{x,t,z_a}\, Max_{z_b(\cdot)}\, Max_{y_a(\cdot)} \{M_1(x, t, z_a, z_b(\bullet), y_a(\bullet)) - \frac{1}{2} R_2\, M_2(x, t, z_a, z_b(\bullet), y_a(\bullet)) - \frac{1}{6} R_3\, M_3(x, t, z_a, z_b(\bullet), y_a(\bullet))\} \quad (9.9')$$

And under CRRA preferences and using Eqn 9.12a, Eqn 9.9 can be approximated as

$$Max_{x,t,z_a}\, Max_{z_b(\cdot)}\, Max_{y_a(\cdot)} \{M_1(x, t, z_a, z_b(\bullet), y_a(\bullet)) - \frac{\gamma}{2} \frac{M_2(x, t, z_a, z_b(\bullet), y_a(\bullet))}{M_1(x, t, z_a, z_b(\bullet), y_a(\bullet))} + \frac{\gamma(1+\gamma)}{6} \frac{M_3(x, t, z_a, z_b(\bullet), y_a(\bullet))}{M_1(x, t, z_a, z_b(\bullet), y_a(\bullet))^2}\} \quad (9.9'')$$

While Eqns 9.9'–9.9" are only approximations, they provide simple formulations that have been commonly used in the empirical evaluation of risk management. First, in situations where $M_3 = 0$, Equations 9.9'–9.9" would reduce to simple mean-variance models. Second, when $M_3 \neq 0$, Equations 9.9'–9.9" show how to introduce skewness effects in risk analysis. Econometric methods have been developed to estimate second and higher moments of a distribution (e.g. Antle, 1983; Antle and Goodger, 1984). The estimation of skewness M_3 gains relevance when the payoff distribution is asymmetric, i.e. when exposure to downside risk differs from exposure to upside risk. Decision makers who are averse to downside risk would be interested in management schemes that increase M_3 and reduce their exposure to downside risk.

9.4.2 Probability-based approach

Although central moments are convenient summary statistics that can be used to assess risk exposure, they may not be sufficient. For example, individuals may be particularly concerned with extreme risk associated with catastrophic events located in the lower tail of the payoff distribution (e.g. the case of crop failure). Such information is not easily captured by central moments, thus motivating the need to evaluate the probability of specific events. A general way to obtain this information is to estimate the distribution function of the stochastic payoff $\pi(x, t, z_a, z_b(\cdot), y_a(\cdot), e)$ given in Eqn 9.2, where $e = (e_p, e_y)$ and risk comes from both price risk e_p and production risk e_y. In this context, it is convenient to proceed in three steps: in a first step, evaluate the marginal distribution of price risk e_p; in a second step, evaluate the marginal distribution of production risk e_y; and in a third step, evaluate the joint distribution of price and production risk.

First, consider price risk $p(e_p)$. As it reflects changes in market conditions between planting time and harvest time, price risk typically varies across outputs depending on supply–demand conditions. For the j-th output, the marginal distribution function of price risk $p_j(e_p)$ is

$$D_{pj}\left(p_{j0}\right) \equiv Prob\left\{p_j\left(e_p\right) \leq p_{j0}\right\},\ j = 1, \ldots, m \quad (9.14)$$

Second, consider the assessment of production risk. It can be done relying on the production function $f(x, t, z_a, z_b(\cdot), y_a(\cdot), e_y)$ in Eqn 9.1, which provides a general representation of the technology frontier under production risk e_y. This suggests that production risk can be evaluated by focusing on the marginal distribution function of $f(x, t, z_a, z_b(\cdot), y_a(\cdot), e_y)$:

$$D_y\left(f_0 \mid x,\ t,\ z_a, z_b\left(\cdot\right),\ y_a\left(\cdot\right)\right) \equiv Prob\left\{f\left(x,\ t,\ z_a,\ z_b\left(\cdot\right),\ y_a\left(\cdot\right),\ e_y\right) \leq f_0\right\} \quad (9.15)$$

Alternatively, when the outputs $y = (y_1, y_2, \ldots, y_m)$ are crops and z_j is the amount of land allocated to the j-th crop, production risk can also be evaluated based on crop yield $Y_j \equiv \frac{y_j}{z_j}$, y_j denoting the production of the j-th crop, $j = 1, \ldots, m$. The yield for the j-th crop can be expressed as $Y_j(x, t, z_a, z_b(\cdot), e_y)$, $j = 1, \ldots, m$. These yield functions would satisfy $z_1\ Y_1(x, t, z_a, z_b(\cdot), e_y) = f(x, t, z_a, z_b(\cdot), y_a(\cdot), e_y)$ and $z_j\ \ Y_j(x, t, z_a, z_b(\cdot), e_y) = y_j(\cdot)$, $j = 2, \ldots, m$. Then, conditional on $(x, t, z_a, z_b(\cdot))$, the distribution of yield risk for the j-th crop is given by

$$D_{Yj}\left(Y_{j0} \mid x,\ t,\ z_a,\ z_b\left(\bullet\right)\right) \equiv Prob\left\{Y_j\left(x,\ t,\ z_a,\ z_b\left(\bullet\right),\ e_y\right) \leq Y_{j0}\right\}, j = 1, \ldots, m \quad (9.16)$$

Third, consider the joint distribution of price risk and production risk when yields are used in risk assessment. Evaluated at points $p_C = (p_{10}, \ldots, p_{m0})$ and $Y_0 = (Y_{10}, \ldots, Y_{m0})$, this joint distribution is

$$D(p_0, Y_0 \mid x,\ t,\ z_a,\ z_b(\bullet)) = Prob\ \{p_j(e_p) \leq p_{j0}, Y_j(x,\ t,\ z_a,\ z_b(\bullet),\ e_y) \leq Y_{j0},\ \ j = 1, \ldots, m\} \quad (9.17a)$$

Since $\pi \equiv \sum_{j=1}^{m} p_j\, Y_j\, z_j - \sum_{i=1}^{n} r_i x_i$, knowing the joint distribution function of (p, Y) is equivalent to knowing the distribution of net income π, conditional on (x, z, t). Using Eqns 9.14 and 9.16, the joint distribution $D(p_0, Y_0 \mid\ x,\ t,\ z_a,\ z_b(\cdot))$ in Eqn 9.17a can be alternatively written as

$$D(p_0,\ Y_0 \mid x,\ t,\ z_a,\ z_b(\bullet)) = C\ [D_{p1}(p_{10}),\ D_{Y1}(Y_{10} \mid x,\ t,\ z_a,\ z_b(\bullet)); \ldots;\ D_{pm}(p_{m0}),\ D_{Y1}(Y_{m0} \mid x,\ t,\ z_a,\ z_b(\bullet))] \quad (9.17b)$$

where $C[D_{p1}, D_{Y1}; \ldots; D_{pm}, D_{Ym}]$ is a copula. The copula in Eqn 9.17b provides formal linkages between the joint risk distribution $D(p_0, Y_0 \mid \cdot)$ and the marginal distributions of price risk $D_{pj}(p_{j0})$ and yield risk $D_{Yj}(Y_{j0} \mid \cdot)$, $j = 1, \ldots, m$ (Nelsen, 1999; Joe, 2014). In general, a copula measures codependency among alternative sources of revenue risk. Codependency can vary across farmers, depending on markets and local agro-climatic conditions. Going beyond correlation measures, codependences can also be affected by the nature of shocks, in which case they would vary across quantiles. Of special interest are the codependency between price and production for a given commodity, codependency between prices for different commodities, and codependency between different risky outputs.

Econometric methods can be used to estimate the marginal distribution functions $D_{pj}(p_{j0})$ and $D_{Yj}(Y_{j0} \mid \cdot)$, $j = 1, \ldots, m$. One possibility is to assume that the associated probability functions belong to a parametric family (e.g. the beta distribution or the gamma distribution) and then estimate the associated parameters. For an application to crop yield risk, see Gallagher (1987). A more flexible approach is to estimate the probability/distribution function using non-parametric methods. This can be done using kernel estimation of the probability density function (e.g. Goodwin and Ker, 1998) or using quantile regression applied to the estimation of the distribution function across all quantiles (e.g. Chavas and Shi, 2015). Similarly, either parametric or non-parametric methods can be used to estimate the copula $C[D_{p1}, D_{Y1}; \ldots; D_{pm}, D_{Ym}]$ in Eqn 9.17b (Nelsen, 1999; Joe, 2014). Such approaches can support an empirical analysis of risk exposure in the presence of multiple sources of risk in a multi-output context. Although estimating distribution functions is empirically more challenging than estimating their moments, knowing the joint distribution function $D(p_0, Y_0 \mid x, t, z_a, z_b(\cdot))$ in Eqns 9.17a–9.17b provides all the information needed to assess risk exposure under general conditions.

9.5 Implications for Risk Management in Agriculture

As discussed in the introduction, most farmers face price risk as well as production risk, and they use risk management schemes to try to reduce their risk exposure (Harwood *et al.*, 1999; Just and Pope, 2002; Kimura and Le Thi, 2011). Price risk and production risk tend to be negatively correlated. This is good for risk-averse farmers: the negative correlation means that production shocks are partially absorbed by opposite price adjustments, leading to reduced exposure to revenue risk (Harwood *et al.*, 1999; Kimura and Le Thi, 2011). This negative correlation between price risk and production risk tends to be stronger in local markets or in dominant production regions (e.g. in the US Corn Belt for corn). And it tends to be weaker in global markets or in more marginal production regions.

Market globalization has several effects on revenue risk (Newbery and Stiglitz, 1981). On the one hand, trade can reduce output price risk. For example, allowing imports into a region facing crop failure reduces the effects of production shortfalls on domestic prices. This can generate substantial benefits from global markets. But market globalization also can increase revenue risk in two ways. First, it can increase price risk by exposing a region to price shocks induced by supply/demand shocks in different regions (Newbery and Stiglitz, 1981). Second, it can reduce the negative correlation between price and production. Indeed, increasing the number of producing regions tends to reduce the effects of local production shocks on induced price adjustments. Thus, under globalization, revenue risk may increase as local production shocks become less absorbed by opposite price adjustments.

9.5.1 Diversification

A management option affecting revenue risk is diversification. As noted above, the effectiveness of diversification depends on the correlation/codependence among risky prospects. Negatively correlated risks increase the ability of diversification to reduce risk exposure. Empirical evidence indicates that such correlations vary among commodities, across space and over time (e.g. Harwood *et al.*, 1999; Kimura and Le Thi, 2011).

Most farms are diversified and produce more than one output. In general, farm diversification can help reduce risk exposure. Few farmers are observed to produce a single output as this would expose them to crop failure (e.g. the catastrophic crop failure that led to the Irish potato famine in 1848–1852). As diseases are typically crop-specific, crop diversification helps reduce disease-generated crop damage. In addition, crop rotation can help reduce pest populations, thus decreasing the risk of facing pest damage. In general, farm diversification reduces the cost of risk, thus providing an incentive for farms to produce multiple outputs. Lin *et al.* (1974) showed that most Californian farms are multi-output enterprises because of risk and risk aversion. Chavas and Di Falco (2012) obtained similar results for Ethiopian farms. But the incentive to diversify can vary across agro-ecosystems. For example, diversification incentives appear to be weaker on Asian irrigated farms than non-irrigated farms. On irrigated paddy land, Kim *et al.* (2014) showed that such farms tend to specialize in rice production for two reasons: (i) irrigation reduces weather risk; and (ii) rice is particularly well adapted to irrigated paddy land. Yet, on non-irrigated land, most farms tend to diversify to reduce revenue risk (Lin *et al.*, 1974; Chavas and Di Falco, 2012). Note that the risk reduction can involve both a lower variance and a higher skewness (i.e. a lower exposure to downside risk). Chavas and Di Falco (2012) and Kim *et al.* (2014) found that, when evaluating the effects of diversification on the cost of risk, the skewness effect can be greater than the variance effect. This indicates that a large part of the risk benefit from farm diversification comes from reducing exposure to downside risk.

Diversification can also go beyond the farm. Part-time farmers can rely on off-farm income to support the farm household (as observed among many small farms in developed countries). And remittances are an important source of income for many rural households in developing countries. Having access to off-farm income can help reduce the cost of risk in several ways: (i) off-farm income may be less risky than farm income; (ii) off-farm income is typically not positively correlated with farm income, thus enhancing the risk-reducing effects of off-farm diversification; and (iii) by increasing mean household income, non-farm income increases the wealth of rural households, which contributes to reducing the cost of risk under DARA preferences (e.g. Chavas *et al.*, 2019).

9.5.2 Managing price risk

Agricultural markets typically exhibit large price volatility for two reasons: (i) food supply is subject to unpredictable weather shocks and pest damages; and (ii) the demand for food is very price-inelastic, implying that supply shocks can get translated into large food price swings. Being price-takers, farmers can face significant price risk. But they have several options for managing price risk. One possibility is forward contracting (e.g. forward sale of a crop to a local elevator). Another possibility is hedging in futures and option markets (Garcia and Leuthold, 2004). A third possibility is to improve information about market conditions, leading to better anticipations of price fluctuations. This stresses the importance of having access to good information (including market information collected and disseminated by government agencies).

9.5.3 Managing production risk

Besides diversification, farmers have many management options to reduce production risk. Farm production decisions can have direct impacts on production risk (Harwood *et al.*, 1999). For example, Just and Pope (1979) showed that fertilizer tends to be 'risk increasing': fertilizer use increases the variance of crop yield. Thus, risk-averse farmers have an incentive to reduce the use of risk-increasing inputs. Other inputs are 'risk reducing.' An example is irrigation, which lessens the adverse effects of rainfall shocks on crop yield (Harwood *et al.*, 1999). Another example is veterinary care, which contributes to reducing the incidence of animal diseases and its adverse impact on livestock productivity. A third example is improved soil management (e.g. high organic matter in a soil tends to improve its water-retention capacity, thus reducing the adverse effects of a drought).

Also, some outputs are riskier than others. As documented in Chavas *et al.* (2019), livestock production is less risky than crop production (as livestock output tends to be less sensitive to weather shocks). This explains why regions facing greater weather risk tend to specialize in livestock (e.g. in the Sahel). Also, drought sensitivity varies across crops. For example, wheat is more drought-resistant than corn. In the United States, wheat is grown in the western Great Plains (where drought risk is higher) and corn is grown in the Corn Belt (where rainfall risk is lower). As documented by Chavas and Shi (2015), crop rotations can reduce production risk. Crop rotations help reduce pest populations, thus lessening the unpredictable adverse effects of pest damage on crop yield. Chavas and Shi (2015) found that crop rotations can be particularly effective at reducing exposure to downside risk. In general, risk-averse farmers have an incentive to implement management strategies that reduce their exposure to risk and downside risk.

Another important option for farmers is to rely on insurance to reduce their risk exposure. This option raises two related puzzles: (i) Given the importance of production risk in agriculture, why are we not observing stronger demands for agricultural insurance from farmers around the world? and (ii) Why does agriculture see limited developments in insurance markets without large government subsidies? Resolving these two puzzles remains the subject of active research. Finally, a last option is to improve information about production conditions, leading to improved anticipation and management of production shocks.

9.6 The Role of Technology

Technology also affects production risk in agriculture. Over the past century, agricultural innovations and improved genetics have generated significant increases in yields and agricultural productivity (Fuglie *et al.*, 2019). Notable examples include the case of hybrid corn (Griliches, 1957), the 'green revolution' (Evenson and Gollin, 2003), and the biotechnology revolution (Qaim, 2009). New technologies have also contributed to reducing the unpredictable effects of pest damage and drought on agricultural production.

For a given farmer, a new technology will be adopted if it is perceived to generate positive net benefit. From our analysis, this means an increase in the certainty equivalent *CE* given in Eqns 9.6 or 9.9. In this context, a new technology is attractive to risk-averse farmers (and thus quickly adopted) when it increases mean income and reduces the cost of risk. As discussed, risk concerns include exposure to both risk and downside risk (e.g. the probability of crop failure). The role of risk becomes particularly relevant when a new technology is perceived to be riskier than a traditional technology, in which case highly risk-averse farmers may fail to adopt it even if it increases mean income (Feder *et al.*, 1985; Foster and Rosenzweig, 2010). This stresses that both expected payoff and risk exposure matter in the evaluation of a new technology and its adoption.

Starting in 1996, many US farmers quickly adopted newly developed genetically modified (GM) seeds for corn, soybean and cotton. Using genetic material from other species, genetically modified crops are better able to control pest damages. As a result, such crops exhibit reduced exposure to both risk and downside risk (e.g. the case of Bt corn that helps reduce yield damages from pest infestation by the European corn borer or root worm). Chavas and Shi (2015) and Chavas (2019) documented how genetically modified seeds have shifted the distribution of corn yield in a way that generates a lower cost of risk due in large part to a reduction in exposure to downside risk. Progress has also been made in developing disease-resistant and drought-tolerant varieties. The role of exposure to downside risk in technology adoption has been explored by Di Falco and Chavas (2009) in Ethiopian agriculture and by Emerick *et al.* (2016) in Indian agriculture.

Although technology can help reduce risk exposure, the outcome of a new technology is typically not perfectly known before it is adopted. Thus, technology adoption always involves some uncertainty (Feder *et al.*, 1985). For a farmer, two types of uncertainty are relevant: (i) Is the new technology adapted to local agro-climatic conditions? and (ii) Does the farmer know how to manage it? For risk-averse farmers, such uncertainties adversely affect technology adoption (Feder *et al.*, 1985). To the extent that risk aversion varies across farmers, this generates the prediction that early adopters would be among farmers exhibiting lower risk aversion, and late adopters would be among those exhibiting higher risk aversion. But two other aspects of technology adoption are also relevant. First, information about a new technology typically varies across individuals (Rogers, 2003). Barham *et al.* (2014a) documented how the ability to learn varies significantly among US farmers. In this context, we expect that the early adopters would be among the quick learners, and the late adopters among the slow learners. Second, the perception of risk depends on how each farmer obtains information about a new technology.

This information can come from multiple sources: from on-farm experimentation (using learning by doing), from seed companies, from extension services, and/or from social networks (including other farmers). On-farm experimentation has been practiced since the beginning of agriculture. Over the past century, most agricultural innovations have come from sources outside the farm (e.g. from geneticists working in research institutions or biotech firms). Extension services have played a role in informing farmers about new technology options (e.g. Jin and Huffman, 2016). The role of social interactions has also been documented (Besley and Case, 1993; Conley and Udry, 2010; Maertens, 2017; Crane-Droesch, 2018). Social learning can help stimulate farm technological adoptions, but its effects are often heterogeneous. This is due to two factors: (i) Each technology may work well only under specific agro-climatic conditions (including rainfall potential and soil fertility); and (ii) Individuals differ in their access to existing social networks and in their ability to learn from them. Early adopters are likely to be farmers who are good at obtaining and processing information from others (Barham *et al.*, 2018). Alternatively, farmers who are less effective at individual learning are more likely to be late adopters; but they can learn from others (including early adopters). The evidence indicates that social networks play an important role in stimulating technology adoption (e.g. BenYishay and Mobarak, 2019).

9.7 Conclusion

This chapter has examined the economics of risk in agriculture, with a focus on the evaluation of the cost of risk under both price risk and production risk. This focus is strengthened by current concerns about climate change and its effects on agricultural risk. Special attention was given to the role of aversion to risk and downside risk (e.g. drought). We discussed the empirical methods available to conduct risk assessment. We explored the implications for farm risk management and technology. The analysis emphasized the many options available to farmers in managing their risk exposure. These include farm diversification (including crop rotation), the use of risk-reducing inputs (such as irrigation), and the adoption of technology that helps reduce risk exposure (such as the adoption of pest-resistant or drought-tolerant varieties).

Our analysis has focused on risk management options at the farm level. First, there is a need to expand the analysis in a dynamic framework and at the market level. Second, the linkages between farm production risk, environmental management and climate change need further investigations. Third, market and policy options are available to manage risk beyond the farm. They include insurance and farm policy. Agricultural policy has played a major role in reducing farm risk, including price support programs that attempt to reduce price risk (e.g. in China), subsidies to farm insurance (e.g. in the United States) and direct income support (e.g. in the European Union). The interactions between farm risk management and farm policy remain imperfectly understood. These issues need further investigations and appear to be good topics for future research.

References

Antle, J.M. (1983) Testing the stochastic structure of production: a flexible moment-based approach. *Journal of Business & Economic Statistics* 1, 192–201.

Antle, J.M. (1987) Econometric estimation of producers' risk attitudes. *American Journal of Agricultural Economics* 69, 509–522.

Antle, J.M. (2010) Asymmetry, partial moments, and production risk. *American Journal of Agricultural Economics* 92, 1294–1309.

Antle, J.M. and Goodger W.J. (1984) Measuring stochastic technology: the case of Tulare milk production. *American Journal of Agricultural Economics* 66, 342–350.

Barberis, N.C. (2013) Thirty years of prospect theory in economics: a review and assessment. *Journal of Economic Perspectives* 27, 173–196.

Barham, B.L., Chavas, J.P., Fitz, D., Rios Salas, V. and Schechter, L. (2014a) Risk, learning, and technology adoption. *Agricultural Economics* 45, 1–14.

Barham, B., Chavas, J.P., Fitz, D., Ríos Salas, V. and Schechter, L. (2014b) The roles of risk and ambiguity in technology adoption. *Journal of Economic Behavior and Organization* 97, 204–218.

Barham, B.L., Chavas, J.P., Fitz, D. and Schechter, L. (2018). Receptiveness to advice, cognitive ability and technology adoption. *Journal of Economic Behavior and Organization* 149, 239–268.

Barseghyan, L., Molinari, F., O'Donoghue, T. and Teitelbaum, J.C. (2018) Estimating risk preferences in the field. *Journal of Economic Literature* 56, 501–564.

BenYishay, A. and Mobarak, A.M. (2019) Social learning and incentives for experimentation and communication. *Review of Economic Studies* 86, 976–1009.
Besley, T. and Case, A. (1993) Modelling technology adoption in developing countries. *American Economic Review* 83, 396–402.
Binswanger, H.P. (1981) Attitudes toward risk: theoretical implications of an experiment in rural India. *Economic Journal* 91, 867–890.
Bocquého, G., Jacquet, F. and Reynaud, A. (2014) Expected utility or prospect theory maximizers? Assessing farmers' risk behaviour from field-experiment data. *European Review of Agricultural Economics* 41, 135–172.
Chambers, R.G. and Quiggin, J. (2000) *Uncertainty, Production, Choice and Agency: The State-Contingent Approach*. Cambridge University Press, New York.
Chambers, G.G., Chung, Y. and Färe, R. (1996) Benefit and distance functions. *Journal of Economic Theory* 70, 407–419.
Chavas, J.P. (2004) *Risk Analysis in Theory and Practice*. Elsevier, New York.
Chavas, J.P. (2008) A cost approach to economic analysis under state-contingent production uncertainty. *American Journal of Agricultural Economics* 90, 435–446.
Chavas, J.P. (2019) Adverse shocks in agriculture: the assessment and management of downside risk. *Journal of Agricultural Economics* 70, 731–748.
Chavas, J.P. and Briec, W. (2019). Efficiency under uncertainty and non-convexity: evaluating the role of probabilities. *Economica* 86, 832–853.
Chavas, J.P. and Di Falco S. (2012) On the role of risk versus economies of scope in farm diversification with an application to Ethiopian farms. *Journal of Agricultural Economics* 63, 25–55.
Chavas, J.P. and Holt, M.T. (1996) Economic behavior under uncertainty: a joint analysis of risk preferences and technology. *Review of Economics and Statistics* 78, 329–335.
Chavas, J.P. and Shi, G. (2015) An economic analysis of risk, management and agricultural technology. *Journal of Agricultural and Resource Economics* 40, 63–79.
Chavas, J.P., Cooper, J. and Wallander, S. (2019) The impact of input and output decisions on agricultural production risk. *Journal of Agricultural and Resource Economics* 44, 513–535.
Conley, T. and Udry, C. (2010) Learning about a new technology: pineapple in Ghana. *American Economic Review* 100, 35–69.
Crane-Droesch, A. (2018) Technology diffusion, outcome variability, and social learning: evidence from a field experiment in Kenya. *American Journal of Agricultural Economics* 100, 955–974.
Debreu, G. (1959) *Theory of Value*. Cowles Foundation, Yale University Press, New Haven, Connecticut.
Di Falco, S. and Chavas, J.P. (2009) On crop biodiversity, risk exposure and food security in the highlands of Ethiopia. *American Journal of Agricultural Economics* 91, 599–611.
Dillon, J. and Scandizzo, P. (1978) Risk attitudes of subsistence farmers in Northeast Brazil: a sampling approach. *American Journal of Agricultural Economics* 60, 425–435.
Dohmen, T., Falk, A., Huffman, D., Sunde, U., Schupp, J. and Wagner, G.G. (2011) Individual risk attitudes: measurement, determinants and behavioral consequences. *Journal of the European Economic Association* 9, 522–550.
Emerick, K., de Janvry, A., Sadoulet, E. and Dar, M.H. (2016) Technological innovations, downside risk, and the modernization of agriculture. *American Economic Review* 106, 1537–1561.
Evenson, R.E. and Gollin, D. (2003) Assessing the impact of the green revolution, 1960 to 2000. *Science* 300, 758–762.
Feder, G., Just, R.E. and D. Zilberman. (1985) Adoption of agricultural innovations in developing countries: a survey. *Economic Development and Cultural Change* 30, 59–76.
Foster, A.D. and Rosenzweig, M.R. (2010) Microeconomics of technology adoption. *Annual Review of Economics* 2, 395–424.
Fuglie, K., Gautam, M., Goyal, A. and Maloney, W. (2019) *Harvesting Prosperity: Technology and Productivity Growth in Agriculture*. World Bank, Washington, DC.
Gallagher, P. (1987) U.S. soybean yields: estimation and forecasting with nonsymmetric disturbances. *American Journal of Agricultural Economics* 69, 796–803.
Garcia, P. and Leuthold, R.M. (2004) A selected review of agricultural commodity futures and options markets. *European Review of Agricultural Economics* 31, 235–272.
Gollier, C. (2001) *The Economics of Risk and Time*. MIT Press, Cambridge.
Goodwin, B.K. and Ker, A.P. (1998) Nonparametric estimation of crop yield distributions: implications for rating group-risk crop insurance contracts. *American Journal of Agricultural Economics* 80, 139–153.

Griliches, Z. (1957) Hybrid corn: an exploration in the economics of technology change. *Econometrica* 25, 501–522.

Halek, M. and Eisenhauer J.G. (2001) Demography of risk aversion. *Journal of Risk and Insurance* 68, 1–24.

Halevy, Y. (2007) Ellsberg revisited: an experimental study. *Econometrica* 75, 503–536.

Harwood, J., Heifner, R., Coble, K., Perry, J. and Somwaru, A. (1999) *Managing Risk in Farming: Concepts, Research, and Analysis*. Agricultural Economic Report No. 774, Economic Research Service, U.S. Department of Agriculture, Washington, DC, 125 pp.

Holt, C.A. and Laury, S.K. (2002) Risk aversion and incentive effects. *American Economic Review* 92, 1644–1655.

Jin, Y. and Huffman, W. (2016) Measuring public research and extension and estimating their impacts on agricultural productivity: new insights from U.S. evidence. *Agricultural Economics* 47, 15–31.

Joe, H. (2014) *Dependence Modeling with Copulas*. Chapman and Hall, CRC Press, New York.

Just, R.E. and Pope, R.D. (1979) Production function estimation and related risk considerations. *American Journal of Agricultural Economics* 61, 276–284.

Just, R.E. and Pope, R.D. (2002) *A Comprehensive Assessment of the Role of Risk in U.S. Agriculture*. Kluwer Academic Publishers, Boston, Massachusetts.

Kahneman, D. and Tversky, A. (1979) Prospect theory: an analysis of decision under risk. *Econometrica* 47, 263–291.

Kim, K., Chavas, J.P., Barham, B. and Foltz, J. (2014) Rice, irrigation and downside risk: a quantile analysis of risk exposure and mitigation on Korean farms. *European Review of Agricultural Economics* 41, 775–815.

Kimura, S. and Le Thi, C. (2011) Farm Level Analysis of Risk and Risk Management Strategies and Policies. OECD, Food, Agriculture and Fisheries Papers, No. 48. OECD Publishing, Paris.

Knight, F.H. (1921) *Risk, Uncertainty and Profit*. Houghton Mifflin, Boston, Massachusetts.

Lin, W., Dean, G. and Moore, C. (1974) An empirical test of utility vs. profit maximization in agricultural production. *American Journal of Agricultural Economics* 56, 497–508.

Liu, E.M. (2013) Time to change what to sow: risk preferences and technology adoption decisions of cotton farmers in China. *Review of Economics and Statistics* 95, 1386–1403.

Maertens, A. (2017) Who cares what others think (or do)? Social learning, social pressures, and cotton farming in India. *American Journal of Agricultural Economics* 99, 988–1007.

Meyer, J. (1987) Two-moment decision models and expected utility. *American Economic Review* 77, 326–336.

Nelsen, R.B. (1999) *An Introduction to Copulas*. Springer, New York.

Newbery, D.M. and Stiglitz, J.E. (1981) *The Theory of Commodity Price Stabilization: A Study in the Economics of Risk*. Clarendon Press, Oxford, UK.

Pratt, J.W. (1964) Risk aversion in the small and in the large. *Econometrica* 32, 122–136.

Prelec, D. (1998) The probability weighting function. *Econometrica* 66, 497–527.

Qaim, M. (2009) The economics of genetically modified crops. *Annual Review of Resource Economics* 1, 665–693.

Quiggin, J. (1993) *Generalized Expected Utility Theory: The Rank-Dependent Expected Utility Model*. Kluwer-Nijhoff, Amsterdam.

Rogers, E.M. (2003) *Diffusion of Innovations*, 5th edn. The Free Press, New York.

Tversky, A. and Kahneman, D. (1992) Advances in prospect theory: cumulative representation of uncertainty. *Journal of Risk and Uncertainty* 5, 297–323.

10 Farmer Risk Preferences and Willingness to Pay for African Rice Landrace Seed: An Experimental Choice Analysis

Aminou Arouna,[1]* Nicholas Tyack[2] and Rachidi Aboudou[1]
[1]*Africa Rice Center (AfricaRice), Abidjan, Côte d'Ivoire;* [2]*University of Saskatchewan, Saskatoon, Canada*

Abstract

Most literature that investigates the impact of farmer risk preferences on agricultural technology adoption has focused on how risk preferences impact the adoption of technologies that enhance productivity but that may be riskier than local technologies. In this work, we elicit the risk preferences of smallholder rice farmers and investigate how these preferences impact how they value having access to seed of farmers' varieties of African rice (*Oryza glaberrima*) maintained in the gene bank of the Rice Biodiversity Center for Africa. These varieties may have lower yields, but they may be more consistent and less risky given their adaptation to the local climatic conditions. We use a Becker–DeGroot–Marschak (BDM) mechanism to elicit farmers' willingness to pay for small amounts of landrace seed (35 g) in Côte d'Ivoire. We find that farmers generally value having access to African rice landraces (with a mean willingness to pay of ~US$0.50), and that this willingness to pay is influenced positively by loss-aversion preferences (but not risk aversion), along with several other factors. This finding is in contrast with past evidence suggesting that loss aversion is connected to slower adoption of novel technologies, and suggests that the impact of risk preferences on technology adoption may depend on the potential ability of technologies to guard against economic losses.

10.1 Introduction

Most literature that investigates the impact that farmers' risk preferences have on their adoption of agricultural technology has focused on how risk preferences impact the adoption of new technologies that enhance productivity (and in some cases, may be riskier than traditional technologies). For example, Liu (2013) finds that Chinese farmers who are more risk- or loss-averse adopted genetically modified Bt cotton varieties later than those who are less risk- or loss- averse. Similarly, in a gender-disaggregated analysis, Magnan *et al.* (2020) find that women's risk aversion and men's loss aversion are negatively correlated with use of improved varieties.

Understanding how risk preferences affect the demand for the seed of varieties that may be more resilient is of interest, given the threat that climate risk poses to smallholder farmers in Africa and elsewhere, and the way these

*E-mail: a.arouna@cgiar.org

 Managing Risk in Agriculture: A Development Perspective (eds A.K. Mishra, S.C. Kumbhakar and G. Lien).
DOI: 10.1079/9781800622289.0010

preferences impact adoption may not be the same as for technologies that do not protect against climate risks. For example, Holden and Quiggin (2017) find that risk aversion positively influenced the likelihood that households adopted drought-tolerant corn varieties but made them less likely to adopt other improved corn varieties.

Although the bulk of the literature examining the role that farmers' risk preferences play in the adoption of new crop varieties has focused on improved varieties, here we analyze how risk preferences impact farmers' demand for landrace seed of African rice (*Oryza glaberrima*), a set of rice varieties that may be more resilient under climatic stress. We elicit the risk preferences of smallholder rice farmers in Côte d'Ivoire and investigate how these preferences impact how they value having access to the seed of farmers' varieties of African rice (*Oryza glaberrima*) maintained in the gene bank of the Rice Biodiversity Center for Africa (RBCA). We contribute to a literature that has focused on how risk preferences may impact the adoption of risk-reducing technologies (Emerick *et al.* 2016; Jin *et al.* 2020), an area of inquiry that will be of particular importance in implementing measures adapting to climate change (e.g. through the promotion of more climate-resilient crop varieties).

We formulate the following two related hypotheses: First, that farmers who are more risk-averse may have a higher demand for seed of African rice landraces, which may be more adapted to growing conditions in West Africa, and second, that farmers who are more loss-averse may also have a higher demand for seed of African rice landraces. These hypotheses are informed by past work that has shown, for example, that risk- and loss-averse individuals may be more likely than individuals who can accept risk and loss to adopt crop varieties demonstrating risk-reducing characteristics (Ward and Singh, 2015; Holden and Quiggin, 2017; Lampe and Würtenberger, 2020). We test these hypotheses by connecting risk preference data collected using the series of questions following Tanaka *et al.* (2010), along with farmers' willingness to pay (WTP) for landrace seed elicited using a Becker–DeGroot–Marschak experiment (Becker *et al.*, 1964).

10.2 Context

10.2.1 Background

AfricaRice manages one of the world's largest collections of the typically low-yielding but drought-tolerant African rice (*Oryza glaberrima*), as well as traditional African landraces of *Oryza sativa* (introduced in the 1500s). Domesticated in West Africa about 3000 years ago and indigenous to the continent, African rice has declined in cultivation and has been replaced in farmers' fields in many cases by the higher-yielding Asian rice (*Oryza sativa*). However, African rice's genetic resources are of interest due to numerous valuable traits in the gene pool, such as tolerance to iron and aluminum toxicity, drought and soil acidity, as well as other qualitative traits related to consumption preferences, such as taste, aroma and cooking qualities (Wang *et al.*, 2014). In particular, varieties of *Oryza glaberrima* are considered more adapted to the climate of West Africa and may offer some protection against severe crop failures under drought conditions.

10.2.2 Institutional setting

This research was carried out in the Vallée du Bandama region, where AfricaRice's main research station is based. AfricaRice is involved in both conserving and distributing tens of thousands of rice accessions as well as contributing to the development of new rice varieties in collaboration with the Africa-wide Rice Breeding Taskforce, which includes AfricaRice breeders and national rice breeders. AfricaRice's gene bank, the RBCA, maintains the world's largest collection of African rice landraces and is the largest collection of rice varieties in Africa, with a total of almost 22,000 accessions. The collection represents a strategic resource for breeding new, advanced rice varieties that are well adapted to growing conditions across Africa. The gene bank also distributes samples widely, with a total of 113,083 samples distributed over the past 25 years to a total of 164 institutions across 57 countries.

10.3 Methodology

This section describes our sample selection and data collection procedures, the methodology used to measure farmers' willingness to pay for African rice landrace seed and how we elicited the risk preferences of the rice farmers in our sample.

10.3.1 Data collection and sample selection

A recently published 'e-registration' of about 8000 rice farmers in Côte d'Ivoire was used to plan the sampling strategy, with a split between more connected villages located closer to the city of Bouaké and roads and villages that were more isolated (Arouna and Aboudou, 2020). Villages with a higher ratio of female-to-male farmers also were targeted, as well as villages reflecting different types of growing ecology (upland, irrigated lowland and rain-fed lowland). Figure 10.1 illustrates the villages' locations on a map.

The dataset includes a total of 569 farmers across 12 villages (six in Hambol, six in Gbeke), with the number of rice farmers per village ranging from 13 rice farmers in Ouanan to 90 rice farmers in Nassoulo. We aimed to survey as many farmers as possible given budgetary constraints and to cover a diverse set of farmers through our sample selection.

An ethics approval was obtained at the Graduate Institute of International and Development Studies following the institution's internal regulations and respecting measures taken by AfricaRice. Safety precautions in response to COVID-19, including social distancing measures, were taken during surveying activities, which were conducted using electronic tablets in outside locations. Masks and disinfectant were purchased for enumerators. Travel occurred strictly within the Vallée du Bandama

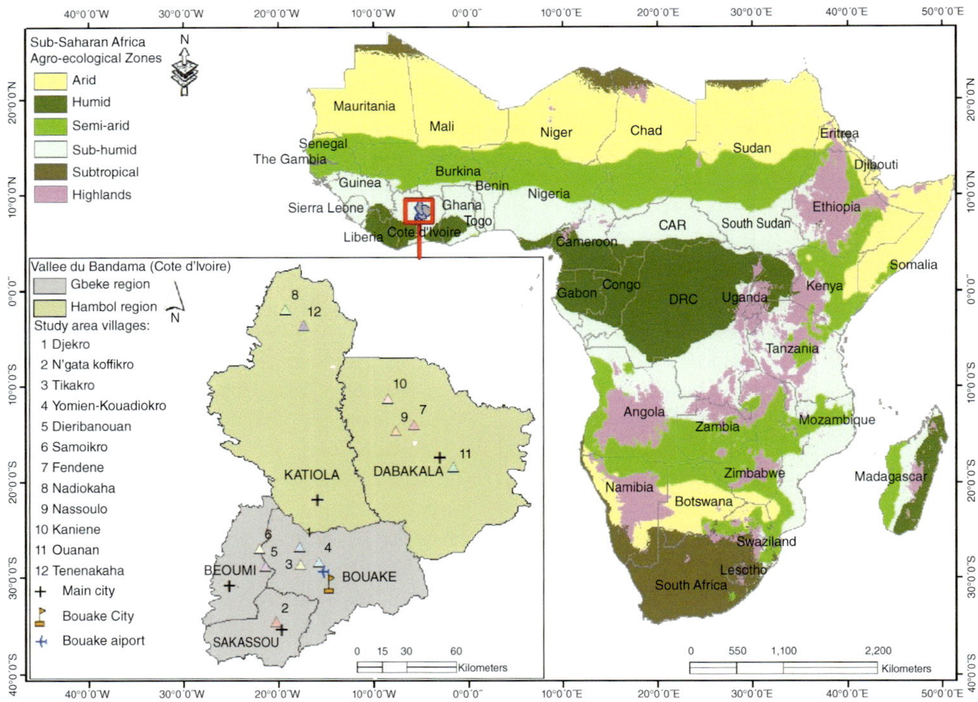

Fig. 10.1. Map of the study villages. Source: Authors' design, 2022.

area around Bouaké with local staff in small groups. Farmers were asked for their consent before being surveyed, with none being forced to participate. During the two periods of data collection, the epidemiological situation was stable and at a low-risk level (around 0.1 cases per 100,000 per day), with active cases clustered in the greater Abidjan area. Survey enumerators were first trained in Bouaké, and then through a pilot conducted before the implementation of the main wave in a village near Bouaké.

10.3.2 The game – Becker–DeGroot–Marschak elicitation of use value

The experimental BDM elicitation of use value took place after several introductory sections of the survey questionnaire. A number of valuation techniques exist, including stated preference techniques (which use respondents' stated willingness to pay for a hypothetical good or service, in which the good or service is not provided and no funds change hands) as well as incentive-compatible experiments (which simulate a hypothetical market but in which consumers face real budget constraints and have the opportunity to physically obtain products). Different approaches may lead to different estimates of willingness to pay (Hamukwala *et al.*, 2018). The Becker–DeGroot–Marschak approach has the advantage in that it avoids the hypothetical bias of stated preference approaches (that is, BDM experiments lead to funds and products changing hands), and thus can be a more rigorous methodology for approximating demand and willingness to pay.

Farmers were randomly assigned to start the experiment with the opportunity of winning either a bag of African rice landrace seed or a bag of advanced rice varieties for Africa (ARICA) seed. The landrace seed was described as:

> A traditional variety (landrace) of African rice (glaberrima) adapted to the ecology of your parcel. The traditional variety is a variety of African rice, not Asian, that was domesticated originally in Africa during the course of the past centuries and that could be better adapted to local conditions.

None of the farmers in our dataset reported that they currently cultivate varieties of African rice. African rice (*Oryza glaberrima*) is rarely cultivated in Africa (present in only a few locations, such as in Senegal and Togo) and thus can be considered novel to most farmers (that is, a market for *Oryza glaberrima* seed or grain can be considered non-existent).

First, farmers were prompted through a ladder elicitation method to think about how much they would be willing to pay for the 35 g bag of seed. A random value of the Franc of the Financial Community of Africa (FCFA) was pulled, and they were asked, 'Would you rather have ___ FCFA or the seed?' If they answered that they would rather have the seed, then the enumerators were instructed to ask the same question but add 25 FCFA to the amount of money. If they answered that they would rather have the money, the enumerators were instructed to ask the same question but subtract 25 FCFA from the previous stated amount of money. This exercise ended once the farmers switched their answers (from seed to money if they originally favored the seed, and from money to seed if they originally preferred the money). At the end of this exercise, they were asked how much they would be willing to pay for the seed (not restricted by the increments of 25 FCFA or the maximum value of 550 FCFA).

After this, the BDM-elicitation dice game was explained to the farmers. The farmers were asked to pick a value they were willing to pay for the bag of landrace or ARICA seed out of the total amount of 550 FCFA offered by the enumerator, associated with a roll of the dice (2 to 12, a potential value resulting from the sum of the two dice rolled) or with zero, if the farmers were not willing to pay anything for the variety. Note that the stake size (approximately US$1) is not large but was deemed to be of economic significance given that the sum was sufficient to purchase 1–2 kg of milled rice in the setting of the villages where the experiment took place, and in the context of large, sometimes food-insecure households. The sum also was large in relation to the small quantity of seed provided. Then, the enumerator explained that if the dice roll was greater than the roll associated with the value the farmers had selected, they would not receive the seed, but rather the full sum of 550 FCFA (approximately US$1). If the dice roll was equal to or less than the value they had selected, they would receive the seed and the 550 minus the value associated with the roll of the dice (and not their willingness-to-pay value). The values used to correspond to dice rolls of 2 to 12 were as follows: 25, 50, 75, 100, 150, 225, 300, 375, 450, 500, 550.

To further help the farmers understand the game, a hypothetical first round was played for a 1 kg bag of fertilizer (though no money or fertilizer changed hands as a result of this practice round). After this practice round, the dice game was played and the farmers received the combination of money and/or seed resulting from the game, before moving to the next seed type (ARICA or landrace).

10.3.3 Elicitation of risk preferences

We elicited farmers' risk preferences using a non-incentivized series of questions that asked farmers to state whether they preferred a more risky or less risky rice variety. Our approach simulated the same set of questions asked in Series 12, and 3 from Tanaka *et al.* (2010). We thus directly tried to simulate farmer decision-making when considering whether to adopt a new variety.

First, for Series 1 and 2, farmers were asked:

> Imagine you are deciding between whether to cultivate two different varieties of rice. One type of rice sometimes will provide very high yields but only if conditions are right. The second type produces more consistent yields year to year. We would like to understand which type of rice you would prefer in different scenarios. The more consistent type of rice will be described first and the less consistent type will be described second.

Farmers were asked to select either Variety A or Variety B for a series of scenarios (the probability of profits from the harvest in FCFA) until they switched from Variety A (the more consistent variety) to Variety B (the 'switching point'). After the farmer switched from Variety A to Variety B, the enumerator moved to the next series of questions (that is, monotonic switching was enforced).

Then, for Series 3 (designed to elicit loss-aversion preferences), farmers were asked a similar set of questions, except with the potential to also lose money (for example, the loss of money spent on agricultural inputs when the harvest failed). Both varieties A and B would lead to losses 50% of the time, but with variety A leading to lower losses (as well as lower gains).

The scenarios are represented in Table 10.1 along with the expected payoff difference in FCFA in the fifth column.

10.3.4 Description of sample

Within our sample of farmers, about one-fifth were females and four-fifths were male. The average age was 43, most farmers (71%) had no education, and 23% had just primary education. The sample was split roughly between the Gbeke and Hambol regions of Côte d'Ivoire. Eighty percent of the sample possessed a phone. On average, farmers were part of a household with approximately 13 members. Farmers on average cultivated rice on about 1 ha of land. About 40% of the sample cultivated rain-fed lowland plots, just under 30% cultivated irrigated lowland plots, and 28% cultivated upland plots.

10.3.5 Estimation strategy

This section describes the estimation strategy of the analysis. We run three regressions, all of which take the BDM-elicited willingness to pay for landrace seed as the dependent variable. These regressions are all run using the ordinary least squares (OLS) method with robust standard errors.

The first set of regressions we run can be simply expressed as in Eqn 10.1. On the left-hand side, the dependent variable is defined as the BDM-elicited willingness to pay for landrace or ARICA seed (in FCFA) of farmer i in village v.

$$WTP_{ive} = \alpha + \beta_i * x + \mu_v + \lambda_e + \varepsilon \qquad (10.1)$$

Here, α is a constant, x is a vector of variables that theoretically may have a significant impact on WTP (including risk preferences), μ_v is a set of village controls, λ_e is a set of enumerator controls (to control for any effect the individual enumerators may have had on WTP), and ε is the error term. The first regression includes our measure of risk aversion, σ, the second our measure of loss aversion, λ, and the third includes both.

Table 10.1. Description of scenarios offered between risky and less risky varieties for Series 1 and 3 (and payoffs in FCFA). (Source: Authors' own work; adapted from Tanaka *et al.* (2010).)

Variety A		Variety B		Expected payoff difference between A and B	Switching point
Series 1					
30%	70%	10%	90%		
40,000	10,000	68,000	5,000	7,700	1
40,000	10,000	75,000	5,000	7,000	2
40,000	10,000	83,000	5,000	6,000	3
40,000	10,000	93,000	5,000	5,200	4
40,000	10,000	106,000	5,000	3,900	5
40,000	10,000	125,000	5,000	2,000	6
40,000	10,000	150,000	5,000	−500	7
40,000	10,000	185,000	5,000	−4,000	8
40,000	10,000	220,000	5,000	−7,500	9
40,000	10,000	300,000	5,000	−15,500	10
40,000	10,000	400,000	5,000	−25,500	11
40,000	10,000	600,000	5,000	−45,500	12
40,000	10,000	1,000,000	5,000	−85,500	13
40,000	10,000	1,700,000	5,000	−155,500	14
Series 3					
50%	50%	50%	50%		
25,000	−4,000	30,000	−21,000	6,000	1
4,000	−4,000	30,000	−21,000	−4,500	2
1,000	−4,000	30,000	−21,000	−6,000	3
1,000	−4,000	30,000	−16,000	−8,500	4
1,000	−8,000	30,000	−16,000	−10,500	5
1,000	−8,000	30,000	−14,000	−11,500	6
1,000	−8,000	30,000	−11,000	−13,000	7

10.4 Results and discussion

We present our main results in this section, consisting of a summary of farmers' willingness to pay for African rice landrace seed, the results of the risk preference elicitation, and finally our main regression results, which investigate the impact of risk preferences on farmers' demand for landrace seed.

10.4.1 Willingness to pay for African rice landrace seed

On average, we find that farmers were willing to pay 257 FCFA (about $0.50) for the 35 g bag of landrace seed. About 60% of farmers received landrace seed, and 40% did not. Figure 10.2 presents a histogram of farmers' willingness to pay for landrace seed. Over eighty farmers, or about 14% of the sample, were not willing to pay for African rice landrace seed.

10.4.2 Distribution of farmers' risk preferences

Figure 10.3 presents a histogram of farmers' risk preferences, from Tanaka *et al.*'s (2010) Series 1 of questions. Most farmers switched from Variety A to Variety B at Question 7, the switching point at which the expected payoff difference between Variety A and Variety B becomes negative. Farmers who switched earlier (for switching point Questions 1–6, when the expected payoff of the more consistent Variety A was still higher than that of Variety B) can be considered those willing to pay a price in terms of expected payoff in order to have the chance to achieve a higher payoff (in 10% of growing seasons).

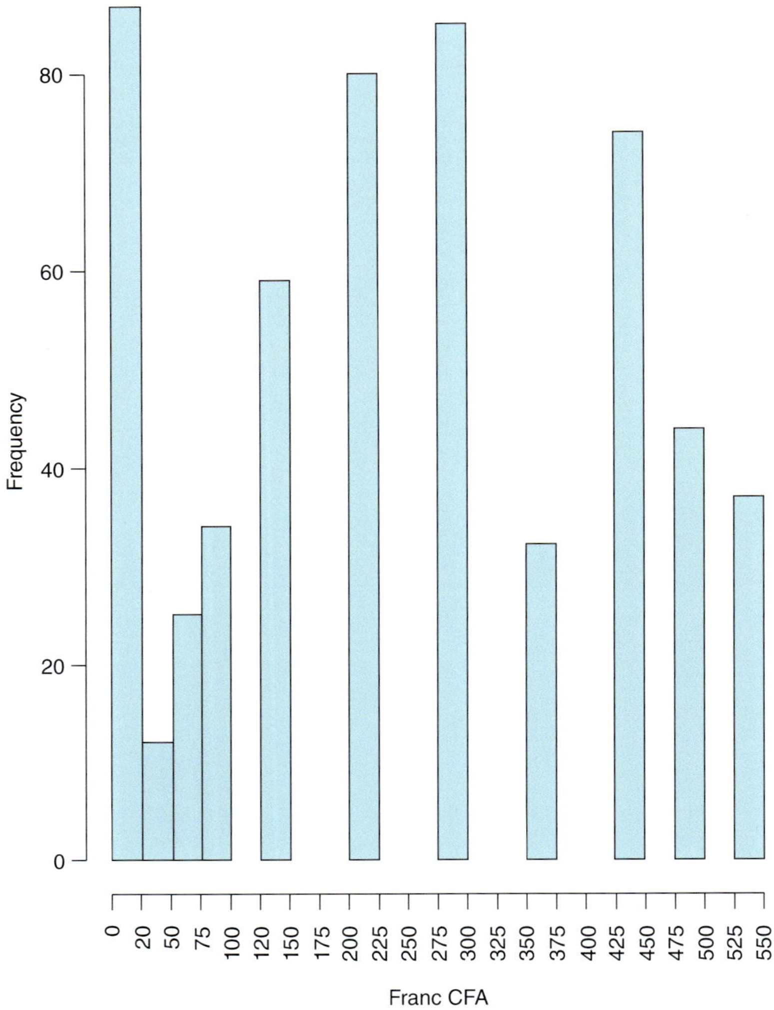

Fig. 10.2. Histogram of farmer willingness to pay (WTP) for landrace seed. Source: Authors' own work.

Those who switched later (switching point Questions 8 through 14) can be considered those willing to pay a price in terms of expected payoff to avoid a lower payoff of 5000 FCFA.

We also present a histogram of farmers' responses to the questions in Series 3 (designed to elicit loss-aversion preferences; Fig. 10.4). We find that most farmers (379) switch at Question 2, while fewer switch at successive questions, with 96 farmers switching at Question 3 and 32 farmers switching at later questions. Sixty-two farmers switch at Question 1.

10.4.3 Impact of risk preferences on farmers' demand for landrace seed

Last, we run three OLS regressions (with robust standard errors) that analyze the impact of risk- and loss-aversion preferences on farmers' willingness to pay for a 35 g bag of African rice landrace seed. For risk-aversion preferences, we use σ, the value function curvature parameter calculated following Tanaka *et al.* (2010). For loss-aversion preferences, we use λ, the loss-aversion parameter approximated following

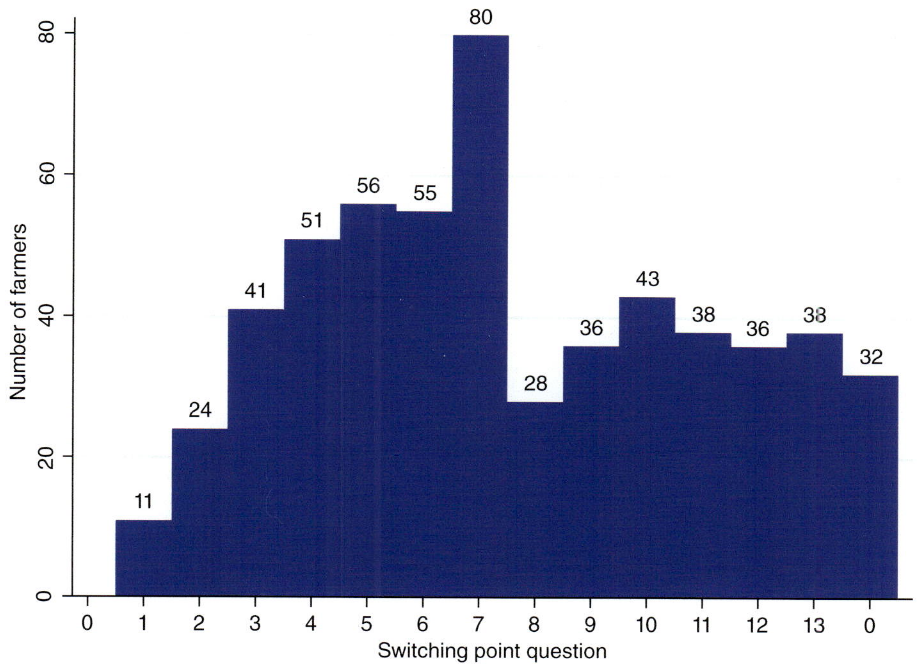

Fig. 10.3. Histogram of selected farmer switching points from Series 1. Source: Authors' own work.

Tanaka *et al.* (2010) and the midpoint technique. Table 10.2 presents our primary results from the three regressions.

We do not find any evidence that farmers' risk-aversion preferences had any significant impact on their willingness to pay for African rice landrace seed (not supporting our first hypothesis, that risk-aversion preferences would positively influence willingness to pay for landrace seed). However, we do find evidence that loss-aversion preferences were positively associated with willingness to pay for African rice landrace seed, with an increase in λ of 1 being associated with a higher willingness to pay of ~41 FCFA. The loss-aversion result is in line with the finding of Lampe and Würtenberger (2020); however, our null finding for the impact of risk aversion on willingness to pay for African rice landrace seed contrasts with the finding of other studies such as Liu (2013) and Holden and Quiggin (2017), who find that risk aversion did have a significant impact on adoption.

We additionally find that several other variables have a significant impact on farmers' willingness to pay for landrace seed. Older farmers and upland farmers were willing to pay less for landrace seed than younger farmers and those in other regions. On the other hand, farmers cultivating more than one crop, who believed their current rice variety to be better than average, and farmers who experienced better-than-average climatic conditions in the past growing season were willing to pay significantly more for African rice landrace seed than farmers who cultivated only one crop and farmers who experienced worse-than-average climatic conditions.

10.5 Conclusion

In this chapter, we use a Becker–DeGroot–Marschak experiment to elicit farmers' willingness to pay for a small amount of African rice (*Oryza glaberrima*) landrace seed, coupled with a (non-incentivized) series of questions related to risk preferences to demonstrate a positive connection between loss aversion and demand for a small quantity of a potentially more adaptive and resilient crop variety. We find no significant relationship between risk aversion and demand for landrace seed. This runs contrary to past evidence that suggests that loss aversion is

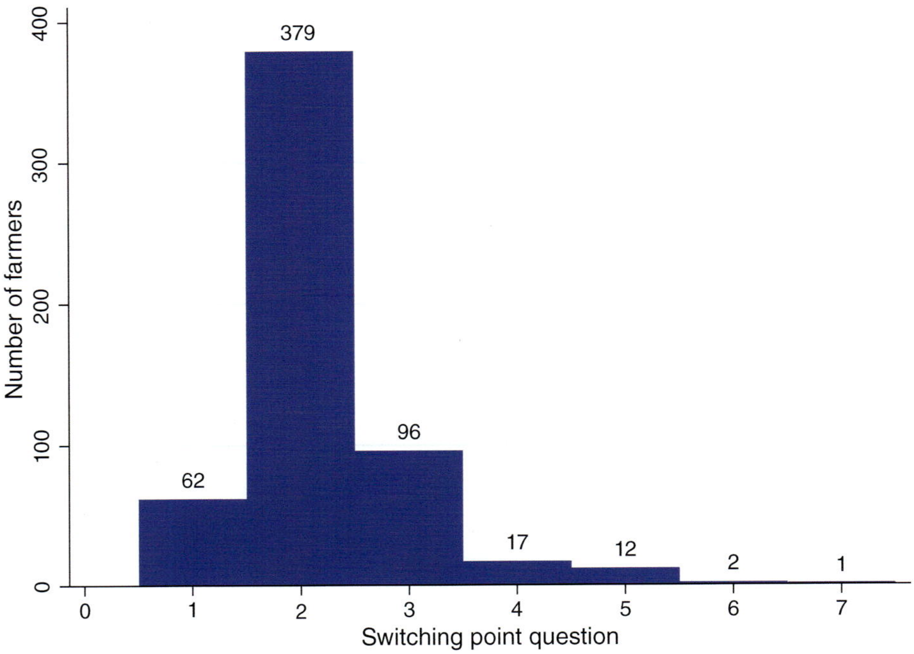

Fig. 10.4. Histogram of selected farmer switching points from Series 3. Source: Authors' own work.

connected to slower adoption of novel technologies (Liu, 2013) and suggests that the impact of risk preferences on technology adoption may depend on the type of technology and its potential ability to guard against losses.

On the other hand, we also find that several other variables play an important role in influencing the demand for African rice landrace seed by the farmers in our sample. For example, we find that upland rice farmers, who are more vulnerable to climatic shocks, were willing to pay less for small amounts of landrace seed (which may perform better in the absence of irrigation) than other farmers. In addition, we find that farmers who experienced better climatic conditions in the previous growing season were also willing to pay more for African rice landrace seed than farmers who experienced poorer climatic conditions. Thus, we provide some evidence that farmers who are more affected by negative climatic shocks may be less likely than less affected farmers to experiment with new varieties that may help protect them against these risks.

There were some limitations to our study. Although the BDM-elicitation of farmers' willingness to pay was incentivized, our risk elicitation did not involve real stakes and thus may have suffered from hypothetical bias. Further work could extend this study by repeating our experiment but with real stakes in the risk elicitation. In addition, our BDM experiment elicited farmer interest for only a very small quantity of seed (suitable only for experimentation). Future studies could investigate how risk aversion and loss aversion impact adoption (not just experimentation) over time.

In coming decades, when climatic shocks and other stresses will become more common and intense, it will become more important to understand how risk preferences and other factors affect smallholder farmers' willingness to experiment with agricultural technologies that provide some downside protection against economic losses. Such information could be valuable for policymakers or extension agents working to promote the adoption of climate-resilient varieties, as depending on their preferences, a different set of farmers may be more willing to adopt new crop varieties (or other technologies/practices) that protect against climate shocks than technologies, such as improved varieties, that provide higher yields but do not protect against shocks such as drought, flood, or pests and diseases.

Table 10.2. Results for impact of risk preferences on WTP for landrace seed (in FCFA). Source: Authors' own work.

VARIABLES	(1) Model 1	(2) Model 2	(3) Model 3
Constant	**205.8*****	**186.4*****	**186.4*****
	(66.48)	(66.63)	(67.22)
Risk aversion (σ)	28.22		0.00828
	(30.78)		(33.33)
Loss aversion (λ)	**100.01*****	**40.79*****	**40.79*****
Favorable climate in last season	(31.89)	(12.59)	(13.77)
		100.5***	**100.50*****
		(31.89)	(30.97)
Female	18.10	17.17	17.17
	(28.27)	(28.22)	(28.16)
Female household head	26.71	33.13	33.13
	(48.08)	(47.99)	(47.97)
Household head	22.82	23.58	23.58
	(24.56)	(24.65)	(24.56)
Age	**−1.802****	**−1.615****	**−1.615****
	(0.764)	(0.757)	(0.765)
Cultivates landrace	-24.06	−19.23	−19.23
	(27.82)	(27.20)	(27.23)
Believes variety better than average	**62.43*****	**57.50*****	**57.50*****
	(17.55)	(17.56)	(17.57)
Upland farmer	**−58.82****	**−57.76****	**−57.76****
	(23.99)	(23.82)	(23.81)
Remembers lost varieties	12.00	14.61	14.61
	(18.20)	(18.03)	(18.04)
Household size	−1.576	−1.598	−1.598
	(1.382)	(1.385)	(1.387)
Cultivates other crops	**52.70***	**53.37***	**53.37***
	(28.91)	(28.97)	(28.96)
Has phone	11.88	13.67	13.67
	(20.41)	(20.49)	(20.51)
Discount rate, two-year period	7.685	−0.154	−0.153
	(14.38)	(14.46)	(14.77)
Income, last month	1.60e-05	1.67e-05	1.67e-05
	(3.09e-05)	(3.01e-05)	(3.01e-05)
Enumerator controls	Yes	Yes	Yes
Village controls	Yes	Yes	Yes
Observations	569	569	569
R-squared	0.170	0.181	0.181

Notes: Robust standard errors in parentheses.
***p < 0.01, **p < 0.05, *p < 0.1

References

Arouna, A. and Aboudou, R. (2020) Dataset of the survey on e-registration and geo-referenced of rice value chain actors for the diffusion of technologies: case of Benin and Côte d'Ivoire. *Data in Brief*, 30, 105642.

Becker, G.M., DeGroot, M.H. and Marschak, J. (1964) Measuring utility by a single-response sequential method. *Behavioral Science* 9, 226–232.

Emerick, K., de Janvry, A., Sadoulet, E. and Dar, M.H. (2016) Technological innovations, downside risk, and the modernization of agriculture. *The American Economic Review* 106, 1537–1561.

Hamukwala, P., Oparinde, A., Binswanger-Mkhize, H.P. and Kirsten, J. (2018) Willingness-to-pay estimates in the Becker–DeGroot–Marschak (BDM) mechanism and the non-hypothetical choice experiment: a case of biofortified maize in Zambia. *Journal of Agricultural Economics* 70, 81–100.

Holden, S.T. and Quiggin, J. (2017) Climate risk and state-contingent technology adoption: shocks, drought tolerance and preferences. *European Review of Agricultural Economics* 44, 285–308.

Jin, J., Xuhong, T., Wan, X., He, R., Kuang, F. and Ning, J. (2020) Farmers' risk aversion, loss aversion and climate change adaptation strategies in Wushen Banner, China. *Journal of Environmental Planning and Management* 63, 2593–2606.

Lampe, I. and Würtenberger, D. (2020) Loss aversion and the demand for index insurance. *Journal of Economic Behavior & Organization* 180, 678–693.

Liu, E.M. (2013) Time to change what to sow: risk preferences and technology adoption decisions of cotton farmers in China. *The Review of Economics and Statistics* 95, 1386–1403.

Magnan, N., Love, A.M., Mishili, F.J. and Sheremenko, G. (2020) Husbands' and wives' risk preferences and improved maize adoption in Tanzania. *Agricultural Economics* 51, 743–758.

Tanaka, T., Camerer, C.F. and Nguyen, Q. (2010) Risk and time preferences: linking experimental and household survey data from Vietnam. *The American Economic Review* 100, 557–571.

Wang, M., Yu, Y., Haberer, G., Marri, P.R., Fan, C. *et al.* (2014) The genome sequence of African rice (*Oryza glaberrima*) and evidence for independent domestication. *Nature* 46, 982–991.

Ward, P.S. and Singh, V. (2015) Using field experiments to elicit risk and ambiguity preferences: behavioral factors and the adoption of new agricultural technologies in rural India. *The Journal of Development Studies* 51, 707–724.

11 Smallholders' Willingness to Pay for Alternative Types of Crop Insurance in China

H. Holly Wang[1]* and Lu Liu[2]
[1]*Purdue University, West Lafayette, USA;* [2]*Chinese Academy of Agricultural Sciences, Beijing, China*

Abstract

Over the past 40 years, crop insurance subsidized by the Chinese government has developed into the world's largest program, growing from only low-coverage crop yield insurance into multiple programs including high-coverage crop yield insurance, price insurance, revenue insurance, and weather-indexed insurance. This chapter introduces the development of China's crop insurance and presents an empirical study of smallholders' preference for alternative types of insurance. Through a survey using choice experiment and contingent valuation methods, we found that given the heavy subsidy, high-coverage yield insurance is preferred the most, followed by price insurance, revenue insurance and weather-indexed insurance.

11.1 Background

Crop insurance has been a popular tool used in most countries to help farmers mitigate their economic risks from farming. Almost all current successful multi-peril crop insurance programs are subsidized by governments because it is believed that moral hazard, adverse selection and systemic risks have prevented the private industry from providing low-cost insurance products. The demanding financial burden on governments has limited such insurance availability mostly to developed countries. Emerging economies have started to follow practices similar to those of developed countries; however, they have begun with meager protection, so the government's subsidy burden is also low.

China is a major country in agricultural production and consumption. Figure 11.1 presents China's crop production from 2005 to 2021. The total crop production in 2021 was 682.85 million tonnes, ranking first in the world, followed by the United States and India. As the top producer of many crops, China's agricultural production has high risks both in the production sector – such as major plant diseases, pests, adverse weather and other natural hazards – and in the market sector, especially price fluctuations. Apart from the traditional risks in these sectors, the 2022 development report of China's agricultural risk management further indicates that non-traditional risks have increased significantly. For example, the COVID-19 pandemic and geographical conflicts cripple crop production and circulation, aggravating crop production's instability and uncertainty (CARMS, 2022).

China's modern crop insurance programs started in 1982 (Tuo, 2018). In the past 10 years,

*E-mail: wanghong@purdue.edu

DOI: 10.1079/9781800622289.0011

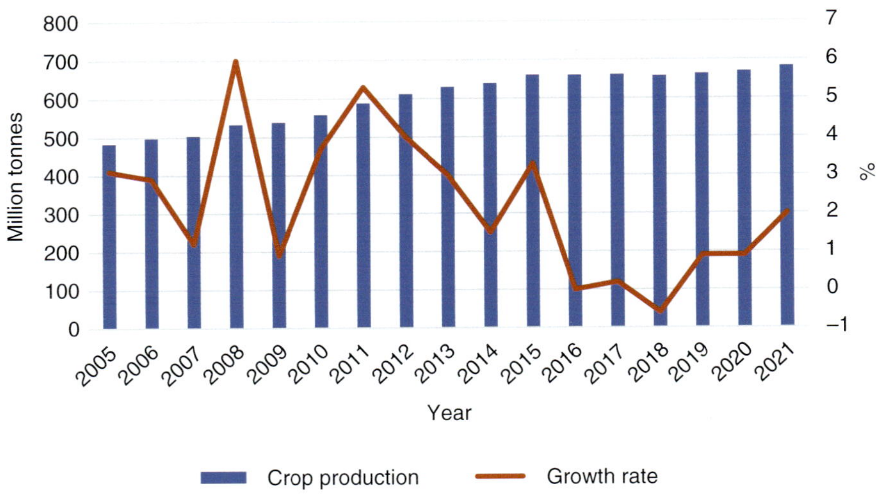

Source: National Bureau of Statistics of China.

Fig. 11.1. China crop production during 2005–2021.

the total agricultural insurance coverage rose from 0.9 trillion yuan (one yuan is about US$0.14) to about 5 trillion yuan. The total premium level was more than 100 billion yuan in 2022, ranked top worldwide. Indemnity payments also rose about ten times over the past decade. The dominant insurance program is multi-peril insurance for major crops. It started from a low level of coverage, equaling only the material costs of crop production such as seed, fertilizer, chemicals, irrigation, machinery service and plastics. This was helpful for small subsistence farms relying on their own labor and land to resume production for the next season. As more specialized farms have emerged in the past few years, with commercial farmers renting land and hiring labor to reach certain scales, crop insurance coverage is being extended to full production cost, including land rent and labor. Currently, crop insurance is transitioning from material cost coverage to full cost coverage, with the latter covering wheat, rice and corn in 13 major grain-producing provinces (MOF, 2021a,b).

In addition to traditional crop insurance that bases its indemnity payment on the loss of crop yield, insurance products have also been piloted basing indemnities on losses from market price, revenue and even weather indices in China. Because China has a relatively short history of providing such insurance with government subsidies, it started with low coverage to foster and test the system, including farmers' participation under low premium costs and the government's low financial obligation. However, Chinese farmers expressed a need for higher coverage. This chapter will: (i) introduce the history and current situation of agricultural insurance for major crops in China; (ii) present our empirical work eliciting farmers' willingness to pay (WTP) for alternative crop insurance; and (iii) compare two methods – contingent valuation and choice experiment – in eliciting WTP and impacting factors.

11.2 Historical and Current Crop Insurance Practices

11.2.1 The development of China's modern crop insurance

Although insurance covering loss related to agriculture was introduced to China more than a century ago, it was quite primitive, e.g. insuring the death of a drafting ox, and on a tiny scale. Modern agricultural insurance started in 1982. However, compared with China's fast-growing economy and insurance industry during that period, agricultural insurance's total volume increased slowly, and its share in the all-property insurance market decreased over the following decade with the absence of government support (Tuo and Wang, 2002).

In 2007, China's central government launched agricultural insurance subsidies for six pilot provinces: Jilin and Inner Mongolia in the northern corn- and wheat-producing region; Jiangsu, Hunan, and Sichuan in the southern rice-producing region; and Xinjiang in the western cotton-producing region. The subsidies were extended to ten provinces in 2008 and have covered all provinces since 2012. In addition to geographic areas, the number of crops covered by the government-subsidized insurance program increased from five field crops – wheat, rice, corn, soybean and cotton – in 2007 to potatoes, oilseeds, lumber, rubber, sugar crops, sows, cows, pigs and many other crops and livestock in 2016 (Boyd *et al.*, 2011; Liu and Tao, 2020). The central government subsidizes the premium from 35% to 50%, depending on the regions and crops, and mandates that provincial and local governments match 25–30%. Farm households may end up paying a tiny portion of the premium (Ye, 2018; Liu and Wang, 2019).

The growth of Chinese agricultural insurance from 2007 to 2021 can be seen in Figures 11.2 and 11.3. Total coverages increased tremendously from 0.17 trillion yuan in 2007 to 4.78 trillion yuan in 2021 (see Fig. 11.2). Correspondingly, the total premium was 5.18 billion yuan and total indemnity was 3.28 billion yuan in 2007, growing to total premium of 81.49 billion yuan and total indemnity of 61.66 billion yuan in 2020 (see Fig. 11.3). The rapid development of agricultural insurance made China rank first in the world in 2021, when the total premium was 97.6 billion yuan.

11.2.2 Existing insurance products

China currently has four types of agricultural insurance products. Because the government subsidizes them, when multiple insurance products are available in the same region, farmers can take only one for production of a particular commodity to rule out double-dipping the subsidies.

First, the most popular product is production cost insurance, which is based on the yield loss caused by natural disasters. It is multi-peril insurance with low coverage but is widely available for both crops and geographic areas. It has a specific yield loss assessment based on when a disaster (such as a flood or frost) happens. Farmers'

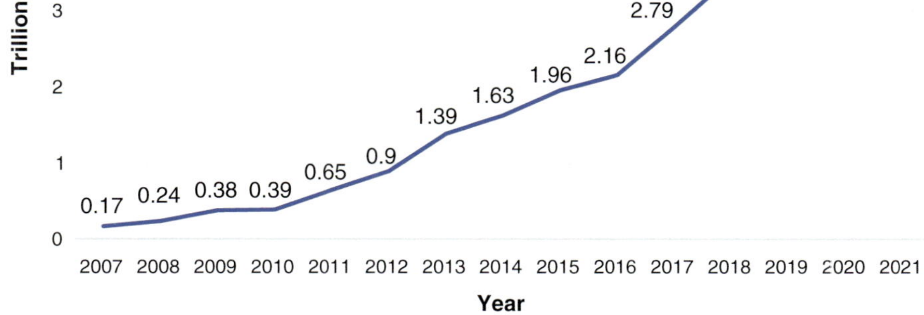

Source: National Bureau of Statistics of China and China Banking and Insurance Regulatory Commission.

Fig. 11.2. The total coverage of agricultural insurance in China during 2007–2021.

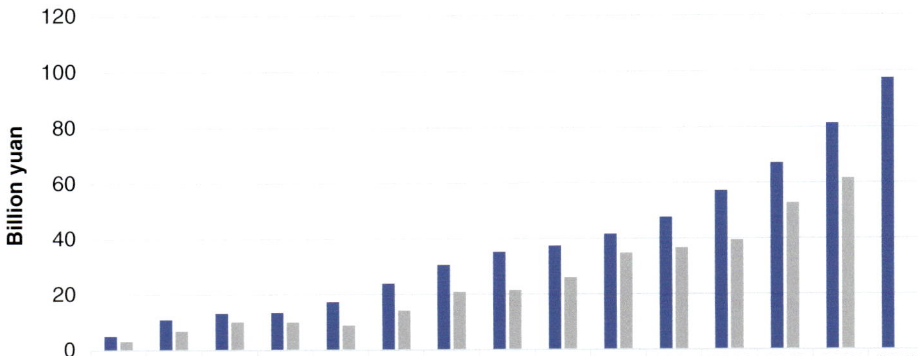

	2007	2008	2009	2010	2011	2012	2013	2014	2015	2016	2017	2018	2019	2020	2021
Total premium	5.18	11.07	13.38	13.57	17.38	24.01	30.67	35.28	37.47	41.71	47.77	57.27	67.25	81.49	97.6
Total indemnity	3.28	6.91	10.19	10.06	8.9	14.22	20.86	21.46	26.01	34.8	36.61	39.43	52.79	61.66	

Year

Source: Tuo (2018), National Bureau of Statistics of China and China Banking and Insurance Regulatory Commission.

Fig. 11.3. The total premium and indemnity of agricultural insurance in China during 2007–2021.

choice of coverage level is based on the region's general yield history and market price. Because farm sizes are tiny in China, in practice this yield loss is based mostly on the village's average yield. In recent decades, Chinese labor and land costs have risen rapidly, partly due to the emergence of larger specialized crop production farms that employ laborers and rent land. Table 11.1 presents the national average of unit production cost and net profit from 2011 to 2020 using corn as an example. The former, which includes material input such as fertilizers, labor and land expenses, increased from 764.2 yuan per mu in 2011 to 1080 yuan per mu in 2020. Labor and land costs, especially, rose much faster than material costs. The government-subsidized cost insurance is transitioning from material cost to total cost products.

Second, target price insurance is an insurance product being piloted by insurance providers and bases indemnities on losses from the market price. For example, the Shanghai government and Anxin Agricultural Insurance Company launched vegetable price insurance in 2011, and the Beijing government and Anhua Agricultural Insurance Company piloted hog price insurance in 2012. In the following years, many price insurance products covering mostly specialty crops were launched in all 31 provinces (MOA 2017; Zhuo, 2022).

The third type of agricultural insurance is weather index insurance. Because agricultural risks are primarily from adverse weather directly and because weather index insurance has advantages in eliminating moral hazard and adverse selection problems, weather index insurance products have been provided in developed and developing countries. In China, Shanghai was the first to pilot such insurance in 2007 for one melon crop, and Jiangxi province followed in 2011 insuring oranges against frost (Liu and Tao, 2020).

Fourth, crop revenue insurance was piloted in 13 Chinese provinces in 2022 (Tuo, 2022). Table 11.1 also shows the downward trend in the average market price from 2011 to 2019, with an increase in 2020. Such inverse changes in the production cost and market price have led to reduced net profits for corn production since 2011 and even negative net profits from 2015 to 2019. To maintain grain production, revenue insurance is called upon. The pilot programs are currently operated by an insurance company or by an insurance company in collaboration with a commodity exchange company. The revenue is based on the futures price on the exchange and on yield, either farm yield or area yield. These products are for major field crops with futures contracts traded on exchanges.

Table 11.1. Cost–profit of corn growing and market prices in China, 2011–2020. (Source: National agricultural cost and income compilation, by National Bureau of Statistics of China, 2016 and 2021.)

	2011	2012	2013	2014	2015	2016	2017	2018	2019	2020
Yield (kg/mu)	472.24	492.55	488.01	499.79	488.81	480.29	501.53	487.02	503.90	502.62
Cost (yuan/mu)	764.2	924.2	1012.0	1063.9	1083.7	1065.6	1026.5	1044.8	1055.7	1080.0
Materials/ services	308.45	344.58	359.71	364.80	376.22	369.55	374.98	383.76	390.20	392.49
Labor expenses	295.49	398.40	455.37	474.68	468.72	458.10	441.20	433.52	424.79	437.96
Land expenses	160.29	181.24	196.96	224.41	238.78	237.94	210.30	227.54	240.68	249.53
Mkt price (yuan/kg)	2.12	2.23	2.18	2.24	1.88	1.54	1.64	1.76	1.79	2.31
Net profit (yuan/mu)	263.1	197.7	77.52	81.82	−134.2	−299.7	−175.8	−163.3	−126.8	107.8

11.2.3 Premium and subsidies

As mentioned earlier, the Chinese government has subsidized farmers' agricultural insurance premiums since 2007. The government realizes that subsidizing agriculture can help boost domestic food supply and transfer welfare to low-income farmers while complying with international trade regulations. So, subsidizing agricultural insurance has become a convenient policy instrument. The Administration Rules of Central Finance Subsidy to Agricultural Insurance Premium (2021) provide specific objectives, rules and procedures for the subsidy (MOF, 2021b). The subsidy covers crops, livestock, forestry and local specialty varieties. The subsidy rates differ depending on regions, commodities and insureds. For example, if the insured is a state-owned enterprise, farmers in poverty may receive higher subsidies. Overall, central government subsidy rates range from 35% to 65% for crops, and the central government also mandates that each province match the premium subsidy for at least 25%.

In addition to the production cost insurance that covered only material costs from the beginning, other types of crop insurance have been initiated since 2017, such as revenue and price insurance by the provincial government. For example, as one of the major crop-producing provinces, Liaoning was among the first ten provinces to introduce government-subsidized crop insurance in 2008. According to the provincial act extending the scope of full production cost insurance and crop revenue insurance for major crops in Liaoning (MOFLN, 2022), total production cost insurance for rice and wheat is available across Liaoning province. For corn, revenue insurance is implemented in six selected counties, and other counties are covered by full production cost insurance. The premium rate also varies across crops and regions with different production risks. Specifically, corn insurance is charged the highest premium rate on average, followed by wheat and rice insurance (see Table 11.2). The subsidy rate for the major crop-producing counties is 45% from the central government. Farmers need to pay only 20% with their own money, which leaves the remaining 35% to provincial and municipal or county governments.

11.2.4 The operation of crop insurance

Commercial insurance companies operate crop insurance. Insurance companies must satisfy certain criteria and get approval from the government's insurance supervision agency. They design agricultural insurance contracts, set premiums, and market and execute the insurance. Farmers or eligible organizations can insure and receive premium subsidies from the government (State Council of the PRC, 2012). Currently, 20 insurers account for 93% of the market share, with five of them specializing in agricultural insurance only. The top ten insurance companies are listed in Table 11.3. With some overlap, they each focused on some particular regions in the past. However, more competition occurs now with multiple insurers working in some regions.

Although the three top specialized agricultural insurance companies are still relatively

Table 11.2. Premium rate of crop insurances for major crops in Liaoning province in 2022. (Source: MOFLN, 2022.)

Target	Cities	Premium Rate (%)
Corn	Shenyang, Anshan, Fushun, Benxin, Dandong, Yingkou, Liaoyang, Tieling, Panjin	6.10
	JinZhou, Fuxin, Chaoyang, Huludao	6.70
Rice	Shenyang, Anshan, Fushun, Benxi, Dandong, Yingkou, Liaoyang, Tieling, Chaoyang, Panjin	4.10
	Jinzhou, Fuxin, Huludao	4.50
Wheat	Shenyang, Anshan, Fushun, Benxi, Dandong, Yingkou, Liaoyang, Tieling, Chaoyang, Panjin	4.20
	Jinzhou, Fuxin, Huludao	4.60

Table 11.3. Top ten insurance companies and their market shares. (Source: Tu and Yuan, 2022.)

Rank	Insurance company	Premium revenue in 2021 (billion)	Rate of change (%)	Market share (%)
1	PICC P&C	42.65	19.30	43.70
2	China Insurance	12.05	13.57	12.35
3	CPIC	10.36	9.78	10.62
4	China Life Insurance	7.95	22.03	8.14
5	Ping An Insurance	3.76	59.34	3.85
6	Guo Yuan Agricultural Insurance	3.60	13.63	3.69
7	Anhua Agricultural Insurance	3.18	4.01	3.26
8	Zhongyuan Agricultural Insurance	2.02	22.92	2.07
9	PAAIC	1.07	23.24	1.10
10	China Continent Insurance	1.05	41.35	1.07

small, as the government subsidy rate reaches about 75%, the overall agricultural insurance industry is growing rapidly. In 2021, the scale of specialized agricultural insurance firms increased by double digits, although at the same time, the overall property insurance scale growth rate was 0.68%.

11.3 Survey and Choice Experiment Design

As Chinese farmers have access to alternative insurance products with government subsidies, they have a choice. We conducted an empirical study on what specific features farmers value for these insurance products, using corn growers in Liaoning, one major corn-growing province, as a case. We employed two valuation methods, choice experiment and contingent valuation, to solicit farmers' valuation for insurance attributes.

11.3.1 Survey methods introduction, regions, time, size

We randomly surveyed 20 farmers in two or three villages from each of the four counties, namely Kangping, Heishan, Panshan and Xiuyan (Table 11.4), in November 2017. These yielded a total of 198 samples. Graduate student enumerators from the Chinese Academy of Agricultural Sciences and Zhejiang University interviewed each farmer for 40 minutes, asking about his or her WTP for certain insurance features through the two methods, choice experiment and contingent valuation.

11.3.2 Descriptive statistics analysis

Additional information was also solicited, such as demographic characteristics and production factors. Table 11.5 shows the descriptive statistics of the sample. About 62% of the respondents were male and almost 80% were family decision

Table 11.4. Survey samples. (Source: Author survey.)

Counties	Date	Num. of obs.
Kangping County, Shenyang City	21 Nov. 2017	50
Heishan County, Jinzhou City	22 Nov. 2017	48
Panshan County, Panjin City	23 Nov. 2017	47
Xiuyan Manchu Autonomous County, Anshan City	25 Nov. 2017	53
In total		198

Table 11.5. Descriptive statistics of variables. (Source: Author survey.)

Covariates	Mean or percentage	Standard deviation
Male (%)	62.12	
Age (years)	53.27	11.49
Decision makers (%)	79.8	
Education above junior high (%)	18.18	
Risk-averse (%)	53.54	
Satisfied with past insurance (%)	68.18	
Acreage 2016 (mu)	14.09	15.98
Yield 2016 (kg/mu)	471.0	165.58
Ratio of production cost to total household expenditure 2016 (%)	16.62	14.43
Ratio of production revenue to total household income 2016 (%)	21.63	20.40
Number of observations	198	

makers, the latter making the data from the survey more convincing. The respondents averaged 53.3 years old, conforming to the reality that young laborers went to cities, leaving older ones farming. Only 18.2% of the respondents received more than 9 years of education. About 53% of the farmers were allegedly risk-averse, and 68.18% were satisfied with past insurance experiences.

Table 11.6 presents the household acreage of corn distribution across the four counties. On average, the acreage of corn was 14 mu per household, indicating there were many smallholders. Only a few households from Kangping, Shenyang, and Heishan, Jinzhou, grow corn on more than 6 ha. The yield per mu in 2016 was 471.0 kg, lower than the national average (Table 11.1). The percentage of corn production cost to total household expenditure was 16.6%, and the percentage of revenue to income was 21.6%, indicating that crop production was only a small part of economic activity for most rural households.

11.3.3 Contingent valuation and choice experiment methods

Contingent valuation methods have long been used to explore users' valuation of a particular product or a particular feature of a product by directly asking only when a product or feature is not in the market so that no observation of the actual accepted price is available (Arrow *et al.*, 1993; Carson *et al.*, 2001; Kahn and Alberini, 2006). New crop insurance or a hypothetical feature of insurance falls in this category. However, contingent valuation is criticized for hypothetical bias, that is, survey respondents may give a value different from their actual WTP for certain reasons. For example, they may overstate the value to make themselves look generous and financially sound. The choice experiment is adopted dominantly over contingent valuation because the former makes it more difficult to do so (Carson *et al.*, 1996; Lusk and Schroeder, 2004; Chang *et al.*, 2009). Instead of directly asking a buyer's WTP, in this method a few alternative insurance products are provided for them to choose from. The alternatives give specific insurance products with particular features, including premium levels. There is also an opt-out choice if none of the provided products is acceptable to the farmer. To reduce hypothetical bias, this is more like an actual situation when a farmer is faced with alternative insurance products.

In this study, we presented four types of crop insurance contracts, i.e. yield insurance,

Table 11.6. Average household corn acreage. (Source: Author survey.)

Acreage (mu)	Kangping, Shenyang	Heishan, Jinzhou	Panshan, Panjin	Xiuyan, Anshan	Num. of households
<5	0	1	12	21	34
5–10	18	5	15	26	64
10–15	11	20	8	4	43
15–20	7	13	9	2	31
20–63	13	6	3	0	22
>63	2	2	0	0	4

revenue insurance, price insurance, and weather index insurance, for respondents to choose in each situation. In Liaoning, yield insurance has been available in the market since 2007, and the three other types of insurance products have been in pilot programs since 2017. Familiarity with these practices makes corn growers in Liaoning good research subjects.

As for coverage, we have three levels from 300 yuan per mu, covering the cost of material inputs and services, to 1000 yuan per mu, equivalent to the average revenue per mu. Given the subsidy from multi-level governments and that farmers may be more concerned with the money they paid but unfamiliar with the policies, we adopted the self-paid premium after deducting subsidies from the actual premium. Lastly, we included 'time to receive indemnity payment' and 'whether the insurance company is state-owned or private' to explore respondents' preference for indemnity services and the possibility of marketization of crop insurance in China, respectively.

Different combinations of these attributes were bundled together to form an insurance product randomly using statistical software. Then, we asked each farmer respondent to make eight choices among given independent hypothetical situations with four types of products: yield insurance (representing the production cost insurance), price insurance, revenue insurance, and weather index insurance. There is also an option not to buy insurance in each situation. Economic and econometric models used to estimate the WTP from the choice experiment observations are not straightforward. We refer readers to Wang *et al.* (2020).

11.4 Results and Discussion

Using contingent valuation, we asked survey respondents the maximum amount they were willing to pay (MWTP) for yield insurance, with increasing coverages from the base 300 yuan to an improved 650 yuan and then 1000 yuan (see Table 11.7). After discarding values greater than the 90th percentile, we present the MWTP for each coverage level in Table 11.7. The average stated MWTP ranges from 6.39 yuan/mu to 16.20 yuan/mu as the coverage level increases from 300 yuan/mu to 1000 yuan/mu. The corresponding minimum and maximum MWTP increases along with the coverage.

The choice experiment study shows that most farmers have positive WTPs for all insurance products (see Table 11.8). We find, on average, that the highest WTP is for yield insurance, 58.04 yuan/mu. The price, revenue, and weather index insurance WTP is 53.88, 51.01, and 43.77 yuan/mu, respectively. These values are under the condition of the government providing 80% of premium subsidies. Not surprisingly, traditional yield (cost) insurance is valued the highest and weather index insurance the lowest. This is not only because farmers are most familiar with and most trust the former, and they are least familiar with and least trust the latter, but also because yield is the primary source of risk for farmers. At the same time, the weather index would bring about basis risk, and the revenue risk is smaller due to the offsetting effect between yield and price in the primary production region. When all insurance is highly subsidized to nearly free to farmers, the advantage of low cost from weather-indexed insurance relative to other insurance is not valuable anymore.

Comparing the WTP obtained from the two methods, the contingent valuation results are smaller than the WTP from the choice experiment. This is expected as the hypothetical bias is much higher from contingent valuation, as shown in criticism. Farmers tend to understate

Table 11.7. Maximum WTP for yield insurance with various coverage levels. (Source: Author survey.)

	MWTP (yuan/mu)				
Coverage (yuan/mu)	Mean	Standard deviation	Min.	Max.	N
300	6.39	3.12	0	20	
650	11.11	4.10	3	27	185
1000	16.20	5.69	6.5	30	

Note: N: number of observations.

Table 11.8. Marginal WTP various crop insurance from choice experiment. (Source: Author survey.)

Attributes	WTP (yuan/mu)	Simulated WTP 95% confidence interval
Types of insurance		
Yield	58.04***	[−85.55, 201.54]
Revenue	51.01***	[−88.59, 190.77]
Price	53.88***	[−101.62, 209.54]
Weather index	43.77***	[−132.02, 218.28]
Coverage		
Yield	0.03***	[−0.03, 0.09]
Revenue	0.03***	[−0.07, 0.13]
Price	0.02***	[−0.07, 0.11]
Weather index	0.02***	[−0.05, 0.09]
Time	14.35***	[−20.99, 49.72]
Company	43.42***	[−49.53, 136.37]

Note: *** denotes statistical significance at 1%.

their WTP in an attempt to pressure the insurance providers to offer low-priced products.

Using the choice experiment method, we also obtained farmers' WTP for specific features of the four types of insurance. The higher coverage will result in higher risk protection value and, in turn, higher WTP. The marginal WTP for each additional yuan of coverage is the highest for revenue insurance, at 3 cents per yuan, followed by yield, price, and weather index insurance, at about 2 cents. These results show that farmers are willing to pay a premium of 2–3% of coverage. Currently, in China, farmers' out-of-pocket premium payment is 1–2% of the coverage level, and thus the result is in line with reality. It shows farmers are willing to pay proportionally for the coverage. The higher the coverage, the higher the premium WTP, assuming the same level of government subsidy. These WTP increases estimated from the two methods are roughly consistent with the contingent valuation on the underestimated side. From coverage of 300–650 yuan and from 650–1000 yuan, the 350 yuan increase will raise the WTP for about 10 yuan, at 3 cents per yuan, and the differences between each adjacent pair of WTP from Table 11.7 are about 5 yuan/mu.

Farmers show a preference for insurance offered by state-owned insurance companies over private ones and are willing to pay 43.42 yuan/mu more for an insurance contract provided by state-owned insurers. The payment time also matters. The shorter processing time means farmers can receive indemnity payments more quickly. They receive indemnity within 1 month after a disaster, which makes farmers willing to pay 14.35 yuan more each month. Factors affecting farmers' WTP are also analyzed for both sets of values obtained from the two methods. Table 11.9 shows the regression results of WTP from the choice experiment methods for each type of insurance product and the WTP for yield insurance with a coverage level of 1000 yuan/mu only elicited directly using contingent valuation.

Other than city dummies, different significant coefficients exist. First, age is an essential factor. Older people tend to have lower WTP for revenue insurance, price insurance, and weather-indexed insurance than younger people. These are newer insurance products, and older farmers tend to be conservative in their insurance demand. The contingent valuation shows that older farmers tend to pay more than younger farmers for yield insurance, but the choice experiment result has an insignificant age effect for yield insurance. Combined effects indicate that older farmers are interested in crop insurance. They prefer the existing type of insurance but are more reluctant to adopt newer types of insurance products.

If farmers have had satisfying insurance experience in the past, they express a higher WTP using the contingent valuation method, but it is not significant for the choice experiments. We can interpret this as showing that the contingent valuation method shows an adverse selection preference. The ratio of production cost to total household expenditure is also significant for yield insurance and price insurance in the choice experiment analysis but not in the contingent valuation. It reveals that these two methods may not always be consistent.

Table 11.9. Factors affecting WTP. (Source: Author survey.)

Variable	WTP for various insurance type				MWTP for yield insurance mu from contingent valuation
	Yield	Revenue	Price	Weather index	
Age	−0.338	−0.447*	−0.586**	−0.752***	0.062*
	(0.261)	(0.244)	(0.267)	(0.247)	(0.037)
Male	1.244	0.033	1.615	5.484	−0.466
	(6.182)	(5.729)	(6.429)	(6.530)	(0.942)
Decision maker	−7.140	0.612	−2.538	−1.351	−1.816
	(8.087)	(7.299)	(8.201)	(7.969)	(1.159)
Jinzhou City	3.136	0.107	3.130	4.709	−2.123*
	(8.434)	(7.494)	(8.129)	(7.400)	(1.105)
Panjin City	22.410**	24.743***	25.342***	22.263**	0.702
	(9.368)	(8.116)	(9.243)	(9.261)	(1.259)
Anshan City	24.764***	29.090***	31.992***	32.071***	0.499
	3.136	0.107	3.130	4.709	(1.285)
Junior high education	3.138	1.005	1.006	−2.464	1.280
	(9.772)	(8.569)	(9.474)	(9.852)	(1.070)
Satisfied with past exp.	4.240	2.982	1.945	−5.754	1.651*
	(6.133)	(5.503)	(6.062)	(6.113)	(0.917)
Risk-averse	−0.526	−0.022	−1.127	3.257	0.216
	(5.998)	(5.289)	(5.912)	(5.947)	(0.838)
Ratio of production cost	20.696*	11.583	22.616**	11.078	−4.045
	(12.280)	(9.507)	(10.764)	(11.478)	(3.064)
Acreage	−0.420	−0.259	−0.437	−0.511	0.029
	(0.352)	(0.323)	(0.383)	(0.316)	(0.044)
Constant	81.640***	74.556***	90.507***	100.709***	14.189***
	(17.768)	(16.234)	(17.232)	(16.503)	(2.180)
R^2	0.119	0.162	0.159	0.171	0.113
N	191				185

Note: *, ** and *** denote statistical significance at 10%, 5% and 1%, respectively. R^2: R-squared; N: Number of observations.

11.5 Summary and Conclusion

As increasing crop production and stabilizing farmers' income are top priorities for the Chinese government. China has fueled the rapid development of the crop insurance industry with heavy subsidies since 2007. Liaoning is one of China's major corn-producing provinces and introduced government-subsidized crop insurance in 2008 as one of the ten pilot provinces. We used choice experiments and contingent valuation methods to explore corn growers' preferences for crop insurance contracts. We interviewed corn growers from Liaoning province in 2017 to elicit their stated preference for yield, revenue, price and weather-index insurance. We extrapolated their WTP for various insurance types, coverage levels, the time needed to receive indemnity, and insurance firm types. Generally, farmers prefer insurance contracts with higher coverage levels and fast indemnity payments from state-owned firms. Their WTP varies across different demographic categories. In particular, the yield insurance covering only the cost of material input in corn production is far from meeting the demand for risk mitigation. WTP for additional coverage exists. Satisfaction with past insurance experience builds farmers' confidence in crop insurance and could facilitate the introduction of new insurance products.

Crop insurance is an attractive tool for farmers to mitigate production risks and market fluctuations. China's crop insurance industry has made enormous strides since 2017. It has grown into the world's top industry, with the annual total premiums reaching 81.49 billion yuan in 2020. Along with the rocketing total premium, new types of crop insurance with various targets, coverage levels, premium and subsidy rates, etc. have been introduced as pilot programs and implemented with official regulations. The time it takes for farmers to receive indemnity has shortened, and the differentiation of premium and subsidy rates across production regions and crops has gradually replaced the uniform rate at the provincial level. Taking Liaoning province as an example, in 2022 yield insurance covered the full production cost for corn, wheat and rice, and revenue insurance for corn was officially implemented in addition to the traditional yield insurance that covered only material costs (MOFLN, 2022). The latter insurance for soybeans, peanuts, potatoes, etc. is also available in the market. Moreover, insurance targeting greenhouses and crops raised in greenhouses, fungi and traditional Chinese medicines is encouraged to support local specialty varieties.

However, plenty of challenges lie ahead for the prosperity of China's crop insurance industry. First, the heavy subsidy imposes a heavy burden on the national financial system. For example, total subsidies in 2017 reached as high as 36.27 billion yuan (Tuo, 2018). The rapid expansion of heavily subsidized crop insurance makes the increasing subsidies unsustainable. It therefore necessitates adjusting subsidy structures, developing commercial crop insurance, and marketizing the crop insurance industry, which is still in the initial stages of exploration. Moreover, no cost–benefit analysis of the subsidy policies has been conducted to date. Second, subsidized crop insurance calls for a more targeted supervisory system on market behavior, solvency, corporate governance and the finance of insurers to ensure fair dealing to protect the rights and interests of the insureds. However, such supervision lags behind the expansion of crop insurance in China. Third, given the frequent occurrences of extreme weather, the development of catastrophe insurance and the re-insurance mechanism should step up to strengthen the resilience of the crop insurance industry.

References

Arrow, K., Solow, R., Portney, P.R., Leamer, E.E., Radner, R. and Schuman, H. (1993) Report of the NOAA panel on contingent valuation. *Federal Register* 58, 4601–4614.

Boyd, M., Pai, J., Zhang, Q., Wang, H.H. and Wang, K. (2011) Factors affecting crop insurance purchases in China: the Inner Mongolia region. *China Agricultural Economic Review* 3, 441–450. DOI: 10.1108/17561371111192301.

CARMS (China Agricultural Risk Management Society) (2022) *The Development Report of China Agricultural Risk Management (2022).* Chinese Academy of Agricultural Sciences, Beijing.

Carson, R.T., Flores, N.E., Martin, K.M. and Wright, J.L. (1996) Contingent valuation and revealed preference methodologies: comparing the estimates for quasi-public goods. *Land Economics* 72, 80–99. DOI: 10.2307/3147159.

Carson, R.T., Flores, N.E. and Meade, N.F. (2001) Contingent valuation: controversies and evidence. *Environmental and Resource Economics* 19, 173–210. DOI: 10.1023/A:1011128332243.

Chang, J.B., Lusk, J.L. and Norwood, F.B. (2009) How closely do hypothetical surveys and laboratory experiments predict field behavior? *American Journal of Agricultural Economics* 91, 518–534. DOI: 10.1111/j.1467-8276.2008.01242.x.

China Banking and Insurance Regulatory Commission (undated) *China Insurance Yearbook, 2008–2022.* CBIRC, Beijing.

Liu, H. and Tao, J. (2020) Policy-oriented agricultural insurance in China: development trend, international comparison and path optimization. *Journal of Huazhong Agricultural University (Social Sciences Edition)* 6, 67–75 and 163–164. (In Chinese).

Liu, L. and Wang, H.H. (2019) Analysis and prospect on farmers' demand for full cost insurance – a case study of maize growers in Liaoning Province. *Agricultural Outlook* 15, 90–95. (In Chinese).

Lusk, J.L. and Schroeder, T.C. (2004) Are choice experiments incentive compatible? A test with quality differentiated beef steaks. *American Journal of Agricultural Economics* 86, 467–482. DOI: 10.1111/j.0092-5853.2004.00592.x.

MOA (Ministry of Agriculture and Rural Affairs of the People's Republic of China) (2017) The Mode of Price Index Insurance for Agricultural Products. Available at: www.moa.gov.cn/ztzl/jrfwnyxdhgflt/sdms_a1012/201710/t20171012_5838724.htm (In Chinese) (accessed December 29, 2022).

MOF (Ministry of Finance of the People's Republic of China) (2021a) The Act of Extending the Scope of Full Production Cost Insurance and Crop Revenue Insurance for the Major Crops. Available at: https://www.gov.cn/zhengce/zhengceku/2021-06/29/content_5621466.htm (accessed December 29, 2022).

MOF (Ministry of Finance of the People's Republic of China) (2021b) The Administration Rules of Central Finance Subsidy to Agricultural Insurance Premium. Available at: https://www.gov.cn/zhengce/zhengceku/2022-01/07/content_5666866.htm (In Chinese) (accessed December 29, 2022).

MOFLN (Ministry of Finance of Liaoning province) (2022) Provincial Act of Extending the Scope of Full Production Cost Insurance and Crop Revenue Insurance for the Major Crops in Liaoning. Available at: https://www.ln.gov.cn/web/zwgkx/lnsrmzfgb/2022n/qk/2022n_dsbq/bmwj/6B6EC58E35A84A1385E46DFC35EFF810/index.shtml (In Chinese) (accessed December 29, 2022).

National Bureau of Statistics of China (undated) *China Statistical Yearbook*. Available at: www.stats.gov.cn/sj/ndsj/ (accessed December 29, 2022).

The State Council of the People's Republic of China (2012) Agricultural Insurance Acts 2012. Available at: https://www.gov.cn/zhengce/2020-12/27/content_5574675.htm (In Chinese) (accessed December 29, 2022).

Tu, Y. and Yuan, Y. (2022) The predicament of agricultural insurance: nearly 40 insurance companies compete to enter the insurance market last year and the premium growth was about 20% while the underwriting profit margin was only 0.4%. *National Business Daily*. Available at: www.nbd.com.cn/articles/2022-06-30/2347119.html (In Chinese) (accessed December 29, 2022).

Tuo, G. (2018) The development of China agricultural insurance from the perspective of policy changes in the past 40 years. *Insurance Studies* 12, 84–87. (In Chinese).

Tuo, G. (2022) Several Problems about the Revenue Insurance for Agricultural Products. Available at: www.agri.cn/V20/SC/jjps/202207/t20220725_7877144.htm (In Chinese) (accessed December 29, 2022).

Tuo, G. and Wang, G. (2002) *China Agricultural Insurance and Rural Social Security System*. Capital University of Economics and Business Press, Beijing (In Chinese).

Wang, H.H., Liu, L., Ortega, D.L., Jiang, Y. and Zheng, Q. (2020) Are smallholder farmers willing to pay for different types of crop insurance? An application of labelled choice experiments to Chinese corn growers. *The Geneva Papers on Risk and Insurance – Issues and Practice.* 45, 86–110. DOI: 10.1057/s41288-019-00153-7.

Ye, C. (2018) Research on improving China's agricultural insurance system. *Journal of Financial Research* 12, 174–188. (In Chinese).

Zhuo, Y. (2022) Stabilizing Income and Price Insurance – crop Insurance Reduces the Price Risks of Agricultural Products. Available at: https://m.picc.com/m/information/content/xwzx/jrbxpl/202206/10/cms20220610_1936099793.html (In Chinese) (accessed December 29, 2022).

12 Role of Risk Aversion and Insurance Adoption in Extreme Climatic Shocks: Evidence from Indian Paddy Growers

Asis Kumar Senapati[1,2*]

[1]*Department of Economics and Development Studies, Central University of Jharkhand, Jharkhand, India;* [2]*Department of Economics, Ravenshaw University, Cuttack, India*

Abstract

This chapter highlights the determinants of the risk attitude of farm households in India. We used a lottery choice experiment to elicit the risk attitudes of farming households. Empirical results suggest that factors such as operator age, education, farm size and winner of the modified Holt–Laury lottery choice experiment significantly influence the risk attitude of Indian farm households. Similarly, the study shows that in flood regions, operator education, farm age, past disaster experience and access to credit significantly influence the adoption of crop insurance. On the other hand, in the drought-prone region, farm age, experience, land ownership and willingness to pay the amount affect crop insurance adoption. Though, theoretically, risk attitude and climatic shocks are associated with insurance adoption, the study found no direct effect of the interaction term on insurance adoption, although it affects it indirectly.

12.1 Introduction

Despite the importance of risk in agricultural production and its consequences for agricultural households, it has been argued that our understanding of the effects of risk and uncertainty is limited (Newbery and Stiglitz, 1981). Two schools of thought discussed risk as a probabilistic phenomenon – the objectivist and the subjectivist. The objectivist school observes probability through repetitive experiments or considers it external. The subjectivists consider probability as the expression of an internal state, i.e. a state of knowledge or confidence (Kaplan and Garrick, 1981). In other words, the knowledge of the probability distribution of an event or the information about the likelihood of an event differentiates risk from uncertainty. In this context, it should be noted that the information about the likelihood of an event can be associated with objective probabilities and subjective probabilities (Savage, 1954). Knight (1921) considers the risk to be attached to objective probabilities, which are true probabilities and not subjective (perceived likelihoods or degrees of belief about outcomes). This view of risk implies that we are concerned with the degree of uncertainty when we talk of risk. When we speak of uncertainty, we refer to the decision maker's state of mind in which they perceive alternative outcomes of a particular action (Roumasset *et al.*, 1979). Kaplan and Garrick (1981) argue that risk is uncertainty plus damages, implying a connotation of loss or damage

*E-mail: asiseco@gmail.com

 Managing Risk in Agriculture: A Development Perspective (eds A.K. Mishra, S.C. Kumbhakar and G. Lien).
DOI: 10.1079/9781800622289.0012

associated with risk, while it is not so in the case of uncertainty. This distinction is implicit in the risk assessment literature (Goldstein *et al.*, 1992; Graham, 2000).

From a measurement of risk perspective, some consider the risk to be synonymous with dispersion, such as the variance of returns for an outcome of interest (Tobin, 1958; Markowitz, 1959; Fama, 1968; Joyce and Vogel, 1970). Others consider risk as the probability of falling below some threshold or critical 'disaster level' (Roumasset, 1976; Scott, 1977). The latter perspective on risk concerns what can be called 'downside risk' or events associated with losses or damages. From this definition of risk, it can be inferred that risk is a combination of uncertainty and damage or loss (Kaplan and Garrick, 1981). An insightful observation made by Kaplan and Garrick (1981) is that for a risk or 'risk scenario' to be necessary for a decision maker, cognizance or awareness is important apart from the probability distribution, that is, the probability of occurrence and consequence of the scenario in terms of the damage or loss it causes.

Accordingly, if the decision maker is aware of the presence of risk, then the risk is considered as 'relative risk,' while with an ignorance of the risk, the risk is regarded as 'absolute risk.' For example, suppose farmers are unaware of some policy decision that could affect the availability of certain inputs, which could be detrimental to crop yields. In that case, we are referring to absolute risk. On the contrary, if farmers know of such a policy decision, we refer to relative risk. In other words, anticipated changes or shocks can be considered a relative risk, while unanticipated shocks can be considered an absolute risk. Rothschild and Stiglitz (1970) describe risk as what the risk-averse decision makers will pay a premium to avoid. Stiglitz states, 'Risk is like love; we have a good idea of what it is, but we can't define it precisely' (cited in Roumasset *et al.*, 1979). Here, it is implied that a risk is an uncertain event attached to a loss. Risk and uncertainty play a significant role in the agricultural decision-making literature (Anderson *et al.*, 1977; Anderson and Dillon, 1992; Just and Pope, 2003). The underlying assumption is that most farmers are 'risk-averse' (Moscardi and de Janvry, 1977; Binswanger, 1980; Antle, 1987). From a developing country perspective, this assumption implies that markets for credit and insurance are non-existent, incomplete or imperfect (Binswanger and Sillers, 1984; Binswanger and Rosenzweig, 1986).

In the expected utility specification, risk aversion arises from the assumption of a strictly concave utility function where the optimal choice is derived from the maximization of expected utility with a specific technology, subject to standard constraints. Note that the standard result of risk aversion has an inverse relationship with adopting riskier technologies or allocating greater resources to riskier prospects, which emerges from the theoretical models of adoption decisions following a neoclassical expected utility framework. Binswanger (1980), Humphrey and Verschoor (2004) and Ihli *et al.* (2013) analyzed the risk behavior of individuals in a developing economy perspective. In contrast, studies like Holt and Laury (2002) and Reynaud and Couture (2012) highlighted individuals' risk behavior in a developed economy.

However, besides the current literature, risk behavior analysis in the agriculture sector has not been researched and is unfamiliar in Odisha, India. Thus, this chapter aims to assess the determinants of risk attitudes of agricultural households and establish the link between risk attitude class and insurance adoption in case of extreme climatic shocks in Indian farm households. The study uses a lottery choice experiment to elicit risk attitudes of farming households in two districts of Odisha, India. The chapter is organized as follows. The next subsection (12.1.1) gives the background. Data and methodology is presented in Section 12.2. Determinants of risk attitude behavior are presented in subsection 12.2.3. The risk attitude and insurance adoption framework is developed in Section 12.3. Results and discussions take place in Section 12.4. Attitude and constraints are highlighted in subsection 12.4.1. Finally, the chapter concludes with some policy implications.

12.1.1 Background

In eastern India, Odisha is a major rice-producing state, but the macro picture shows a very dismal story so far as technology adoption is concerned. Agriculture is the predominant sector of Odisha, followed by the service sector in terms of contribution to net state domestic product (NSDP). Its share has always been above 30%, with little fluctuations in some years (Economic Survey of Odisha, 2008–2009). Agriculture employs more than 65% of the workforce, and 80% of the population depends on it directly or indirectly. So, these two points manifest that agriculture is still the dominant sector in Odisha.

However, the agrarian structure of the state is peculiar in that it has cultivable land of 61.65 lakh hectares, out of which about 50.1% is irrigated and the rest is yet to be irrigated. Thus, 50% of cultivable lands are still rain-fed, where weather plays a major role. So, the problems, prospects and potentials involving agricultural development vary from one type of agriculture to another. Since agricultural production is still backward and depends so much on climatic conditions and weather factors like rainfall, the present study assumes importance. It verifies that agriculture is not that developed in Odisha though it is an important sector.

Crop protection insurance is one of the most important risk management alternatives for farmers. It protects them by covering crop losses caused by extreme weather shocks and acts as a cushion to boost their self-confidence and enable them to continue farming despite having a bad year due to natural disasters. It assists farmers in accessing credit markets for financing their crop production, as most of the crop protection insurance schemes are linked with the credit structure of a bank. It certainly reduces farmers' risk and uncertainty behavior while undertaking farm production decisions, which is considered the most important feature of the decline in the growth rate of agriculture in India. Risk-averse farmers invest less in agricultural activities like purchasing new high-yield variety seeds, chemical fertilizer, new modern equipment and better quality seeds, which is detrimental to the overall food security problems of the country. So, to minimize the risk, agricultural insurance is crucial (Senapati, 2020a). Government interventions can provide incentives for farmers to manage risk personally. Crop insurance is now available in all the state's regions (Senapati, 2020c).

12.2 Data and Methodology

We collected a sample of 400 farm households from two districts of Odisha. Cuttack and Bolangir districts were selected out of the 30 districts of Odisha, in accordance with their definition in Chand *et al.* (2011) as the two most irrigated and rain-fed districts in Odisha. We selected farmers in Cuttack and Bolangir based on their land records. Experiments were conducted from May to August 2016. We applied the modified Holt–Laury lottery method and the lottery choice procedures adopted from Holt and Laury (2002) and Ihli *et al.* (2013), respectively, to measure risk attitudes. We formed several groups for experiments. Quiz tests were conducted to establish the awareness of farmers about climate change ecology and agriculture in general. Sessions were conducted both in the morning and in the afternoon in schools or temples. Since Odia was the familiar language, we asked questions in Odia, later translated to English for report writing. Most farmers were familiar with playing cards, so we used them with different payoffs for our experiments, explaining the experiment's rules by putting appropriate cards in boxes classified as risky and safe options.

Senapati's (2020b) paper highlights the detailed experiment process. To motivate our participants, we arranged some refreshments after the experiment. We then asked the participants to draw a card from their preferred boxes. We adopted an incentive mechanism developed by Humphrey and Verschoor (2004). Numbers were assigned to all the groups and from these groups members were selected at random. They then exercised their choice to win a prize for their group. In this way a group would win the prize money, not an individual participant. See Appendix A for a detailed understanding of how experiments were conducted in the study area.

12.2.1 Attitude toward risk using Expected Utility Approach

Using von Neumann and Morgenstern's (1944) utility, we identified the risk attitude class of the sample respondents. We gave sample respondents ten choice tasks and two boxes (safe vs risky option) to exercise their choice. Box A contained heart and diamond cards with a payoff of Indian rupees (INR) 25 and INR 30. Similarly, Box B contained spade and club cards with a payoff of INR 1 and INR 55.

Case (a): risk attitude class based on Box A

Here we have classified the risk attitude class based on the payoff given in Box A. A farmer is classified as risk-averse if his/her expected value of the utility is less than the utility of the expected value. From our experiment, we found:

$$E\left[U(X)\right]=P_H X_H+P_D X_D=(0.5)(\text{INR }25)+(0.5)(\text{INR }30)=\text{INR }27.5 \quad (12.1)$$

where $U(X)=X^{1/2}$, $X_H=\text{INR } 25$, and X_D = INR 30, P_H and P_D = 0.5 and H = hearts; D = diamonds.

$$E[U(X)]=P_H U(X_H)+P_D U(X_D)=(0.5)X_H^{1/2}+(0.5)X_D^{1/2}=(0.5)(25)^{1/2}+(0.5)(30)^{1/2}=5.23 \quad (12.2)$$

$$U\left[E(X)\right]=U(X)^{1/2}=U(27.5)^{1/2}=5.24 \quad (12.3)$$

A farmer is classified as risk-loving if their expected value of the utility is more than the utility of the expected value. Thus, from our experiment, we found:

$$E\left[U(X)\right]=P_H U(X_H)+P_D U(X_D)>U\left[E(X)\right] \quad (12.4)$$

$$E[U(X)]=P_H U(X_H)+P_D U(X_D)=(0.5)X_H^2+(0.5)X_D^2=(0.5)(25)^2+(0.5)(30)^2=762.5 \quad (12.5)$$

$$U\left[E(X)\right]=U(X)^2=U(27.5)^2=756.25 \quad (12.6)$$

Similarly, a farmer is classified as risk-neutral if their expected value of the utility is exactly equal to the utility of the expected value. Mathematically,

$$E\left[U(X)\right]=P_H U(X_H)+P_D U(X_D)=U\left[E(X)\right] \quad (12.7)$$

$$E[U(X)]=P_H U(X_H)+P_D U(X_D)=(0.5)(25)+(0.5)(30)=27.5 \quad (12.8)$$

$$U\left[E(X)\right]=U(27.5)=27.5$$

Case (b): risk attitude class based on Box B

Here we have classified the risk attitude class based on the payoff given in Box B. A farmer is classified as risk-averse if their expected value of the utility is less than the utility of the expected value. Mathematically,

$$E[U(X)]=P_S X_S+P_C X_C=(0.5)(\text{INR }1)+(0.5)(\text{INR }55)=\text{INR }28 \quad (12.9)$$

where $U(X)=X^{1/2}$, $X_S=\text{INR }1$, and X_C = INR 55, P_s and P_s = 0.5 and S = spades; C = clubs.

$$E[U(X)]=P_S U(X_S)+P_C U(X_C)=(0.5)X_S^{1/2}+(0.5)X_C^{1/2}=(0.5)(1)^{1/2}+(0.5)(55)^{1/2}=4.208 \quad (12.10)$$

$$U\left[E(X)\right]=U(X)^{1/2}=U(28)^{1/2}=5.291$$

A farmer is classified as risk-loving if their expected value of the utility is more than the utility of the expected value. Mathematically,

$$E[U(X)]=P_S U(X_S)+P_C U(X_C)=(0.5)X_S^2+(0.5)X_C^2=(0.5)(1)^2+(0.5)(55)^2=1{,}513 \quad (12.11)$$

$$U\left[E(X)\right]=U(X)^2=U(28)^2=784 \quad (12.12)$$

Similarly, a farmer is classified as risk-neutral if their expected value of the utility is exactly equal to the utility of the expected value. Mathematically,

$$E\left[U(X)\right]=P_S U(X_S)+P_C U(X_C)=(0.5)(1)+(0.5)(55)=28 \quad (12.13)$$

$$U\left[E(X)\right]=U(X)=U(28)=28 \quad (12.14)$$

12.2.2 The Holt–Laury lottery method and modification

Holt and Laury (2002) provided ten choice tasks to exercise the safe and risky options. Ihli *et al.* (2013) used the same approach, and I have also adopted the same procedure with minor modifications. Table 12.1 highlights the Holt–Laury lottery choice payoff matrix for the sample farm household in irrigated and rain-fed regions. There are ten choice tasks, and the monetary units are held constant. The expected

Table 12.1. Payoff matrix of the modified HL lottery method, Bolangir and Cuttack districts, Odisha, India. (Source: Author's own work.)

Choice task	Box A	Box B	Expected value A	Expected value B	CRRA ranges[a]	Risk aversion class[b]
1	With 1 card worth INR 30 and 9 cards worth INR 25	With 1 card worth INR 55 and 9 cards worth INR 1	25.5	6.4	$-\infty<r<-1.71$	Extremely RL
2	With 2 cards worth INR 30 and 8 cards worth INR 25	With 2 cards worth INR 55 and 8 cards worth INR 1	26	11.8	$-1.71<r<-0.95$	Highly RL
3	With 3 cards worth INR 30 and 7 cards worth INR 25	With 3 cards worth INR 55 and 7 cards worth INR 1	26.5	17.2	$-0.95<r<-0.49$	Very RL
4	With 4 cards worth INR 30 and 6 cards worth INR 25	With 4 cards worth INR 55 and 6 cards worth INR 1	27	22.6	$-0.49<r<-0.15$	RL
5	With 5 cards worth INR 30 and 5 cards worth INR 25	With 5 cards worth INR 55 and 5 cards worth INR 1	27.5	28	$-0.15<r<0.14$	RN
6	With 6 cards worth INR 30 and 4 cards worth INR 25	With 6 cards worth INR 55 and 4 cards worth INR 1	28	33.4	$0.14<r<0.41$	Slightly RA
7	With 7 cards worth INR 30 and 3 cards worth INR 25	With 7 cards worth INR 55 and 3 cards worth INR 1	28.5	38.8	$0.41<r<0.68$	RA
8	With 8 cards worth INR 30 and 2 cards worth INR 25	With 8 cards worth INR 55 and 2 cards worth INR 1	29	44.2	$0.68<r<0.97$	Very RA
9	With 9 cards worth INR 30 and 1 card worth INR 25	With 9 cards worth INR 55 and 1 card worth INR 1	29.5	49.6	$0.97<r<1.37$	Highly RA
10	With 100% of INR 30 and 0% worth of INR 25	With 100% of INR 55 and 0% worth of INR 1	30	55	$1.37<r<\infty$	Extremely RA

Note: Prizes are displayed in Indian Rupees (Rs = INR).

[a]Coefficient of relative risk aversion assuming a power risk utility function.

[b]RL, RN and RA represent risk-loving, risk-neutral and risk-avoiding, respectively (Senapati, 2020b).

value changes as farm households shift from one study to another. Here, one has to exercise ten choices from Box A or Box B. Until choice task four, the expected value of A is greater than that of B, and from choice task six onwards, the expected value of B is greater than that of A. If the farm household switches from the safe box to risky box, then one can determine the corresponding risk attitude class. If they initially switch to B until they choose task four, then they are a risk-taking person. Similarly, if a farm household switches after choice task six to option B, then they will be known as a risk-averse person. If they switch exactly at choice task five, then they are risk-neutral. The detailed procedure is highlighted in the appendix.

Table 12.2 highlights the risk attitude class obtained from the modified Holt–Laury lottery choice experiment method. We found that of the total sample, 259 farmers, or nearly 65%, had a risk-averse attitude. Eighty-one farmers were classified as risk-loving farmers in the study region, which is close to 21%. Similarly, we found that almost 15% of farmers were risk-neutral in both study districts.

Table 12.2. Classification of risk attitude class, Bolangir and Cuttack districts, Odisha, India. (Source: Author's field survey from May to August 2016.)

Classification	Frequency	Percentage (%)
Risk-loving	81	20.25
Risk-neutral	60	15.00
Risk-avoiding	259	64.75
Total	400	100

12.2.3 Determinants of risk attitude behavior of the study region

It is essential to see how socio-demographic variables such as gender, age, education, household size and marriage affect the farm household's risk attitude. Similarly, socio-economic variables such as credit, disaster experience, farm size and group membership play an important role in influencing the risk attitude of a farmer. Table 12.3 highlights variables and their impact on risk attitude.

Table 12.4 depicts the results of the ordered-probit model assuming risk attitude (modified Holt–Laury (HL) lottery) as the response variable. It shows that a high value of modified HL lottery implies the risk-loving attitude of the farm household. There is no multicollinearity among the explanatory variables, as the mean variance inflation factor (VIF) is less than 5. In this model, factors such as age, education, farm size and winner of the lottery choice experiment are negative and significantly influence the risk attitude of the farm household. As the farmer's age increases, they become more risk-averse, which is in line with Nielsen

Table 12.3. Factors affecting farmers' risk attitude behavior, Bolangir and Cuttack districts, Odisha, India. (Source: Author's field survey from May to August 2016.)

Variables	Impact observed from the literature
Socio-demographic variables	
Gender	Female farm households are more risk-averse than male (Croson and Gneezy, 2009).
Age	As age increases, farmers become more risk-averse (Nielsen *et al.*, 2013).
Education	As education level increases, farmers become more risk-averse (Harrison *et al.*, 2007).
Household size	Large families are more risk-averse than small families (Miyata, 2003).
Marriage	Married farmers are more risk-averse than unmarried farmers
Socio-economic variables	
Farm size	Large farmers are more risk-averse than others (Wiki *et al.*, 2004)
Farmer group membership	Farmers belonging to some kind of union are less risk-averse (Barham and Chitemi, 2009)
Access to credit	Access to credit makes farmers risk-averse (Eswaran and Kotwal, 1989)
Past disaster experience	As the disaster experience of the farmer increases, they become less risk-averse

Table 12.4. Parameter estimated from ordered probit regression, Bolangir and Cuttack districts, Odisha, India. Dependent variable: risk attitude (modified HL lottery). (Source: Author's field survey from May to August 2016.)

Variables	Coefficient
Age	−0.018* (0.011)
Gender	0.105 (0.395)
Education	−0.427* (0.24)
Household size	0.029 (0.042)
Farm size	−0.072*** (0.03)
Farmer group membership	−0.003 (0.28)
Off-farm income source	0.073 (0.271)
District	−0.334 (0.251)
Farming experience	−0.009 (0.012)
Past disaster experience	0.289 (0.518)
Access to credit	−0.032 (0.267)
Marriage	0.25 (0.52)
Winner modified HL lottery	−5.157*** (0.427)
LR Chi^2	503.14***
Pseudo R^2	0.71
Cut 1	−3.491 (0.955)
Cut 2	−1.29 (0.91)

Notes: Standard errors are indicated in parentheses.
*,** and *** indicate statistical significance at 1%, 5% and 10%, respectively.

et al. (2013). As the level of education of the farmers increases, they become relatively more risk-averse, in line with Harrison *et al.* (2007), who obtained a similar association between risk attitude and level of education. Farmers with a large acreage of land are more risk-averse than those with fewer acres of land, in line with Wik *et al.* (2004). The farmers who participated in the experiment and won are more risk-averse than others.

12.3 Risk Attitude and Insurance Adoption in Agriculture: Theoretical Model

Crop insurance adoption by farmers is used as the dependent variable. It is a dichotomous choice variable substituted as 1 and 0. Rahman (2008) used the binary probit model to analyze various factors related to crop diversification adoption tendency in the case of Bangladesh. Zhou *et al.* (2008) and Pfeiffer and Lin (2014) followed a similar mechanism to analyze various determinants of water-saving technology adoption in the case of China. Saqib *et al.* (2016) and Wang *et al.* (2016) also used the probit framework in the case of Pakistan and China to analyze various socio-economic determinants of technology adoption. Gaurav (2015) explored the risk and insurance association using panel data in India. Senapati (2020a) used the probit model to explore determinants of crop insurance adoption behavior in Odisha, India. In this chapter, I have used the binary logistic regression following Tan *et al.* (2020) to derive the odds ratio of farmers adopting crop insurance as follows:

$$Y_i = \delta X_i + U_i \tag{12.15}$$

The specific form of the model is demonstrated as:

$$P(Y_i \geq 0 \mid X_i) = \ln\left(\frac{p_i}{1-p_i}\right) = \delta_0 + \sum \delta_i X_i + U_i \tag{12.16}$$

This chapter's econometric model includes shock, risk attitude and control variables. Incorporating the above three variables, the model can be rewritten as:

$$Y_i = \delta_0 + \beta_i SHOCK + \Sigma \gamma_i RISK + \Sigma \alpha_i X_i + U_i \tag{12.17}$$

In Eqn 12.17, the shock represents the covariate shocks such as flood and drought. Risk attitude class variables denote the risk-taking, risk-avoiding and risk-neutral class, β captures shock, and γ is the risk-attitude behavior effect of the sample respondent. α represents the parameter estimates of control variables affecting the adoption of crop insurance. For testing whether the risk behavior class has any relation to mitigating the covariate shocks, the interaction term between shocks and risk class is introduced as:

$$\begin{aligned} Y_i = \delta_0 + \beta_i SHOCK + \Sigma \gamma_i RISK \\ + \Sigma \lambda (SHOCK * RISK) + \Sigma \alpha_i X_i + U_i \end{aligned} \tag{12.18}$$

where λ represents the interaction effect between covariate shocks such as flood and drought along with the risk behavior class of the sample respondent, which can influence the crop insurance adoption decision. Other variables have been defined above.

12.4 Results and Discussion

This section presents the descriptive information of several factors and the ordered probit regression results to measure the risk attitude. Additionally, a binary logistic regression is used to obtain the odd log ratio of the determinants of crop insurance adoption. Table 12.5 shows the description and summary statistics of the variables used in the empirical analysis.

Table 12.5 also shows descriptive information about the sample respondents who actively participated in the experiment in the two districts of Odisha. We found that, out of the total sample, insurance adoption in the Bolangir district is 55%. We found 42% of farmers have a risk-averse attitude in production and technology adoption decisions. The total area used for crop production is 2.82 ha, whereas land used for rice is 2.41 ha. Almost 97% of farmers perceived past disasters as droughts in the study region. We found that 75% of farmers who had insured their crop previously were interested in renewing. The average insured amount is INR 42,990. Similarly, the average premium paid per hectare is INR 1,074.75. In Bolangir, 92% of farmers produce crops other than rice. About 63% of farmers have adequate access to credit in their area. Almost one person in the family works as a non-agricultural laborer. On average, farmers own 2.77 livestock in the study region.

On the other hand, we found that, out of the total sample, insurance adoption in the Cuttack district is 74%. We found 28% of farmers have risk-averse attitudes in production and technology adoption decisions. The total land area used for the production of all crops is 2.96 ha, while land used for rice is 2.55 ha. About 96% of farmers perceived past disasters as floods in the study region. We

Table 12.5. Descriptive information about variables in Bolangir and Cuttack districts, Odisha, India. (Source: Author's field survey from May to August 2016.)

			Bolangir district		Cuttack district	
Variables	Variables description	Unit	Mean	SD	Mean	SD
Insurance	Insurance adoption	Yes = 1, No = 0	0.55	0.49	0.74	0.44
Risk aversion	Lottery choice experiment	High = 0, Low = 1	0.42	0.49	0.28	0.45
Production	Total land area used for production	Per hectare	2.82	3.06	2.96	3.35
	Land area used for rice production	Per hectare	2.41	2.93	2.55	2.76
	Dummy variable: experienced a disaster in previous years	Yes = 1, No = 0	0.965	0.184241	0.96	0.18
Crop insurance environment	Dummy variable: already insured previously	Yes = 1, No = 0	0.75	0.43	0.74	0.43
	Insured amount	INR	42,990	34,366.92	44,475	37,163.76
	Premium amount	INR/hectare	1,074.75	859.17	1,111.87	929.09
Education	Schooling years	Year	6.29	4.43	5.74	5.14
Risk management	Dummy variable: produces crop other than rice	Yes = 1, No = 0	0.92	0.27	0.85	0.35
	Dummy variable: credit accessibility	Yes = 1, No = 0	0.63	0.48	0.85	0.36
	Number of non-agricultural laborers in the family	Person	0.87	1.25	1.45	1.47
	Number of livestock	Number	2.775	1.82	2.93	2.21

found that 74% of farmers who had insured their crops previously were interested in renewing. The average insured amount is INR 44,475. Similarly, the average premium paid per hectare is INR 1111.87. In Cuttack, 85% of farmers produce crops other than rice. Some 85% of farmers have adequate access to credit in their area. Almost one and a half people in the family work as non-agricultural labor. On average, farmers own 2.93 livestock in the study region.

Table 12.6 shows that the LR Chi^2 is highly significant, and the value of Pseudo R^2 is 0.65. There is no multicollinearity among the explanatory variables, as the mean VIF is less than 5. The regression results show that out of 16 explanatory variables, only four variables, farm age, farm experience, land ownership and willingness to pay an amount, are statistically significant. The odd log ratio of insurance adoption declines by 0.168 as land ages. The interest to insure declines if the farm is very old, since the probability of getting a high yield is low. The odd log ratio of insurance adoption again declines by 0.068 as the experience of the farmer increases. Experience in handling farm business is a crucial variable as farmers are well acclimatized with basic knowledge of farm production. They used to avoid taking insurance and adopted other risk management strategies. The odd log ratio of insurance adoption increases by 0.209 if the farmer owns the land. When the farmer owns the land, their interest to insure crops increases compared to tenants in the study region. Similarly, the odd log ratio of insurance adoption increases by 0.001 if the farmer's willingness to pay the amount increases. Risk aversion is not a significant factor in deciding insurance adoption in drought regions. However, the interaction effect between the shock and risk attitude of the farmer is insignificant.

The last two columns of Table 12.6 show the results for Cuttack district farmers. The LR Chi^2 is highly significant, and the value of Pseudo R^2 is 0.13. There is no multicollinearity among the explanatory variables, as the mean VIF is less than 5. The regression results show that out of 16 explanatory variables, only four

Table 12.6. Logistic regression results for Bolangir and Cuttack districts, Odisha, India Dependent variable: crop insurance adoption. (Source: Authors' calculation based on survey data.)

	Bolangir district		Cuttack district	
Variables	Coefficient	Standard error	Coefficient	Standard error
Awareness	0.035	0.037	−0.105	0.144
Age	0.103	0.107	−0.005	0.131
Gender	−1.165	143.22	−0.188	0.131
Education	0.023	0.019	0.059**	0.025
Farm group membership	0.025	0.047	0.066	0.067
Farm age	−0.168**	0.076	0.305***	0.093
Farm experience	−0.068*	0.05	0.017	0.071
Land ownership	0.209***	0.046	−0.055	0.081
Credit	0.017	0.04	0.318***	0.104
Extension contact	−0.025	0.056	−0.118	0.074
Risk aversion	0.825	143.22	0.004	0.223
Information access	−0.029	0.04	0.011	0.067
Past disaster experience	−0.019	0.08	0.33**	0.166
Willingness to pay amount	0.001***	0.0005	−0.0001	0.0002
Drought	1.145	143.22	−0.182	0.141
Drought*risk	−0.945	143.23	0.016	0.234
Constant	0.271	2.118	0.001	0.005
LR Chi^2	147.88***		28.46**	
Pseudo R^2	0.65		0.13	
Number of observations	200		200	

Note: *,** and *** indicate statistical significance at 1%, 5% and 10%, respectively.

variables, education, farm age, credit and past disaster experience, are statistically significant. The odd log ratio of insurance adoption increases by 0.059 as the farmer becomes more educated. The interest in insurance increases if the level of education is very high, since educated farmers value insurance more than their uneducated counterparts.

The odd log ratio of insurance adoption again increases by 0.305 as land becomes old. The reason is that farmers are highly uncertain about the yield due to the farm's age, which induces them to insure their crops. The odd log ratio of insurance adoption increases by 0.318 if the farmer has access to credit facilities. The reason is that, in the study region, insurance for farmers who borrowed money is compulsory, whereas for non-borrowers it is optional, as per the government-sponsored crop insurance scheme, the *Pradhan Mantri Fasal Bima Yojana*, introduced in 2016. Similarly, I found the odd log ratio of insurance adoption increases by 0.33 if the farmers have experienced disaster earlier. In the study region, most farmers had experienced either floods or cyclones yearly. Again, in flood regions, risk aversion is not a significant factor in deciding insurance adoption. Similarly, the interaction effect between the farmer's shock and risk attitude is also insignificant.

We found most of the farmers who insured rice are also interested to insure again due to their past disaster experience. In Bolangir district, drought is a regular phenomenon. Rising temperatures and crop damages due to insect and pest attack are worrisome factors for the sample farm household. Another problem we found was that they have much less contact with extension officers and much less access to information. Access to market for various agricultural inputs is also lacking in the study region as the market is far away from their home. In the Cuttack district, farmers are also interested in renewing their insurance due to the significant increase in floods in the region. We found most of the farmers in this region are well educated and aware of the insurance program. Their access to credit is also very high. Distance to market is not a factor here as most of the farmers avail inputs nearby their home. Again, the only worrisome factor is not having enough contact with agricultural extension officers. They lack proper guidance regarding fertilizer use and type of seed to be used for better production.

12.4.1 Assessment of farmers' attitudes and their response to the crop insurance program

The present study has examined the attitude and response of sample farmers toward crop insurance schemes, which can be used as a reference by the insurance companies in two geographically distinct locations. The study used the mean score using a Likert scale to analyze the data. Following the Likert scale, the study tried to examine the attitudes of sample households to an insurance scheme and identify various constraints to crop insurance participation. Table 12.7 shows that sample households in the Bolangir district perceived farming as risky. Due to this, farmers have positive attitudes toward insurance, but they do not perceive insurance as reducing their worries and stress completely. The results further show that sample respondents believed that the government should bear the entire burden when there is a disaster resulting in huge crop losses. For this reason, they perceived that agricultural insurance should be made mandatory for all farmers.

Farmers in the rain-fed region rank farm business and insurance as important factors in their production decisions. Small and marginal farmers in the area are not getting enough benefits from crop insurance programs. On the other hand, in the irrigated Cuttack region, farmers rank the government's responsibility for recovering crop losses as the most important factor for their production decisions, followed by the significance of insurance adoption. They also reported that small and marginal farmers are not getting enough subsidies from the government.

Table 12.8 highlights the constraints faced by the farmers in the study region. We found lack of information and lengthy bureaucracy controls are less important. We found claim settlement problems, awareness issues, relatively high premium rates and the perception of farmers to be constraints to crop insurance adoption decisions. In the irrigated Cuttack region, they did not perceive that insurance would completely reduce their worries and stress. Therefore, they

Table 12.7. Attitude of farmers toward crop insurance program in Bolangir and Cuttack districts, Odisha, India. (Source: Author's field survey from May to August 2016.)

	Bolangir district					Cuttack district				
	Incidence					Incidence				
Statements	SA	A	D	SD	Mean score	SA	A	D	SD	Mean score
Farming is a risky business	64	107	18	11	2.6	113	79	8	0	2.94
Agricultural insurance is beneficial	105	55	36	4	2.75	135	46	5	14	2.92
Crop insurance lowers stress and uncertainty	42	105	40	13	2.4	21	38	88	53	1.78
Government is responsible for recovering losses in agriculture	153	47	0	0	3.14	167	33	0	0	3.19
Agricultural insurance should be mandatory	87	95	15	3	2.77	56	87	22	35	2.35
Small and marginal farmers are not getting enough help from insurance	65	81	37	17	2.47	116	51	11	22	2.75
Loss and damage from shocks are beyond the scope of insurance	15	21	75	89	1.51	9	49	106	36	1.79
Losses in crop production are due to invisible power	115	61	13	11	2.83	25	76	81	18	2.12

Note: SA, A, D and SD indicate strongly agree, agree, disagree and strongly disagree, respectively.

did not favor mandatory crop insurance policies. The results further show sample respondents perceived that the government should bear the entire burden when there is a disaster resulting in huge crop losses.

Table 12.8 highlights the constraints, such as lack of knowledge on the advantages of crop insurance, lack of proper information about the availability of agricultural insurance and giving less priority to insurance, were viewed as less important by the sample respondents. Sample households identified the first two constraints as less important because they are very much aware of the crop insurance program in their region. They also believed crop insurance is not a priority compared to other important needs such as educating their children, huge health expenditures and managing the farm through several other risk-reducing strategies. Thus, the constraints identified by this study could be responsible for the low participation rates of sample farmers in the crop insurance program.

In the rain-fed region, we found farmers have several constraints to insurance adoption, such as less compensation to recover the crop losses due to drought, delay in claim settlement and high premium rates. They also reported that even though they are aware of insurance, due to poverty they cannot insure their crop. Their beliefs in predestination and destiny again create significant problems for insurance adoption. On the other hand, in the irrigated Cuttack region, farmers mostly highlighted lower compensation amounts and high bureaucratic controls as the major constraints, followed by problems related to claim settlement and high premium rates in insurance adoption. Due to low income, most of the farmers are interested in migrating to other regions in search of jobs, since income from rice production is relatively low due to the frequent occurrence of floods in this region.

12.5 Conclusion

This chapter shows the significance of risk attitude class and socio-demographic and socio-economic determinants in insurance adoption decisions. We used household-level data in irrigated and rain-fed regions of Odisha, India. The study used an ordered probit model and lottery choice experiment method and found that factors such as operator age, educational attainment, farm size and winner of the modified HL lottery play an important role in highlighting the risk attitude class of the farm household.

Though no significant direct association exists between risk attitude and climatic shocks

Table 12.8. Challenges faced in adoption of crop insurance in Bolangir and Cuttack districts, Odisha, India. (Source: Author's field survey from May to August 2016.)

Statements	Bolangir district					Cuttack district				
	Incidence				Mean score	Incidence				Mean score
	SA	A	D	SD		SA	A	D	SD	
Problems related to claim settlement	106	55	23	16	2.71	133	59	8	0	3.02
Lack of awareness	93	39	65	3	2.59	27	30	101	42	1.84
High premium rate	145	31	22	2	2.99	136	61	3	0	3.05
Lack of information about the availability of crop insurance	78	43	62	17	2.42	13	55	43	89	1.63
Insurance is not a priority to farmers compared to other needs	111	33	12	44	2.54	7	18	95	80	1.46
Beliefs of farmers in predestination and destiny	162	15	15	8	3.04	183	17	0	0	3.26
Difficulty in implementing insurance policy	107	30	49	14	2.62	97	81	15	7	2.78
Low income of the farmers	172	13	10	5	3.13	166	28	6	0	3.16
Late payment of compensation	191	9	0	0	3.29	191	9	0	0	3.29
Lower compensation amounts	180	16	4	0	3.23	152	42	6	0	3.11
Bureaucratic control	26	87	66	21	2.16	170	26	0	4	3.17

Note: SA, A, D and SD indicate strongly agree, agree, disagree and strongly disagree, respectively.

with insurance adoption, the effect can be observed indirectly. Similarly, using binary logistic regression, the study found the level of education, age of the farm, past disaster experience and access to credit play an important role in crop insurance adoption behavior in the flood-affected region of Odisha. On the other hand, operator age, experience, land ownership and willingness to pay for insurance affect the crop insurance adoption behavior in the drought-affected region of Odisha. The chapter recommends that policymakers obtain sufficient knowledge of risk attitude class, general buying attitude and the constraints farm households face in the different ecosystems before designing and offering formal insurance programs.

Acknowledgments

I would like to thank the editor and anonymous reviewers for their constructive comments and suggestions. The earlier draft was presented at SJSOM, IIT Bombay in an International workshop on "Reimagining risk and vulnerability in Indian Agriculture" from 10th March to 11th March, 2023. The author has benefited from discussions with Prof. Sarthak Gaurav and Prof. Phanindra Goyari for the conceptual development of risk and vulnerability in India, and is grateful to all participants in the experiment from both the regions of Odisha. However, the usual disclaimer applies.

References

Anderson, J.R. and Dillon, J.L. (1992) *Risk Analysis in Dryland Farming Systems. Farming Systems Management Series (2).* Food and Agriculture Organization, Rome.

Anderson, J.R., Dillon, J.L. and Hardaker, J.B. (1977) *Agricultural Decision Analysis*. The Iowa State University, Ames, Iowa.

Antle, J.M. (1987) Econometric estimation of producers' risk attitudes. *American Journal of Agricultural Economics* 69, 509–522.

Barham, J. and Chitemi, C. (2009) Collective action initiatives to improve marketing performances: lessons from farmer groups in Tanzania. *Food Policy* 34, 53–59.

Binswanger, H.P. (1980) Attitudes toward risk: experimental measurement in rural India. *American Journal of Agricultural Economics* 62, 395–407.

Binswanger, H.P. and Rosenzweig, M.R. (1986) Behavioral and material determinants of production relations in agriculture. *Journal of Development Studies* 22, 503–539.

Binswanger, H.P. and Sillers, D.A. (1984) Risk aversion and credit constraints in farmers' decision making: a reinterpretation. *Journal of Development Studies* 20, 5–21.
Chand, R., Raju, S.S. and Pandey, L.M. (2011) Growth crisis in agriculture: severity and options at national level and state level. In: Balakrishnan, P. (ed.) *Economic Reforms and Growth in India*. Orient Black Swan, New Delhi, pp. 283–300.
Croson, R. and Gneezy, U. (2009) Gender differences in preferences. *Journal of Economic Literature* 47, 448–474.
Economic Survey of Orissa (2008–2009) Government of Odisha, Bhubaneswar, India.
Eswaran, M. and Kotwal, A. (1989) Credit as insurance in agrarian economies. *Journal of Development Economics* 31, 37–53.
Fama, E.F. (1968) Risk, return, and equilibrium: some clarifying comments. *The Journal of Finance* 23, 29–40.
Gaurav, S. (2015) Are rainfed agricultural households insured? Evidence from five villages in Vidarbha, India. *World Development* 66, 719–736.
Goldstein, B.L., Demak, M., Northridge, M. and Wartenberg, D. (1992) Risk to groundlings of death due to airplane accidents: a risk communication tool. *Risk Analysis* 12, 339–341.
Graham D. (2000) Making sense of risk. *Risk Analysis* 20, 302–306.
Harrison, G.W., Lau, M.I. and Rutstrom, E.E. (2007) Estimating risk attitudes in Denmark: a field experiment. *The Scandinavian Journal of Economics* 109, 341–368.
Holt, C.A. and Laury, S.K. (2002) Risk aversion and incentive effects. *The American Economic Review* 92, 1644–1655.
Humphrey, S.J. and Verschoor, A. (2004) Decision-making under risk among small farmers in East Uganda. *Journal of African Economics* 13, 44–101.
Ihli, H.J., Chiputwa, B., Bauermeister, G.F. and Mußhoff, O. (2013) Measuring risk attitudes of smallholder farmers in Uganda: how consistent are results of different methods. *Proceedings of the Second International Agricultural Risk, Finance, and Insurance Conference (IARFIC), June 16–18, Vancouver, Canada*.
Joyce, J.M. and Vogel, R.C. (1970) The uncertainty in risk: is variance unambiguous? *The Journal of Finance* 25, 127–134.
Just, R.E. and Pope, R.D. (2003) Agricultural risk analysis: adequacy of models, data, and issues. *American Journal of Agricultural Economics* 85, 1249–1256.
Kaplan, S. and Garrick, B.J. (1981) On the quantitative definition of risk. *Risk Analysis* 1, 11–27.
Knight, F.H. (1921) *Risk, Uncertainty and Profit.* Houghton Mifflin, Boston, Massachusetts.
Markowitz, H.M. (1959) *Portfolio Selection: Efficient Diversification of Investments*. Wiley, New York.
Miyata, S. (2003) Household's risk attitudes in Indonesian villages. *Applied Economics* 35, 573–583.
Moscardi, E. and De Janvry, A. (1977) Attitudes toward risk among peasants: an econometric approach. *American Journal of Agricultural Economics* 59, 710–716.
Newbery, D. and Stiglitz, J. (1981) *The Theory of Commodity Price Stabilisation: A Study in the Economics of Risk*. Clarendon Press, Oxford, UK.
Nielsen, T., Keil, A. and Zeller, M. (2013) Assessing farmers' risk preferences and their determinants in a marginal upland area of Vietnam: a comparison of multiple elicitation techniques. *Agricultural Economics* 44, 255–273.
Pfeiffer, L. and Lin, C-Y.C. (2014) Does efficient irrigation technology lead to reduced groundwater extraction? Empirical evidence. *Journal of Environmental Economics and Management* 67, 189–208.
Rahman, S. (2008) Determinants of crop choices by Bangladeshi farmers: a bivariate probit analysis. *Asian Journal of Agriculture and Development* 5, 29–41.
Reynaud, A. and Couture, S. (2012) Stability of risk preference measures: results from a field experiment on French farmers. *Theory and Decision* 73, 203–221.
Rothschild, M. and Stiglitz, J.E. (1970). Increasing risk: I. A definition. *Journal of Economic Theory* 2, 225–243.
Roumasset, J.A. (1976) *Rice and Risk: Decision Making Among Low-income Farmers*. North-Holland, Amsterdam.
Roumasset, J.A., Boussard, J.M. and Singh, I. (1979) *Risk, Uncertainty, and Agricultural Development*. Southeast Asian Regional Centre for Graduate Study and Research in Agriculture (SEARCA) and Agricultural Development Council (ADC), Laguna, Philippines.
Saqib, S.E., Ahmad, M.M., Panezai, S. and Rana, I.A. (2016) An empirical assessment of farmers' risk attitudes in flood-prone areas of Pakistan. *International Journal of Disaster Risk Reduction* 18, 107–114.
Savage, L. (1954) *The Foundations of Statistics*, 2nd edn. Wiley, New York.
Scott, J.C. (1977) *The Moral Economy of the Peasant: Rebellion and Subsistence in Southeast Asia*. Yale University Press, New Haven, Connecticut.
Senapati, A.K. (2020a) Do farmers value insurance against extreme droughts and floods? Evidence from Odisha, India. *Global Business Review* 21, 1–24.
Senapati, A.K. (2020b) Evaluation of risk preferences and coping strategies to manage with various agricultural risks: evidence from India. *Heliyon* 6, 1–13.

Senapati, A.K. (2020c) Insuring against climatic shocks: evidence on farm households' willingness to pay for rainfall insurance product in rural India. *International Journal of Disaster Risk Reduction* 42, 1–12.

Tan, Y., Qian, L., Sarkar, A., Nurgazina, Z. and Ali, U. (2020) Farmer's adoption tendency towards drought shock, risk-taking networks and modern irrigation technology: evidence from Zhangye, Gansu, PRC. *International Journal of Climate Change Strategies and Management* 12, 431–448.

Tobin, J. (1958) Liquidity preference as behavior towards risk. *Review of Economic Studies* 25, 65–86.

von Neumann, J. and Morgenstern O. (1944) *Theory of Games and Economic Behaviour*. Princeton University Press, Princeton, New Jersey.

Wang, N., Gao, Y., Wang, Y. and Li, X. (2016) Adoption of eco-friendly soil-management practices by smallholder farmers in the Shandong province of China. *Soil Science and Plant Nutrition* 62, 185–193.

Wik, M., Kebede, T., Bergland, O. and Holden, S. (2004) On the measurement of risk aversion from experimental data. *Applied Economics* 36, 2443–2451.

Zhou, S., Herzfeld, T., Glauben, T., Zhang, Y. and Hu, B. (2008) Factors affecting Chinese farmers' decisions to adopt water-saving technology. *Canadian Journal of Agricultural Economics* 56, 51–61.

Appendix

Q. How are you going to win these prizes?

Ans: To win these prizes, you will first have to choose between two Boxes, Box A and Box B for each of the 10 rows.

Q. How do these two boxes differ?

Ans: Each box contains 10 cards. The two boxes contain cards of a different suit (spades, hearts, diamonds and clubs) with a different value. We draw only one card from the selected box, which will be the prize. If you choose Box A, you can win a prize of INR 30 (diamond card) or a prize of INR 25 (heart card). And if you choose Box B, you can win a prize of INR 55 (club card) or a prize of INR 1 (spade card). We are going to ask you:

Q. Which of these two boxes you prefer?

Ans: Note that with Box A the difference between the prizes is small, while it is large in the case of Box B.

In addition, in Box A the prize of INR 30 is smaller than the prize of INR 55 in Box B, and the prize of INR 25 in Box A is greater than the prize of INR 1 in Box B.
Thus, you will choose between Box A and Box B in 10 rows, one after another.

Box A:

- Box A contains nine heart cards and one diamond card (see Fig.12.1). Each heart cards is worth INR 25 and the diamond card is worth INR 30.
- If this Box is selected and the diamond card is subsequently drawn, you will win INR 30. In the case that one of the heart cards is drawn, you will win INR 25.
- So, if we pick a card from the box, it may be a heart or a diamond. But, it is more likely that we will pick one of the heart cards because there are more heart cards than diamond cards in the box.

Box B:

- Now, let's look at Box B. What is different about it? Well, this box contains nine spade cards and one club card (see Fig. 12.2). Each spade card is worth INR 1 and the club card is worth INR 55.
- If this box is selected and the club card is subsequently drawn, you will win INR 55. In the case that one of the spade cards is drawn, you will win INR 1.
- So, if we pick a card from the box, it may be a spade or a club. But, it is more likely that we will pick one of the spade cards because there are more spade cards than club cards in the box.

13 Attitudes toward Risk and Farmers' Willingness to Pay for Drought-Tolerant Maize and Weather Index Insurance in West Africa

Tahirou Abdoulaye,[1]* Essossinam Ali,[2] Abel-Gautier Kouakou[1] and Tesfamicheal Wossen[3]

[1]*International Institute of Tropical Agriculture, Bamako, Mali;* [2]*University of Kara, Kara, Togo;* [3]*International Institute of Tropical Agriculture, Nairobi, Kenya*

Abstract

In West Africa, climate change is seen as a major risk to agricultural production and food systems. Several innovations have been developed and proposed to farmers to help them reduce the risk and better adapt to climate change. Among the climate-smart agricultural technologies available, drought-tolerant crops and weather index insurance have received attention in recent years. Understanding farmers' decision-making under risk will improve the ability to set up development programs that will lead to the adoption of these innovations. This chapter uses empirical work to show farmers' willingness to pay for these risk-reducing innovations, including drought-tolerant maize in Nigeria and weather-based index insurance in Togo. We conclude that in both cases there is a potential market for risk-reducing climate-smart agricultural innovations in West Africa.

13.1 Introduction

Agricultural production is prone to several risks that affect nutrition, food security and livelihoods. In Africa, especially West Africa, climate change is seen as a major risk to agricultural production and food systems (World Bank, 2022). The stochastic nature of weather, pest and other biotic and abiotic shocks, coupled with the intrinsic feature of agriculture (the lag between production decision and output realization), make agriculture inherently risky. This is particularly important as crop production in many developing countries depends largely on seasonal rainfall. Understanding risk-mitigating and coping mechanisms available to smallholders in developing countries, where formal insurance schemes are still largely unavailable, is thus crucial for designing risk-mitigating strategies.

Among the climate-smart agricultural technologies available, drought-tolerant crops and weather index insurance (WII) have received attention in recent years in West Africa. This is possibly explained by the implementation of (research) initiatives that intend to help rural communities adapt to abiotic stress factors (Burke and Lobell, 2010; FAO, 2015) and develop

*E-mail: t.abdoulaye@cgiar.org

DOI: 10.1079/9781800622289.0013

hydrometeorological information services in West Africa, including the Sahel region (World Bank, 2022). In addition, regional institutions such as the Permanent Interstates Committee for Drought Control in the Sahel (CILSS) are being strengthened to provide these agricultural technologies (World Bank, 2022).

However, successfully adopting these agricultural technologies requires understanding farmers' decision-making under risk, since it is an inevitable part of many farming activities in West Africa. Evidence of the effects of risk and risk aversion on farmers' technology adoption is well-documented (e.g. Foster and Rosenzweig, 2010; Liebenehm and Waibel, 2014). Likewise, studies that assess the weight of evidence for the two major received theories of choice under risk, expected utility theory (EUT) and prospect theory (PT), are well-documented (e.g. Kahneman and Tversky, 1979; Harrison *et al.*, 2010). In this chapter, we contribute to both fields of literature by highlighting the usefulness of both EUT and PT for analyzing farmers' willingness to pay (WTP) for agricultural technologies in West Africa. We present the cases of drought-tolerant maize in northern Nigeria and WII in northern Togo.

The rest of the chapter is organized into two case studies. We begin in Section 13.2 with the first case study, which elicits farmers' valuation of drought-tolerant maize in northern Nigeria, considering heterogeneity in preferences for the stress-tolerance traits that exist across risky situations under gain and loss domains. Section 13.3 presents the second case study, in which we look at farmers' WTP for WII in northern Togo. In each case study, we present the motivation of the study, the methods used and the results and discussion. Section 13.4 concludes.

13.2 Case Study 1: Attitudes toward Risk and Farmers' WTP for Drought-Tolerant Maize in Northern Nigeria

Maize is Africa's most important food crop, cultivated on an estimated 35 million hectares and supporting more than 300 million people on the continent (e.g. Fisher *et al.*, 2015). However, given that maize is predominantly grown under rain-fed systems, it is susceptible to the increasing intensity and frequency of drought shocks induced by climate change. The occurrence of these shocks has been reported to increase the risk of crop failure, which negatively affects the income, food and nutrition security of millions of smallholder farmers and their families (Fisher *et al.*, 2015).

Considerable research efforts have been directed at developing adaptation strategies against abiotic (and biotic) stress factors that are limiting maize productivity. For example, the Drought Tolerant Maize for Africa (DTMA) project aimed to develop and deploy improved drought-tolerant maize varieties in sub-Saharan Africa (Burke and Lobell, 2010). However, the adoption of drought-tolerant maize varieties depends on farmers' acceptance and demand, showing the importance of identifying market potentials for drought-tolerant traits. Market potentials for improved maize varieties are determined by farmers' preferences for or valuation of the traits, as reflected by the price premiums (or discount) they are willing to pay (Ouma *et al.*, 2007; Bello and Abdulai, 2016). Although there have been efforts in maize research in the development and dissemination of new varieties in sub-Saharan Africa, few studies have examined preferences and willingness to pay for maize crop traits (e.g. De Groote *et al.*, 2011; Kassie *et al.*, 2017). Most of these studies are centered on determining consumers' preferences in the context of the food safety domain (e.g. De Groote and Kimenju, 2008; De Groote *et al.*, 2011, 2016).

To our knowledge, the work of Kassie *et al.* (2017) is the only study that has consistently estimated the implicit prices farmers are willing to pay for drought tolerance in maize compared with other intrinsic traits (like texture and cob size), using a hypothetical choice experiment framework. We add to this literature by employing an incentive-compatible non-hypothetical choice experiment to elicit farmers' valuations of the different traits of stress-tolerant maize varieties, including the drought-tolerant trait, and examine the effects of attitudes to risk therein. We hypothesize that the magnitude of farmers' valuations of drought-tolerant trait varieties and the potential acceptability of these varieties may likely vary across attitudes to risk, all things being equal.

13.2.1 Method and data

The methods used in this research include modeling farmers' WTP and risk attitudes, data collection and data analysis.

Modeling farmers' WTP and attitudes toward risk

To model farmers' WTP for drought-tolerant maize, we employ a Mixed Multinomial Logit (MMNL) model with error correction (EC; Train and Weeks, 2005). We use the MMNL-EC model because, in addition to allowing for preference heterogeneity across sample farmers, the model addresses the systematic effects across utilities between alternatives caused by the opt-out option in an experimental design (Scarpa and Alberini, 2005). Similarly, the estimation in the WTP space provides greater coherence in the results when a welfare measure such as WTP for an attribute is the central focus of the research, as in our case. The WTP specification also allows for the welfare values to be estimated directly, such that appropriate distributions for the values are assumed from the onset and the coefficients directly represent the WTP values for the attributes.

For specification in a preference space, the utility function of individual n for selecting alternative j in choice situation t is a function of the price K_{njt} and the non-price x_{njt} attributes. The utility function is specified as follows:

$$U_{njt} = \beta_{nk} K_{njt} + \sum\nolimits_{j=1}^{J} \beta_{nj} x_{njt} + \eta_{njt} + \varepsilon_{njt} \quad (13.1)$$

where scalar β_{nk} and vector β_{nj} vary randomly over farmers; η_{njt} is an error component distributed normally but with zero mean, and ε_{njt} is an unobserved random term that is assumed to be independent and identically distributed and Gumbel-distributed. Having drawn a distinction between price and non-price attributes, the WTP for an attribute j can be obtained from the estimated coefficients in Eqn 13.1. The marginal WTPs are computed as the ratio of the partial derivative of the utility function in terms of the attributes of interest to the partial derivative of the utility function in terms of the price variable (Bliemer and Rose, 2010). However, turning to utility specification in the WTP space, the utility function is re-parameterized such that the coefficients can be directly interpreted as marginal WTP effects (Train and Weeks, 2005; Scarpa *et al.*, 2008). The basic specification in the WTP space can be expressed as follows:

$$U_{njt} = \beta_{nk}\left[K_{njt} + \sum\nolimits_{j=1}^{J} \beta'_{nj} x_{njt} + \eta_{njt} \right] + \varepsilon_{njt} \quad (13.2)$$

where $\beta'_{nj} = \beta_{nj} / \beta_{nk}$ denote the WTP values and β_{nk} is formulated as a normalizing constant.

To model farmers' attitudes toward risk, we use the cumulative prospect theory (CPT; Tversky and Kahneman, 1992). It extends the EUT in terms of loss aversion and probability distortion (Tanaka *et al.*, 2010), allowing for a distinction between gains and losses in attitudes toward risk.

We express the utility function over wealth w for a two-outcome gamble as follows:

$$\varphi(w) = \begin{cases} (w - w_0)^{\delta^+} & \text{if } w \geq w_0 \\ -\theta\left[(-w + w_0)^{\delta^-}\right] & \text{if } w < w_0 \end{cases} \quad (13.3)$$

where δ^+ and δ^- represent the curvature of the value function in gains and losses, respectively; w_0 is the reference point; and θ is a loss-aversion parameter. In practice, the gains and losses are compared to a reference point that defines the gain and loss domains and can accommodate different weightings. We set reference point $w_0 = 0$ to identify whether a farmer is facing a gain or loss and to facilitate comparison of farmers' attitudes. Using the Tversky and Kahneman (1992) specification of the probability weighting function, we transformed the probabilities and applied Prelec's (1998) specification as a robustness check. The probability weighting function, $\Pi(v)$, is $= \frac{v^{\mu^+}}{[v^{\mu^+} + (1-v)^{\mu^+}]^{1/\mu^+}} \text{ if } w \geq 0$ and $\frac{v^{\mu^-}}{[v^{\mu^-} + (1-v)^{\mu^-}]^{1/\mu^-}} \text{ if } w < 0$, where μ^+ and μ^- are probability sensitivity parameters in gains and losses, respectively. If $\mu < 1$, the weighting function is inverse S-shaped and the farmer is risk-averse, whereas if $\mu > 1$, the weighting function is S-shaped and the farmer is risk-loving.

Data collection

The study was conducted in the predominantly maize-producing areas of northern Nigeria. Data were collected through a household survey conducted between June and September 2018. A multi-stage stratified sampling was used to select the surveyed farmers. First, we stratified northern Nigeria based on agro-ecological zones (AEZs) into two strata, dry and moist savannas. Second, we randomly selected three states and four sub-locations (villages) in the dry savanna agro-ecology – Katsina, Kano and Kaduna – and from the moist savanna agro-ecology, we sampled sub-locations from two states – Niger and Kwara. Finally, we randomly sampled a proportionate number of respondents from the lists of smallholder maize farmers provided by extension agents in each sub-location, making our total sample size 600 respondents. It is important to note that we sampled sub-locations for the International Institute of Tropical Agriculture (IITA) on-farm agronomic trials (OFAR) and non-OFAR clusters that share similar population characteristics along the dimension that could affect the outcome variables. The survey was carried out using face-to-face interviews with a questionnaire. In addition to collecting preference data using the randomized controlled experiments (RCEs) and lotteries (see next section), we collected data on farmers' plot-level agronomic activities and on farmers' characteristics, including demographic, economic, cognitive and attitudinal variables.

A. MAIZE TRAIT PREFERENCE CHOICE EXPERIMENT In the non-hypothetical choice experiment, farmers were presented with four choice scenarios. Each scenario consisted of an opt-out option and two experimentally generated alternatives that were described by stress tolerance and price, both four-level attributes. For each choice scenario, the participants had to choose one of the three options or alternatives that best reflected their preferences. This procedure is similar to farmers' typical purchasing decisions when buying seed inputs in the market. The non-hypothetical nature was introduced by including economic incentives in the design in the following way. At the end of the experiment, once participants had completed the four choice tasks, one of the tasks was randomly drawn as binding, and participants were asked to pay the corresponding price. If the farmer chose the opt-out option in the randomly drawn choice task, he or she would not purchase any maize seed and would not pay anything. We did not have such a case in our experiment. We adopted the design by Lusk and Schroeder (2004) by letting all the maize varieties be available in each choice set. Table 13.1 shows the attributes and attribute levels used in the experiment.

B. RISK PREFERENCE CHOICE EXPERIMENTS In the risk preference lottery choice experiment, farmers were presented with a series of choices between two lotteries with varying probabilities and outcomes in the gain and loss domains. At the end of the experiment, once participants had made the choices, one of the choices was then randomly drawn as binding, and it was played and paid. Table 13.2 shows the lottery choice experiment in gain and loss domains.

Generally, the idea behind our elicitation strategies is to reveal the amount needed to identify a farmer's switching point that provides information about his or her preferences. For example, a risk-neutral farmer without distortion of probabilities considers only the expected value of the lottery when making his or her choices. The neutral switching point is the reversal of the expected value between the two lotteries in a series. Farmers who switched before the neutral switching point are considered risk-seeking, whereas farmers who switched after the neutral switching point are considered risk-averse. Similarly, to control for framing effects, we designed the nine series so that the row position of the neutral switching point was not systematically the middle row but instead varied across the series (Bougherara *et al.*, 2017)

After the experiments, we asked farmers to complete a survey, which, in addition to asking

Table 13.1. Attributes and attribute levels. (Source: Bello *et al.*, 2020.)

Attributes	Description	Attribute levels
Stress tolerance	Improved maize varieties with resistance to biotic and abiotic factors	None (conventional), *Striga*, drought, *Striga* plus drought
Price	Purchasing price of maize varieties (naira/kg)	₦250, ₦380, ₦400, ₦420

Table 13.2. Lottery choice experiment (risk in gain/loss domain). (Source: Bello *et al.*, 2020.)

Lottery choice experiment (risk in gain domain)						Lottery choice experiment (risk in loss domain)					
Lottery A		Lottery B		Choice		Lottery A		Lottery B		Choice	
3/10	7/10	1/10	9/10			1/10	9/10	3/10	7/10		
₦200	₦100	₦270	₦75	A	B	₦ −200	₦ −75	₦ −150	₦ −100	A	B
₦200	₦100	₦280	₦75	A	B	₦ −210	₦ −75	₦ −150	₦ −100	A	B
₦200	₦100	₦350	₦75	A	B	₦ −350	₦ −75	₦ −150	₦ −100	A	B
₦200	₦100	₦390	₦75	A	B	₦ −390	₦ −75	₦ −150	₦ −100	A	B
₦200	₦100	₦430	₦75	A	B	₦ −430	₦ −75	₦ −150	₦ −100	A	B
₦200	₦100	₦450	₦75	A	B	₦ −480	₦ −75	₦ −150	₦ −100	A	B
₦200	₦100	₦480	₦75	A	B	₦ −520	₦ −75	₦ −150	₦ −100	A	B
₦200	₦100	₦520	₦75	A	B	₦ −540	₦ −75	₦ −150	₦ −100	A	B
₦200	₦100	₦600	₦75	A	B	₦ −580	₦ −75	₦ −150	₦ −100	A	B
₦200	₦100	₦700	₦75	A	B	₦ −620	₦ −75	₦ −150	₦ −100	A	B
₦200	₦100	₦900	₦75	A	B	₦ −700	₦ −75	₦ −150	₦ −100	A	B
₦200	₦100	₦1,200	₦75	A	B	₦ −780	₦ −75	₦ −150	₦ −100	A	B
₦200	₦100	₦2,200	₦75	A	B	₦ −850	₦ −75	₦ −150	₦ −100	A	B
₦200	₦100	₦3,000	₦75	A	B	₦ −1000	₦ −75	₦ −150	₦ −100	A	B

about demographic and farm characteristics, included retrospective questions about the farmers' input use and output. We also asked the farm households whether they had been affected by any shocks in the past four production seasons and ranked shocks by severity. This is because exposure to shocks may affect farmers' preferences and demand.

13.2.2 Empirical estimation strategy

Our empirical estimation strategy builds on the RCEs and binary choice lotteries described in Tables 13.1 and 13.2. Respondents were presented with up to four choice tasks each. For the analysis, we used a sample of 1200 observations from 300 respondents, both in the OFAR trial and non-OFAR treatments, as well as a pooled sample of 2400 observations from the 600 respondents. In addition, we used 500 Halton draws per respondent and per random term for this simulation-based estimation (Halton, 1960). Given that the estimation of MMNL models in the WTP space is well documented (e.g. Scarpa *et al.*, 2008), in our empirical estimation strategy we mostly focus on the estimation of risk under gain and loss domains. In the analysis, the stress-tolerant quality attributes were dummy-coded to allow for a straightforward interpretation of the model coefficients. In contrast, for the price attribute, actual values were used (Hensher *et al.*, 2015).

Let's assume that farmer i has a choice scenario g: a dichotomous choice between lottery A and lottery B. Hence, the utility elicited in choice scenario g is expressed as:

$$Z^l = \Pi(v_g^l).\varphi(w_{g-}^l) + \Pi(1 - v_g^l).\varphi(w_{g\downarrow}^l), l \in \{A, B\} \quad (13.4)$$

where v_g^l and $1 - v_g^l$ are the high and low probabilities of high w_{g-}^l and low $w_{g\downarrow}^l$ outcomes, respectively. The farmer i compares the expected utility from choosing lottery A with the expected utility from choosing lottery B. The farmer selects lottery B if $I^* = Z^B - Z^A > 0$. I^* is a latent variable (unobservable) that captures the net utility from the decision and can be expressed as a function of the farmer's observable choices in the following latent variable model:

$$I_i^* = \Delta Z_i + \vartheta_i \quad I_i = 1\left[I_i^* > 0\right] \quad (13.5)$$

where Z_i is expected utility and I_i is a binary indicator variable that equals 1 if farmer i chooses lottery B, 0 otherwise. We specify the Fechner error as $\Gamma = \vartheta/\sigma$, with $\Gamma \sim N(0, \sigma)$. The probability of choosing lottery B can be represented as:

$$\begin{aligned} \Pr(\mathrm{B})\,(I_i = 1) &= \Pr(I^* + \vartheta_i > 0) \\ &= \Pr(\Gamma > -I^*/\sigma) = F(I^*/\sigma) \end{aligned} \quad (13.6)$$

where F is the distribution function for ϑ_i. Assumptions made on the functional form of F (for instance, if we choose the normal distribution function, then the maximum likelihood [ML] method helps estimate the parameters) help derive the five parameters using the ML estimator: risk aversion δ, loss aversion θ, probability distortion in gains μ^+, probability distortion in losses μ^- and the standard deviation of the Fechner error term σ. The log-likelihood function of the ML estimator is specified as:

$$\ln\left(L\left(\delta,\theta,\mu^+,\mu^-,\sigma\right)\right) = \sum_i \left\{ \left[\ln\left(F\left(\frac{I^*}{\sigma}\right)\right)\right].\mathrm{I}(I_{ig}=1) + \left[\ln\left(1-F\left(\frac{I^*}{\sigma}\right)\right)\right].\mathrm{I}(I_{ig}=0) \right\} \quad (13.7)$$

13.2.3 Results and discussion

In this section, we present the results on the extent to which risk aversion influences farmers' valuation of the stress-tolerance traits of maize as well as our findings on whether farmers' involvement in OFAR trials influences their preferences for the stress-tolerance traits. Table 13.3 presents the ML estimates for the choice model in WTP using the unrestricted specification. Restricted models are with loss-aversion specification $\delta+ = \delta-$, and unrestricted models have the specification $\delta+ \neq \delta-$. We notice that our sampled farmers are willing to pay a premium for the drought-tolerant maize, as the implicit price of the drought-tolerant trait is significantly different from zero (at 1% level). This suggests that farmers' preferences for drought-tolerance attributes in maize are significantly high. In addition, from the Cholesky matrix, we observe that the estimates of the diagonal elements are statistically significant. This provides evidence of substantial taste (unobserved) heterogeneity for the stress-tolerance attributes, including the drought-tolerant one, among the sampled farmers in both gain and loss domains.

To further explore this heterogeneity in preferences among the farmers, in Table 13.3, we present the interactions of drought-tolerance traits with farmers' risk aversion ($\delta+$, $\delta-$), in the gain and loss domains. The interaction with risk attitude in our results validates the existence of domain-specific effects in farmers' preferences for drought-tolerance traits across treatments.

Starting with OFAR trial farmers, our results in Table 13.3 (unrestricted specification) show that their interaction with risk for the drought-tolerance trait is positive and statistically significant in the gain domain but negative, albeit insignificant, in the loss domain. These findings suggest that, in the gain domain, the drought-tolerant trait attracts higher premiums among farmers with higher risk aversion. We also observe that without domain specificity, more literate farmers located in dry savannas and who have participated in OFAR trials are more likely than less literate farmers located in moist savannas, and who have not participated in such trials, to pay higher premiums for all the stress-tolerance traits. Preference for maize varieties with *Striga*-only traits and combined *Striga*-drought traits is higher among less-experienced maize farmers with a lower level of maize yield who have experienced production shocks within the past four seasons.

In the case of farmers who have not participated in OFAR trials, we also find a domain-specific preference for the drought-tolerance trait, as the interaction terms for risk are positive in the gain domain and negative in the loss domain, suggesting that, in the gain domain, the drought-tolerant trait attracts higher premiums among farmers who have higher risk aversion. In the loss domain, the drought-tolerant trait attracts higher premiums among farmers with lower risk aversion.

Generally, our results show the considerable market potential for stress-tolerance traits in maize varieties in northern Nigeria. Farmers who participated in OFAR trials are willing to pay a significantly higher premium for the drought-tolerance traits than farmers who did not participate in such trials. Of the sampled farmers, those who participated in trials are willing to pay on average ₦8.60 as a premium for stress-tolerance traits, including drought tolerance, relative to the conventional alternative in the gain and loss domains, and the sampled farmers who did not participate are willing to pay on average ₦17.76 as a premium for such traits. The preference for the identified stress traits, including the drought-tolerance one, is especially pronounced among farmers in dry savanna AEZs relative to those in the moist savanna.

Table 13.3. Maximum likelihood estimates in WTP space – unrestricted specification. (Source: Bello *et al.*, 2020.)

	OFAR		Non-OFAR	
	Gain	Loss	Gain	Loss
#Number of responses	300	300	300	300
#Number of observations	35,700	35,700	35,700	35,700
Log-likelihood	−1,303.8	−1,150.9	−1,596.4	−1,602.5
Variable	**Estimates**	**Estimates**	**Estimates**	**Estimates**
Implicit prices for stress attributes				
MEAN				
Striga	.721*** (.090)	.996*** (.113)	.854** (.340)	.445** (.226)
Drought	1.792*** (.646)	3.960*** (.620)	.270*** (.026)	.386** (.178)
Str_Drt	1.351*** (.065)	4.166*** (.582)	.729*** (.108)	.625** (.288)
Constant	−2.643*** (.341)	−1.960*** (.149)	−2.427*** (.479)	−3.929*** (.245)
STANDARD DEVIATION				
Striga	.496*** (.043)	.443** (.214)	.570** (.233)	.704*** (.027)
Drought	.296* (.158)	.356*** (.033)	.689** (.289)	.621 (.798)
Str_Drt	.993* (.578)	.835*** (.099)	.121** (.054)	-.662*** (.025)
Implicit prices for shifts in sensitivities				
Striga * risk	.974** (.446)	−.163** (.074)	3.522*** (1.144)	−3.893*** (.619)
Drought * risk	.304** (.120)	−.192 (.193)	.792** (.337)	−.707*** (.027)
Str_Drt * risk	.588* (.307)	−.770 (.889)	1.760*** (.245)	1.114*** (.245)
Other interaction terms	YES	YES	YES	YES

Notes: ***, ** and * mean significance at 1%, 5% and 10%, respectively.
Figures in parenthesis are standard errors.
Restricted models are with loss aversion specification $\delta + = \delta -$, while unrestricted models have the specification $\delta + \neq \delta -$.

Furthermore, we observe significant preference ordering that is domain-specific in our estimates. For example, in the loss domain, both farmers who participated in trials and those who did not participate are willing to pay the highest premium for maize varieties with multiple stress traits (i.e. *Striga*-drought trait), followed by drought-only varieties, and then maize with only the *Striga*-resistant trait. However, in the gain domain, farmers who participated in trials express the highest preference for varieties with drought tolerance only, followed by varieties with combined *Striga*-drought traits and then *Striga*-only maize varieties. On the other hand, for farmers who did not participate in trials, *Striga*-only varieties are valued highest, followed by Striga-drought traits and drought-only varieties.

13.3 Case Study 2: Farmers' WTP for WII in Northern Togo

Here we present our second case study focusing on farmers' WTP for WII in northern Togo.

WII can play an essential role in risk pooling and enhance developing countries' resilience in the face of climate change (Glauber, 2004; Bhattacharya and Osgood, 2014; Broberg, 2019; Budhathoki *et al.*, 2019; Nordlander *et al.*, 2020). WII can be an effective distribution channel for farm credit facilities, advisory services, and other agricultural risk management solutions. It is seen as one of the most efficient instruments that could be used in climate risk management in developing countries (Miranda, 1991; Richards *et al.*, 2004; Chantarat *et al.*, 2007; Horton and Keith, 2019; Shirsath *et al.*, 2019; Ali *et al.*, 2020). However, despite the advantages that the WII products offer in protecting farmers against natural hazards and adverse weather events (Miranda and Glauber, 1997; Barnett and Mahul, 2007; Giné and Yang, 2009), neither public nor private insurance companies currently exist in many developing countries. The lack of local empirical evidence to provide local solutions to climate stress affects food security and agricultural development.

This study seeks to estimate the WTP for WII in a semi-subsistence farming setting in northern Togo. Specifically, it: (i) assesses farmers' perceptions of WII policy interventions; (ii) evaluates the effect of attribute-based insurance contracts on farmers' WTP for such insurance; and (iii) estimates the average amount that farmers would be willing to pay for such insurance. The study contributes to understanding how insurance could be used as a climate risk management strategy. This study sets itself apart from previous works by focusing on cereals such as maize and sorghum, well-known staple crops produced in the West African subregion for subsistence.

13.3.1 Method and data

Modeling farmers' WTP

Methods from the stated preferences approach, including the contingent valuation method (CVM) and the choice modeling approach (CMA), are widely used for the valuation of hypothetical goods (Hanemann, 1984; Blamey *et al.*, 1999; Hanley *et al.*, 2001; Tesfaye *et al.*, 2019). Our current model uses the choice modeling approach described below:

$$U_{ij} = V_{ij}\left(X_{ij}\right) + \varepsilon_{ij} \tag{13.8}$$

U_{ij} is the utility function of farmer *i* when choosing alternative *j* within a finite choice set $\mathcal{S}$. This utility function can be decomposed into two components: the deterministic component (V_{ij}), which captures observable characteristics like respondents' socio-economic characteristics, and the stochastic component (ϵ_{ij}), which is the individual's unobservable characteristics that might influence his or her choice. X_{ij} is a vector of observable characteristics that could influence the utility function of the individual *i* when choosing the alternative *j*. Assuming that V_{ij} is linear in its parameters, the utility function of individual observable characteristics for choosing the jth alternative could be written as:

$$U_{ij} = aX_{ij} + \varepsilon_{ij} \tag{13.9}$$

The decision of farmer *i* is stochastic because of the unobserved components associated with choices being made. Assuming that farmer *i* is an expected utility maximizer (a von Neumann–Morgenstern utility function), between two alternative options *j* and *t* within the choice set (*j* and *t* are elements of $\mathcal{S}$ and $j \neq t$), farmer *i* will choose the option *t* rather than any other alternative *j*, if and only if the expected utility from choosing option *t* is greater than choosing option *j*. The choice of alternative *t* implies that:

$$Prob(U_{it} > U_{ij}) \\ = Prob[(V_{it} - V_{ij}) > (\varepsilon_{ij} - \varepsilon_{it}), \text{for all } t \neq j] \quad (13.10)$$

Assuming the independence between alternative choices, then the probability of choosing alternative *t* over alternative *j* can be represented by a conditional logit structure:

$$Prob\{t / S\} \\ = Prob[(V_{it} > V_{ij}), \forall t \neq j] = \frac{exp(\omega V_{it})}{\sum_j exp(\omega V_{ij})} \quad (13.11)$$

where ω is a scale parameter. If we consider a log-likelihood function (and estimation), it is possible to compute farmers' WTP for compensating variation in welfare (WTPCVW) for each attribute, using the status quo option as the baseline option of welfare measurement:

$$WTPCVW = b_y^{-1}[\ln \sum_i exp(V_i^1) - ln \sum_i exp(V_i^0)] \quad (13.12)$$

V_i^0 is the utility from the baseline option, and V_i^1 is the utility from the alternative state. b_y is the coefficient of the attribute's cost. It also represents the value of the marginal utility of income. If we assume that before and after the policy option the choice set includes a single solution, the welfare measured by Eqn 13.11 will be reduced to Eqn 13.13.

$$WTPCVW = \frac{1}{b_y}[V_i^1 - V_i^0] \quad (13.13)$$

From Eqn 13.12 the implicit prices could be derived since the deterministic part of the individual utility function can be expressed as a linear function of all the observed characteristics.

Data collection

The study was conducted in northern Togo. Data were collected through a household survey. Enumerators conducted face-to-face interviews and collected information about farmers' preferences for drought, flood and high-temperature insurance, WTP for them, and preferences on hypothetical insurance attributes such as loans, access to drought-tolerant seeds (DTS) and the use of climate information in decision-making. Data were also collected on agricultural households' socio-economic characteristics, plots and perceptions of climate change.

A multi-stage sampling technique was used to select a sample of 704 agricultural households. First, three regions in northern Togo – Centrale, Kara and Savanes – were selected on the basis of their rainfall characteristics. These regions are characterized by three AEZs and have one wet season and one dry season each year, compared with two wet seasons and two dry seasons in the southern regions of Togo. Second, we randomly selected three districts in each region. Third, we randomly selected five villages in each district. Finally, the surveyed farm households were randomly selected using the list of farm households provided by the villages' administrators.

INSURANCE CHOICE EXPERIMENT Four attributes were considered in the design of hypothetical WII contracts, as shown in Table 13.4. The first attribute is the insurance premium. Three different premium levels were offered, based on the maximum payout per hectare: US$17.50 (12.5% of the maximum payout), US$14 (10%) and US$7 (5%). The second attribute means that an insured farmer will receive DTS when signing the WII contract. The third attribute concerns loans. The problem of access to loans in the agricultural sector has been a common and critical issue in low-income countries. Therefore, making loans available could improve agricultural production. To this end, insured farmers would get a loan of US$60 in which repayment plus a fixed interest (maximum of 10% of the loan amount) was made after the harvest. Attribute four offers weather-related information that can be taken from a satellite and given to the insured farmers through mobile phones in the local language.

The selected insurance packages were displayed on four cards, each showing two insurance packages plus the status quo option (not interested in the proposed insurance contract). An example of a card is given in Table 13.5. With the assumption that the individual is rational in

Table 13.4. Attribute and attribute levels for hypothetical WII contracts. (Source: Ali *et al*., 2020.)

Attributes	Possible levels		
Premium (with a maximum payout of US$140.00 per hectare)	US$17.50	US$14.00	US$7.00
The insured will get drought-tolerant seeds (DTS) when signing the contract	Yes	No	
Get loan (US$60.00) and repayment plus fixed interest of US$6.00 is made after the harvest, but the insured farmers must be organized by group or cooperative	Yes	No	
The insurer guarantees the provision of weather information through mobile phones in the insured's local language	Yes	No	

Table 13.5. Example of a card given to respondents. (Source: Ali *et al*., 2020.)

Weather index-based attributes	Package 1	Package 2	I am not interested in such packages
Premium (with a maximum pay out of $140.00 per hectare)	$17.50	$14.00	
The insured will get DTS when signing the contract	No	Yes	
The insurer guarantees the provision of weather information through mobile phones in the insured local language	Yes	No	
Choose the preferred insurance package	☐	☐	☐

his or her choice, and following an explanation of what the WII is and how it functions, those interested were asked to specify their preferred insurance package. If the respondent chose the status quo option, it implied that his or her utility level from participating in the new WII program was less than his or her utility level from not participating in the program and vice versa.

13.3.2 Results and discussion

The analysis found that 99.43% of farmers were willing to enroll in the insurance program. Considering drought, flood and high-temperature insurance as different insurance products, the results show that about 99% of respondents were likely to participate in the proposed drought insurance program, 48.86% were interested in flood insurance, and 5.26% in high-temperature insurance. We also found that 97.3% of farmers were willing to insure maize crops against drought and 24.43% were willing to insure sorghum crops against drought. Only 20.45% of farmers would accept to insure rice against drought. Besides the selected crops, 25% of respondents expressed their desire to insure cotton and 10% opted for soybeans.

Table 13.6 shows the results from the conditional logit regression. We found that the insurance premium has a negative effect on the probability of participating in WII. Therefore, offering insurance as a stand-alone product seemed unattractive. We also found that providing DTS and loans for agricultural activities and access to weather information increases the probability of participation in the WII markets. The results of the attributes-based specification (Model 1) are also similar to those considering the full interaction with socio-economic characteristics (Model 2). For instance, we found that households with access to credit value DTS more highly than access to loans or weather information and therefore are more likely to enroll in the WII program.

The magnitude of the premium that farmers are willing to pay for alternative WII packages, along with estimates of market shares, are shown in Table 13.7. Package 3 is by far the most preferred insurance contract. It provides DTS to farmers when signing the contract, as well as a loan of US$60 (requiring repayment after harvest with a fixed interest rate of 10%,

Table 13.6. Econometric results from conditional logit. (Source: Ali *et al*., 2020.)

	Model 1			Model 2		
Variables	Odds ratio	Coefficient	Z	Odds ratio	Coefficient	Z
Premium	0.99***	−0.00***	−8.40	0.99***	−0.00***	−3.40
Access to DTS	13.01***	2.56***	14.82	11.69***	2.45***	14.06
Access to loan	3.66***	1.29***	5.89	3.81***	1.33***	5.27
Weather information supply	17.25***	2.84***	16.92	17.11***	2.83***	14.37
Access to DTS*FNFI credit				1.68***	0.52***	2.65
Premium*Farm size				1.00***	0.00***	3.43
Premium*Access to climate information				1.00***	0.00***	2.61
Premium*Household asset				1.00	0.00	1.53
Access to DTS*No formal education				0.49***	−0.70***	−2.76
Premium*Primary				0.99***	−0.00***	−3.08
Premium*Age				0.99**	−0.00**	−2.06
Pseudo R^2			0.46			0.48
Number of households	704					
Choice sets per household	9					
Number of observations	6336					
Wald-statistics			831			826

Note: ***$p < 0.01$; **$p < 0.05$.

and under the condition that the insured are organized in a group of five to ten people). The insurer provides weather information through TV, radio and mobile phones in the insureds' local language. About 61% of farmers preferred package 3. The average estimated insurance premium to be paid is about US$11 (Table 13.7). This result shows that combining WII with other risk management tools would enhance the insurance take-up in developing countries characterized by low adaptive capacities.

Table 13.7. Market shares and willingness to pay for WII packages. (Source: Ali *et al.*, 2020.)

Packages	P1	P2	P3
Premium (US$)	7.0	7.0	17.5
DTS	No	No	Yes
Loan	Yes	Yes	Yes
Weather information	No	Yes	Yes
Market share	2.13%	24.29%	61.22%
WTP (US$)	0.15	1.17	10.71

13.4 Conclusion

Using case studies that elicit farmers' WTP for drought-tolerant maize in northern Nigeria and WII in northern Togo, this chapter highlights the usefulness of both EUT and PT for analyzing farmers' WTP for agricultural technologies in West Africa.

In the case of Nigeria, based on a PT framework, we show that, in the gain domain, the drought-tolerant maize attracts a higher premium among farmers with higher risk aversion, whereas, in the loss domain, it attracts a higher premium among farmers who have lower risk aversion. In addition, the sample farmers are willing to pay on average ₦8.60–17.76 as a premium for stress-tolerance traits, including drought tolerance, relative to the conventional alternative in the gain and loss domains. The preference for the identified stress traits, including the drought-tolerance one, is especially pronounced among farmers in dry savanna AEZs relative to those in the moist savanna.

Using an EUT framework in the Togo study, we show that offering insurance as a standalone product is not attractive. However, providing DTS and loans for agricultural activities positively affects the probability of farmers participating in the WII market. Moreover, access to weather information increases the probability of participation in the WII market.

These findings show that, although both theories and innovations are distinct, they still provide similar conclusions in this context of climate change. Involving different countries and agro-ecology, our case studies using different approaches support the idea that there is a potential market for climate-smart agricultural technologies in West Africa.

References

Ali, E., Egbendewe, Y.G., Abdoulaye, T. and Sarpong, D.B. (2020) Willingness to pay for weather index-based insurance in semi-subsistence agriculture: evidence from northern Togo. *Climate Policy* 20, 534–547.

Barnett, B.J. and Mahul, O. (2007) Weather index insurance for agriculture and rural areas in lower-income countries. *American Journal of Agricultural Economics* 89, 1241–1247.

Bello, M. and Abdulai, A. (2016) Impact of *ex-ante* hypothetical bias mitigation methods on attribute non-attendance in choice experiments. *American Journal of Agricultural Economics* 98, 1486–1506. DOI: 10.1093/AJAE/AAV098.

Bello, M., Abdoulaye, T., Abdulai, A., Wossen, T. and Menkir, A. (2020) *The Roles of Risk and Ambiguity in Measuring Farmers Preference for Stress Tolerance in Maize in Northern Nigeria*. Working paper. International Institute of Tropical Agriculture, Ibadan, Nigeria.

Bhattacharya, H. and Osgood, D.E. (2014) Weather index insurance and common property resources. *Agricultural and Resource Economics Review* 43, 438–450. DOI: 10.1017/S1068280500005530.

Blamey, R., Gordon, J. and Chapman, R. (1999) Choice modelling: assessing the environmental values of water supply options. *Australian Journal of Agricultural and Resource Economics* 43, 337–357. DOI: 10.1111/1467-8489.00083.

Bliemer, M.C.J. and Rose, J.M. (2010) Construction of experimental designs for mixed logit models allowing for correlation across choice observations. *Transportation Research Part B: Methodological* 44, 720–734. DOI: 10.1016/J.TRB.2009.12.004.

Bougherara, D., Gassmann, X., Piet, L. and Reynaud, A. (2017) Structural estimation of farmers risk and ambiguity preferences: a field experiment. *European Review of Agricultural Economics* 44, 782–808. DOI: 10.1093/ERAE/JBX011.

Broberg, M. (2019) Parametric loss and damage insurance schemes as a means to enhance climate change resilience in developing countries. *Climate Policy* 20, 693–703. DOI: 10.1080/14693062.2019.1641461.

Budhathoki, N.K., Lassa, J.A., Pun, S. and Zander, K.K. (2019) Farmers interest and willingness-to-pay for index-based crop insurance in the lowlands of Nepal. *Land Use Policy* 85, 1–10. DOI: 10.1016/J.LANDUSEPOL.2019.03.029.

Burke, M. and Lobell, D. (2010) Food security and adaptation to climate change: what do we know? *Advances in Global Change Research* 37, 133–153. DOI: 10.1007/978-90-481-2953-9_8/COVER.

Chantarat, S., Barrett, C.B., Mude, A.G. and Turvey, C.G. (2007) Using weather index insurance to improve drought response for famine prevention. *American Journal of Agricultural Economics* 89, 1262–1268. DOI: 10.1111/J.1467-8276.2007.01094.X.

De Groote, H. and Kimenju, S.C. (2008) Comparing consumer preferences for color and nutritional quality in maize: application of a semi-double-bound logistic model on urban consumers in Kenya. *Food Policy* 33, 362–370. DOI: 10.1016/J.FOODPOL.2008.02.005.

De Groote, H. Kimenju, S.C. and Morawetz, U.B. (2011) Estimating consumer willingness to pay for food quality with experimental auctions: the case of yellow versus fortified maize meal in Kenya. *Agricultural Economics* 42, 1–16. DOI: 10.1111/J.1574-0862.2010.00466.X.

De Groote, H., Narrod, C., Kimenju, S.C., Bett, C., Scott, R.P.B. *et al.* (2016) Measuring rural consumers' willingness to pay for quality labels using experimental auctions: the case of aflatoxin-free maize in Kenya. *Agricultural Economics* 47, 33–45. DOI: 10.1111/AGEC.12207.

FAO (2015) *Climate Change and Food Security: Risks and Responses*. FAO, Rome. Available at: https://www.fao.org/3/i5188e/I5188E.pdf (accessed January 18, 2023).

Fisher, M., Abate, T, Lunduka, R.W., Asnake, W., Alemayehu, Y. and Madulu, R.B. (2015) Drought tolerant maize for farmer adaptation to drought in sub-Saharan Africa: determinants of adoption in eastern and southern Africa. *Climatic Change* 133, 283–299. DOI: 10.1007/S10584-015-1459-2.

Foster, A.D. and Rosenzweig, M.R. (2010) Microeconomics of technology adoption. *Annual Review of Economics* 2, 395–424.

Giné, X. and Yang, D. (2009) Insurance, credit, and technology adoption: field experimental evidence from Malawi. *Journal of Development Economics* 89, 1–11. DOI: 10.1016/J.JDEVECO.2008.09.007.

Glauber, J.W. (2004) Crop insurance reconsidered. *American Journal of Agricultural Economics* 86, 1179–1195. DOI: 10.1111/J.0002-9092.2004.00663.X.

Halton, J.H. (1960) On the efficiency of certain quasi-random sequences of points in evaluating multi-dimensional integrals. *Numerische Mathematik* 2, 84–90. DOI: 10.1007/BF01386213/fulltext.html.

Hanemann, W.M. (1984) Welfare evaluations in contingent valuation experiments with discrete responses. *American Journal of Agricultural Economics* 66, 332–341. DOI: 10.2307/1240800.

Hanley, N., Mourato, S. and Wright, R.E. (2001) Choice modelling approaches: a superior alternative for environmental valuation? *Journal of Economic Surveys* 15, 435–462. DOI: 10.1111/1467-6419.00145.

Harrison, G.W., Humphrey, S.J. and Verschoor, A. (2010) Choice under uncertainty: evidence from Ethiopia, India and Uganda. *The Economic Journal* 120, 80–104. DOI: 10.1111/J.1468-0297.2009.02303.X.

Hensher, D.A., Rose, J.M. and Greene, W.H. (2015) *Applied Choice Analysis*. Cambridge Books, Cambridge University Press, Cambridge, UK. Available at: https://ideas.repec.org/b/cup/cbooks/9781107465923.html (accessed: January 18, 2023).

Horton, J.B. and Keith, D.W. (2019) Multilateral parametric climate risk insurance: a tool to facilitate agreement about deployment of solar geoengineering? *Climate Policy* 19, 820–826. DOI: 10.1080/14693062.2019.1607716.

Kahneman, D. and Tversky, A. (1979) Prospect theory: an analysis of decision under risk. *Econometrica* 47, 263–292. DOI: 10.2307/1914185.

Kassie, G.T., Abdulai, A., Greene, W.H., Shiferaw, B., Abate, T. *et al.* (2017) Modeling preference and willingness to pay for drought tolerance (DT) in maize in rural Zimbabwe. *World Development* 94, 465–477. DOI: 10.1016/J.WORLDDEV.2017.02.008.

Liebenehm, S. and Waibel, H. (2014) Simultaneous estimation of risk and time preferences among small-scale cattle farmers in West Africa. *American Journal of Agricultural Economics* 96, 1420–1438. DOI: 10.1093/AJAE/AAU056.

Lusk, J.L. and Schroeder, T.C. (2004) Are choice experiments incentive compatible? A test with quality differentiated beef steaks. *American Journal of Agricultural Economics* 86, 467–482. DOI: 10.1111/J.0092-5853.2004.00592.X.

Miranda, M.J. (1991) Area-yield crop insurance reconsidered. *American Journal of Agricultural Economics* 73, 233–242. DOI: 10.2307/1242708.

Miranda, M.J. and Glauber, J.W. (1997) Systemic risk, reinsurance, and the failure of crop insurance markets. *American Journal of Agricultural Economics* 79, 206–215. DOI: 10.2307/1243954.

Nordlander, L., Pill, M. and Romera, B.M. (2020) Insurance schemes for loss and damage: Fools gold? *Climate Policy* 20, 704–714. DOI: 10.1080/14693062.2019.1671163.

Ouma, E., Abdulai, A. and Drucker, A. (2007) Measuring heterogeneous preferences for cattle traits among cattle-keeping households in East Africa. *American Journal of Agricultural Economics* 89, 1005–1019. DOI: 10.1111/J.1467-8276.2007.01022.X.

Prelec, D. (1998) The probability weighting function. *Econometrica* 66, 497–528.

Richards, T.J., Manfredo, M.R. and Sanders, D.R. (2004) Pricing weather derivatives. *American Journal of Agricultural Economics* 86, 1005–1017.

Scarpa, R. and Alberini, A. (2005) *Applications of Simulation Methods in Environmental and Resource Economics*. Springer, Dordrecht, Netherlands.

Scarpa, R., Thiene, M. and Train, K. (2008) Utility in willingness to pay space: a tool to address confounding random scale effects in destination choice to the Alps. *American Journal of Agricultural Economics* 90, 994–1010. DOI: 10.1111/J.1467-8276.2008.01155.X.

Shirsath, P., Vyas, S., Aggarwal, P. and Rao, K.N. (2019) Designing weather index insurance of crops for the increased satisfaction of farmers, industry and the government. *Climate Risk Management* 25, 100189. DOI: 10.1016/J.CRM.2019.100189.

Tanaka, T., Camerer, C.F. and Nguyen, Q. (2010) Risk and time preferences: linking experimental and household survey data from Vietnam. *American Economic Review* 100, 557–571. DOI: 10.1257/AER.100.1.557.

Tesfaye, A., Hansen, J., Kassie, G.T., Radeny, M. and Solomon, D. (2019) Estimating the economic value of climate services for strengthening resilience of smallholder farmers to climate risks in Ethiopia: a choice experiment approach. *Ecological Economics* 162, 157–168.

Train, K. and Weeks, M. (2005) Discrete choice models in preference space and willingness-to-pay space. In: Scarpa, R. and Alberini, A. (eds) *Applications of Simulation Methods in Environmental and Resource Economics*, Vol 6. Springer, Dordrecht, Netherlands, pp. 1–16. DOI: 10.1007/1-4020-3684-1_1.

Tversky, A. and Kahneman, D. (1992) Advances in prospect theory: cumulative representation of uncertainty. *Journal of Risk and Uncertainty* 5, 297–323.

World Bank (2022) *Addressing Food Insecurity and Boosting the Resilience of Food Systems in West Africa*. World Bank, Washington, DC. Available at: https://www.worldbank.org/en/news/press-release/2021/11/18/addressing-food-insecurity-and-boosting-the-resilience-of-food-systems-in-west-africa (accessed: January 18, 2023).

Section III

Climate Change, Risk and Risk Management

Another issue confronting smallholders and farmers in Western economies is climate change risk. The variability in weather conditions and climate risk are hard to predict, but they have serious implications for farming businesses and poor households in particular. Thus, producers need to account for climate risk when choosing risk management strategies in farming. Having formed an empirical framework for measurement of risk attitudes and risk preferences using various methods described in Section II, we move forward to a better understanding of the tools that smallholders can use to mitigate risk. These tools include contract farming, input usage, efficiency and productivity. Using farm-level studies from developing, emerging and other economies, the final section of this book details emerging climate change risks, perception of climate risks, adaptation mechanisms, natural disasters and risk management tools that farmers use.

Chapter 14 argues that most farmers in developing countries face negative cash flows during planting and growing periods due to various reasons, including weather variability, poor crop harvests and low profitability. Thus, they need credit to engage in contract farming. The authors investigate finance utilization from different sources and contract fulfillment among chili farmers in Indonesia. Chapter 15 discusses the emerging climate change risks and calls for climate resilience assessments of smallholder farming systems in developing countries. To that end, the chapter argues for a context-specific framework for assessing the climate resilience of semi-arid farming systems in India and evaluating the framework's relevance.

Chapter 16 discusses perception of risk as a prerequisite for adopting adaptation strategies for mitigating the effects of climate change. Production systems need to adopt climate-smart agriculture and mitigation practices to help foster climate resilience, such as improved cultivars that are resistant to pests and climate conditions (droughts/excessive rainfalls). The chapter assesses farmers' risk perceptions of climate change and how they relate to the use of agronomic practices to combat the effects of weather variability.

Natural disasters like earthquakes significantly affect rural populations and farming systems. Disaster's indirect effects on savings and production choices are less understood in the literature. Risk preferences play a significant role in determining economic options, such as savings and production choices after natural disasters. Chapter 17 uses lottery games in Peru to assess the effects of natural disasters on farmers' risk aversion in the affected and unaffected areas.

Recent decades have witnessed changes in precipitation and regional and global temperatures. To become resilient to climate change, rural communities, particularly farming communities, need help bolstering their adaptive capacities. Using household survey data collected from China, Chapter 18 assesses the impacts of climate change adaptations on wheat farmers' exposure to risk.

Several technological and agronomic options can partially offset climate change's adverse effects on agriculture and agriculture-based livelihoods. Crop diversification is a risk management strategy used by farmers in developing and emerging economies. Crops differ in their response to climate change, and risk-averse farmers often choose a portfolio of crops with low-correlated returns. Chapter 19 evaluates crop diversification's potential for improving system productivity and reducing its sensitivity to climate change. The authors use a spatially disaggregated district-level panel dataset and apply a moment-based specification of the stochastic production function.

The intensity and frequency of extreme weather events have risen due to anthropogenic climate change. Climate change is more apparent in some regions than in others. Temperatures in China, for example, have risen faster than the average temperature globally. Volatile weather and adverse weather events increase production risks and reduce yields and farm profitability. The use of microcredit in response to climate change also has attracted considerable attention recently. Consequently, Chapter 20 shows linkages between extreme weather events and apple farmers' demand for credit in six counties of China's Shaanxi province.

Agricultural practices evolve to match the prevailing climate; thus, climate change will significantly affect the farming sector. Timely adaptation is vital for efficient, economically sustainable production in the future. Farmers are well motivated to seek a better understanding of changes in weather patterns relevant to them. However, when farmers use weather/climate information to make production decisions they may either be depending on subjective climate information based on first-hand experiences and feelings or on objective data as might best be measured by official statistical sources. Chapter 21, using a natural experiment in the United States, investigates how human perceptions and beliefs about climate change adapt after experiencing an extreme weather event.

14 Finance Utilization and Chili Contract Fulfillment

Eliana Wulandari[1*], Miranda P.M. Meuwissen[2], Maman H. Karmana[1] and Alfons G.J.M. Oude Lansink[2]
[1]*Universitas Padjadjaran, West Java, Indonesia;*
[2]*Wageningen University & Research, Wageningen, Netherlands*

Abstract

This study investigates the relationship between finance utilization from different finance providers and contract fulfillment rates in Indonesian chili farms. The contract fulfillment rate was calculated as the ratio of the quantity of chili delivered to the trader to the quantity of chili defined in the contract. Furthermore, the contract fulfillment rate was regressed on the variables reflecting the availability of finance and socio-economic factors, using a censored regression model. Results from personal interviews with 102 chili farmers who had a written contract with a trader show that the farmers generally had a low contract fulfillment rate, i.e. 57%. The contract fulfillment rate was positively associated with in-kind finance utilization from farmers' associations, but negatively associated with subsidized credit utilization from banks and flexible payments of inputs to agricultural input kiosks. The findings suggest that policymakers and value chain actors may stimulate in-kind finance programs offered by farmers' associations to improve contract fulfillment and thereby food supply continuity.

14.1 Introduction

Contract farming has advantages for farmers, such as a guaranteed market and farm technical assistance (Brambilla and Porto, 2011). These advantages improve productivity and income (Sharma, 2008). For firms further downstream in the chain, contract farming provides benefits with regard to assurance and stability of raw produce (Key and Runsten, 1999; Singh, 2002). Furthermore, contract farming benefits farms in reducing investment cost, conflicts with landowners and labor issues (Saptana *et al.*, 2006).

An increasing number of agribusiness firms and smallholders are involved in contract farming in Indonesia. Previous studies on contract farming in Indonesia have shown that it helps farmers increase their farm efficiency (Fauziyah, 2010) and productivity (Manzilati *et al.*, 2010). Productivity increases because farmers need to work more carefully and follow the planting procedures detailed in the contract. Furthermore, farmers get more useful knowledge from the practice of contract farming compared to focusing solely on extension programs (Manzilati *et al.*, 2010).

Key and Runsten (1999) have shown that production costs from crops under contractual arrangements are much higher than the costs of traditional crops, since the contract often requires specific use of inputs and skilled labor. Hence, their study suggests that finance is very important to help contract farmers produce

*E-mail: eliana.wulandari@unpad.ac.id

 Managing Risk in Agriculture: A Development Perspective (eds A.K. Mishra, S.C. Kumbhakar and G. Lien).
DOI: 10.1079/9781800622289.0014

crops that follow the requirements in the contract. Furthermore, the majority of farmers face negative cash flows during planting and growing periods (Bozoglu and Ceyhan, 2007). Therefore, they need finance to purchase inputs prior to harvesting (Angelucci and Conforti, 2010).

Finance can be accessed from different sources; each can be expected to provide specific agricultural inputs. For instance, farmers can seek credit from banks, money lenders, relatives, cooperatives (Pham and Lensink, 2007; Armendariz and Labie, 2011), micro finance institutions (MFIs; Kaino, 2005; Bastin and Matteucci, 2007) and development programs (Pham and Lensink, 2007). Farmers also can obtain financial assistance through government subsidies. Subsidies are provided for agricultural inputs, such as fertilizers and seeds, and for interest on credit (Dorward and Chirwa, 2011; Wulandari *et al.*, 2017). Despite the important role of finance in fulfilling the contract requirements, to the best of our knowledge, the effect of finance utilization from different sources on contract fulfillment has not yet been studied.

This study aims to analyze the role of finance sources in contract fulfillment rates in Indonesian agriculture. Finance providers are banks, MFIs, farmers' associations, traders, agricultural input kiosks and other sources such as relatives and friends. This study focuses on chili farms because contract farming plays an important role in the chili value chain (Rudiyanto, 2014; Tsurayya and Kartika, 2015). Through contracts, farmers are assured of prices (Oelviani, 2013). Chili contracts specify the price paid to farmers and the local market conditions under which the price can be changed. If the price in the local market is higher than the contract price, then the price is adjusted accordingly. Apart from prices, contracts typically specify the chili variety and the number of chili plants that need to be grown. Contracts also specify the quantity of chili that has to be delivered (i.e. fixed at 600 g per plant).

The results of this study show that the farmers generally had a low contract fulfillment rate. Finance utilization in the form of in-kind finance provided by a farmers' association seemed to play an important role in stimulating the continuity of food supplies, indicated by a positive relation to the contract fulfillment rate. In contrast, subsidized credit utilization from a bank and flexible payments of inputs to an agricultural input kiosk had a negative association with the contract fulfillment rate. Education also had a positive association with the contract fulfillment rate. The results provide insight into the role of different finance sources in contract fulfillment; these insights are useful for policymakers, finance providers and supply chain actors in Indonesian agriculture. For policymakers, the results of this study can help in designing policy to enhance the availability of finance to stimulate the continuity of food supplies. For finance providers and supply chain actors, such as traders and farmers' associations, this study provides insight into the effect of their financial provisions on supply chain continuity.

The rest of this chapter is organized as follows. Section 14.2 describes the sources of finance available to farmers in Indonesia. Section 14.3 describes the methods used and an explanation of how the contract fulfillment rate is calculated. Results on contract fulfillment rates and the effects of finance utilization on contract fulfillment rates are described and discussed in Section 14.4. Section 14.5 provides conclusions and policy implications.

14.2 Sources of Finance

Indonesian farmers can access finance from many different sources. Credit is supplied by banks, MFIs and other sources, such as friends or relatives (Machmud and Huda, 2011). In-kind finance is provided by farmers' associations and traders, whereas agricultural input kiosks offer a scheme of flexible payments for inputs (Wulandari *et al.*, 2017). A full description of the finance sources available to farmers in Indonesia is provided in Wulandari *et al.* (2017).

14.2.1 Banks

Farmers can access finance from banks. They can also apply for finance under a subsidized credit program offered by banks in collaboration with the Indonesian government; an example is the program of microcredit loans (*KUR-Kredit Usaha Rakyat*). The KUR is designed to help farmers obtain credit, especially farmers who have no collateral (Agricultural Ministry of Indonesia, 2013).

Banks favor farmers who have contracts with a firm because contracts can act as a guarantee (with a so-called *alvalis*) to obtain credit from banks (Tutik *et al.*, 2014).

14.2.2 Micro finance institutions

Farmers can also apply for commercial credit offered by MFIs. Farmers are required to register as members of an institution. Credit from such institutions benefits small farmers who cannot obtain credit from banks.

14.2.3 Farmers' associations

Farmers' associations channel government aid to farmers who are members of the associations. The Indonesian government provides associations with cash aid through the rural agribusiness development program (*PUAP-Pengembangan Usaha Agribisnis Perdesaan*), which targets farmers' associations to motivate farmers to increase production. The government also provides them with seed and fertilizer, farm equipment, and subsidies to purchase seed and fertilizer. The associations then distribute the finance obtained from the government to their members. Finance is provided in-kind, i.e. by distributing agricultural inputs to their members. In return, some farmers' associations require their members to sell their produce to traders.

14.2.4 Traders

Some traders provide in-kind finance to farmers with a sales contract, which is set by the traders and farmers together. In-kind finance includes seeds, fertilizers and pesticides. The contract specifies the rules for production, the amount of produce to be delivered and prices. In case of non-performance, farmers' contracts are terminated (Saptana *et al.*, 2006).

14.2.5 Agricultural input kiosks

Some agricultural input kiosks provide finance to farmers by selling inputs, such as seeds, fertilizers and pesticides, for a flexible payment. This means that farmers can pay for the input after 1 week and 1 month of the purchase. Some kiosks allow farmers to pay after harvesting, as long as the farmers agree to sell the harvest to the kiosk.

14.2.6 Others

Relatives and friends are other finance providers. Informal sources of finance are traditionally important for rural people, as they supply short-term credit for urgent needs, which are usually not met by the banks and MFIs (Duong and Izumida, 2002).

14.3 Research Methods

14.3.1 Survey and data

To measure the contract fulfillment rate and analyze its relation to finance utilization and other socio-economic variables, structured questionnaires for farmers were prepared and pre-tested to evaluate consistency and clarity. The questions covered four main areas: (i) socio-economic characteristics of farmers; (ii) finance obtained from different sources during the growing season; (iii) quantity of chili that farmers agreed to supply; and (iv) quantity of chili delivered to the trader.

The survey was conducted from January 2014 to August 2015. Data were collected in person from farmers who had a written contract with a trader or with a farmers' association. A total of 102 farmers were surveyed in West Java and Central Java.

Within Java, study sites were selected based on the importance of chili production. The two provinces in Java with the largest area were selected as study sites. According to data from the Central Bureau of Statistics in Indonesia (BPS, 2013), chili production occurs mainly in the provinces of West Java and Central Java. Farmers were randomly selected prior to the interviews, and oral consent was obtained from the farmers because some participants had little schooling. Oral consent also was used by Reyes-Garcia *et al.* (2016).

14.3.2 Analytical methods

The contract fulfillment rate was calculated as the ratio of the quantity of chili (kilograms) actually delivered to the trader to the quantity of chili (kilograms) defined in the contract. The fulfillment rate, therefore, is between 0 and 1. Next, the contract fulfillment rate was regressed on the variables related to the availability of finance and socio-economic factors, using a censored regression model. Censored regression is a useful tool for obtaining unbiased estimation of a regression model with a limited dependent variable (Tobin, 1958), i.e. in this case the rate cannot be less than 0. The generic form of a censored model (Verbeek, 2004) is

$$y_i^* = x_i' \beta + \varepsilon_i, i = 1, 2, \ldots\ldots, N \text{ (102 farmers)} \quad (14.1)$$

The dependent variable is y_i^*, the contract fulfillment rate for farmer *i*, the independent variables (x_i') are variables reflecting finance and socio-economic variables, and ε_i is the error term. Finance is measured by dummy variables that take the value 1 if the farmer obtained finance from commercial credit (*commercial credit*) and subsidized credit (*subsidized credit*) from banks, commercial credit from micro finance institutions (*mfi*), in-kind finance from farmers' associations (*farmers' association*) and traders (*trader*), flexible payment of inputs to agricultural input kiosks (*kiosk*) and finance from other sources of finance (*other*), such as from relatives and friends.

If a farmer relied on multiple sources of finance, each source was accounted for. For instance, if a farmer obtained finance from a farmers' association, an agricultural input kiosk and a friend, then the dummy variables *farmers' association*, *kiosk* and *other* took the value 1. The model also included socio-economic factors: age, education, farming experience, farm size, quality and intercropping. Age (*age*) is measured as the age of the farmer in years. Education (*education*) is measured as the number of years of the farmer's formal education, and farming experience (*farming experience*) is the number of years the farmer has experience in chili production. Farm size (*farm size*) is measured as the number of hectares the farmer used for producing crops. Quality (*quality*) is represented by a dummy variable, which takes the value 1 if the farmer faced at least one rejected delivery because the quality did not meet the standard required by the contract. Intercropping (*intercropping*) is a dummy variable, which takes the value 1 if the farmer planted any other crops jointly with chili on his or her farm.

Finance is expected to increase the contract fulfillment rate because it facilitates better access to inputs and technologies. Finance in the form of commercial and subsidized credit from banks, as well as commercial credit from MFIs, can be used for the following activities: buying irrigation equipment that can increase production (Wossen *et al.*, 2014), investing in agricultural equipment or renovating old buildings (Brummer and Loy, 2000) or obtaining more inputs or better quality inputs (Girabi and Mwakaje, 2013). All these activities are expected to increase the contract fulfillment rate. In-kind finance from farmers' associations (Lamprinopoulou *et al.*, 2006) and traders (Schipmann and Qaim, 2011) and flexible payment of inputs to agricultural input kiosks may help farmers obtain inputs that are used throughout the growing season, such as fertilizers and pesticides. Better access to these inputs can increase farm production and subsequently the contract fulfillment rate. Finally, finance from other sources, including family and friends, may help farmers purchase agricultural inputs during planting and growing periods, as the majority of farmers are faced with negative cash flows during these periods (Bozoglu and Ceyhan, 2007).

However, the effect on contract fulfillment of finance utilization from all sources of finance may be obscured when farmers exhibit opportunistic behavior and deliberately fail to comply with the contract requirements. This occurs when farmers sell output to the market in situations when the market price is higher than the contract price (Gow and Swinnen, 2001). Furthermore, the effect of in-kind finance utilization may be obscured when the in-kind finance is not used to benefit the crop. Gow and Swinnen (2001) showed that farmers were selling inputs obtained through in-kind finance rather than using them on their land. Similarly, the effect of credit utilization may be obscured if farmers do not use the credit for their crops. Evidence shows that some farmers used credit for other purposes, such as renovating a house or buying a motorcycle (Rizal and Zulfa, 2012).

The socio-economic variables age, education, farming experience and farm size are all expected to increase the contract fulfillment rate. Older farmers are expected to have more financial resources than younger farmers (Lucas and Pabuayon, 2011), which may enable them to more easily obtain inputs that can increase production, leading to an increase in their contract fulfillment rate. More educated farmers are expected to make better use of new technology than less educated farmers (Ghimire and Huang, 2016). Lastly, more experienced farmers and those with larger farms have better opportunities than less experienced farmers and those with smaller farms to adapt to climate change by using crop diversification (Gebrehiwot and van der Veen, 2013), which may lead to increased production and a better contract fulfillment rate.

Before carrying out the regression, all independent variables, except for the dummy variables, were standardized to prevent scale effects of the β coefficients. Standardization was performed by subtracting the variable-specific mean from each variable and dividing the result by the standard deviation (UCLA, 2015). Homoskedasticity was tested using the Breusch–Pagan test, and multicollinearity was checked by calculating the variance inflation factors (VIF) for each variable. Following Rook *et al.* (1990), no multicollinearity was found in the model as all VIF values were well below 10. Furthermore, the presence of reverse causality was checked especially in the variables reflecting finance utilization. For instance, in the case of a contract with a farmers' association, a higher contract fulfillment rate might in turn lead to better access to finance from the farmers' association.

The definitions and summary statistics for all variables are presented in Table 14.1. The table shows that a large group (52%) of chili farmers obtained subsidized credit from a bank, whereas only few obtained commercial credits from a bank (5%) or from an MFI (2%). Furthermore, 13% of the chili farmers obtained in-kind finance from a farmers' association and 27% from a trader. Flexible payment of inputs to an agricultural input kiosk had been obtained by 13% of the chili farmers, and one-fifth of the chili farmers obtained finance from other finance sources. Farmers averaged 40 years old with 7 years of formal education and 10 years of farming experience. Table 14.1 also shows that 12% of the respondents had had at least one delivery rejected. Furthermore, a large group (69%) of the farmers grew other crops together with chili during the 2013 growing season. Table 14.1 also shows that the average contract fulfillment rate was 57%.

14.4 Results

The results of the censored regression of contract fulfillment rate on the variables reflecting finance utilization and socio-economic factors are presented in Table 14.2. The results show that the relationship between the contract fulfillment rate and finance utilization varied across finance providers. The utilization of subsidized credit negatively associated with the contract fulfillment rate, implying that the contract fulfillment rate was, *ceteris paribus*, lower with finance utilization from this source. A possible explanation is that subsidized credit might target a special program for smaller farmers who may have limited experience in chili production. These farmers may be less skilled in producing chili according to the quantity and quality requirements specified by the contract. Another explanation is that the subsidized credit may have been used for other activities. Riaz *et al.* (2012) found that farmers in Pakistan did not use credit for crop purposes but for other activities, such as household needs or construction of a house, thereby not benefitting overall farm production.

As expected, finance utilization from a farmers' association had a positive association with the contract fulfillment rate. This result suggests that the contract fulfillment rate is higher when in-kind finance is obtained from farmers' associations. By being a member of an association, a farmer can benefit from the provision of inputs (Lamprinopoulou *et al.*, 2006), which enables farmers to increase their production and fulfill their contracts. However, finance utilization from a trader had a negative association with the contract fulfillment rate, although the association is not statistically significant. The negative association might be because of a side-selling problem. Farmers might behave opportunistically because there were no strict rules that applied sanctions in a contract when farmers did not sell their products as agreed in the contract (Prihadi *et al.*, 2010).

Table 14.1. Definitions and summary statistics of sample respondents (n = 102). (Source: Authors' calculation.)

Variable		Variable definition	Sample mean
Bank	Commercial credit	1: obtained commercial credit from bank in 2013	0.05
	Subsidized credit	1: obtained subsidized credit from bank in 2013	0.52
Micro finance institution		1: obtained commercial credit from MFI in 2013	0.02
Farmers' association		1: obtained in-kind finance from farmers' association in 2013	0.13
Trader		1: obtained in-kind finance from trader in 2013	0.27
Kiosk		1: obtained flexible payment of inputs to agricultural input kiosk in 2013	0.13
Others		1: obtained finance from other finance sources in 2013	0.20
Age		Age (years)	40
Education		Education (years)	7
Farming experience		Chili production experience (years)	10
Farm size		Farm size (hectares)	1.16
Quality		1: has faced at least one rejected delivery because of quality issues	0.12
Intercropping		1: grew other crops on their chili farms	0.69
Contract fulfillment rate		Ratio of the quantity (kg) of chili actually delivered to the quantity (kg) of chili defined in the contract for the growing season of 2013	0.57

Table 14.2. Coefficients from the censored regression of contract fulfillment rate on variables representing finance utilization and socio-economic factors. (Source: Authors' calculation.)

Variables		Coefficients	Standard errors
Bank	Commercial credit[d]	0.046	0.165
	Subsidized credit[d]	−0.252	0.116[b]
MFI[d]		0.290	0.236
Farmers' association[d]		0.375	0.131[a]
Trader[d]		−0.188	0.119
Kiosk[d]		−0.307	0.108[a]
Others[d]		0.072	0.086
Age (years)		−0.030	0.036
Education (years)		0.070	0.041[c]
Farming experience (years)		0.013	0.037
Farm size (hectares)		−0.053	0.040
Quality[e]		−0.057	0.125
Intercropping[f]		−0.006	0.078
Constant		0.720	0.119

Notes: [a]Significant at 1%, [b]significant at 5% and [c]significant at 10% level.

[d]Dummy variable representing whether the respondent had obtained finance from the finance provider during 2013; 1 if yes, 0 otherwise.

[e]Dummy variable representing whether the respondent has faced at least one rejected delivery because of quality issues; 1 if yes, 0 otherwise.

[f]Dummy variable representing whether the respondent grew other crops on their chili farms; 1 if yes, 0 otherwise.

Finance in the form of flexible payments to an agricultural input kiosk had a negative association with the contract fulfillment rate, which indicates that farmers who obtained finance from the kiosk had, *ceteris paribus*, a lower contract fulfillment rate. This might be because farmers with limited experience in chili production used the kiosk as a last option to obtain finance, i.e. they had difficulty obtaining finance from formal finance sources, such as banks (Supriatna, 2009).

Most socio-economic factors had no statistically significant relation with the contract fulfillment rate. Education, which had a positive relation with the contract fulfillment rate, was the only exception. The positive association is consistent with the prior expectation that more educated farmers have better skills than less educated farmers to use new technology (Ghimire and Huang, 2016), which may increase production and hence the contract fulfillment rate. Reverse causality tests were performed to investigate the relation between statistically significant variables and the contract fulfillment rate. Reverse causality was found between the contract fulfillment rate and finance utilization from commercial banks, farmers' associations and kiosks. Hence, although the relations with finance utilization from these sources were positive, the direction of the causality is unknown.

14.5 Conclusion and Recommendations

This study analyzed the role of finance utilization from different finance providers in the contract fulfillment of chili farmers in Indonesia. Sources of finance included in the study were banks, MFIs, farmers' associations, traders, agricultural input kiosks and other sources. The results show that the farmers in the sample generally had a low contract fulfillment rate, with an average of 57%. Finance utilization from a farmers' association was positively associated with the contract fulfillment rate. In contrast, subsidized credit utilization from a bank and flexible payments of inputs to an agricultural input kiosk had a negative association with the contract fulfillment rate. Finally, education had a positive association with the contract fulfillment rate.

The results of this study imply that in-kind finance provided by farmers' associations seems to play an important role in stimulating the continuity of food supplies, and thus policies may need to be introduced to focus on enhancing the availability of finance from this finance provider. Although this study was limited to chili farms, the results can be generalized to other types of farms in Indonesia that have similar characteristics and operate under similar types of contracts, i.e. farms that produce annual crops under contracts that specify the quantity of crops to be delivered. An example is potato contracts between farmers and traders (Silvia *et al.*, 2015). The results also can be generalized to countries with conditions similar to those of Indonesia, i.e. countries that have an institutional environment of agricultural value chains similar to Indonesia's and that have a similar farm structure. Examples of countries for which the results may be relevant include countries in South, Southeast and East Asia and in sub-Saharan Africa (Eastwood *et al.*, 2010).

Funding

The author (EW) received financial support for conducting the research from the Netherlands Fellowship Programme (NFP).

Acknowledgments

The authors would like to thank the farmers and finance providers for participating in this study. Further, the authors are grateful to the agricultural officers and agricultural networks for providing contact details for farmers.

References

Agricultural Ministry of Indonesia (2013) *Pedoman teknis kredit usaha rakyat sektor pertanian [Technical guidelines of micro credit loans for agriculture]*. Directorate General of Horticulture, Agricultural Ministry of Indonesia, Jakarta, Indonesia.

Angelucci, F. and Conforti, P. (2010) Risk management and finance along value chains of small island developing states: evidence from the Caribbean and the Pacific. *Food Policy* 35, 565–575.

Armendariz, B. and Labie, M. (2011) *The Handbook of Microfinance*. World Scientific Publishing, Singapore.

Bastin, A. and Matteucci, N. (2007) Financing coffee farmers in Ethiopia: challenges and opportunities. *Savings and Development* 31, 251–282.

Bozoglu, M. and Ceyhan, V. (2007) Measuring the technical efficiency and exploring the inefficiency determinants of vegetable farms in Samsun Province, Turkey. *Agricultural Systems* 94, 649–656.

BPS (Central Bureau of Statistics in Indonesia) (2013) Available at: http://www.bps.go.id (accessed January 18, 2023).

Brambilla, I. and Porto, G. (2011) Market structure, outgrower contracts, and farm output. Evidence from cotton reforms in Zambia. *Oxford Economic Papers* 63, 740–766.

Brummer, B. and Loy, J.P. (2000) The technical efficiency impact of farm credit programmes: a case study of northern Germany. *Journal of Agricultural Economics* 51, 405–418.

Dorward, A. and Chirwa, E. (2011) The Malawi agricultural input subsidy programme: 2005/06 to 2008/09. *International Journal of Agricultural Sustainability* 9, 232–247.

Duong, P.B. and Izumida, Y. (2002) Rural development finance in Vietnam: a microeconometric analysis of household surveys. *World Development* 30, 319–335.

Eastwood, R., Lipton, M. and Newell, A. (2010) Farm size. In: Evenson R. and Pingali P. (eds) *Handbook of Agricultural Economics Vol.* 4. Elsevier, Amsterdam, pp. 3323–3397.

Fauziyah, E. (2010) Analisis efisiensi teknis usaha tani tembakau: suatu kajian dengan menggunakan fungsi produksi frontier stokhastik [*Technical efficiency analysis of tobacco: A stochastic frontier function*]. *Embryo* 7, 1–7.

Gebrehiwot, T. and van der Veen, A. (2013) Farm level adaptation to climate change: the case of farmer's in the Ethiopian highlands. *Environmental Management* 52, 29–44.

Ghimire, R. and Huang, W-C. (2016) Adoption pattern and welfare impact of agricultural technology: empirical evidence from rice farmers in Nepal. *Journal of South Asia Development* 11, 113–137. DOI: 10.1177/0973174116629254.

Girabi, F. and Mwakaje, A.E.G. (2013) Impact of microfinance on smallholder farm productivity in Tanzania: the case of Irambia District. *Asian Economic and Financial Review* 3, 227–242.

Gow, H.R. and Swinnen, J.F.M. (2001) Private enforcement capital and contract enforcement in transition economies. *American Journal of Agricultural Economics* 83, 686–690.

Kaino, T. (2005) Rural credit markets in Myanmar: a study of formal and non-formal lenders. *Asian Journal of Agriculture and Development* 4, 1–15.

Key, N. and Runsten, D. (1999) Contract farming, smallholders, and rural development in Latin America: the organization of agroprocessing firms and the scale of outgrower production. *World Development* 27, 381–401.

Lamprinopoulou, C., Tregear, A. and Ness, M. (2006) Agrifood SMEs in Greece: the role of collective action. *British Food Journal* 108, 663–676.

Lucas, M.P. and Pabuayon, I.M. (2011) Risk perceptions, attitudes, and influential factors of rainfed lowland rice farmers in Ilocos Norte, Philippines. *Asian Journal of Agriculture and Development* 8, 61–77.

Machmud, Z. and Huda, A. (2011) SMEs' access to finance: an Indonesia case study in selected East Asian economies. In: Harvie, C., Oum, S. and Narjoko, D. (eds) *Small and Medium Enterprises (SMEs) Access to Finance in Selected East Asian Economies*. ERIA research project report 2010-14, ERIA, Jakarta, pp. 261–290.

Manzilati, A., Yustika, A.E., Krisnamurthi, B. and Suman, A. (2010) Implikasi kontrak usaha tani dengan perusahaan terhadap keberlanjutan usaha tani (studi pada komoditi jagung) [*The implications of farming contracts with companies on the sustainability of farming (a study in the commodity of corn)*]. *Wacana* 13, 439–457.

Oelviani R. (2013) Penerapan metode analytic hierarchy process untuk merumuskan strategi penguatan kinerja sistem agribisnis cabai merah di Kabupaten Temanggung [*The implementation of an analytic hierarchy process to design strategy to strengthen the performance of the red cayenne agribusiness system in Temanggung Regency*]. *Informatika Pertanian* 22, 11–19.

Pham, T. and Lensink, R. (2007) Lending policies of informal, formal and semiformal lenders: evidence from Vietnam. *Economics of Transition* 15, 181–209.

Prihadi, N., Darusman, D., Nugroho, B. and Wijayanto, N. (2010) Kelembagaan kemitraan industri pengolahan kayu bersama rakyat untuk membangun hutan di Pulau Jawa [*Institutional partnership between wood-processing industries and communities on forest development in Java Island*]. *Jurnal Penelitian Sosial dan Ekonomi Kehutanan* 7, 127–138.

Reyes-Garcia, V., Pyhala, A., Diaz-Reviriego, I., Duda, R., Fernandez-Llamazares, A. *et al.* (2016) Schooling, local knowledge and working memory: a study among three contemporary hunter-gatherer societies. *Plos One* 11, 1–18. DOI: 10.1371/journal.pone.0145265.

Riaz, A., Khan, G.A. and Ahmad, M. (2012) Utilization of agriculture credit by the farming community of Zarai Tariqiati Bank Limited (ZTBL) for agriculture development. *Pakistan Journal of Agricultural Sciences* 49, 557–560.

Rizal, S. and Zulfa, E. (2012) Evaluasi kegiatan program Kemitraan dan Bina Lingkungan (PKBL) Bank Mandiri [*The evaluation of the partnership and community development programme of Mandiri Bank*]. *EPIGRAM* 9, 61–67.

Rook, A.J., Dhanoa, M.S. and Gill, M. (1990) Prediction of the voluntary intake of grass silages by beef cattle 2. Principal component and ridge regression analyses. *Animal Production* 50, 439–454. DOI: 10.1017/S000335610000492X.

Rudiyanto, A.A. (2014) Pola kemitraan Koperasi Sejahtera Abadi dalam meningkatkan keuntungan petani cabai [*The partnership of Sejahtera Abadi Cooperative in increasing the profitability of chili farmers*]. *JEJAK Journal of Economics and Policy* 7, 173–183. DOI: 10.15294/jejak.v7i1.3596.

Saptana, Sunarsih and Indraningsih, K.S. (2006) Mewujudkan keunggulan komparatif menjadi keunggulan kompetitif melalui pengembangan kemitraan usaha hortikultura [*Realizing comparative advantage into competitive advantage through developing horticultural business partnerships*]. *Forum Penelitian Agro Ekonomi* 24, 61–76.

Schipmann, C. and Qaim, M. (2011) Supply chain differentiation, contract agriculture, and farmers' marketing preferences: the case of sweet pepper in Thailand. *Food Policy* 36, 667–677.

Sharma, V.P. (2008) India's agrarian crisis and corporate-led contract farming: socio-economic implications for smallholder producers. *International Food and Agribusiness Management Review* 11, 25–48.

Silvia, H., Syamsun, M. and Kartika, L. (2015) Analisis strategi peningkatan daya saing komoditas kentang di Kabupaten Karo, Sumatera Utara [*Strategy analysis for increasing the competitiveness of the potato commodity in Karo Regency, North Sumatera*]. *Jurnal Ilmu Pertanian Indonesia* 20, 164–170.

Singh, S. (2002) Contracting out solutions: political economy of contract farming in the Indian Punjab. *World Development* 30, 1621–1638.

Supriatna, A. (2009) Pola pelayanan pembiayaan sistem kredit mikro usaha tani di tingkat pedesaan [*Patterns of micro-financial services for farming system activities in rural areas*]. *Jurnal Litbang Pertanian* 28, 111–118.

Tobin, J. (1958) Estimation of relationships for limited dependent variables. *Econometrica* 26, 24–36.

Tsurayya, S. and Kartika, L. (2015) Kelembagaan dan strategi peningkatan daya saing komoditas cabai Kabupaten Garut [*The institutions and strategies for improving the competitiveness of the chili commodity in Garut District*]. *Journal Manajemen dan Agribisnis* 12, 1–13. DOI: 10.17358/JMA.12.1.1.

Tutik, Noor, I. and Hayat, A. (2014) Peran stakeholder dalam pengembangan tebu rakyat: studi pada wilayah kerja Pabrik Gula Redjosarie Kabupaten Magetan [*The role of stakeholders in the development of people's cane: study in Redjosarie Sugar Factory Working Area, Magetan Regency*]. *Journal Administrasi Publik* 2, 823–829.

UCLA (University of California Los Angeles). (2015) How Do I Standardize Variables in Stata? Available at: https://stats.oarc.ucla.edu/stata/faq/how-do-i-standardize-variables-in-stata/ (accessed July 19, 2015).

Verbeek M. (2004) *A Guide to Modern Econometrics*, 4th edn. Wiley, Chichester, UK.

Wossen, T., Berger, T., Swamikannu, N. and Ramilan, T. (2014) Climate variability, consumption risk and poverty in semi-arid northern Ghana: adaptation options for poor farm households. *Environmental Development* 12, 2–15.

Wulandari, E., Meuwissen, M.P.M., Karmana, M.H. and Oude Lansink, A.G.J.M. (2017) Access to finance from different finance provider types: farmer knowledge of the requirements. *PloS One* 12, e0179285. DOI: 10.1371/journal. pone.0179285.

15 Assessing the Climate Resilience of Semi-Arid Farming Systems in India: Framework and Application

Arjuna Srinidhi[1,2*], Saskia E. Werners[1,3], Fulco Ludwig[1], Marcella D'Souza[2] and Miranda P.M. Meuwissen[1]
[1]*Wageningen University & Research, Wageningen, Netherlands;* [2]*Watershed Organisation Trust, Pune, India;* [3]*United Nations University, Bonn, Germany*

Abstract

Semi-arid regions in India are characterized by poor natural resources, smallholder farmers and an increasing frequency of extreme weather events. The vulnerability of semi-arid farming systems to climate change and the state of agrarian distress in India underscore the need to assess their climate resilience. We observe weaknesses in existing frameworks for assessing resilience, such as a bias toward assessing a system's status quo and the lack of clarity between assessing resilience to specific stresses and a system's more general resilience attributes. The need for context-specific tools and frameworks has also been highlighted in literature on climate resilience. To address these gaps, we develop a context-specific framework to assess the climate resilience of semi-arid farming systems in India and evaluate its relevance. The Climate Resilience In Semi-arid India (CRISI) framework, with customized system functions, indicators, resilience capacities and attributes, has been developed by combining insights from existing literature, from the local knowledge and expertise of the research team and from a case study application. Considering emerging climate change risks, the case study further shows us there is an urgent need to build the resilience capacities of adaptability and transformability of the farming system. CRISI's inclusiveness in identifying stakeholder needs, its systemic understanding of resilience and its provision of information for decision-making make it useful for the farming community, practitioners and policymakers in assessing and building climate resilience.

15.1 Introduction

Dryland regions, consisting of arid, semi-arid and sub-humid climates, cover about 55% of India's areal extent. The semi-arid regions in particular are dynamic, influenced by changes in temperature and precipitation patterns as well as anthropogenic land-use changes (Ramarao *et al.*, 2019). Even small changes in temperature and rainfall patterns can impact natural resources such as water availability, soil health, forest cover and pastureland – and in turn, the agrarian livelihoods of the vulnerable communities who live in these regions (Kuchimanchi *et al.*, 2019; Ramarao *et al.*, 2019). Agriculture and allied livelihoods employ more than 50% of India's population (Birthal *et al.*, 2017), and more than 85% of the farming community

*E-mail: arjuna.srinidhi@wotr.org.in

DOI: 10.1079/9781800622289.0015

consists of small and marginal farmers (GoI, 2016). More than half of these small and marginal farmers are located in the resource-poor arid and semi-arid regions, adding to their vulnerability and poor adaptive capacities (Behera and France, 2016).

Given the complexity of the challenges such farmers face, several studies recommend using the lens of resilience to assess a farming system's sustainability and its ability to cope with shocks (Folke *et al.*, 2010; Bullock *et al.*, 2017; Meuwissen *et al.*, 2019). A farming system is a unit of analysis that is above individual farms and consists of a network of farms, farmers and other actors interacting in a specific agro-ecological context (Giller, 2013; Dixon and Stringer, 2015). Many frameworks exist to assess the resilience of socio-ecological systems and farming systems (Resilience Alliance, 2010; Walker and Salt, 2012; UNDP Drylands Development Centre, 2013; Nemec *et al.*, 2014; Quinlan *et al.*, 2016). They vary from more conceptual (e.g. Cabell and Oelofse, 2012) to data-intensive, quantitative approaches (e.g. FAO, 2016). O'Connell *et al.* (2015) reviewed and described some of the weaknesses in existing frameworks for assessing resilience, such as who is included in the process, a narrow understanding of resilience that focuses on the status quo and a lack of clarity between assessing resilience to specific stresses and a system's more general resilience attributes.

In the context of farming systems, Meuwissen *et al.* (2019) adapt earlier frameworks specifically paying attention to the importance of multiple value chains, relationships between farm and non-farm populations and the role of the enabling environment. One aspect that all resilience assessment frameworks recommend is adapting and operationalizing the steps and indicators to the local context. Dixon and Stringer (2015) review tools and frameworks that focus on the climate resilience assessments of smallholder farming systems in sub-Saharan Africa and highlight the need for context-specific tools and frameworks that analyze local power relations, capture existing vulnerability and promote participation, flexibility and learning. Here, we understand climate resilience as the capacity of a system to dynamically respond to, recover from and even thrive in changing climate conditions while continuing to maintain essential functions, identities and structures (Rockefeller Foundation, 2009; Dixon and Stringer, 2015).

The objective of our research is to develop a context-specific framework to assess the climate resilience of semi-arid farming systems in India and evaluate the framework's relevance. We develop the framework by combining insights from existing literature, from the local knowledge and experience of the research team and from a case study application. We refer to our framework as the Climate Resilience In Semi-arid India (CRISI) framework (phonetically similar to *Krisi* or *Krishi*, which translates to 'agriculture' in Hindi). In semi-arid India, where farming depends on water harvested within a catchment area, we use the watershed boundary to demarcate the farming system. The scale at which we apply the CRISI framework is the micro-watershed. Micro-watersheds are the smallest hydrologic unit in the hierarchal system of watersheds, typically 500–1000 ha in India (Symle *et al.*, 2014). The technically correct term 'micro-watershed' is used interchangeably with the more generic term 'watershed' throughout the chapter.

We apply the framework to assess the resilience of Kalamkarwadi, a micro-watershed in the state of Maharashtra, to provide additional insights to refine the framework. We evaluate the framework's relevance for assessing the climate resilience of farming systems in semi-arid India against three criteria: (i) identification of all relevant stakeholders and their needs; (ii) following a systems perspective while assessing resilience (as opposed to a disciplinary/sectoral perspective); and (iii) providing information for decision-making processes. We find that the CRISI framework adequately meets these three criteria and believe it has application potential in generating evidence-based insights that can guide climate change adaptation and climate-resilient development policies at the state and national levels, not just in semi-arid regions of India but in other parts of the world with similar socio-ecological contexts.

This chapter is organized into five sections. After this introduction, we discuss in Section 15.2 how the CRISI framework was developed. Section 15.3 presents the application of the framework to the Kalamkarwadi case study and the insights that contributed to making the framework context-specific. In Section 15.4 we evaluate the framework's relevance for assessing the climate resilience of farming systems in the context of semi-arid India. Our final discussion and conclusions are presented in Section 15.5.

15.2 CRISI – A Context-Specific Framework to Assess the Climate Resilience of Farming Systems in Semi-Arid India

To develop a climate resilience assessment framework specific to semi-arid regions in India, we followed a two-stage process. The first stage involved identifying the broad steps and generic functions, and resilience capacities and attributes from existing literature on the resilience of farming systems. Literature on the impact assessment of watershed development (WSD) interventions in India (e.g. Bharucha *et al.*, 2014; Taylor and Bhasme, 2020; Shah *et al.*, 2021) and the local knowledge and experience of the research team also contributed to this first stage. The second stage involved applying the framework in a case study and using the insights generated to refine the context-specific list of system functions, indicators, and resilience capacities and attributes. This application of the CRISI framework was done through a participatory rural appraisal (PRA) process that involved semi-structured discussions with a multi-stakeholder group and focus group discussions (FGDs) with marginalized sections of the local community. The multi-stakeholder discussions included the local community as well as researchers and development professionals who have been working with the Watershed Organisation Trust (WOTR), the partner non-governmental organization involved in the research. Quantitative data available at the village council and from project baseline and endline reports also contributed to the assessment of the case study.

The CRISI framework consists of the following six steps as shown in Table 15.1. These steps are: (1) system description; (2) challenges; (3) system functions; (4) resilience capacities; (5)

Table 15.1. Steps of the Climate Resilience In Semiarid India (CRISI) framework. (Source: Authors' analysis.)

Step	Actions to be undertaken in the assessment
1. System description	a) Identify stakeholders, governance structures and power dynamics within the farming system. b) Describe the context and explore the stakeholders' views of the farming system, including what they value and why. c) Discuss the scope of the climate resilience-building interventions and the time and spatial scale.
2. Challenges	a) Explore stakeholders' views on existing stresses and vulnerabilities, particularly climate-related, and what they expect, including short-term and long-term views. b) Describe the environmental, economic, social and institutional issues and stresses.
3. System functions	a) Discuss the different functions provided by the farming system and appropriate indicators to reflect their functions. b) Assess the performance of these indicators in response to (specific) known stresses over the assessment period.
4. Resilience capacities	a) Define the four resilience capacities (*Anticipation*, *Robustness*, *Adaptability* and *Transformability*) in the local context and discuss relevant indicators for each. b) Assess the capacities by factoring in the story behind the performance of system function indicators (3b) and the contribution of any strategies adopted to cope with climate stresses on the farming system.
5. Resilience attributes	a) Discuss the list of resilience attributes identified for semi-arid farming systems and their relevance to the local context. b) Assess the status of these attributes over the assessment period.
6. Reflection	a) Discuss the outcomes of the resilience assessment with the community, reflecting on its accuracy and making revisions when required. b) Over the assessment period, reflect on the adequacy of the resilience-building measures and the need for additional adaptation or transformative changes.

resilience attributes; and (6) reflection. The first five steps are based on a holistic understanding of resilience by Meuwissen *et al.* (2019) that pays attention to the farming system as a whole and emphasizes the importance of assessing both resilience to specific stresses and the more general resilience attributes of a system. With respect to Step 4, Meuwissen *et al.* (2019) conceptualize resilience through the capacities of robustness, adaptability and transformability. Considering the increasing frequency of extreme weather events being experienced in semi-arid India, we add a fourth capacity, anticipation, as suggested by Mathijs and Wauters (2020). We also add a sixth and final step, reflection, based on insights from the Resilience, Adaptation Pathways and Transformation Assessment (RAPTA; see O'Connell *et al.*, 2015). The RAPTA framework has been tested in development contexts and suggests that this final step helps to improve the quality of the resilience assessment, in adaptive management of the resilience-building measures, and also to assess the need for additional adaptation or transformative changes in building resilience (Grigg *et al.*, 2015; Maru *et al.*, 2017).

The first two steps of system description and challenges focus on outlining the system boundary, the relevant stakeholders, governance structures, climate-related stresses and other existing socio-economic vulnerabilities. Step 3 addresses the functions of the farming system. In the context of semi-arid India, we identify ten functions (Table 15.2). Eight of these are based on the public and private functions of farming systems used by Paas *et al.* (2021) in the context of European farms. In the context of semi-arid India, two more functions – social organization and equity (standard of living) – were added based on the analysis of semi-arid watersheds by Kerr *et al.* (2002). With regard to equity, we use two perspectives based on principles of 'fairness' (Jones, 2009) and the research team's experience in the local context: (i) standard of living through access to employment, income and assets; and (ii) decision-making and power dynamics in local governance. We include the former perspective as a system function in Step 2 (Table 15.2) and include the latter as a resilience attribute in Step 5. The PRA exercise during the application of the framework in the case study helped arrive at appropriate indicators for the system functions.

Table 15.2. Farming system functions and indicators used in the CRISI framework.

No.	Farming system functions	Indicators in local context / What to look for
1	Social organization[a]	No. of community-based institutions; membership and participation; readiness for voluntary labor
2	Economic viability[b]	Income from the farming system (crop and livestock); total per capita income (including non-farm income)
3	Food and nutrition security[b]	Total food production (cereals, vegetables, milk, etc.); data on malnutrition and anemia; diversity in diet
4	Animal health and welfare[b]	Total livestock ownership; production of meat and milk; health and well-being of animals
5	Other bio-based resources[b]	Production quantity from sources such as non-timber forest produce, from common lands; diversity of resources; income from such sources
6	Health of ecosystem[b]	Soil quality; forest cover; water-table depth
7	Biodiversity of habitat[b]	Diversity of crops, other plants, trees, animals and wildlife
8	Attractiveness of area[b]	Waste management; composting; roadside plantation; public toilets
9	Quality of life[b]	No. of families and months of migration under distress; ownership of TVs, smartphones, internet access; working hours (men and women); returning families; perception of village youth
10	Equity (standard of living)[a]	Access to employment; quality of the house; number of households with assured income; number of households with BPL (below poverty line) cards

Note: [a]Based on Kerr *et al.* (2002) and [b]based on Paas *et al.* (2021).

In Step 4, we expand the possible range of resilience-building interventions that can be assessed by incorporating the four resilience capacities of: (1) anticipation – the capacity to identify potential risks and take proactive steps; (2) robustness – the capacity to withstand stresses and (un)anticipated shocks; (3) adaptability – the capacity to change the composition of farming system inputs, outputs and risk management strategies in response to shocks and stresses (without changing the system's structure and feedback mechanism); and (4) transformability – the capacity to significantly change the farming system's internal structure and feedback mechanism in response to either severe shocks or enduring stress that makes business as usual impossible. The PRA exercise during application of the framework in the case study helped arrive at appropriate indicators for each of the resilience capacities (Fig. 15.1). The indicators provide guidance on what to look for in the local context.

Step 5 aims to discuss the system's capacity to deal with new and/or unforeseen stresses to the system, also known as general resilience. To assess this, Meuwissen *et al.* (2019) describe resilience attributes as 'individual and collective competencies that along with an enabling (or constraining) environment affect the general resilience of a system.' An initial list of resilience attributes that combined all the indicators and attributes mentioned in Cabell and Oelofse (2012) and Paas *et al.* (2021) was used as a starting point. The PRA exercises during the application of the framework in Kalamkarwadi helped introduce new attributes relevant to the local context and remove those that were deemed redundant or not applicable. The final list of attributes is presented in Table 15.3.

Step 6 is an iterative step of reflection, in which we focus on reviewing the assessment results from Steps 1 to 5 and improving the assessment's accuracy. We discuss issues such as the adequacy of the resilience-building measures and the need for additional adaptation or transformative changes. This step also gives stakeholders the chance to reflect on indicator dynamics over the time frame of the assessment.

15.3 Applying the CRISI Framework: The Kalamkarwadi Case Study

We applied the CRISI framework to assess the resilience of Kalamkarwadi, a micro-watershed in the state of Maharashtra, to validate the customization made and to provide additional insights to refine the framework. The farming system in the watershed of Kalamkarwadi, located in the Parner block of the Ahmednagar district of Maharashtra, was chosen as the site for the case study (Fig. 15.2). This case was selected based on its semi-arid ecosystem, the history of WSD interventions, data availability and consent from the local community. Falling in the rain-shadow region of the Western Ghats, this area receives about 500 mm of rain annually. About 90% of the rainfall occurs during the monsoon season from June to September. The

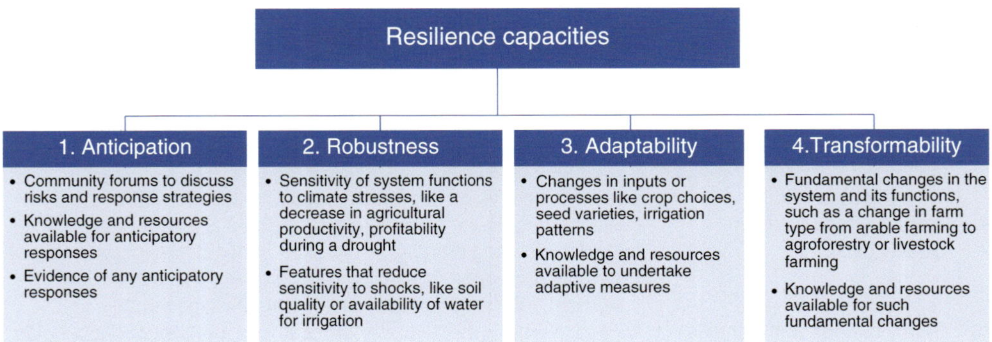

Source: Authors' analysis.

Fig. 15.1. The four resilience capacities and indicators in the local context for each capacity.

Table 15.3. Resilience attributes and indicators.

	Resilience attributes	Indicators in local context / What to look for
1	Reasonable profitability[a]	Economic viability in the long run; evidence of profits; extent of reliance on subsidies; type and size of investments
2	Social self-organization[a]	Community-based organizations that are inclusive and actively involved in decision-making
3	Ecological self-regulation[a]	Reliance on ecosystem services and common property resources in place of external inputs
4	Appropriate connectedness[a]	Connectedness between components of the system, such as crops and livestock, large farmers and landless laborers
5	Functional diversity[a]	Diversity in the ecosystem services provided within the watershed and on the farm; a variety of water resources; forest produce and other bio-based resources
6	Optimal redundancy[a]	Duplication or redundancy in the system, such as multi-cropping, a mix of livestock breeds, alternate irrigation sources, alternate livelihood sources
7	Spatial and temporal heterogeneity[a]	Diversity in the landscape, farm types, soil types and different cropping patterns in farms and over time (seasons/years)
8	Exposure to disturbance[a]	Small disturbances that test and increase the resilience of a system, such as exposure to droughts, leading to a shift to drought-resistant crops
9	Reflectivity and shared learning[a]	People and institutions that learn from past experiences and each other; specific changes or new practices introduced after significant shocks
10	Human capital building[a]	Education; diversified skills in the local community; adopting new practices and technology
11	Diverse policies[b]	Appropriate state and national policies (incentives, pricing, access to markets, etc.) that help farming systems become more profitable and sustainable in the long-term
12	Infrastructure and information for innovation[b]	Existing infrastructure; the adoption of new knowledge and cutting-edge technologies such as weather-based farm advisories through smartphone apps
13	Support for rural life[b]	Livelihoods attractive to all sections of the community; a lifestyle that is deemed as dignified by the community
14	Access to credit, insurance and other financial safety nets[c]	Credit from institutional lenders with viable terms, interest rates; awareness of and access to insurance; compensation for damages not covered by insurance
15	Equity (decision-making and power dynamics)[c]	Participation of marginalized sections of society (women, tribal, elderly, landless, etc.) in project planning, execution and general decision-making
16	Governance arrangements that support transformation[c]	Community-based institutions with discussions on climate, markets, disasters (pandemics like COVID); changes or preparedness for change among such institutions and community

Note: [a]attributes from Cabell and Oelofse (2012), [b]attributes introduced by Paas *et al.* (2021) and [c]new attributes added to the CRISI framework based on authors' analysis.

district experiences significant rainfall variability – varying from less than 200 mm to more than 900 mm annually.

Social Centre, a non-governmental organization based in Ahmednagar, Maharashtra, implemented a WSD project between 1996 and 2001. Considering the long-term impacts of climate change and the need to assess the sustainability of resilience-building interventions, data availability for more than 20 years was a key

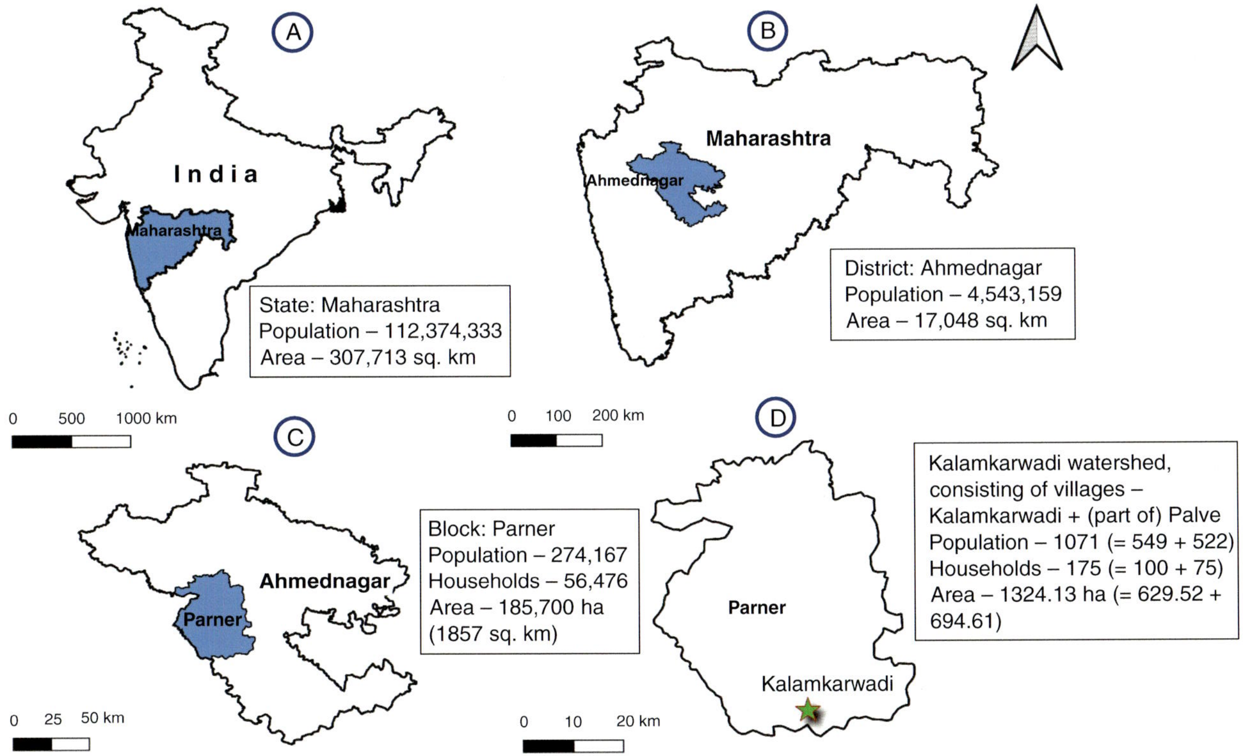

Source: Government of India, Census 2011.

Fig. 15.2. Demographic details of the Kalamkarwadi watershed and its location vis-à-vis the country, India (A); the state, Maharashtra (B); the district, Ahmednagar (C) and the block, Parner (D).

consideration in selecting the case study. Care was taken to obtain consent from the local community for the research and ensure the case study had no unusual or unique characteristics, such as special caste and class demographics or proximity to urban areas. The Kalamkarwadi watershed satisfied these criteria and qualified as a representative WSD project in semi-arid Maharashtra.

Applying the CRISI framework to assess the resilience of Kalamkarwadi involved three visits between October and November 2020. The first visit involved an introductory and rapport-building meeting in the village, where the study's purpose was explained and consent for documenting all the discussions and use in research was sought. The system boundary for the study was discussed, and a detailed list of all relevant stakeholders was created. In the next two visits, we followed a PRA process (see Narayanasamy, 2009) for the assessment in Kalamkarwadi. The appraisal exercises consisted of semi-structured discussions with a multi-stakeholder group as well as FGDs with marginalized sections of the local community. The multi-stakeholder discussions had two to three members representing each stakeholder group, such as large farmers, small farmers, women, youth and elderly members of the local community, as well as researchers and development professionals who have been working with the WOTR for several years and experts in the areas of agriculture, land and water resource management. The FGDs were undertaken with members of marginalized groups like women and smallholder farmers to ensure their opinions and issues around the local power dynamics were appropriately captured. Quantitative data available at the village council and from project baseline and endline reports also contributed to the assessment of the case study.

Step 1: System description. The selected watershed consists of the entire village of Kalamkarwadi and a part of the Palve village. The total area of the watershed is 1324.13 ha, with a population of 1071 in 175 households. The Kalamkarwadi village has a larger proportion of households and agricultural lands within this watershed and lends its name to the watershed. Kalamkarwadi is distinct with a very homogeneous demographic in terms of caste – most families are *Marathas*. About 50% of the families are small and semi-medium farmers (< 4 ha), about 30% are medium farmers (4 to 8 ha) and the rest are large farmers (> 8 ha). The watershed has no landless families. The soils are light and highly permeable and have very low depth. Agriculture and livestock are the primary sources of livelihood in the watershed. In terms of local governance bodies, there are several active ones – a functioning village development committee (VDC), three self-help groups (SHGs) with 15 to 20 women in each, a dairy cooperative and a youth group.

WSD interventions were carried out under the Indo-German Watershed Development Programme (IGWDP) between 1996 and 2001. As documented in project completion reports and validated through interactions with the local community, the WSD project contributed to several positive impacts, such as increased water availability, reduced land erosion, better soil quality, an increase in the cropped area of more than 25% during the *Kharif* (monsoon) and 50% during *Rabi* (winter) season, increases of about 135% in grain yields and 42% in fodder yields and an improvement in milk production from about 705 l/day to 2024 l/day (Social Centre, 2002). Traditional crops in Kalamkarwadi included sorghum, pearl millet, wheat and pulses. After WSD, cropping patterns changed to include more cash crops, including vegetables, soybean and horticulture. These changes resulted in a significant improvement in income levels (Social Centre, 2002). Since 2017, the WOTR has been implementing a new project with the aim of climate-proofing the watershed. This project includes some maintenance work on older watershed structures, training programs on organic farming and improved agronomic practices to improve soil health, water budgeting and crop planning.

Together with the community and other stakeholders involved in the assessment, we decided that the time frame used in the assessment of Kalamkarwadi would cover 25 years, from the start of WSD in 1996 to 2020. The resilience of the system was evaluated for five specific time slices within the assessment period: 1996, 2001, 2009, 2018 and 2020. These five time slices correspond to the start of the WSD project, the completion of the WSD project, a year of normal rain, a drought year and the year of most recent available data, respectively.

STEP 2: CHALLENGES. Before implementation of WSD activities, poor agriculture and livestock productivity meant that a large proportion of the limited household income was used for purchasing food, grain and fodder. Migration for work was common, with at least one person from each household seeking work in larger villages and towns. Drinking water was a perpetual challenge during the summer months, and the village had to rely on water supply by tankers.

Interactions with the local community during our field visits and a comparison of geographic information system images show that from 2001 to 2020, the WSD's positive impacts have largely been sustained. However, an analysis of local rainfall data shows significant challenges of drier conditions and an increasing frequency of droughts. Further, the annual average rainfall has declined from about 580 mm in 1986 to about 480 mm in 2020 (Department of Agriculture, 2021). Such a shift toward drier and more arid conditions is corroborated by local experiences and climate change impact assessments across central India (Ramarao *et al.*, 2019; Krishnan *et al.*, 2020).

Although soil health and access to water resources have improved, many farmers have lands in upper parts of the watershed that are unproductive. Most WSD activities focus on common lands in the village and haven't led to investments in private lands, like land levelling. The PRA exercise also highlighted challenges in managing the watershed resources due to differences between the watershed's geographical boundaries and the administrative boundaries of the two villages and the village council (*gram panchayat*).

STEP 3: FUNCTIONS OF THE WATERSHED. To assess the watershed functions, the stakeholders in the PRA rated the indicators in Table 15.2 on a scale of 1 (poor) to 5 (excellent), indicating their relative performance over the assessment period. The scoring considers data from different sources for the list of indicators for each function (Table 15.2) and the consensus of the group involved in the PRA. The scoring was grounded in available information that included both qualitative and quantitative data. Qualitative data on indicators such as readiness for voluntary labor, diversity in diet, perception of village youth, waste management system and migration scenario were sought on the basis of recall. Quantitative data on indicators such as community-based institutions, total crop and livestock production, income, soil quality, malnutrition and BPL cards were obtained from project completion reports and from information available with the village council. Such data were discussed in the multi-stakeholder group before arriving at the final score for each function.

Although the absolute value of the scores may be contested, we believe the direction of the scores over time provides the most value. The discussions during the appraisal exercise showed a significant improvement in the number of community-based organizations – like the VDC and women's SHGs – and community engagement since the initiation of WSD activities. Assessment results for each of the functions at the five time slices are shown in Fig. 15.3. No significant quantities of other bio-based resources were reported in Kalamkarwadi (and hence they are excluded from Fig. 15.3).

An overall improvement across all indicators is observed from 1996 to 2020, driven by improved natural resources and better agricultural productivity after the WSD interventions. However, a dip in performance can be observed during the drought year 2018 for all indicators except social organization and equity (standard of living) – which increased even in 2018. Social organization increased because the challenges of the drought motivated the community to collaborate and find solutions such as purchasing water through tankers. The factors that led to an increase in equity (standard of living) are discussed in Step 6.

STEP 4: RESILIENCE CAPACITIES. We started the assessment of the resilience capacities by first defining anticipation, robustness, adaptability and transformability in the local context and discussing relevant indicators for each (Table 15.3). Given their experiences of frequent droughts and the recent COVID-19 pandemic, anticipation was a capacity that farmers could quickly relate to and helped validate its inclusion. We then referred to the performance of the watershed functions (Fig. 15.3) and sought the stories behind the performance. For instance, we explored why food and nutrition security dropped during the drought in 2018 but bounced back rapidly during the good rains of 2020. The appraisal exercise showed that although water-intensive cash crops have brought in better profits, they have

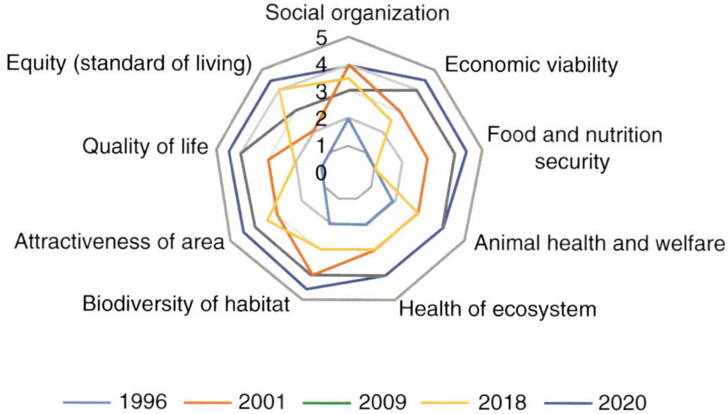

Source: Primary data collected in this study.

Fig. 15.3. Perceived performance of the watershed functions in Kalamkarwadi.

increased the farmers' vulnerability to droughts – an indicator of the system's lack of robustness.

Based on the indicators mentioned in Fig. 15.3, we sought evidence for each resilience capacity in Kalamkarwadi. Several WSD interventions had a positive influence on robustness and anticipation. For instance, we saw a positive influence on anticipation through: (i) the establishment of the VDC, SHGs, dairy cooperative and youth groups providing forums to discuss risks; and (ii) the exposure, training and capacity-building measures in the recent climate-proofing project contributing to an improved understanding of the increasing frequency of extreme weather events and the need to take proactive actions. A positive influence on robustness was seen through: (i) the WSD interventions such as drainage line treatments and area treatments that decreased soil erosion and improved the water table; and (ii) the introduction of borewells for drinking water, which helped the community during droughts. There were, however, some unintended consequences – such as the shift toward water-intensive cash crops and cross-bred cattle, which have increased the risk of heat stress and droughts. Although the community coped with buying tanker water and fodder during the 2018 drought, there is recognition that these changes have reduced its adaptability to a drier and more variable local climate. We found limited evidence that contributed to transformability. The introduction of biogas units linked to toilets was transformative in the behavioral changes it brought about (through its contribution to improvement in quality of life and a more integrated farm–livestock system), but the current lack of alternatives to the dominant crop–livestock farming system suggests that the capacity for transformability is not very strong yet.

STEP 5: RESILIENCE ATTRIBUTES. A list of resilience attributes derived from literature on the resilience of farming systems served as a starting point for discussions on unforeseen stresses that went beyond the commonly experienced climate risks like droughts and unseasonal rain. The appraisal exercise helped refine the list of attributes to the semi-arid context, as discussed in Section 15.2 (Table 15.3). The attributes were then rated on a scale of 1–5, indicating the degree to which each attribute was present during the assessment period. Appraisal participants were asked to give a score of 1 to the poorest form of the attribute they could imagine, such as relying on a single crop on a small piece of land (indicating the lack of diversity or redundancy), and to give a score of 5 to the best possible form of the attribute, such as multi-cropping and a mixed crop and livestock system. The scores as well as comments from stakeholders that support the relative scores of the attributes are presented in Table 15.4.

STEP 6: REFLECTION. The final step involved reflecting on the results of the resilience assessment from earlier steps, making revisions where required

Table 15.4. Performance of the resilience attributes of Kalamkarwadi. (Source: Primary data collected in this study.)

	Resilience attributes	Scores (1–5: low–high)	Indicators that support the relative score assigned
1	Reasonable profitability	3	Income from crops and livestock has increased, but so have input costs and losses due to extreme weather events.
2	Social self-organization	4	The VDC and the SHGs have sustained even after projects that initiated them have been completed.
3	Ecological self-regulation	3	Water resources are very sensitive to rainfall and suffer during a drought, such as in 2018. Needed to be supplemented with purchased water.
4	Appropriate connectedness	5	Success of the biogas linked to toilets has brought about a lot of connectedness between crops, livestock and community household needs.
5	Functional diversity	4	Mixed crop–livestock systems and inter or mixed cropping provide diversity as well as some redundancy.
6	Optimal redundancy	4	In addition to the points under functional diversity, increased sowing area and some non-farm income for smallholders have helped.
7	Spatial and temporal heterogeneity	5	Farmers often have fragmented land holdings and leave some of the less productive land parcels fallow. In years of good rainfall, these provide the opportunity for additional yield. Crop rotation provides temporal heterogeneity.
8	Exposure to disturbance	3	Exposed to recurrent droughts but has not to led to major changes in the crop choices or the predominant crop–livestock system in the village.
9	Reflectivity and shared learning	4	The success of all the watershed work has involved many demonstrations and learning from each other's experiences.
10	Human capital building	3	Farmers see the need for training related to higher value chain agricultural produce, marketing and diversifying livelihoods, as these skills are currently inadequate.
11	Diverse policies	3	Do not find the current policies very supportive of agriculture, especially regarding market access and minimum support price during public procurement.
12	Infrastructure and information for innovation	2	New technology like weather-based apps for farmers have been introduced but not yet very widely adopted.
13	Support for rural life	3	Moderately happy with rural life and would prefer if children got a better education and went to cities.
14	Access to credit, insurance and other financial safety nets	2	When institutional credit was not available, farmers did have access to informal sources of credit but at much higher interest rates.
15	Equity (decision-making and power dynamics)	4	The efforts to make the VDC participatory and inclusive have been sustained even after the completion of the watershed activities.
16	Governance arrangements that support transformation	1	Governance arrangements for business-as-usual scenarios (like purchasing water through tankers during a drought or digging a borewell) seem adequate. However, no specific vision for long-term planning or major transformational actions was observed (like investing in other areas of the agri-value chain or non-farm livelihoods).

and reflecting on the adequacy of the climate resilience-building measures. For instance, while discussing the performance of the function of equity (standard of living), the appraisal exercise had initially suggested a dip in the score during the 2018 drought, in line with the overall reduction in income and increased losses during the year. However, while reviewing the results, the multi-stakeholder group realized that the losses sustained by the better-off farmers with higher investments were much more significant than those sustained by the marginal farmers. Some of the marginal farmers depended on income from other sources – such as non-farm labor – and suffered relatively lower losses. Therefore, equity (standard of living) had gone up during 2018 while the performance of most other functions had decreased. This insight also shows that an alternate source of income contributes to functional diversity and optimal redundancy in the system (Table 15.4), leading to improved resilience for smallholder farmers.

Reflecting on the adequacy of the resilience-building measures, the appraisal stakeholders agreed that WSD interventions had contributed to several important gains – like improvements in soil quality, water table, vegetative cover, agriculture and livestock productivity, community engagement and equity in access to employment and quality of life. These have contributed to an increase in resilience capacities like robustness and anticipation. However, there were limits to focusing on WSD interventions alone as a development strategy. Over the years, cropping patterns in Kalamkarwadi have changed to include more cash crops that are also water intensive. Although they have led to better income levels in years of good rainfall, the losses during drought years – such as in 2018 – were substantial. In terms of livestock, the type and quantity of assets have changed, from 92 cross-bred and 59 indigenous cows in 1996 to 304 cross-bred and 41 indigenous cows in 2020 and a significant increase in milk production over the years.

Here again, the low rainfall in 2018 impacted the milk output and led to the distress sale of cattle in some cases. These experiences highlight that focusing on WSD interventions alone is not adequate and there is an urgent need to focus on the resilience capacities of adaptability and transformability while also improving resilience attributes like access to credit, insurance and other financial safety nets; infrastructure and information for innovation; and governance arrangements that support transformation. Exposure, training and capacity-building measures in recent years have improved understanding about the limits of WSD in the local community. These measures are also slowly motivating the adoption of innovative practices like locale-specific and weather-based agro-advisories, water budgeting and crop planning in the ongoing climate-proofing project.

15.4 Evaluating the Relevance of the CRISI Framework

In this section, we evaluate the framework's relevance for assessing farming systems' climate resilience in the context of semi-arid India. Similar questions about the relevance of frameworks and indicators have been asked in the domain of climate change adaptation (Dinshaw *et al.*, 2014; Pringle and Leiter, 2018; Hallegatte and Engle, 2019). Considering the similarities in challenges between assessing adaptation and assessing resilience to climate change, we draw inspiration from the Adaptation Metrics Mapping Evaluation (AMME) Programme (see IPAM, 2021) and identify the following three criteria to evaluate the CRISI framework: (i) identifying all relevant stakeholders and their needs; (ii) following a systems perspective while assessing resilience (as opposed to a disciplinary/sectoral perspective); and (iii) providing information for decision-making processes. We find that the CRISI framework adequately meets these three criteria, as discussed below.

15.4.1 Identifying all relevant stakeholders and their needs

Step 1 of the CRISI framework explicitly focuses on identifying stakeholders, governance structures and power dynamics within the farming system, and Step 2 seeks to identify stakeholders' views on existing stresses and vulnerabilities and what they expect, including short-term and long-term views. Identifying the relevant stakeholders was also done in a participatory manner with the involvement of the local community as a part of the first step. The appraisal exercise then consisted of a diverse group that included large farmers, small farmers, women, youth and elderly members of the local community; development professionals who have been working with the WOTR for several years; and experts in the areas of agriculture, land and water resource management. Although the effort was to try to reach a consensus on various aspects of the assessment, differing opinions, especially from the marginalized sections of the group, were separately recorded and appropriately captured in the final scores. For instance, the smallholder farmers were concerned about risks during shorter time frames like an agricultural season (3 months–1 year) and dissatisfied with access to credit, insurance and other financial safety nets (low score in Table 15.4). The CRISI framework thus enables identification of the needs of a diverse and relevant set of stakeholders.

15.4.2 Following a systems perspective while assessing resilience

Step 1 of the CRISI framework begins with a discussion about the stakeholders' views of the farming system, including what they value and why. Using such a participatory process to define the system, as opposed to an expert-driven identification of the system and its sub-components, is key to adopting a systems perspective (Darnhofer *et al.*, 2012; Sriskandarajah *et al.*, 2016). Another key characteristic of a systems perspective is adopting an inter- and transdisciplinary approach as opposed to a disciplinary approach (Gibbon, 2012). The CRISI framework is based on the frameworks by Meuwissen *et al.* (2019) and the RAPTA, which are both embedded in a systems perspective.

This can be seen in the comprehensive list of functions, capacities and attributes used in CRISI that cover all five types of capital (physical, natural, social, human and financial) that are often cited to demonstrate a holistic systems approach to sustainability (DFID, 1999; Porritt, 2007). Therefore, the CRISI framework is not restricted to assessing the impact of issues related only to irrigation, agricultural productivity or income but rather looks at the resilience of the farming system from a much more holistic and interconnected perspective. The final step on reflection is also important from a systems perspective as it gave the opportunity to reflect on backward and forward linkages of interventions and their consequences. For instance, some of the unintended consequences discussed in Step 4 of the case study, such as WSD prompting the shift to water-intensive cash crops and cross-bred cattle, came to light during this step.

15.4.3 Providing information for decision-making processes

Applying the CRISI framework to assess the Kalamkarwadi watershed provided numerous insights for decision-making, such as farmers' current preferences for water-intensive cash crops and hybrid livestock breeds and the lack of incentives to change. Also, given the vulnerability of current livelihoods to changing rainfall patterns, the need for diversifying livelihoods and investing in other areas of the agricultural value chain was an important insight. In addition, a very practical aspect of CRISI from a decision-making perspective is its spatial scale of application. Participants in the appraisal found the selection of the micro-watershed a useful unit for analysis as this is also the scale at which WSD projects are typically implemented (Symle *et al.*, 2014). Another aspect important to decision-making is one of timescales. Applying the CRISI in Kalamkarwadi shows the value of considering longer timescales during assessments, since issues such as changing rainfall patterns and the need for additional land management interventions became explicit only when the discussions covered a timescale of more than 20 years.

15.5 Discussion and Conclusion

To arrive at a framework for assessing the climate resilience of semi-arid farming systems in India, we developed the CRISI framework based on existing literature on resilience assessments, the local knowledge and experience of the research team and insights from application of the framework to a case study. We also evaluated the framework's relevance against three criteria – identifying all relevant stakeholders and their needs, following a systems perspective while assessing resilience and providing information for decision-making processes – and found that the framework adequately meets them. The CRISI framework's strengths lie in its holistic understanding of resilience, its focus on participatory approaches and its customized lists of functions, indicators and attributes that help operationalize it for assessing the resilience of farming systems in semi-arid India.

The framework assesses the farming system's resilience to specific climate stresses and other forms of uncertainty, risk and externalities. Droughts are a recurrent and specific climate stress faced by communities in semi-arid India, but other forms of uncertainty in weather – like a delayed onset of the monsoon, dry spells, unseasonal rains at different times of the year and heat waves – and market fluctuations are being experienced with increasing frequency (Kuchimanchi *et al.*, 2019; Singh *et al.*, 2019). The CRISI framework factors in the drivers of vulnerability such as income, food security, the health of the ecosystem, equity and quality of life. It also factors in the need to anticipate challenges and transform the system through appropriate resilience capacities, resilience attributes and a focus on reflective learning.

In assessing the resilience of semi-arid farming systems in India, the CRISI framework makes several improvements over other resilience assessment frameworks. It goes beyond impact assessments that rely on biophysical and economic indicators, focusing on issues such as equity and prompting discussions around what would enhance resilience in the future. It broadens the range of resilience-building activities that can be assessed, highlights the need for a diverse set of stakeholders and encourages assessments over longer time frames. The bias toward focusing on the status quo during the assessment is reduced by referring to quantitative data from earlier years, such as geographic information system images and information on crop sown areas, during the PRA exercises to bring more objectivity into the assessment of the earlier time period. Step 6 of the framework also brings in critical reflection about the system's resilience vis-à-vis emerging challenges, such as climate change, and the need to build the resilience capacities of adaptability and transformability of the farming system.

Through the case study, we demonstrated the application of the CRISI framework at the micro-watershed scale, and we believe that it has potential to be used in the work of government agencies, non-governmental organizations and research institutes that focus on semi-arid farming systems in India. Several government programs in India are aimed at improving natural resources to reduce the impact of droughts in semi-arid regions, such as the Integrated Watershed Management Projects (IWMP) under the Pradhan Mantri Krishi Sinchayee Yojana (PMKSY) scheme and the soil and water conservation work carried out under the Mahatma Gandhi National Rural Employment Guarantee Act (MGNREGA). The CRISI framework has application potential in generating evidence-based insights that can guide such programs in the planning, monitoring and evaluation stages and in turn contribute to shaping future agriculture policy in India.

In applying the CRISI framework, one limitation that we encountered was that this process was data-intensive and some participants in the appraisal exercises found the framework, with its ten system functions, four resilience capacities and 16 resilience attributes, quite cumbersome and time-consuming. However, there was consensus among the participants that its usefulness for decision-making would be lost by trying to simplify the framework. In the case of Kalamkarwadi, we had the benefit of existing baseline and endline data, such as geographic information system images and surveys on agriculture and livestock production. We were able to supplement this quantitative data with three meetings for additional qualitative data. If such extensive baseline and endline data were not available, applying the CRISI framework would require additional time for the primary data collection. A workaround for this could be relying on secondary data from local administrative offices, with checks on the quality and reliability of the data. It also must be noted that although applying the CRISI framework identifies interventions that have contributed to the current resilience of the farming system, it cannot establish a more certain attribution. Considering the complexity of issues that farming systems face, several other socio-economic factors likely also contribute to the system's resilience.

We see two specific areas of further research related to the CRISI framework – in socio-ecological systems of larger or smaller scales and in different development contexts. Although the CRISI framework has some scale-independent indicators (i.e. they do not depend on the geographical size or population) such as soil quality, water table depth and per capita income, other indicators, such as the number of institutions and total food production, are very specific to the size of the micro-watershed and its population. Challenges related to smallholder farmers in India, including the specific challenges faced in semi-arid contexts, apply in other types of farming systems in India as well as in other developing countries (Behera and France, 2016). The COVID-19 pandemic has further shown that issues around equity, the resilience capacity of anticipation and attributes such as reflective and shared learning, access to financial safety nets and governance arrangements that support transformation apply even in developed societies.

While assessing case studies from different contexts, it would be important to keep in mind that the ratings of the indicators might not be directly comparable. However, we believe it is still possible to use the CRISI framework for such comparative assessments by focusing on the direction of change (for indicators of the functions, capacities and attributes) rather than the absolute values, and by focusing on the stories behind the performance of the indicators. In an upcoming publication, Srinidhi *et al.* (2023) demonstrate the application of the CRISI framework in such comparative assessments by focusing on the direction of change (increasing or decreasing) and the relative change over the time period of the assessment (comparing the difference between the performance of indicators before and after intervention).

References

Behera, U. and France, J. (2016) Integrated farming systems and the livelihood security of small and marginal farmers in India and other developing countries. *Advances in Agronomy* 138, 235–282.

Bharucha, Z.P., Smith, D. and Pretty, J. (2014) All paths lead to rain: explaining why watershed development in India does not alleviate the experience of water scarcity. *The Journal of Development Studies* 50, 1209–1225. DOI: 10.1080/00220388.2014.928699.

Birthal, P.S., Negi, D.S. and Devesh, R. (2017) Enhancing farmers' income: who to target and how? *Policy Paper – National Centre for Agricultural Economics and Policy Research 30*. National Centre for Agricultural Economics and Policy Research, Indian Council of Agricultural Research, New Delhi.

Bullock, J.M., Dhanjal-Adams, K.L., Milne, A., Oliver, T.H., Todman, L.C. *et al.* (2017) Resilience and food security: rethinking an ecological concept. *Journal of Ecology* 105, 880–884. DOI: 10.1111/1365-2745.12791.

Cabell, J.F. and Oelofse, M. (2012) An indicator framework for assessing agroecosystem resilience. *Ecology and Society* 17. DOI: 10.5751/ES-04666-170118.

Darnhofer, I., Gibbon, D. and Dedieu, B. (2012) Farming systems research: an approach to inquiry. In: Darnhofer, I., Gibbon, D. and Dedieu, B. (eds) *Farming Systems Research into the 21st Century: The New Dynamic*. Springer, Dordrecht, Netherlands, pp. 3–31.

Department of Agriculture (2021) Rainfall Recording and Analysis, Department of Agriculture Maharashtra State. Available at: https://maharain.maharashtra.gov.in/ (accessed February 22, 2021).

DFID (1999) *Sustainable Livelihoods Guidance Sheets*. Department for International Development, London.

Dinshaw, A., Fisher, S., McGray, H., Rai, N. and Schaar, J. (2014) Monitoring and evaluation of climate change adaptation: methodological approaches. In: *OECD Environment Working Papers 74*, OECD Publishing, Paris. DOI: 10.1787/19970900.

Dixon, J.L. and Stringer, L.C. (2015) Towards a theoretical grounding of climate resilience assessments for smallholder farming systems in sub-Saharan Africa. *Resources* 4, 128–154.

FAO (2016) *Rima-II. Resilience Index Measurement and Analysis*. FAO, Rome.

Folke, C., Carpenter, S.R., Walker, B., Scheffer, M., Chapin, T. *et al.* (2010) Resilience thinking: integrating resilience, adaptability and transformability. *Ecology and Society* 15. DOI: 10.5751/ES-03610-150420.

Gibbon, D. (2012) Methodological themes in farming systems research and implications for learning in higher education. In: Darnhofer, I., Gibbon, D. and Dedieu, B. (eds) *Farming Systems Research into the 21st Century: The New Dynamic.* Springer, Dordrecht, Netherlands, pp. 95–115.

Giller, K.E. (2013) Can we define the term 'farming systems'? A question of scale. *Outlook on Agriculture* 42, 149–153. DOI: 10.5367/oa.2013.0139.

GoI (2016) *State of Indian Agriculture 2015–2016*. Department of Agriculture CFW, Ministry of Agriculture and Farmers Welfare, New Delhi, pp. 253.

Grigg, N., Abel, N., O'Connell, D. and Walker, B. (2015) *Resilience Assessment Desktop Case Studies in Thailand and Niger*. CSIRO, Canberra, pp. 38.

Hallegatte, S. and Engle, N.L. (2019) The search for the perfect indicator: reflections on monitoring and evaluation of resilience for improved climate risk management. *Climate Risk Management* 23, 1–6.

IPAM (2021) *AMME Framework: Adaptation Metrics Mapping Evaluation. V 1.1. International Platform on Adaptation Metrics (IPAM)*, Rabat, Morocco.

Jones, H. (2009) *Equity in Development: Why it is Important and How to Achieve*. Overseas Development Institute, London.

Kerr, J., Pangare, G. and Pangare, V.L. (2002) Watershed development projects in India: an evaluation. *Research Report of the International Food Policy Research Institute 127*. IFPRI, Washington, DC. DOI: 10.2499/0896291294rr127.

Krishnan, R., Sanjay, J., Gnanaseelan, C., Mujumdar, M., Kulkarni, A. and Chakraborty, S. (2020) *Assessment of Climate Change over the Indian Region: A Report of the Ministry of Earth Sciences (MoES), Government of India*. Springer Nature, Singapore.

Kuchimanchi, B.R., Nazareth, D., Bendapudi, R., Awasthi, S. and D'Souza, M. (2019) Assessing differential vulnerability of communities in the agrarian context in two districts of Maharashtra, India. *Climate and Development* 11, 918–929. DOI: 10.1080/17565529.2019.1593815.

Maru, Y., O'Connell, D., Grigg, N., Abel, N., Cowie, A. *et al.* (2017) *Making 'Resilience', 'Adaptation' and 'Transformation' Real for the Design of Sustainable Development Projects: Piloting the Resilience, Adaptation Pathways and Transformation Assessment (RAPTA) Framework in Ethiopia*. CSIRO, Canberra, pp. 87.

Mathijs, E. and Wauters, E. (2020) Making farming systems truly resilient. *EuroChoices* 19, 72–76. DOI: 10.1111/1746-692X.12287.

Meuwissen, M.P.M., Feindt, P.H., Spiegel, A., Termeer, C.J.A.M., Mathijs, E. *et al.* (2019) A framework to assess the resilience of farming systems. *Agricultural Systems*, 176, 102656. DOI: 10.1016/j.agsy.2019.102656.

Narayanasamy, N. (2009) *Participatory Rural Appraisal: Principles, Methods and Application*, SAGE Publications, India, New Delhi.

Nemec, K.T., Chan, J., Hoffman, C., Spanbauer, T.L., Hamm, J.A. *et al.* (2014) Assessing resilience in stressed watersheds. *Ecology and Society*, 9, 34. DOI: 10.5751/ES-06156-190134.

O'Connell, D., Walker, B., Abel, N. and Grigg, N. (2015) *The Resilience, Adaptation and Transformation Assessment Framework: From Theory to Application*. CSIRO, Canberra, pp. 96.

Paas, W., Coopmans, I., Severini, S., van Ittersum, M., Meuwissen, M. *et al.* (2021) Participatory assessment of sustainability and resilience of three specialized farming systems. *Ecology and Society*, 26. DOI: 10.5751/ES-12200-260202.

Porritt, J. (2007) *Capitalism as if the World Matters*. Routledge, Abingdon, UK.

Pringle, P. and Leiter, T. (2018) Pitfalls and potential of measuring climate change adaptation through adaptation metrics. In: Christiansen, L., Martinez, G. and Naswa, P. (eds) *Adaptation Metrics: Perspectives on Measuring, Aggregating and Comparing Adaptation Results*. UNEP DTU Partnership, Copenhagen, pp. 29–48.

Quinlan, A.E., Berbés-Blázquez, M., Haider, L.J. and Peterson, G.D. (2016) Measuring and assessing resilience: broadening understanding through multiple disciplinary perspectives. *Journal of Applied Ecology*, 53, 677–687. DOI: 10.1111/1365-2664.12550.

Ramarao, M.V.S., Sanjay, J., Krishnan, R., Mujumdar, M., Bazaz, A. and Revi, A. (2019) On observed aridity changes over the semiarid regions of India in a warming climate. *Theoretical and Applied Climatology* 136, 693–702. DOI: 10.1007/s00704-018-2513-6.

Resilience Alliance (2010) *Assessing Resilience in Social-Ecological Systems: Workbook for Practitioners, Version 2.0.* The Resilience Alliance, Stockholm.

Rockefeller Foundation. (2009) Building climate change resilience. *Rockefeller Foundation White Paper*. Rockefeller Foundation, New York.

Shah, S.H., Harris, L.M., Johnson, M.S. and Wittman, H. (2021) A drought-free Maharashtra? Politicising water conservation for rain-dependent agriculture. *Water Alternatives* 14, 573–596.

Singh, C., Solomon, D., Bendapudi, R., Kuchimanchi, B., Iyer, S. and Bazaz, A. (2019) What shapes vulnerability and risk management in semi-arid India? Moving towards an agenda of sustainable adaptation. *Environmental Development* 30, 35–50. DOI: 10.1016/j.envdev.2019.04.007.

Social Centre (2002) *Kalamkarwadi Watershed: An Overview*. Social Centre, Ahmednagar, India, p. 6.

Srinidhi, A., Werners, S.E., Dadas, D., D'Souza, M., Ludwig, F. *et al.* (2023) Retrospective climate resilience assessment of semi-arid farming systems in India. *International Journal of Water Resources Development.* DOI: 10.1080/07900627.2023.2207680.

Sriskandarajah, N., Givá, N. and Hansen, H.P. (2016) Bridging divides through spaces of change: action research for cultivating the commons in human-inhabited protected areas in Nicaragua and Mozambique. In: Hansen, H.P., Steen Nielsen, B., Sriskandarajah, N., and Gunnarsson, E. (eds) *Commons, Sustainability, Democratization*. Routledge, Abingdon, UK, pp. 139–166.

Symle, J., Lobo, C., Milne, G. and Williams, M. (2014) Watershed development in India: an approach evolving through experience. *Agriculture and Environmental Services Discussion Paper 4*, World Bank Group, Washington, DC. DOI: 10.13140/RG.2.1.1120.2164.

Taylor, M. and Bhasme, S. (2020) Between deficit rains and surplus populations: the political ecology of a climate-resilient village in South India. *Geoforum* 126, 431–440.

UNDP Drylands Development Centre (2013) *Community Based Resilience Analysis (CoBRA) Conceptual Framework and Methodology*. UNDP, New York, pp. 24.

Walker, B. and Salt, D. (2012) *Resilience Practice: Building Capacity to Absorb Disturbance and Maintain Function*. Island Press, Washington, DC.

16 Risk Perceptions of Climate Change, Risk Preferences and Cacao Farming

Alexis H. Villacis,[1]* Ashok K. Mishra,[1] Victor Barrera[2] and Juan Dominguez[3]
[1]Arizona State University, Tempe, USA; [2]National Institute for Agricultural Research of Ecuador (INIAP), Quito, Ecuador; [3]Escuela Superior Politécnica del Litoral (ESPOL), Guayaquil, Ecuador

Abstract

In this study, we investigate cacao farmers' risk perceptions of climate change and how they relate to the use of agronomic practices that will help them combat the effects of weather variability. We measure risk preferences and risk perceptions using a nationally representative survey and framed field experiments conducted in Ecuador – the world's largest producer of fine and flavor cacao. We find that cacao farmers, overall, have reasonable perceptions of high risk of climate change. More importantly, we find that relationships between climate change risk perceptions and potential agronomic mitigation activities occur for the case of fertilization practices only. That is, individuals who perceive a greater risk of climate change have higher expenditures on fertilization activities. Finally, we show that these findings are robust to the influence of experimentally elicited risk preference parameters, i.e. prospect theory or DOSPERT parameters. As the perception of risk is a necessary prerequisite for deciding on actions to adapt to climate risk, these results have important policy implications for the development of programs aimed at mitigating the effects of climate change on the cacao sector and the global chocolate industry.

16.1 Introduction

The effects of climate change will impact the cacao sectors of producing countries in West Africa and Latin America. The main producing countries, such as Côte d'Ivoire, Ghana and Ecuador, among others, will see their national economies affected by changes in droughts and rainfall patterns (Amfo and Ali, 2020). These changes will disturb the land suitability for cacao production, increasing land pressure and competition from other crops, causing deforestation and increasing the occurrence of plant pests and diseases (Ruf *et al.*, 2015; Lahive *et al.*, 2019; Ceccarelli *et al.*, 2021). Ultimately, there will be consequences for the livelihoods of smallholders – who depend on income from cacao production – and the global chocolate industry.

Production systems need to adopt climate-smart agriculture and mitigation practices to help foster climate resilience while sustainably increasing cacao productivity to meet an increasing global demand for chocolates. Some of these practices include the adoption of efficient irrigation systems to offset the effects of droughts and maintain optimal humidity levels in the plantations, the adoption of phytosanitary control practices and pruning to manage the spread of cacao pests and diseases caused by excessive rainfalls, the adoption of improved cultivar

*E-mail: alexis.villacis@asu.edu

 Managing Risk in Agriculture: A Development Perspective (eds A.K. Mishra, S.C. Kumbhakar and G. Lien).
DOI: 10.1079/9781800622289.0016

varieties that are resistant to pests and climate conditions (droughts/excessive rainfalls), the adoption of fertilizers to offset potential losses of soil nutrients, and increased knowledge of climate change impacts on the crop and the sector, among others.

Since the perception of risk is a necessary prerequisite for the adoption of adaptation strategies for mitigating the effects of climate change (Maddison, 2007; Villacis *et al.*, 2021), the adoption by cacao producers of the practices mentioned above depends partially on how they perceive the risks associated with climate change. The relationship between risk perceptions and actions to adapt to climate change highlights the role of psychological constraints in the development of the global cacao sector. Understanding this relationship can help to design customized and more effective policies and programs aimed at mitigating the cacao sector's vulnerability to the effects of climate change.

In this chapter, we examine cacao farmers' risk perceptions of climate change and how they relate to the use of agronomic practices that will help combat the effects of weather variability. To do so, we conducted a nationally representative survey and framed field experiments with cacao farmers in Ecuador. Ecuador ranks third in the world for producing cacao beans and first for producing fine and flavor cacao (ICCO, 2021). Most of its cacao production comes from small farmers cultivating 1–5 ha (USDA FAS, 2015) with low productivity levels (Villacis *et al.*, 2022a). Given that the perceived risks of climate change can be influenced by the distinction between personal (i.e. self-regarding) levels of risk judgments and societal (i.e. other-regarding) levels of risk judgments (van der Linden, 2015, 2017), we assess risks perceptions over different domains. These include judgments of the climate risks perceived over input costs, output prices, management practices and geographical place. As it is also possible that climate change risk perceptions are influenced by risk preferences (Villacis *et al.*, 2021), we also account for the role of this behavioral trait by using the frameworks of prospect theory (Kahneman and Tversky, 1979; Tversky and Kahneman, 1992) and the domain-specific risk-taking (DOSPERT) scale (Weber *et al.*, 2002; Blais and Weber, 2006).

We present three key results. First, we find that farmers have reasonable perceptions of high risk of climate change over all the domains assessed. Second, we find that relationships between climate change risk perceptions and potential mitigation activities occur for the case of fertilization practices only. That is, individuals who perceive a greater risk of climate change have higher expenditures on fertilization activities, regardless of the risk perception domain. We do not, however, find evidence of relationships between climate change risk perceptions and expenditures on irrigation, phytosanitary control, pruning and plantation renovation activities. Third, we show that these findings are robust to the influence of experimentally elicited risk preference parameters, i.e. prospect theory or DOSPERT parameters. These findings are important because they shed light on how farmers perceive the current threats posed by climate change and how they adapt. By showing the detachment between climate change and specific agronomic practices, we also highlight where policies and programs seeking to foster cacao climate resilience should target future efforts.

The rest of this chapter is organized as follows. Section 16.2 describes the theoretical framework and discusses climate change risk perceptions, prospect theory and the DOSPERT scale as proxies for farmers' risk preferences. Section 16.3 discusses the survey and data and presents descriptive statistics. Section 16.4 presents the empirical framework used to model the associations between climate change risk perceptions, agronomic practices performed by cacao farmers and risk preferences. Section 16.5 presents the results from the estimation of the empirical model. Section 16.6 summarizes and discusses the implications of the findings.

16.2 Theoretical Framework

16.2.1 Risk perceptions of climate change

Climate change risk perceptions are influenced by the ability to evaluate changes in long-term weather conditions, which are difficult for people to perceive (Weber, 2010). They also are affected by psychological factors (van der Linden, 2017); thus, the literature incorporates five determinants of perceptions of risk of climate change:

socio-demographic, cognitive, experiential and socio-cultural factors and risk preferences (Helgeson *et al.*, 2012; van der Linden, 2015; Villacis *et al.*, 2021). Given the great heterogeneity of these determinants, climate change risk perceptions vary significantly among individuals, regions and countries, with most low- and middle-income countries having greater climate change risk perceptions than the developed world.

16.2.2 Prospect theory

Climate change phenomena involve imprecise or low probabilities of events; thus, risk preferences have been argued to influence the perception of risk associated with weather phenomena. Notably, the framework of prospect theory has been shown to help capture the influence of status quo bias and probability distortion in the context of climate change (Villacis *et al.*, 2021).

The prospect theory utility function is defined as:

$$PT(x, p; y, 1-p) = \begin{cases} v(y) + w(p)\left[v(x) - v(y)\right]; \\ x > y > 0 \text{ or } x < y < 0 \\ w(p)v(x) + w(1-p)v(y); \\ x < 0 < y \end{cases} \quad (16.1)$$

where $PT(x, p; y, 1-p)$ is the expected value over binary prospects $(x; y)$, with corresponding probabilities $(p; 1-p)$; $v(x)$ is the value function defined as a piecewise function that assigns different values for gains $(x > 0)$ and losses $(x < 0)$, such that:

$$v(x) = \begin{cases} x^{\sigma}; & x \geq 0 \\ -\lambda(-x)^{\sigma}; & x < 0 \end{cases} \quad (16.2)$$

σ represents the curvature of the value function for gains and losses and defines risk aversion. λ represents the curvature below 0 relative to the curvature above 0 and establishes the degree of loss aversion. The function $w(p)$ represents probability weighting. We use the weighting function of Prelec (1998), where α determines the curvature of the probability weighting function, and it is the proxy for probability distortion:

$$w(p) = \frac{1}{\left[\exp(\ln(1/p))\right]^{\alpha}} \quad (16.3)$$

An individual is risk-loving if $\sigma > 1$, risk-neutral if $\sigma = 1$ and risk-averse if $\sigma < 1$. The individual is said to be more sensitive to losses than to gains when $\lambda > 1$, and less sensitive to losses than to gains when $\lambda < 1$. If $\alpha = 1$, the probability weighting function $w(p)$ is linear, and the agent has no probability distortion. If $\alpha < 1$, $w(p)$ departs from linear probability weighting, and the agent will overweight small probabilities and underweight large probabilities (Tversky and Kahneman, 1992). If $\alpha > 1$, the agent underweights small probabilities and overweights large probabilities. Prospect theory probability weighting and loss aversion (α and λ) relate closely to climate change and its behavioral implications. They account for status quo bias, the overweighting of low-probability extreme weather events, and the further behavioral impact of economic losses caused by climate-related crop damages (Villacis *et al.*, 2021).

16.2.3 DOSPERT scale

As risk taking has been argued to be domain-specific, we assess risk taking in six domains using the DOSPERT scale (Weber *et al.*, 2002; Blais and Weber, 2006). The domains investigated include gambling, investing, ethical choices and behaviors relating to health/safety, social interaction and recreation. DOSPERT questions prompt participants to rate the risk associated with different domain-specific activities. The scale in each question ranges from 1 (not at all risky) to 7 (extremely risky). The numerical values of the responses are added to obtain a domain-specific single score that can vary from 6 to 42, with 42 being the most risk-averse.

16.3 Data and Descriptive Statistics

16.3.1 Study area and survey

In partnership with the Ecuadorian Institute of Agricultural Research (INIAP), we collected data for this study between August and December of 2021 in five coastal provinces of Ecuador (i.e. Esmeraldas, Guayas, Los Ríos, Manabí and Santo Domingo de Los Tsáchilas). According to Ecuador's Corporación Financiera Nacional

(National Finance Corporation; CFN, 2018), these provinces account for 80% of the total cacao area planted in the country. We randomly selected 25 villages, where we implemented household surveys and framed field experiments to collect data on household and farm characteristics and elicit individual risk preferences. During the past decade, a national extension program led by INIAP promoted the adoption of pruning practices in the cacao plantations of these villages. The surveys and framed field experiments involved 523 farmers, who received monetary compensation for their participation in this study (see more details of the logistics of the study in Villacis *et al.* (2022b)).

Summary statistics for each of the main variables of interest are shown in Table 16.1. The average participant is about 49 years old, has finished elementary school and completed some high school years, earns between US$800 and US$72,000 per year and cultivates cacao on about 2.5 ha. Expenditures on fertilization, phytosanitary control, irrigation, pruning and farm renovations are measured as yearly average expenses and include inputs and labor costs.

16.3.2 Descriptive analysis of risk perceptions of climate change

We measured climate change risk perceptions along personal (i.e. self-regarding) levels of risk judgments and societal (i.e. other-regarding) levels of risk judgments (Villacis *et al.*, 2021). We asked participants about the climate change perceived risks over crop management practices, output prices, input costs and geographic location. Following Weber *et al.* (2013), we assessed risk perceptions using psychometric methods, i.e. from answers to attitudinal statements on Likert scales. For example, participants were asked if they thought that 'the effects of climate change on the output prices' would be *not important, slightly important, moderately important, important* or *very important*. Risk perceptions took a value of 1 if an individual answered *not important*; a value of 2 if *slightly important*; a value of 3 if *moderately important*; a value of 4 if *important*; and a value of 5 if *very important*.

We use this type of assessment with the objective of omitting positive or negative connotations and to capture the spirit of risk as it relates to the variance of outcomes, regardless of whether these represent gains or losses to the individual (Harrison and Rutström, 2008; Mata *et al.*, 2018). For example, two different individuals can say that the effects of climate change on the output prices are very important, being the case that one thinks the cacao price will decrease while the other thinks the price will increase. Overall risk perceptions of climate change were reasonably high over all domains, with input costs having a mean ($\overline{y}$) of 4.33 and a standard deviation (*SD*) of 1.00; output price with $\overline{y} = 4.40$ and $SD = 0.96$; crop management

Table 16.1. Summary statistics: household and farm characteristics. (Source: Authors' own elaboration.)

	(1)	(2)	(3)	(4)	(5)
VARIABLES	Mean	Std. Dev.	Min.	Max.	Obs.
Age	49.47	15.63	17	87	523
Household size	4.08	1.72	1	10	523
Gender (female = 1)	0.27	0.44	0	1	523
Years of education	10.37	4.13	0	17	523
Area of cacao cultivation (hectares)	2.56	2.58	0.1	30	523
Years of cacao farming experience	19.87	15.52	1	70	523
Annual income (US$)	8,915.68	9,133.60	800	72,000	523
Plantation age (years)	18.26	16.85	1	80	523
Fertilization expenditures (US$)	256	835	0	16,200	523
Phytosanitary control expenditures (US$)	302	1,281	0	17,550	523
Irrigation expenditures (US$)	83	131	0	800	523
Pruning expenditures (US$)	624	1,156	0	18,000	523
Farm renovation expenditures (US$)	195	563	0	5,000	523

practices with $\overline{y} = 4.41$ and $SD = 0.86$; and geographic location with $\overline{y} = 4.33$ and $SD = 0.91$ (see Panel A of Table 16.2).

16.3.3 Descriptive analysis of risk preferences parameters

We conducted framed field experiments (Harrison and List, 2004) consisting of incentivized lottery games to elicit prospect theory risk preference parameters (Kahneman and Tversky, 1979; Tversky and Kahneman, 1992). We used the lottery game employed by Villacis *et al.* (2021), which is based on the experimental design of Tanaka *et al.* (2010). This lottery game illustrates a farm composed of ten equally sized lots with particular payoffs over different years from using two types of technologies, either Seed A or Seed B. Participants are told that at the end of each year, only one random lot out of the ten will survive, thus, they need to decide at which year they will 'switch' from Seed A to Seed B. Participants received \$5 for showing up to the experimental sessions and a bonus that ranged from −\$4 to \$4. The bonus was applied based on their actual choices during the lottery games (see Villacis *et al.* (2021, 2022b) for more details on the experimental procedures and the incentivized payments).

Tanaka's design consists of three series or price lists; thus, participants reported their answers in each series. To estimate the structural parameters of the prospect theory utility function, we used the 'mid-point' approximation method. This method helps determine unique values of prospect theory structural parameters from plays in the lottery games. Results show the average participant: (i) is relatively risk-averse (average of $\sigma = 0.62$); (ii) distorts probability information and tends to overweight low probabilities (average of $\alpha = 0.78$); and (iii) is more sensitive to losses than to gains, at a magnitude of roughly five to one (average of $\lambda = 5.55$) (see Panel B of Table 16.2).

Using the survey instrument, we also measured risk attitudes using the DOSPERT scale to evaluate participants' risk-taking behavior. Panel C of Table 16.2 reports risk taking in five domains, including ethical and financial choices and behaviors relating to health safety, recreation and social interaction (Weber *et al.*, 2002; Blais and Weber, 2006). Out of 42 possible points in each domain, the social domain had the highest mean value across individual farmers

Table 16.2. Summary statistics: risk perceptions of climate change and risk preferences. (Source: Authors' own elaboration.)

	(1)	(2)	(3)	(4)	(5)
VARIABLES	Mean	Std. Dev.	Min.	Max.	Obs.
	Panel A: Risk perceptions of climate change				
Over input costs	4.33	1.00	1	5	523
Over output price	4.40	0.96	1	5	523
Over management practices	4.41	0.86	1	5	523
Over geographical place	4.33	0.91	1	5	523
	Panel B: Risk preferences – prospect theory				
Risk aversion	0.62	0.45	0	2	523
Probability distortion	0.78	0.35	0	1	523
Loss aversion	5.55	4.33	0.12	11.79	523
	Panel C: Risk attitude – DOSPERT scale				
Ethical	15.29	7.12	6	42	523
Financial	18.39	6.26	6	42	523
Health safety	19.14	6.88	6	42	523
Recreational	14.60	7.77	6	42	523
Social	22.44	6.44	6	42	523

Notes: Risk perceptions of climate change take a value of 1 if an individual answered not important; a value of 2 if slightly important; a value of 3 if moderately important; a value of 4 if important; and a value of 5 if very important. DOSPERT scale measures risk-taking behavior in five domains, with 42 possible points in each domain.

(mean = 22.44), and the recreational domain had the lowest mean value (mean = 14.60).

16.4 Empirical Framework

Following Villacis *et al.* (2021), we use an ordered probit model to examine the relationships between risk perceptions of climate change, risk preferences and crop management practices. Risk perceptions of climate change are modeled as the dependent variable, and risk preferences and crop management practices as independent variables. The ordered probit model, including individual-specific controls, is given by:

$$y_i^* = \mathbf{X}_i'\beta + \mathbf{R}_i'\psi + \mathbf{C}_i'\gamma + u_i \quad i = 1,\ldots,n \qquad (16.4)$$
$$u_1 \sim \mathrm{N}(0,1)$$

i indexes participant *i*, $i = 1, \ldots, n$, where *n* is the study size; $y_i^*\left(-\infty < y_i^* < +\infty\right)$ is the underlying latent variable representing respondent *i's* propensity to agree with a certain risk perception; $\mathbf{X}_i$ is a vector containing individual-specific controls, including cognitive, experiential, socio-demographic and socio-cultural factors that help explain risk perceptions of climate change; $\mathbf{R}_i$ is a vector containing individual-specific risk attitudes, namely the DOSPERT domain scales (ethical, financial, health safety, recreational and social) or the prospect theory risk preference parameters (e.g. risk aversion, probability distortion and loss aversion); $\mathbf{C}_i$ is a vector containing individual-specific farm management practices, namely expenditures on farm renovation, fertilization, phytosanitary control, irrigation and pruning activities; β, ψ and γ are vectors of parameters, *not* including an intercept. The relationship between the unobserved (latent) y_i^* and the observed variable $\boldsymbol{y}_i$ (individual *i's* response to the survey question concerning climate change risk perceptions) is given by:

$$\begin{aligned} y_i &= 1 \text{ if } -\infty < y_i^* < \kappa_1 \\ y_i &= 2 \text{ if } \kappa_1 < y_i^* < \kappa_2 \\ y_i &= 3 \text{ if } \kappa_2 < y_i^* < \kappa_3 \\ y_i &= 4 \text{ if } \kappa_3 < y_i^* < \kappa_4 \\ y_i &= 5 \text{ if } \kappa_4 < y_i^* < +\infty \end{aligned} \qquad (16.5)$$

where the parameters $\kappa_1, \ldots, \kappa_4$ are the threshold parameters, also known as 'cut-points.' We assess risk perceptions of climate change over the four domains discussed above: cacao farming management practices, output prices (price of cacao), input prices (production costs) and geographical location (village).

We expect the individual-specific farm management practices (expenditures on farm renovation, fertilization, phytosanitary control, irrigation and pruning activities) to correlate positively with climate change risk perceptions. That is, after controlling for other factors, individuals with higher climate change risk perceptions are more likely to spend more resources on their farming activities (Maddison, 2007) than individuals with lower risk perceptions. These relationships can be argued as adaptation strategies to the effects of climate change. Finally, as we estimate model specifications using different risk measures, we test for the joint significance of these parameters using a likelihood ratio test.

16.5 Results and Discussion

16.5.1 Empirical model estimates

In this section, we present results from our empirical model. First, Tables 16.3 and 16.4 present results from estimating Eqn 16.4 using prospect theory and DOSPERT parameters, respectively. Finally, in Table 16.5, we present the marginal effects of our statistically significant results.

Ordered probit estimates

We begin by studying the ordered probit estimates of the relationship between cacao farmers' risk perceptions of climate change and the use of agronomic practices that will help combat the effects of weather variability. Table 16.3 reports results from estimating Eqn 16.4 with the farmer's climate change risk perceptions as the dependent variable. Columns (1), (3), (5) and (7) serve as a baseline regression that estimates the relationship between farmers' risk perceptions of climate change and expenditures on agronomic practices with no risk preference variables included. In these columns, we find evidence of a relationship between farmers' climate change risk perceptions and fertilizer expenditures. The coefficient on fertilizer expenditures is positive

Table 16.3. Ordered probit estimates – prospect theory risk preferences parameters as controls. (Source: Authors' own elaboration.)

	(1)	(2)	(3)	(4)	(5)	(6)	(7)	(8)
VARIABLES	Over input costs	Over input costs	Over output price	Over output price	Over management practices	Over management practices	Over geographical place	Over geographical place
Fertilization expenditures	0.023**	0.023**	0.025**	0.026**	0.030**	0.031**	0.025**	0.026***
	(0.010)	(0.010)	(0.010)	(0.011)	(0.012)	(0.013)	(0.010)	(0.009)
Phytosanitary control expenditures	−0.003	−0.003	−0.003	−0.003	−0.008**	−0.008***	−0.003	−0.003
	(0.003)	(0.003)	(0.003)	(0.003)	(0.003)	(0.003)	(0.002)	(0.002)
Irrigation expenditures	0.057	0.058	0.014	0.013	0.062	0.063	0.063	0.064
	(0.047)	(0.047)	(0.051)	(0.052)	(0.062)	(0.063)	(0.049)	(0.049)
Pruning expenditures	0.001	0.000	−0.001	−0.001	−0.005	−0.004	0.000	0.001
	(0.007)	(0.007)	(0.005)	(0.005)	(0.006)	(0.005)	(0.005)	(0.004)
Farm renovation expenditures	−0.007	−0.007	−0.014	−0.014	0.017*	0.017*	−0.009	−0.009
	(0.011)	(0.011)	(0.012)	(0.012)	(0.010)	(0.010)	(0.014)	(0.014)
Pt risk aversion		−0.025		0.113		0.101		0.198**
		(0.098)		(0.114)		(0.069)		(0.088)
Pt probability distortion		−0.017		0.104		−0.001		−0.034
		(0.162)		(0.140)		(0.160)		(0.211)
Pt loss aversion		−0.009		0.005		0.008		0.015
		(0.009)		(0.014)		(0.012)		(0.012)
Additional controls	Yes	Yes	Yes	Yes	Yes	Yes	Yes	Yes
Test of joint significance								
LR $chi^2(3)$		0.47		1.21		0.65		2.63
Prob > chi^2		0.92		0.75		0.88		0.45
Observations	523	523	523	523	523	523	523	523

Notes: This table reports ordered probit estimates with climate change risk perceptions as the dependent variable. Expenditures on fertilization, phytosanitary control, irrigation, pruning and farm renovation are measured as yearly average expenses in US$100 and include inputs and labor costs. The risk aversion, probability distortion and loss aversion variables are prospect theory (Pt) risk preference parameters. Threshold parameters not reported. Additional controls included: age, household size, number of children in the household, gender, association membership, education level, area of cacao farming, experience in cacao farming, plantation age and income. Test of joint significance reports the joint significance of prospect theory risk preference parameters. Asterisks *, ** and *** denote significance at the 10%, 5% and 1% levels, respectively. Standard errors are clustered at the village level in parentheses.

and statistically significant across all domains of climate risk. In contrast, the coefficients of the other expenditures on agronomic practices, in general, are not statistically significant. One exception is the negative relationship between phytosanitary control expenditures and the climate risk perceived over crop management practices.

Next, in columns (2), (4), (6) and (8), we include the prospect theory risk preference parameters in our main regression specification, estimating the relationship between cacao farmers' risk perceptions of climate change and the use of agronomic practices. Similar to the results in columns (1), (3), (5) and (7), and although in these columns we account for the farmer's risk preferences with the prospect theory parameters, we again find evidence for a relationship between the farmer's climate change risk perceptions and fertilizer expenditures. The coefficient on fertilizer expenditures remains positive and statistically significant, while the coefficients of the other expenditures on agronomic practices, in general, are not statistically significant. Again, the exception is the negative relationship between phytosanitary control expenditures and the climate risk perceived over crop management practices.

We generally find no relationships between cacao farmers' risk perceptions of climate change and prospect theory risk preferences. One exception is the positive relationship between risk aversion and the climate risk perceived over the geographical place. We test for the joint significance of the prospect theory parameters using a likelihood ratio test, and the results lead us to fail to reject the null hypothesis of no joint significance across all domains of climate risk.

As a robustness check, we use DOSPERT parameters measuring risk-taking behavior in different domains as an additional measure of risk preferences. Columns (1), (3), (5) and (7) in Table 16.4 show the baseline regression (same as Table 16.3) that estimates the relationship between farmers' risk perceptions of climate change and expenditures on agronomic practices with no risk preference variables included. Interestingly, when we include the DOSPERT variables in our main regression specification, we continue to find strong evidence for a positive relationship between farmers' climate change risk perceptions and fertilizer expenditures. Results in columns (2), (4), (6) and (8) show that even though in these columns we account for the farmer's risk preferences with DOSPERT parameters, we again find evidence for a relationship between the farmer's climate change risk perceptions and fertilizer expenditures. Similar to the previous exercise, the coefficients of the other expenditures on agronomic practices, generally, are not statistically significant, except for the negative relationship between phytosanitary control expenditures and the climate risk perceived over crop management practices.

In contrast to the previous exercise, we do find relationships between cacao farmers' risk perceptions of climate change and DOSPERT risk parameters. Specifically, we find that risk-taking behavior related to ethical choices negatively correlates with the climate risk perceived over input costs and output price. Similarly, we find that risk-taking behavior related to health safety choices positively correlates with the climate risk perceived over input costs, crop management practices and geographic location. We test for the joint significance of the DOSPERT parameters using a likelihood ratio test, and results show they are jointly significant at the 5% significance level for the climate risk perceived over input costs, at the 1% significance level for the risk perceived over output price and at the 10% significance level for the risk perceived over crop management practices. Nonetheless, we fail to reject the null hypothesis of no joint significance for the case of climate risk perceived over the geographical place.

Ordered probit marginal effects

Table 16.5 reports the ordered probit marginal effects calculated at mean values of the independent variables. We report the marginal effects on the probability of perceiving the risks of climate change as being not important ($y' = 1$), slightly important ($y' = 2$), moderately important ($y' = 3$), important ($y' = 4$) or very important ($y' = 5$). For brevity, we report only the statistically significant coefficients. Individuals with higher fertilizer expenditures stated higher climate change risk perception over all the domains investigated (input costs, output price, crop management practices and geographical place). Empirically, an increase of US$100 in fertilization-related expenditures is associated

Table 16.4. Ordered probit estimates – DOSPERT scale parameters as controls. (Source: Authors' own elaboration.)

	(1)	(2)	(3)	(4)	(5)	(6)	(7)	(8)
VARIABLES	Over input costs	Over input costs	Over output price	Over output price	Over management practices	Over management practices	Over geographical place	Over geographical place
Fertilization expenditures	0.023**	0.023**	0.025**	0.026**	0.030**	0.033***	0.025**	0.026***
	(0.010)	(0.010)	(0.010)	(0.012)	(0.012)	(0.013)	(0.010)	(0.010)
Phytosanitary control expenditures	−0.003	−0.002	−0.003	−0.002	−0.008**	−0.007**	−0.003	−0.002
	(0.003)	(0.004)	(0.003)	(0.003)	(0.003)	(0.003)	(0.002)	(0.003)
Irrigation expenditures	0.057	0.058	0.014	0.005	0.062	0.063	0.063	0.064
	(0.047)	(0.046)	(0.051)	(0.049)	(0.062)	(0.059)	(0.049)	(0.045)
Pruning expenditures	0.001	−0.000	−0.001	−0.002	−0.005	−0.005	0.000	−0.000
	(0.007)	(0.007)	(0.005)	(0.006)	(0.006)	(0.006)	(0.005)	(0.005)
Farm renovation expenditures	−0.007	−0.007	−0.014	−0.015	0.017*	0.016*	−0.009	−0.009
	(0.011)	(0.010)	(0.012)	(0.010)	(0.010)	(0.009)	(0.014)	(0.014)
DOSPERT ethical		−0.034**		−0.032**		−0.015		−0.016
		(0.017)		(0.016)		(0.012)		(0.014)
DOSPERT financial		0.004		0.020		−0.006		0.005
		(0.011)		(0.017)		(0.010)		(0.013)
DOSPERT health safety		0.028***		0.011		0.030***		0.021*
		(0.010)		(0.012)		(0.007)		(0.012)
DOSPERT recreational		−0.001		−0.015		−0.013		−0.006
		(0.017)		(0.018)		(0.008)		(0.012)
DOSPERT social		0.002		0.013		0.006		−0.002
		(0.015)		(0.016)		(0.014)		(0.014)
Additional controls	Yes	Yes	Yes	Yes	Yes	Yes	Yes	Yes
Test of joint significance								
LR $chi^2(5)$		12.59		16.7		9.55		5.27
Prob > chi^2		0.028**		0.005***		0.089*		0.385
Observations	523	523	523	523	523	523	523	523

Notes: This table reports ordered probit estimates with climate change risk perceptions as the dependent variable. Expenditures on fertilization, phytosanitary control, irrigation, pruning and farm renovation are measured as yearly average expenses in US$100 and include inputs and labor costs. The DOSPERT variables represent psychometric scales that assess risk taking in different domains. Threshold parameters not reported. Additional controls included: age, household size, number of children in the household, gender, association membership, education level, area of cacao farming, experience in cacao farming, plantation age and income. Test of joint significance reports the joint significance of the DOSPERT parameters. Asterisks *, ** and *** denote significance at the 10%, 5% and 1% levels, respectively. Standard errors are clustered at the village level in parentheses.

Table 16.5. Ordered probit marginal effects. (Source: Authors' own elaboration.)

VARIABLES	(1) Over input costs	(2) Over output price	(3) Over management practices	(4) Over geographical place
Marginal effect on probit ($y' = 1$)				
Fertilization expenditures	−0.001**	−0.001**	−0.001**	−0.001***
	(0.001)	(0.001)	(0.000)	(0.000)
DOSPERT ethical	0.002*	0.002**	0.000	0.001
	(0.001)	(0.001)	(0.000)	(0.001)
DOSPERT health safety	−0.001**	−0.001	−0.001**	−0.001
	(0.001)	(0.001)	(0.000)	(0.001)
Marginal effect on probit ($y' = 2$)				
Fertilization expenditures	−0.002**	−0.002*	−0.002*	−0.001***
	(0.001)	(0.001)	(0.001)	(0.000)
DOSPERT ethical	0.002**	0.002**	0.001	0.001
	(0.001)	(0.001)	(0.001)	(0.001)
DOSPERT health safety	−0.002***	−0.001	−0.001***	−0.001
	(0.001)	(0.001)	(0.000)	(0.001)
Marginal effect on probit ($y' = 3$)				
Fertilization expenditures	−0.002*	−0.001*	−0.004**	−0.003**
	(0.001)	(0.000)	(0.002)	(0.001)
DOSPERT ethical	0.003*	0.001**	0.002	0.002
	(0.002)	(0.000)	(0.001)	(0.001)
DOSPERT health safety	−0.003**	−0.000	−0.003***	−0.002**
	(0.001)	(0.000)	(0.001)	(0.001)
Marginal effect on probit ($y' = 4$)				
Fertilization expenditures	−0.004**	−0.006**	−0.007***	−0.005***
	(0.002)	(0.003)	(0.002)	(0.002)
DOSPERT ethical	0.006*	0.007*	0.003	0.003
	(0.003)	(0.004)	(0.003)	(0.003)
DOSPERT health safety	−0.005***	−0.003	−0.006***	−0.004*
	(0.001)	(0.003)	(0.002)	(0.002)
Marginal effect on probit ($y' = 5$)				
Fertilization expenditures	0.009**	0.010**	0.013**	0.010***
	(0.004)	(0.005)	(0.005)	(0.004)
DOSPERT ethical	−0.013**	−0.012**	−0.006	−0.006
	(0.007)	(0.006)	(0.005)	(0.005)
DOSPERT health safety	0.011***	0.004	0.011***	0.008*
	(0.004)	(0.005)	(0.003)	(0.005)
Observations	523	523	523	523

Notes: This table reports ordered probit marginal effects calculated at mean values of independent variables. Climate change risk perceptions are the dependent variable. Expenditures on fertilization are measured as yearly average expenses in US$100 and include inputs and labor costs. The DOSPERT variables represent psychometric scales that assess risk taking in different domains. Threshold parameters not reported. Additional controls included: age, household size, number of children in the household, gender, association membership, education level, area of cacao farming, experience in cacao farming, plantation age and income. Asterisks *, ** and *** denote significance at the 10%, 5% and 1% levels, respectively.

with being: (i) 0.9% more likely to perceive the risk of climate change over *input costs* as very important ($y' = 5$); (ii) 1% more likely to perceive the risk of climate change over *output price* as very important ($y' = 5$); (iii) 1.3% more likely to perceive the risk of climate change over *crop management practices* as very important ($y' = 5$); and (iv) 1% more likely to perceive the risk of climate change over the *geographical place* as very important ($y' = 5$). This result suggests that farmers who use fertilizers perceive climate change risk as more threatening to all outcomes associated with their farming practices, markets and the local area.

Interestingly, the marginal effects of the health safety risk-taking behavior suggest that farmers who are more risk-averse in this specific domain stated higher climate change risk perceptions over input costs, crop management practices and geographical place. Empirically, an increase of one unit in the DOSPERT scale of this specific domain is associated with being: (i) 1.1% more likely to perceive the risk of climate change over *input costs* as very important ($y' = 5$); (ii) 1.1% more likely to perceive the risk of climate change over *crop management practices* as very important ($y' = 5$); and (iii) 0.8% more likely to perceive the risk of climate change over the *geographical place* as very important ($y' = 5$). This result suggests that farmers who are more risk-averse over health safety issues perceive climate change risk as more threatening to *input costs, crop management practices* and the *geographical place*. We found no relationships between this specific risk domain and the stated climate change risk perceptions over output price.

Finally, the marginal effects of the ethical risk-taking behavior suggest that farmers who are more risk-averse in this specific domain stated lower climate change risk perceptions over input costs and output price. Empirically, an increase of one unit in the DOSPERT scale of this particular domain is associated with being: (i) 1.3% less likely to perceive the risk of climate change over *input costs* as very important ($y' = 5$); and (ii) 1.2% less likely to perceive the risk of climate change over *output price* as very important ($y' = 5$). This result is puzzling and suggests that farmers who are more risk-averse over ethical issues perceive climate change risk as less threatening to *input costs* and *output price*. We found no relationships between this specific risk domain and the stated climate change risk perceptions over crop management practices and geographical place.

16.6 Summary and Concluding Remarks

The perception of a risk is a prerequisite for adopting adaptation strategies for mitigating the effects of climate change. Thus, adopting practices aimed at mitigating the effects of climate change in the cacao sector depend, in part, on how farmers perceive the risks associated with weather variability. Recent evidence shows that past shocks and future risks are perceived as important across production, market, institutional, financial and personal risk domains (Feyisa *et al.*, 2023). In this chapter, we focus on climate change risk perceptions among cacao producers from Ecuador – the largest producer of fine and flavor cacao in the world – and investigate their relationships with the use of agronomic practices that will help farmers combat the effects of climate change.

Our analysis leads to three key findings. First, we find that farmers have high risk perceptions of climate change over the four domains assessed: input costs, output price, crop management practices and geographic location. Thus, farmers in our study area are aware of the multi-dimensional impacts of climate change over their lives. Second, we find that relationships between climate change risk perceptions and potential mitigation activities occur only for the case of fertilization practices. Fertilization, particularly in cacao farming, appears to be the factor that the average farmer associates the most with the risks and effects of climate change. Third, we show that these associations are robust to the influence of risk preferences as represented by prospect theory risk preference parameters or DOSPERT scale parameters. We do not, however, find evidence of relationships between climate change risk perceptions and expenditures on irrigation, phytosanitary control, pruning and plantation renovation activities.

As consumption of chocolate has steadily increased over recent decades – especially in emerging economies – the demand for cacao beans has constantly exceeded the global supply. Unless producing countries keep increasing their supply of beans, chocolate products could become more expensive. Climate change can potentially limit cacao production worldwide, as it is expected to hit hardest in regions around the equator, which are the only areas where cacao beans can be grown. Thus, our findings presented here are important for policymakers. Policy interventions that aim to help mitigate the effects of climate change can focus their efforts on the agronomic practices that currently lack association with the risks of climate change. Thus, they can be a valuable addition to farmers' current activities. Providing free/subsidized agricultural inputs and other initiatives can help the cacao sector adapt to the effects of climate change.

To the extent that policymakers are committed to developing the cacao sector, proactive policies must provide integral development initiatives to foster this sector's resilience. Given that the effects of climate change will be heterogeneous, the 'one size fits all' policies should be discouraged. For example, irrigation access can be a possible solution in places where the impacts will be felt in the form of increased droughts. Meanwhile, places that will experience more rainfall, and thus humidity, will need efforts focused on drainage initiatives and extensive phytosanitary control to combat pests and diseases associated with higher humidity levels.

The limitations of our study suggest several directions for future work. First, we did not explore the possibility that new areas could become more suitable for cacao production. As new areas within current producing countries and new countries have the potential to become high-yielding cacao regions, it will be important to know if farmers in those areas would be willing to become cacao producers. Moreover, there is the question of whether this change will be associated with their climate change risk perceptions. Second, we did not explore the possibility that farmers with higher climate change risk perceptions would be willing to undergo training on climate-smart agriculture and mitigation practices and, thus, become more resilient to climate impacts. Finally, although beyond the scope of this study, unfair wages and child labor have been long-standing issues in the cacao sector, and climate change can exacerbate them. Future research can investigate the effects of climate change on these two specific issues affecting the global cacao sector.

References

Amfo, B. and Ali, E.B. (2020) Climate change coping and adaptation strategies: how do cocoa farmers in Ghana diversify farm income? *Forest Policy and Economics* 119, 102265. DOI: 10.1016/j.forpol.2020.102265.

Blais, A-R. and Weber, E.U. (2006) A domain-specific risk-taking (DOSPERT) scale for adult populations. *Judgment and Decision Making* 1, 33–47.

Ceccarelli, V., Fremout, T., Zavaleta, D., Lastra, S., Imán Correa, S. *et al.* (2021) Climate change impact on cultivated and wild cacao in Peru and the search of climate change-tolerant genotypes. *Diversity and Distributions* 27, 1462–1476. DOI: 10.1111/ddi.13294.

CFN (2018) Ficha Sectorial: Cacao y Chocolate. Corporacióon Financiera Nacional, Quito. Available at: https://www.cfn.fin.ec/wp-content/uploads/2018/04/Ficha-Sectorial-Cacao.pdf (accessed November 20, 2022).

Feyisa, A.D., Maertens, M. and de Mey, Y. (2023) Relating risk preferences and risk perceptions over different agricultural risk domains: insights from Ethiopia. *World Development* 162, 106137. DOI: 10.1016/j.worlddev.2022.106137.

Harrison, G.W. and List, J.A. (2004) Field experiments. *Journal of Economic Literature* 42, 1009–1055. DOI: 10.1257/0022051043004577.

Harrison, G.W. and Rutström, E. E. (2008) Risk aversion in the laboratory. In: Cox, J.C. and Harrison, G.W. (eds) *Risk Aversion in Experiments* (Research in Experimental Economics, Vol. 12). Emerald Group Publishing, Bingley, UK, pp. 41–196. DOI: 10.1016/S0193-2306(08)00003-3.

Helgeson, J., van der Linden, S. and Chabay, I. (2012) In: Wals, A.E.J. and Corcoran, P.B. (eds) *Learning for Sustainability in Times of Accelerating Change*. Wageningen Academic Publishers, Wageningen, Netherlands, pp. 329–346.

ICCO (2021) *May 2021 Quarterly Bulletin of Cocoa Statistics*. The International Cocoa Organization, Abidjan. Available at: https://www.icco.org/may-2021-quarterly-bulletin-of-cocoa-statistics/ (accessed November 20, 2022).

Kahneman, D. and Tversky, A. (1979) Prospect theory: an analysis of decision under risk. *Econometrica* 47, 263–291.

Lahive, F., Hadley, P. and Daymond, A.J. (2019) The physiological responses of cacao to the environment and the implications for climate change resilience. A review. *Agronomy for Sustainable Development* 39, 5. DOI: 10.1007/s13593-018-0552-0.

Maddison, D. (2007) *The Perception of and Adaptation to Climate Change in Africa*. The World Bank Development Research Group Sustainable Rural and Urban Development Team, Policy Research Working Paper 4308. World Bank, Washington, DC. DOI: 10.1596/1813-9450-4308.

Mata, R., Frey, R., Richter, D., Schupp, J. and Hertwig, R. (2018) Risk preference: a view from psychology. *Journal of Economic Perspectives* 32, 155–172. DOI 10.1257/jep.32.2.155.

Prelec, D. (1998) The probability weighting function. *Econometrica* 66, 497–527. DOI: 10.2307/2998573.

Ruf, F., Schroth, G. and Doffangui, K. (2015) Climate change, cocoa migrations and deforestation in West Africa: what does the past tell us about the future? *Sustainability Science* 10, 101–111. DOI: 10.1007/s11625-014-0282-4.

Tanaka, T., Camerer, C.F. and Nguyen, Q. (2010) Risk and time preferences: linking experimental and household survey data from Vietnam. *American Economic Review* 100, 557–571. DOI: 10.1257/aer.100.1.557.

Tversky, A. and Kahneman, D. (1992) Advances in prospect theory: cumulative representation of uncertainty. *Journal of Risk and Uncertainty* 5, 297–323. DOI: 10.1007/BF00122574.

USDA FAS (2015) *Ecuador Cacao Update and Outlook*. US Department of Agriculture, Foreign Agricultural Service, Washington, DC. Available at: https://www.fas.usda.gov/data/ecuador-ecuador-cocoa-update-and-outlook (accessed November 20, 2022).

van der Linden, S. (2015) The social-psychological determinants of climate change risk perceptions: towards a comprehensive model. *Journal of Environmental Psychology* 41, 112–124. DOI: 10.1016/j.jenvp.2014.11.012.

van der Linden, S. (2017) *Determinants and Measurement of Climate Change Risk Perception, Worry, and Concern* (SSRN Scholarly Paper ID 2953631). Social Science Research Network, Rochester, New York. DOI: 10.2139/ssrn.2953631.

Villacis, A.H., Alwang, J.R. and Barrera, V.H. (2021) Linking risk preferences and risk perceptions of climate change: a prospect theory approach. *Agricultural Economics* 52, 863–877. DOI: 10.1111/agec.12659.

Villacis, A.H., Alwang, J.R., Barrera, V. and Dominguez, J. (2022a) Prices, specialty varieties, and postharvest practices: insights from cacao value chains in Ecuador. *Agribusiness* 38, 426–458. DOI: 10.1002/agr.21730.

Villacis, A.H., Bloem, J.R. and Mishra, A.K. (2022b) *Farm Investments, Risk Preferences, and Aspirations*. Agricultural and Applied Economics Association, Anaheim, California. DOI: 10.22004/ag.econ.322520.

Weber, E.U. (2010) What shapes perceptions of climate change? *WIREs Climate Change* 1(3), 332–342. DOI: 10.1002/wcc.41.

Weber, E.U., Blais, A-R. and Betz, N.E. (2002) A domain-specific risk-attitude scale: measuring risk perceptions and risk behaviors. *Journal of Behavioral Decision Making* 15, 263–290. DOI: 10.1002/bdm.414.

Weber, M., Weber, E.U. and Nosić, A. (2013) Who takes risks when and why: determinants of changes in investor risk taking*. *Review of Finance* 17, 847–883. DOI: 10.1093/rof/rfs024.

17 Natural Shocks and Farm Household Behavior: The Consequences of an Earthquake in Pisco, Peru

Marrit van den Berg[1] and Ricardo Fort[2]
[1]*Wageningen University, Wageningen, Netherlands;* [2]*Group for the Analysis of Development (GRADE), Lima, Peru*

Abstract
Natural disasters have indirect effects that may affect the victims for a substantial time after the direct impact. This chapter studies the impact of the Pisco earthquake on Peruvian cotton farmers more than a year after the earthquake hit. We find that incomes were only half of what they would have been without the earthquake. Damaged plots had not been recovered, and farmers seemed to cultivate less land and use lower amounts of inputs than they did before the earthquake. Risk aversion increased substantially, including for farmers who were not directly affected by the earthquake. However, preferences for insurance did not change.

17.1 Introduction

Numerous earthquakes happen every year around the world. In 2021 alone, 159 significant earthquakes (magnitude > 6) occurred (USGA, undated). The most infamous recent earthquake was arbitrarily the Haiti earthquake of August 14, 2021, which killed more than 2200 people and damaged tens of thousands of houses. In 2010, the same country was hit by an even more devastating earthquake that killed between 200,000 and 250,000 people and caused damages of between US$7.2 million and US$8.1 million (Cavallo *et al.*, 2010). Taking a broader perspective, between 2000 and 2019, 7348 natural disaster events were recorded worldwide by the Emergency Events Database (EM-DAT), affecting more than 4 billion people (many on more than one occasion), claiming approximately 1.23 million lives and causing about US$3 trillion in economic losses (CRED-UNDRR, 2020).

In the first days or weeks after they occur, major disasters receive wide media coverage. We are updated regularly about the number of casualties and the relief efforts. However, after a while, the media turns silent, and somewhat later most relief workers leave. This does not mean that the disaster is over for those who were affected. Disasters are often uninsured and assistance may be limited, especially in lower- and middle-income countries (LMICs). Often, disasters destroy not only houses and other domestic assets, but also businesses, thus limiting the capacity to generate income. Such damage to businesses may severely hinder the rate of recovery.

The indirect effects of disasters are less understood than the direct impacts in terms of casualties and destruction (Karim and Noy,

*E-mail: marrit.vandenberg@wur.nl

 Managing Risk in Agriculture: A Development Perspective (eds A.K. Mishra, S.C. Kumbhakar and G. Lien).

DOI: 10.1079/9781800622289.0017

2016). These indirect effects occur through damage to inputs used in production or infrastructure or from the fact that reconstruction pulls resources away from production. The indirect impacts may be larger than the direct impacts (Anttila-Hughes and Hsiang, 2013), and they are often most severe for the poor (Karim and Noy, 2016), who may lose a larger share of their assets (Morris *et al.*, 2002) and face a greater percentage reduction in income (Dercon *et al.*, 2005; Lopez-Calva and Juarez, 2009). Poor households may also lack the capacity to rebuild their asset base over time (Carter *et al.*, 2007), but this does not necessarily mean they revert to low-income livelihood strategies and fall into permanent poverty (Van den Berg, 2010).

A relatively recent line of research studies the consequences disasters have on preferences, such as risk aversion. Risk preferences play a significant role in determining economic choices such as savings (Dohmen *et al.*, 2011; Bommier and Grand, 2019), investment (Zimmerman and Carter, 2003; Caggese, 2012) and production choices (Moschini and Hennessy, 2001; Tesfaye and Tirivayi, 2020). If natural hazards indeed affect risk preferences, they may affect livelihoods not only directly by destroying assets, but also indirectly by affecting the speed of recovery through changing the degree of risk aversion. However, the existing literature, which mainly identifies risk attitudes using lottery games with given probabilities, finds mixed effects of natural disasters, ranging from increased risk aversion, to no effects, to increased love of risk (Purcell, 2021).

This chapter adds to the evidence on the effects of natural disasters by analyzing the consequences of the Pisco earthquake that hit Peru on August 15, 2007. The impact of the earthquake was substantial, with 519 casualties and more than 1300 injured. The estimated direct economic losses were about US$300 million. We used a panel dataset of 162 farm households interviewed in both 2006 and 2009 – before and after the earthquake. Some 72 of these households were cotton farmers from the Pisco valley, which was heavily affected by the earthquake. The rest were cotton farmers from carefully selected control valleys. The contribution is twofold. First, we combine an analysis of disaster consequences for crop production and income with an analysis of the effects on risk preferences in a single chapter. Second, we use panel data from before and after the earthquake. With the exception of Hanaoka *et al.* (2018), Kahsay and Osberghaus (2018) and Purcell (2021), existing studies are based on cross-sectional data and thus assume that disasters are not correlated with unobservables such as choice of residence.

The structure of the chapter is as follows. Section 17.2 discusses the existing literature on natural hazards and risk aversion. Section 17.3 summarizes the sample selection and data collection process and specifies the empirical strategies used. The empirical results are presented in Section 17.4, after which Section 17.5 concludes.

17.2 Natural Hazards and Risk Preferences

Natural disasters can have devastating consequences and are often traumatic experiences for those involved. It is therefore plausible that they would influence how people perceive risk and uncertainty. It is likely that people who experience a certain shock update their perceptions based on the likelihood of such shocks. They also may perceive the world as a riskier place in general, and they may even change their attitudes toward risk. If this is indeed the case, such changes would affect the speed of recovery. Changes in risk attitudes would affect entrepreneurial behavior, which facilitates recovery and growth. As aid money is often infused just after a disaster occurs, even short-lived preference changes could have large consequences. It is, however, not immediately obvious whether a disaster would make people more risk-averse and hinder recovery or make them more risk-tolerant and stimulate investment and growth.

The psychological literature provides evidence that hazard experience makes people more worried and fearful (Weinstein, 1989; Cutchin *et al.*, 2008) and that worry leads to more risk-averse choices (Raghunathan and Pham, 1999; Lerner and Keltner, 2001). On the other hand, individuals may react emotionally in the aftermath of a shock and exhibit more risk-loving behavior (Eckel *et al.*, 2009; Cameron and Shah, 2015). The expression of these changes likely depends on the nature of the shock: psychological evidence of diminishing sensitivity suggests that, in high-risk environments, people may not respond to the addition of a small independent risk (Kahneman and Tversky, 1979).

The economic literature also predicts ambiguous effects of natural disasters on risk behavior. Disasters destroy wealth, which has been shown to be negatively associated with risk aversion. In addition, natural disasters are likely to result in an increased perceived riskiness of the environment. Information about the likelihood of hazard occurrence is not perfect, and people are likely to update their expectations after a disaster hits. This will add to background risk perception. Under specific assumptions about the utility function, background risk makes people less willing to take other independent risks (Gollier and Pratt, 1996). In this case, an increased perceived likelihood of earthquakes, for example, would decrease willingness to invest in risky assets or in rainfall-dependent agriculture (Cameron and Shah, 2015). It would also increase the demand for insurance, even if this insurance does not cover earthquake damage. Interestingly, Quiggin (2003) shows that the opposite holds for preferences displaying constant risk aversion, i.e. when neither a fixed increment to all risky outcomes nor a scalar multiplication of outcomes affects preference rankings. In this case, aversion to a given risk is always reduced by the presence of independent background risk. That is, the occurrence of a natural disaster would decrease risk aversion in other domains.

Whether natural disasters change risk preferences is thus ultimately an empirical question that may depend on the type and severity of shock and the nature of the population affected. This is reflected in the diversity of findings in the literature. Most studies based their conclusions on choices in lottery games with a given (mostly 50–50) probability to separate risk aversion from risk perception, but this is not the case for all studies, as detailed below.

Several studies found that natural disasters increased risk aversion. Cameron and Shah (2015) concluded that people living in Javanese villages that had experienced an earthquake or a flood in the last three years were more risk-averse than those living in similar villages without such hazard experience. Similarly, Cassar *et al.* (2017) observed higher risk aversion for tsunami victims in Thailand. Beine *et al.* (2020) observed that respondents in Tirana, Albania, whom they reached just after a series of two large earthquakes hit the city, made more risk-averse choices than those they had reached just before the earthquakes. Van den Berg *et al.* (2009) concluded that natural hazards increased risk aversion among farmers in both Nicaragua and Peru. Finally, Bourdeau-Brien and Kryzanowski (2020) found that natural disasters in general lead to more risk-averse behavior in US investors. They based their analysis on US municipal bond transactions, contrary to the other studies, which used lottery games. Hence, they may measure changes in risk perceptions as well as risk aversion.

However, some other studies concluded that natural disasters made people more risk-tolerant. Using lottery games, Eckel *et al.* (2009) found that, in the direct aftermath of Hurricane Katrina in New Orleans, evacuated people were more risk-loving than a control group. Similarly, Hanaoka *et al.* (2018) found that Japanese earthquake victims expressed higher risk tolerance. Interestingly, whereas Eckel *et al.* found that the effect had disappeared after a year, Hanaoka *et al.* observed a persistent effect after 5 years, although only for men. Using different approaches, a higher risk tolerance was observed for Australian flood victims by Page *et al.* (2014) and for German storm victims by Kahsay and Osberghaus (2018). Page *et al.* used the choice between a scratch card (a lottery with generally unknown probabilities) and a fixed sum of money, and Kahsay and Osberghaus used a survey question on preparedness to assess risk in general, which has been shown to be correlated with experimental measures of risk attitudes. In both cases, one could argue that these methods do not fully separate risk attitudes from risk perceptions. However, it is unlikely that experiencing a major shock has resulted in lower risk perceptions, so we can only conclude that the increased willingness to take risks is due to a change in risk attitudes.

There may seem to be a pattern in these studies, with increased risk aversion in mainly lower- and middle-income countries and increased risk-seeking in high-income countries. However, in the United States, natural disasters were found to increase risk aversion for investors (Bourdeau-Brien and Kryzanowski, 2020) and to increase risk-seeking by poor African Americans

(Eckel *et al.*, 2009). The findings of Bchir and Willinger (2013) confirm that the story is more complex than being just wealth-related. Using 50–50 lotteries, they found that risk-seeking increased in areas exposed to lahars (volcanic flow) risk compared to unexposed areas in Peru, but only for the poor. Risk preferences seemed unaffected for higher income categories.

Two more studies illustrate that disaster responses can differ even within a single country. Using the same survey-based measure as Brown *et al.* (2018), Kahsay and Osberghaus (2018) found that one ethnic group was more risk-averse after exposure to a cyclone in Fiji, whereas another was not, suggesting a key role for culture. Finally, Purcell (2021) revealed heterogeneity between types of shock. Using lottery game data from Indonesia, she concluded that especially unexpected shocks with a high death toll and high destruction increased risk aversion, whereas less traumatic shocks did not.

17.3 Data and Methodology

17.3.1 Data collection

On August 15, 2007, an earthquake of 7.9 degrees on the Richter scale hit the central coast of Peru with terrible consequences in terms of deaths, injuries and material damages. The year before, a survey was carried out among 437 rural households in six valleys of the Peruvian coast as part of an impact evaluation of a government irrigation program. One of these valleys was the valley of Pisco, which was at the epicenter of the earthquake. The other valleys did not suffer direct damage. Data collected in the survey covered income and production data for the year 2005. In addition, the questionnaire included sections to measure risk and insurance preferences.

In 2009, we did a follow-up survey to assess the consequences of the earthquake. We decided to focus on producers of cotton, by far the dominant crop in Pisco. The 2006 survey contained information for 85 households in Pisco valley, all agricultural producers. For 75 of them, cotton was the main product. Independent of the earthquake, these farmers were affected by heavily fluctuating prices, especially for Tangüis cotton – the dominant variety in the area. Our intention was to re-interview the cotton farmers in the Pisco valley and comparable farmers in other valleys. We selected four control valleys that had a substantial number of Tangüis cotton farmers and that were otherwise similar to Pisco: Cañete, Chancay Huaral, Irchim and Santa. Within these valleys, we intended to trace all farmers who indicated that they cultivated cotton in 2005. We succeeded in re-interviewing most affected and potential control farmers: 72 out of 75 (96%) in Pisco and 107 out of 115 (93%) in the control valleys. The income data and production data collected in this survey covered the year 2008, which was the year after the earthquake, not the earthquake year itself.

The questionnaires used were mostly identical between the two survey rounds. They were extensive and included several sections on household composition, income-generating activities, assets, risk and shocks. The activity sections covered detailed questions on resource allocation, inputs, outputs and prices over the previous calendar year, which allowed calculation of annual incomes. The 2009 questionnaire included an additional section on damage done by the earthquake. This additional section was used only in the affected area.

In addition to the more standard survey questions, the surveys included sections to measure risk and insurance preferences. We used a hypothetical lottery game to measure risky choice behavior. The participants were presented with the choice between five crops, each with two possible incomes per hectare: for good years and for bad years. The options provide a range from risk-free to very risky, with payoffs increasing in good years and decreasing (but less so) in bad years (see Table 17.1). Unfortunately, we did not specify the probability of a year being good or bad, so both differences in risk attitudes and differences in risk perceptions could explain differences in choices between farmers. The insurance questions reflected the government's interest in weather insurance. The insurance section started with a question on knowledge of agricultural insurance schemes, then provided an elaborate explanation of a weather insurance scheme with payouts based on regional average damages, and finally probed for willingness to pay.

Table 17.1. Choices in the risk game. (Source: Authors' own work.)

Choice	Bad year (S./)	Good year (S./)	Risk index
A	1000	1000	1: risk-free
B	770	1350	2: low risk
C	540	1700	3: intermediate risk
D	310	2050	4: moderate risk
E	80	2400	5: high risk

Note: S./ represents the Peruvian sol.

17.3.2 Pre-earthquake descriptive statistics

To get insight into the characteristics of our sample household and the comparability of our treatment and control groups, we looked at the pre-earthquake data for three sets of variables (see Table 17.2). The first set involves variables of interest regarding income and crop production that we expect could possibly be affected by the earthquake even a year later. Key variables are total income and its main components, crop income and wage income. To go deeper into the determinants of crop income, we also consider cultivated area, input use intensity, area under cotton and cotton yields. The second set of variables involves the hypothetical risk behavior indicators: the choice in the risk game and the knowledge of, interest in and willingness to pay for weather insurance. The third set of variables involves other factors that characterize the household, such as the characteristics of the household head, household composition and some farm and income characteristics that are unlikely to be affected by the earthquake (farm size) or helpful for baseline characterization but not interesting enough to be outcome indicators (number of crops, whether a household is growing maize and whether a household has off-farm income).

In 2005, average incomes were almost 15,000 soles, or about US$4400. Crop production contributed 75% and wage income 17% of income, indicating that non-agricultural self-employment and livestock played only a limited role in the study area. Some 60% of households earned some form of off-farm income. Most households were headed by males, and the household heads averaged 57 years old with 7 years of education. Only 6% of the sample was aware of weather insurance. However, after an explanation, 71% were interested. These characteristics did not differ significantly between the households in Pisco and the controls.

The two groups of households were different, however, on other aspects. Households in Pisco were on average larger than households in the control areas, though the dependency ratio was the same. They also owned more land and cultivated larger areas with slightly more inputs per hectare. This implied a larger area under cotton than in the control area. Yet, households in the control areas obtained higher cotton yields than the Pisco households and were more likely than the Pisco households to cultivate other crops, such as maize, though for these households cotton also was the dominant crop. Surprisingly, respondents in Pisco opted for more risky hypothetical crops and had a lower willingness to pay for insurance. We have no explanation for these unexpected differences.

17.3.3 Empirical strategy

Despite our focus on cotton households in similar valleys, the households in the control valleys were not fully comparable with the households in Pisco, even before the earthquake. To account for these differences, we applied several pseudo-experimental methods to identify the impact of the earthquake. Below, we first describe how we analyzed impact for the impact and production variables. Next, we turn to the risk variables. We use a different approach for the latter variables due to the relatively large difference in risk choices between the two groups of households at baseline.

As a starting method for the income and production outcomes, we took first difference in differences (DID) to control for possible time-invariant differences between the households in Pisco and those in the control valleys:

$$\Delta Y_i = \beta_0 + \beta_1 E_i + \varepsilon_i \qquad (17.1)$$

where ΔY_i is a vector of the outcome indicators for household i and E indicates whether the

Table 17.2. Descriptive statistics for 2005 (before the earthquake). (Source: Authors' own work.)

	All		Other		Pisco[1]		Comparison
	Mean	SD	Mean	SD	Mean	SD	Pr(|T|>|t|)
Income and production							
Total income (S./)	14,759	14,724	13,981	16,674	15,932	11,176	0.35
Crop income (S./)	11,002	15,641	10,214	17,919	12,189	11,400	0.37
Wage income (S./)	2,438	5,320	2,192	46,450	2,809	6,212	0.48
Cultivated area (ha)	4.0	3.0	3.4	3.4	4.7	2.2	0.00
Input use (S./ha)	1,097	466	1,047	502	1,174	398	0.06
Area under cotton (ha)	3.3	2.3	2.5	2.0	4.5	2.2	0.00
Cotton yield (kg/ha)	2,915	988	3,146	1,061	2,569	747	0.00
Risk behaviour							
Choice risk game	3.1	1.7	2.8	1.7	3.5	1.6	0.00
Knows insurance	0.06	0.24	0.08	0.28	0.03	0.17	0.10
Interested in insurance	0.71	0.46	0.67	0.47	0.76	0.43	0.20
Willingness to pay	98	52	105	55	88	46	0.07
Controls							
Male household head (yes = 1)	0.9	0.3	0.9	0.3	0.9	0.3	0.32
Age head (years)	56.7	14.5	57.8	13.9	55.0	15.4	0.22
Education head (years)	7.0	4.6	7.0	4.5	7.1	4.9	0.90
Household size	4.3	2.0	3.9	1.8	4.8	2.3	0.01
Dependency ratio	0.3	0.3	0.3	0.3	0.3	0.3	0.67
Farm size	4.2	3.2	3.7	3.6	5.0	2.4	0.00
Number of crops	1.3	0.6	1.4	0.7	1.1	0.3	0.00
Growing maize (yes = 1)	0.1	0.3	0.1	0.3	0.0	0.1	0.00
Off-farm income (yes = 1)	0.6	0.5	0.6	0.5	0.6	0.5	0.43
N	178		107		71		178

Notes: [1]Household from Pisco was dropped from the dataset. This household had 29 members in 2008, which caused balancing problems in the PSM-DID analyses. The other results were not affected by exclusion of this household.
N: number of observations; S./ : the Peruvian sol.

household was exposed to the earthquake. To be specific, E takes the value 1 for households in the valley of Pisco and 0 for households in the other valleys.

Next, we added baseline control (X_0) and outcome (Y_0) variables to account for observed heterogeneity (DID-COV):

$$\Delta Y_i = \beta_0 + \beta_1 E_i + \beta_2 X_{0i} + \beta_3 Y_{0i} + \varepsilon_i \quad (17.2)$$

As a robustness check, we also estimated a model without baseline outcome indicators. The results were very similar, so we report only the results for the full model.

In addition, we used a different approach to impact assessment: propensity score matching (PSM). This is a non-parametric method that generates estimates of impact by matching treatment and control households based on a set of characteristics that are not affected by the treatment, in this case the earthquake. The first step of PSM involves a binary regression of the treatment on the household characteristics to generate the propensity score, or probability to be in the treatment. The second step then involves matching treatment households to non-treatment households using a matching algorithm. The matched sample is restricted to the common support, i.e. the overlapping region of the propensity score for treated and control households. This method results in unbiased impact estimates if the covariates are balanced after matching and there are no differences in unobservable characteristics between treatment and control. To guarantee that the matching variables were unaffected by the earthquake, we used the baseline controls and outcomes. We used probit regression and kernel matching for our analyses.

Finally, we combined the above described parametric and non-parametric methods (DID-PSM). By using the inverse propensity scores as weights in the first difference regression, the estimates become doubly robust: when either the outcome model or the selection model is properly specified, the outcomes are unbiased. Formally, this model is defined as follows:

$$\Delta Y_i = \beta_0 + \beta_1 w_i E_i + \beta_2 X_{i0} + \beta_3 Y_{i0} + \varepsilon_i \quad (17.3)$$

where w_i are the inverse of the propensity score weights. To increase the internal validity of the estimand, we present a final set of results in which we restrict (Eqn 17.3) to the common support of the propensity score for treated and control groups (DID-PSM common support).

Whereas balancing was easily achieved for the income and production variables, after the elimination of a single outlier (see Fig. 17.1), the baseline difference in risk choices was sufficiently large to prevent proper balancing. We therefore decided to restrict our analysis to comparison of means between groups and periods for the risk variables. An additional advantage of this approach is that it switches focus from impact to trends over time. Whereas the earthquake did not affect the production capacity of households in neighboring valleys, the occurrence of such a major disaster close by may still have affected their risk preferences. However, we would expect this effect to be smaller than for the households in the earthquake area. As risk and insurance preferences may depend on personal characteristics such as age, gender and education, we analyzed the results of the risk game and the insurance questions only for those households that had the same informant in both survey years.

17.4 Results

17.4.1 Direct earthquake damage

Before we discuss the effect of the 2007 earthquake on our 2008 indicators, it is important to know the extent of the direct effects. Not surprisingly given the strength of the earthquake, all survey households in the Pisco valley were affected.

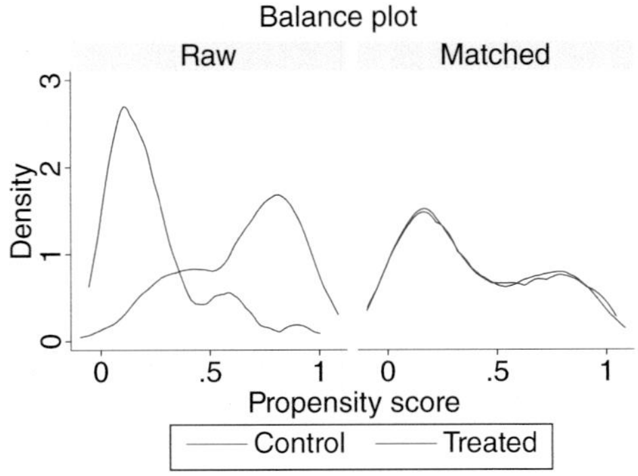

Source: Authors' own work.

Fig. 17.1. Balancing plot for PSM.

All respondents reported damage to their home, with an average value close to mean annual incomes for 2005. Moreover, 39% reported losses of household assets. Damage to income-generating activities was less widespread but still substantial. Some 29% reported yield losses, with an average value of 40% of the pre-earthquake total incomes or just over half of pre-earthquake crop incomes. Some 15% reported losses to non-farm self-employment. Losses of agricultural farm structures and farm animals, which would directly affect future income-generating capacity, were reported by only about 4% of households each. Similarly, only 6% reported loss of crop land. Yet many households reported damage to plots: 22% of plots were reported as more than just a little damaged, and none of these had recovered much. In about half the cases, the damage involved de-levelling. For the rest there were reports of cracking of the ground, destruction of canals or flooding.

17.4.2 Income and crop production

The different impact estimates give qualitatively similar results (Table 17.3). The main differences are the smaller effects measured without controlling for baseline household characteristics and the larger effects for the PSM. Significance levels are mostly unaffected by the estimation method, except for cultivated area. Below, we will focus on the DID-PSM estimates on the common support, which is our preferred method.

While incomes in the control valleys increased by 4499 soles, or 32%, incomes decreased substantially for households affected by the earthquake. The estimated negative effect of

Table 17.3. Parameter estimates. (Source: Authors' own work.)

	Trend (no earthquake)	DID	DID-COV	PSM	DID-PSM	DID-PSM (common support)
All households						
Total income	4,499**	−5,924*	−13,023***	−16,238***	−13,506**	−13,491**
(S./)	(1,975)	(2,371)	(3,737)	(5,736)	(5,710)	(5,723)
Crop income	455	−4506*	−10,153**	−15,221**	−11,544**	−11,535**
(S./)	(1,906)	(2,439)	(3,715)	(7,706)	(5,552)	(5,569)
Labor income	2051***	−448	−930	1136	−277	−270
(S./)	(1,042)	(1,207)	(952)	(1,714)	(1,370)	(1,373)
N	178	178	178	176	178	168
Cultivating households						
Cultivated area (ha)	1.94***	−0.92**	−1.25	−2.07**	−1.10	−1.10
	(0.33)	(0.45)	(0.79)	(1.00)	(0.94)	(0.95)
Grows cotton	−0.44***	0.14	−0.04	−0.02	0.00	0.00
	(0.06)	(0.11)	(0.10)	(0.12)	(0.11)	(0.11)
Cotton area (ha)	−0.82**	−0.24	0.20	0.51	0.27	0.27
	(0.30)	(0.67)	(0.55)	(0.90)	(0.72)	(0.72)
Inputs/ha	901***	−299	−210	−333	−407	−406
	(131)	(184)	(271)	(431)	(336)	(338)
N	139	139	139	130	139	130
Cotton households						
Cotton yields	364	−531*	−1,010***	−1,185***	−1,473***	−1,245***
(Kg/ha)	(216)	(291)	(258)	(436)	(214)	(229)
N	86	86	86	81	86	81

Notes: Standard errors in parentheses; *** $p < 0.01$, ** $p < 0.05$, * $p < 0.1$. N (number of observations).

the earthquake was about double the income increase experienced in the neighboring valleys. As a result, incomes in Pisco are now lower than in the control valleys, whereas without the earthquake, they would have been substantially higher. In fact, incomes of the Pisco farmers would have been almost double without the earthquake: their average income was 14,507 soles with an estimated treatment effect of −13,491. This negative income effect is due to foregone crop income. The earthquake did not affect the positive trend for labor income.

Some 22% of households stopped cultivating crops between 2005 and 2008. This share was similar between Pisco and the other valleys and therefore seems unrelated to the earthquake. The majority of these households were classified as non-cultivating agricultural households and still in possession of their land. Only a few households (3%) were reclassified as non-agricultural without farm land.

For those households still engaged in crop production, the cultivated area and the value of inputs per hectare almost doubled. We did not find a significant impact of the earthquake on these variables using our preferred estimates. However, some less-preferred estimates suggest that the increase in cultivated area was less in Pisco than in other valleys. The direction of our preferred effect was also negative, and the size of the coefficient was substantial. Its lack of significance could be the result of lack of power due to the relatively small size of the sample. Also, the earthquake coefficient for input intensity is consistently negative. This could point to a lower capacity to invest due to redirection of money to, for example, reconstruction or repair of the home, or a lower willingness to invest due to lower land quality. Yet the coefficient is not significant for any of the methods.

Many households stopped growing cotton, and the average area under cotton decreased. Real cotton prices were similar between the two surveys, but we expect that the notable price instability played an important role in this reduction. If the earthquake had increased risk aversion throughout the sample, this would have strengthened this response. We found no additional decrease in cotton area in Pisco. Finally, the earthquake is estimated to have resulted in a substantial decrease in cotton yields. Cotton yields in the control valleys had increased compared to 2005, but yields in Pisco had decreased. This outcome is probably at least partly due to the reported loss in soil quality caused by the earthquake.

17.4.3 Risk and insurance choices

As indicated above, the earthquake may have caused increased risk aversion in both Pisco and the nearby valleys, although we would expect any effect to be largest for the directly affected households in Pisco. Looking at the choices in the risk game and preferences for insurance, we find only limited evidence for increased risk aversion due to the earthquake (see Table 17.4).

Between the two survey rounds, awareness of weather insurance increased from 5% to almost a quarter of respondents, with no significant differences between Pisco and the other valleys. The interest in purchasing insurance increased slightly between the two years in both areas, but this increase was not significant in either area or the combined area. Also, the average amount that people were willing to pay for insurance did not change significantly over time. This suggests that people did not become more inclined to purchase weather-based crop insurance after the earthquake.

We do, however, find strong trends in the responses for the hypothetical risk game. The alternative selected after the earthquake was on average almost one option less risky than the before-earthquake choice for Pisco households, and half an option less risky for control households. The difference between these changes, however, was not statistically significant, and the larger effect for Pisco could be a correction of the unexpected higher risk taking before the earthquake. Contrary to the insurance data, the responses in the risk game suggest that the earthquake decreased the willingness to engage in risky behavior, though not necessarily more for directly affected households.

To further explore the surprising difference in risk choice between Pisco and the other valleys before the earthquake, we ran ordered probit and logit regressions of the risk choice on personal and household characteristics and a set of valley dummies to search for an explanation. We ran the regressions using both the 2005 data only and the two-period panel. When using the panel, we used the 2005 data for the potentially endogenous

Table 17.4. Risk and insurance choices. (Source: Authors' own work.)

	All	Other	Pisco	Difference	DID-COV
Risk taking					
Choice risk game (1 = no risk, 5 = very risky)					
2005	3.08	2.88	3.59	0.81***	
	[1.69]	[1.67]	[1.62]	(0.28)	
2008	2.43	2.30	2.66	0.36*	
	[0.24]	[1.02]	[1.28]	(0.19)	
Difference	−0.65***	−0.49**	−0.94***	−0.44	−0.43
	(0.17)	(0.20)	(0.28)	(0.34)	(0.34)
Weather insurance					
Awareness					
2005	0.05	0.07	0.02	−0.06	
				(0.04)	
2008	0.24	0.25	0.22	−0.03	
				(0.08)	
Difference	0.18***	0.18***	0.20***	0.02	0.02
	(0.04)	(0.05)	(0.06)	(0.08)	(0.08)
Interested in purchasing					
2005	0.75	0.71	0.82	0.11	(0.07)
2008	0.81	0.75	0.91	0.16**	
				(0.07)	
Difference	0.06	0.04	0.09	0.05	0.05
	(0.05)	(0.06)	(0.07)	(0.10)	(0.10)
Willingness to pay					
2005	73.40	73.37	73.46	0.08	
	[62.18]	[67.04]	[53.34]	(10.57)	
2008	71.40	69.79	74.18	4.39	
	[53.08]	[57.98]	[43.70]	(9.01)	
Difference	−2.00	−3.58	0.73	4.31	7.52
	(6.68)	(6.43)	(9.30)	(13.90)	(13.73)
N	150	95	55		

Notes: ***p < 0.01, **p < 0.05, *p < 0.1; table reports mean, (standard errors) and [standard deviations] N (number of observations).

household characteristics in 2008. The only significant coefficients are for the Irchim dummy, which implies less risky choices in this valley (all regressions), and the time and treatment dummy in the panel regression, which indicate that risk aversion was higher in 2008, and especially so in the Pisco area. The Pisco dummy was not significant in any of the regressions. The regressions thus support the descriptive results, but provide no clear explanation of the differences between valleys before the earthquake.

17.5 Conclusion

On August 15, 2007, a severe earthquake hit the valley of Pisco in Peru. We found that the year after the earthquake, incomes of households that experienced direct damage were still negatively affected: incomes were estimated to be about half of what they would have been without the earthquake. This was due to decreased crop incomes and because cotton yields were lower than they would have been without the earthquake. Our data suggest two possible reasons. First, farmers indicated that their damaged plots had not recovered much. Second, we found weak evidence that farmers increased cultivated area and the use of external inputs less than they would have without the earthquake. The size of the estimated coefficients is substantial, and the lack of significance (in our preferred estimates) could be caused by lack of power due to the relatively small sample size. We believe that these results provide a lower bound to the real impact of the earthquake, as the control group consists of households living in valleys that may have been affected indirectly by the earthquake.

We found evidence of lower willingness to take risk for both directly affected households and similar households in nearby valleys. Unfortunately, the lottery game used did not allow us to assess whether this was due to changes in risk aversion or in risk perceptions. What we do know is that we did not measure an increased perceived likelihood of earthquakes: the lottery game reflected agricultural choices with harvest levels in years with good and bad weather. Hence, we measured how experiencing one specific shock translated into choices regarding other shocks. This is in accordance with findings in an emerging literature on the effect of natural disasters on risk aversion which uses abstract lottery games to measure risk aversion. In this context, the lack of a significant trend in the interest in and willingness to pay for crop insurance is surprising. Several other studies found a positive effect of risk aversion on uptake of or interest in insurance (Fahad *et al.*, 2018; Yanuarti *et al.*, 2019; Ali *et al.*, 2021; Zhang *et al.*, 2021). The high interest in crop insurance even before the earthquake could explain part of the results, but this does not explain why we do not find an increased willingness to pay.

The effect of the earthquake on risk aversion was almost double for the farmers who were directly affected than for those who were indirectly affected, but this difference was not statistically significant. We argue that the occurrence of such a major disaster nearby may have affected the preferences of our control group as well. Such a reaction is not unheard of. Goebel *et al.* (2015) even found that Germans became more risk-averse after Japan's Fukushima disaster, presumably through the perceived risk of meltdown of domestic reactors. The increase in risk aversion in our study area likely contributed to the drop in income and the move away from cotton production, the dominant crop in Pisco before the earthquake. Although real prices for cotton did not decrease between the two survey periods, the instability of prices before and during the survey period made the crop less attractive, especially for those with high levels of risk aversion.

In conclusion, the indirect effects of the Pisco earthquake were substantial among the rural population, which was reflected in much lower incomes. The underlying reasons appear to be damage to agricultural fields, diversion of resources to repair damage to houses and a higher risk aversion, though the limited sample size make it difficult to give a definite answer. The results, however, do imply that the negative effects of the earthquake largely outlived the international media attention and that further assistance beyond direct emergency aid would have been justified. Programs to reclaim damaged agricultural land could have helped limit the drop in crop incomes. In addition, further assistance may have prevented the decrease in willingness to engage in risky behavior.

References

Ali, W., Abdulai, A., Goetz, R. and Owusu, V. (2021) Risk, ambiguity and willingness to participate in crop insurance programs: evidence from a field experiment*. *Australian Journal of Agricultural and Resource Economics* 65, 679–703.

Anttila-Hughes, J. and Hsiang, S. (2013) Destruction, disinvestment, and death: economic and human losses following environmental disaster. Available at SSRN: 2220501. DOI: 10.2139/ssrn.2220501.

Bchir, M.A. and Willinger, M. (2013) Does the exposure to natural hazards affect risk and time preferences? Some insights from a field experiment in Perú. Working Papers 13-04, LAMETA, University of Montpellier, France.

Beine, M.A., Charness, G., Dupuy, A. and Joxhe, M. (2020) Shaking things up: on the stability of risk and time preferences. *CESifo Working Paper No. 8187.* Munich Society for the Promotion of Economic Research, Munich, Germany, pp. 54.

Bommier, A. and Grand, F.L. (2019) Risk aversion and precautionary savings in dynamic settings. *Management Science* 65, 1386–1397.

Bourdeau-Brien, M. and Kryzanowski, L. (2020) Natural disasters and risk aversion. *Journal of Economic Behavior & Organization* 177, 818–835.

Brown, P., Daigneault, A.J., Tjernström, E. and Zou, W. (2018) Natural disasters, social protection, and risk perceptions. *World Development* 104, 310–325.

Caggese, A. (2012) Entrepreneurial risk, investment, and innovation. *Journal of Financial Economics* 106, 287–307.

Cameron, L. and Shah, M. (2015) Risk-taking behavior in the wake of natural disasters. *Journal of Human Resources* 50, 484–515.

Carter, M.R., Little, P.D., Mogues, T. and Negatu, W. (2007) Poverty traps and natural disasters in Ethiopia and Honduras. *World Development* 35, 835–856.

Cassar, A., Healy, A. and Von Kessler, C. (2017) Trust, risk, and time preferences after a natural disaster: experimental evidence from Thailand. *World Development* 94, 90–105.

Cavallo, E.A., Powell, A. and Becerra, O. (2010) *Estimating the Direct Economic Damage of the Earthquake in Haiti* (IDB-WP-163). Inter-American Development Bank, Washington, DC, pp. 16. Available at: https://publications.iadb.org/publications/english/document/Estimating-the-Direct-Economic-Damage-of-the-Earthquake-in-Haiti.pdf (accessed July 15, 2022).

CRED-UNDRR (2020) *The Human Cost of Disasters: An Overview of the Last 20 Years (2000–2019).* Centre for Research on the Epidemiology of Disasters, Brussels; United Nations Office for Disaster Risk Reduction, Geneva, Switzerland. Available at: https://www.preventionweb.net/files/74124_humancostofdisasters20002019reportu.pdf (accessed July 15, 2022).

Cutchin, M.P., Martin, K.R., Owen, S.V. and Goodwin, J.S. (2008) Concern about petrochemical health risk before and after a refinery explosion. *Risk Analysis* 28, 589–601.

Dercon, S., Hoddinott, J. and Woldehanna, T. (2005) Vulnerability and shocks in 15 Ethiopian villages, 1999–2004. *Journal of African Economies* 14, 559–585.

Dohmen, T., Falk, A., Huffman, D., Sunde, U., Schupp, J. and Wagner, G.G. (2011) Individual risk attitudes: measurement, determinants, and behavioral consequences. *Journal of the European Economic Association* 9, 522–550.

Eckel, C.C., El-Gamal, M.A. and Wilson, R.K. (2009) Risk loving after the storm: a Bayesian-network study of Hurricane Katrina evacuees. *Journal of Economic Behavior & Organization* 69, 110–124.

Fahad, S., Wang, J., Hu, G., Wang, H., Yang, X., Shah, A.A. *et al.* (2018) Empirical analysis of factors influencing farmers crop insurance decisions in Pakistan: evidence from Khyber Pakhtunkhwa province. *Land Use Policy* 75, 459–467.

Goebel, J., Krekel, C., Tiefenbach, T. and Ziebarth, N.R. (2015) How natural disasters can affect environmental concerns, risk aversion, and even politics: evidence from Fukushima and three European countries. *Journal of Population Economics* 28, 1137–1180.

Gollier, C. and Pratt, J.W. (1996) Risk vulnerability and the tempering effect of background risk. *Econometrica* 64, 1109–1123.

Hanaoka, C., Shigeoka, H. and Watanabe, Y. (2018) Do risk preferences change? Evidence from the Great East Japan earthquake. *American Economic Journal: Applied Economics* 10, 298–330.

Kahneman, D. and Tversky, A. (1979) Prospect theory, an analysis of decisions under risk. *Econometrica* 47, 263–291.

Kahsay, G.A. and Osberghaus, D. (2018) Storm damage and risk preferences: panel evidence from Germany. *Environmental and Resource Economics* 71, 301–318.

Karim, A. and Noy, I. (2016) Poverty and natural disasters – a qualitative survey of the empirical literature. *The Singapore Economic Review* 61, 1640001.

Lerner, J.S. and Keltner, D. (2001) Fear, anger, and risk. *Journal of Personality and Social Psychology* 81, 146–159.

Lopez-Calva, L.F. and Juarez, E.O. (2009) Evidence and policy lessons on the links between disaster risk and poverty in Latin America. MPRA Paper 18342, University Library of Munich, Germany.

Morris, S.S., Neidecker-Gonzales, O., Carletto, C., Munguía, M., Medina, J.M. and Wodon, Q. (2002) Hurricane Mitch and the livelihoods of the rural poor in Honduras. *World Development* 30, 49–60.

Moschini, G. and Hennessy, D.A. (2001) Uncertainty, risk aversion, and risk management for agricultural producers. *Handbook of Agricultural Economics* 1, 87–153.

Page, L., Savage, D.A. and Torgler, B. (2014) Variation in risk seeking behaviour following large losses: a natural experiment. *European Economic Review* 71, 121–131.

Purcell, H. (2021) *The Heterogeneous Impacts of Natural Disasters on Risk Preferences in Indonesia.* University of Pennsylvania Population Center Working Paper (PSC/PARC), 2021-73, Philadelphia, Pennsylvania.

Quiggin, J. (2003). Background risk in generalized expected utility theory. *Economic Theory* 22, 607–611.

Raghunathan, R. and Pham, M.T. (1999) All negative moods are not equal: motivational influences of anxiety and sadness on decision making. *Organizational Behavior and Human Decision Processes* 79, 56–77.

Tesfaye, W. and Tirivayi, N. (2020) Crop diversity, household welfare and consumption smoothing under risk: evidence from rural Uganda. *World Development* 125, 104686.

USGA (undated) Earthquake Hazards Program. US Geological Survey, Reston, Virginia. Available at: https://www.usgs.gov/programs/earthquake-hazards (accessed July 15, 2022).

Van den Berg, M. (2010) Household income strategies and natural disasters: dynamic livelihoods in rural Nicaragua. *Ecological Economics* 69, 592–602.

Van den Berg, M., Fort, R. and Burger, K. (2009) Natural hazards and risk aversion: experimental evidence from Latin America. *Contributed Paper prepared for presentation at the International Association of Agricultural Economists Conference, Beijing, China, August 16–22, 2009*.

Weinstein, N.D. (1989) Effects of personal experience on self-protective behavior. *Psychological Bulletin* 105, 31–50.

Yanuarti, R., Aji, J.M.M. and Rondhi, M. (2019) Risk aversion level influence on farmer's decision to participate in crop insurance: a review. *Agricultural Economics* 65, 481–489.

Zhang, R., Fan, D., Lai, G., Wu, J. and Li, J. (2021) Rank-dependent preferences, social network and crop insurance uptake: field experimental evidence from rural China. *Agricultural Finance Review* 81, 636–656.

Zimmerman, F.J. and Carter, M.R. (2003) Asset smoothing, consumption smoothing and the reproduction of inequality under risk and subsistence constraints. *Journal of Development Economics* 71, 233–260.

18 Adaptation to Climate Change and Risk Exposure: Insights from Smallholder Wheat Farmers in China

Wanglin Ma[1]*, Hongyun Zheng[2] and Puneet Vatsa[1]
[1]Lincoln University, Christchurch, New Zealand; [2]Huazhong Agricultural University, Wuhan, China

Abstract

We analyze the impacts of adopting climate change adaptations (CCAs) on the risk exposure of wheat farmers using data collected from smallholder wheat farmers in China. A moment-based production function is used to calculate four risk exposure indicators: variance (variability of farm revenue), semi-variance, skewness (exposure to downside risk) and kurtosis. A two-stage predictor substitution approach is used to address the endogeneity of adopting climate change adaptations. The results show that the number of adaptations is determined mainly by family size, off-farm-worker ratio, asset ownership and insurance. The greater the number of adaptations, the lower the farmers' risk exposure as measured by variability, semi-variability, kurtosis and skewness of farm revenue. These findings call for practical interventions to encourage climate change adaptations for the wheat sector, not only in China but also in other countries, as climate change threatens global agriculture.

18.1 Introduction

About 43% of the global population lives in rural regions. These regions are also home to about 500 million smallholder farming households, many of which live in abject poverty (Ritchie, 2021). Improving agricultural performance is crucial to lifting the world's poor out of poverty and feeding the global population, which is projected to reach 9.7 billion by 2050. However, climate change continues to challenge the sustainability of agricultural production and human development. Recent decades have witnessed concerning changes in precipitation and regional and global temperatures. Furthermore, climate-induced natural disasters such as droughts and floods have intensified and become more frequent, threatening crop yields and leading to higher food prices and increased food insecurity, especially in the world's vulnerable regions (Challinor *et al.*, 2014; World Bank, 2020; Bora, 2022; Weerasekara *et al.*, 2022). These are exigent problems, pointing to the need to help rural communities become resilient to climate change by bolstering their adaptive capacities.

Agronomists, government agencies and non-government organizations in different countries have proposed and promoted a wide range of climate change adaptations (CCAs) among smallholder farmers to sustain farm productivity, reduce vulnerability, and enhance agricultural systems' resilience to climate change. These

*E-mail: wanglin.ma@lincoln.ac.nz

DOI: 10.1079/9781800622289.0018

strategies include, for example, adjustments in sowing time; adoption of drought-tolerant varieties; shifts to new crops; better soil, water, and fertilizer management; crop diversification; minimum tillage; and irrigation. A growing number of studies have investigated the effects of the adoption of CCAs, which helps us better understand the effectiveness of adaptation strategies. These studies can be classified into two broad strands based on how the adaptations are measured and the econometric approaches applied to estimate the adaptations' effects.

In the first strand of studies, researchers measured CCAs as a binary decision. Because such adaptations are not random and smallholder farmers choose to adopt or not adopt these strategies, studies have employed different approaches to address the inherent selection bias. These include the propensity score matching (PSM) model (Fentie and Beyene, 2019; Faisal *et al.*, 2021; Jamil *et al.*, 2021), the augmented inverse probability weights (AIPW) estimator (Zhu *et al.*, 2021), the endogenous treatment effect (ETE) model (Onyeneke, 2021) and the endogenous switching regression (ESR) model (Gorst *et al.*, 2018; Ojo and Baiyegunhi, 2020). For example, by estimating the PSM model, Wongnaa and Babu (2020) found that CCAs helped farmers achieve significantly higher farm productivity and incomes in cacao production in Ghana. Zhu *et al.* (2021) employed the AIPW model to analyze data collected from banana farmers in China. They found that CCAs contributed to higher banana yields. Ojo and Baiyegunhi (2020) estimated the ESR model and found a positive link between CCAs and net income from rice production in south-west Nigeria.

The second strand of studies is devoted to multiple adaptation strategies and uses models such as multinomial endogenous switching regression (MESR) to allow the treatment variable to have more than two choices (Di Falco and Veronesi, 2013; Ho and Shimada, 2021; Issahaku *et al.*, 2021; Oparinde, 2021; Shahzad *et al.*, 2021). For example, by estimating data from Ethiopia, Di Falco and Veronesi (2013) found that a portfolio of CCAs significantly increased farm net revenues, and changing crop varieties increased net revenues when coupled with water conservation strategies or soil conservation strategies. Ho and Shimada (2021) found that various types of responses to climate change improved rice yield, profitability and income; they also reduced chemical fertilizer use. Oparinde (2021) estimated the impacts of CCAs (i.e. construction of bore-holes, adjustment of stocking time and creation of embankments) on the welfare of fish farmers in Nigeria. The results showed that adjusting fish stocking time increased the food security index by 106.1%, creating embankments increased the index by 16.9% and the stocking time adjustment reduced the poverty gap by 87.2%.

Some studies have also investigated the impacts of CCAs on farmers' exposure to risk (Wang *et al.*, 2018; Amondo *et al.*, 2019; Issahaku and Abdulai, 2020; Shahzad and Abdulai, 2020; Issahaku *et al.*, 2021; Sarr *et al.*, 2021). For example, Issahaku and Abdulai (2020) investigated climate-smart practices such as crop choice and soil and water conservation measures. They found that farmers' adoption of crop choice and soil and water conservation increased crop revenues and reduced production risks, with the largest impact on crop revenues coming from joint adoption. Sarr *et al.* (2021) found that the adoption of climate-friendly agriculture (captured by a modified, rain-fed variant of the system of rice intensification) enhanced rice yields and reduced the downside risk of crop failure in Tanzania. Issahaku *et al.* (2021) estimated the impacts of the adoption of adaptation strategies comprising irrigation, soil conservation and enhanced cropping calendar management on farm performance and the multi-dimensional poverty index in Ghana. They found that adaptation strategies implemented in combination are more effective in increasing output and reducing the probability of crop failure than those implemented in isolation.

This chapter contributes to the literature by estimating the impacts of CCAs on wheat farmers' exposure to risk, using household survey data collected from China. It is interesting to explore Chinese wheat farmers' climate change adaptation behaviors for the following reasons. In 2020, the total output of wheat produced in China was 134 million metric tonnes, making it the world's largest wheat-producing country (FAOSTAT, 2022). Wheat yields are sensitive to climatic variables. Gupta *et al.* (2017) found that a 1°C increase in average daily maximum and minimum temperatures lowered wheat yields by between 2–4% in India. It is reported that wheat yields are expected to decline by nearly

20% in Malaysia, Vietnam and the United States due to climate change (World Bank, 2020). Therefore, understanding Chinese wheat farmers' CCAs and exploring the associations between these adaptations and risk exposure would help provide valuable insights for designing appropriate measures to minimize the effects of climate change on China's wheat sector, making it more profitable, productive and sustainable.

This study's two primary contributions to the literature on CCAs are these. First, we measure *the number of adaptations* using a count variable. This measurement approach is different from previous studies that measure the adoption of CCAs as a binary variable (Jamil *et al.*, 2021; Onyeneke, 2021) or as multiple packages (Oparinde, 2021). Very few studies have considered the number of CCAs (Kpadonou *et al.*, 2017; Zakaria *et al.*, 2020; Ojo *et al.*, 2021). Second, we employ a two-stage predictor substitution model to address the endogeneity of the number of adaptations used. Because farmers *choose* whether to adopt CCAs, the number of adaptations they use is potentially endogenous, and failure to address the endogeneity would generate biased estimates.

The rest of this chapter is structured as follows: In Section 18.2, we detail the estimation strategies used in the analysis. Data and descriptive statistics are introduced in Section 18.3. We present and discuss the empirical results in Section 18.4. The conclusion and policy implications are presented in Section 18.5.

18.2 Estimation Strategies

18.2.1 Moments of production function

Following previous studies (Abro *et al.*, 2018; Bozzola and Smale, 2020; Sarr *et al.*, 2021; Ma *et al.*, 2022), we used a flexible moment-based production function to disentangle the impact of the adoption of CCAs on the wheat farmers' exposure to risk. Specifically, we assumed a continuous and twice differentiable production function as follows:

$$Y_i = f\left(X_i;\beta_i\right) + \varepsilon_i \tag{18.1}$$

where Y_i refers to the wheat output of farmer *i*. In this chapter, the output is defined as the gross farm revenue from wheat production rather than yields, to take quality differences into account. X_i represents a vector of inputs, and β_i refers to the corresponding parameters to be estimated. The inputs considered include land, seeds, fertilizers, pesticides, labor and other inputs (e.g. machinery and irrigation). ε_i is an error term. The production function in Eqn 18.1 can be estimated by either the Cobb–Douglas functional form or the translog functional form (Mukasa, 2018; Ma *et al.*, 2022). A likelihood ratio test suggested that the translog functional form would be more suitable than the Cobb–Douglas functional form. Therefore, we estimated a translog function using an ordinary least squares (OLS) regression. The translog function is as follows:

$$\begin{aligned}\ln\left(Y_i\right) = \beta_0 + \sum\nolimits_{m=1}^{6} \beta_m \ln(X_{im}) + \\ 0.5\sum\nolimits_{m=1}^{6} \beta_{mm} \ln(X_{im})^2 + \\ \sum\nolimits_{n=1}^{6}\sum\nolimits_{m=1}^{6} \beta_{nm} \ln(X_{in})\ln(X_{im}) + \varepsilon_i\end{aligned} \tag{18.2}$$

where $\ln(Y_i)$ is the natural logarithm of the gross revenue of farmer *i*. $\ln(X_{im})$ is the natural logarithm of inputs *m* for farmer *i*, and $\ln(X_{im})^2$ is its squared term. $\ln(X_{in})$ $\ln(X_{im})$ is the interaction term between inputs *n* and *m* ($n \neq m$). β_0 is a constant. β_m, β_{mm} and β_{nm} are parameters to be estimated. After estimating the moment-based production function, the error term is predicted as follows:

$$(\hat{\varepsilon}_i)^j = E[Y_i - f(X_i;\beta_i)]^j, j = 2,3,4 \tag{18.3}$$

where $\hat{\varepsilon}_i$ is the predicted error term after estimating the production function in Eqn 18.2, capturing the uncertainties after controlling for the effects of other factors. It provides a basis for estimating relevant moments of the gross revenue of wheat production, conditional on the production inputs X_i (Wang *et al.*, 2018). In particular, the extent of risk exposure is calculated by the second (variance; j = 2), third (skewness; j = 3) and fourth (kurtosis; j = 4) moments of the distributions of gross revenue. To study the relevance of CCAs to the most vulnerable farmers, we also considered a semi-variance indicator to focus on gross revenue realization below a specified benchmark, which is the expected value in our study (Bozzola and Smale, 2020). The four moments (variance, semi-variance, skewness and kurtosis) capture the distributions of farm revenue and reflect the production risk wheat farmers face; they are used as the dependent variables in our econometric estimations.

18.2.2 Two-stage predictor substitution (2SPS) model

Although approaches such as PSM, AIPW, ETE and ESR models have been employed to address the endogeneity of CCAs in previous studies (Ojo and Baiyegunhi, 2020; Jamil *et al.*, 2021; Onyeneke, 2021; Zhu *et al.*, 2021), they require a binary endogenous treatment variable. The MESR model requires the endogenous treatment variable to have different options (Ho and Shimada, 2021; Oparinde, 2021; Setsoafia *et al.*, 2022). These approaches are not appropriate to our case because our endogenous variable measures the number of CCAs adopted, a count variable. In this chapter, we employ a two-stage predictor substitution (2SPS) model to estimate the impact of the number of adaptations on wheat farmers' exposure to risk (Terza *et al.*, 2008; Ying *et al.*, 2019). The first-stage equation of the 2SPS model is as follows:

$$CCAs_i = \alpha_i Z_i + \gamma_i IV_i + \xi_i \tag{18.4}$$

where $CCAs_i$ represents the number of CCAs. Z_i is a vector of control variables. IV_i is the instrumental variable used for the 2SPS model identification. α_i and γ_i are parameters to be estimated. ξ_i is an error term.

A variable representing the distance to the nearest input market is employed as the IV. Following previous studies (Pizer, 2016; Liu *et al.*, 2021; Zheng *et al.*, 2021), we use a falsification test to check the validity and effectiveness of the employed IV. The test results, which are not presented here for the sake of simplicity, confirm our IV's appropriateness. Because $CCAs_i$ in Eqn 18.4 captures a count variable, we employ a Poisson regression model to estimate it.

In the second stage, the 2SPS model estimates a risk exposure function as follows:

$$R_i = \beta_i \widehat{CCAs_i} + \eta_i Z_i + \psi_i \tag{18.5}$$

where R_i refers to a vector of outcome variables (variance, semi-variance, skewness and kurtosis) that capture risk exposure. $\widehat{CCAs_i}$ denoted the number of CCAs, predicted after estimating Eqn 18.4. To address the endogeneity, we use the predicted rather than the observed value of *CCAs* (Wooldridge, 2015). Z_i is defined earlier. ψ_i is an error term. An OLS regression model helps estimate Eqn 18.5.

18.3 Data Source and Descriptive Statistics

18.3.1 Data source

We used data from a household survey of 558 smallholder farmers in China, collected between June and July 2019. Despite experiencing rapid growth in both yields and total output of wheat in the past few years, China's wheat industry has suffered considerable losses due to frequent extreme weather events, such as rising temperatures, droughts and floods. In 2020, about 20 million hectares of croplands were damaged, accounting for about 11.9% of the total sown area of crops in China (NBS, 2020). Moreover, the Chinese agricultural sector comprises many smallholder farms, with an average cultivated land area per household of only 0.69 ha.

We employed a stratified sampling strategy to select rural smallholder farmers from three major wheat-growing provinces: Shandong, Henan and Anhui. Wheat production is an essential livelihood source for local farmers in the selected provinces. After selecting three provinces in the first step, we randomly chose two cities from each selected province in the second step, namely Linyi and Zaozhuang in Shandong, Xinyang and Zhumadian in Henan and Suzhou and Huaibei in Anhui. Thereafter, two to three towns were randomly chosen from each selected city, and then two to three villages were randomly chosen within each selected town. In the final step, about 10–30 households were randomly selected within each village, resulting in a sample of 558 households.

We considered common agricultural practices in China and the data used in previous studies to gather information on farmers' CCAs (Cholo *et al.*, 2019; Amadu *et al.*, 2020; Issahaku and Abdulai, 2020; Wongnaa and Babu, 2020; Zakaria *et al.*, 2020; Ho and Shimada, 2021; Ojo *et al.*, 2021). The questions pertained to eight aspects: (i) whether farmers adopted drought-tolerant wheat varieties (DTWVs); (ii) whether they planted more than one wheat variety; (iii) whether they adopted crop rotation; (iv) whether they used water-saving irrigation technology; (v) whether they adopted organic soil amendments (e.g. organic fertilizer or farmyard manure); (vi) whether they used crop residue turnover technology; (vii) whether they used minimum tillage technology; and (viii) whether they adopted an integrated crop–livestock system. Each adopted strategy was assigned a value of 1 if the farmer adopted the corresponding strategy and 0 otherwise; summing the 1s across each farmer yielded a count variable between 0 and 8, representing the number of CCAs. By analyzing eight adaptations, we present a more detailed perspective on how farmers are coping with climate change than previous studies (Bozzola and Smale, 2020; Diallo *et al.*, 2020; Kimathi *et al.*, 2021; Ahmed, 2022), which have considered only one or a few technologies.

In the questionnaire, output (e.g. gross revenue) and inputs (e.g. cultivated land area, expenditures on seeds and fertilizers) of wheat production for the 2018–2019 cropping season were also collected. We prepared questions to obtain farmers' individual characteristics (e.g. age, gender and education level), household- and farm-level characteristics (e.g. family size, farm size and crop insurance purchase behaviors) and locational information (e.g. provincial dummies). Enumerators fluent in Mandarin and main local languages administered the face-to-face interviews with the selected farmers. All enumerators were employed from an agriculture-centric university and received training before the formal survey to try to ensure the quality of the collected data.

18.3.2 Descriptive statistics

Table 18.1 presents the summary statistics of the variables used for the empirical analysis, including the eight CCAs adopted by wheat farmers, output and input variables, risk exposure indicators and variables that explain farmers' CCAs and risk exposure. The upper part of Table 18.1 shows that the adoption rates of different adaptations vary. Specifically, most farmers adopted crop rotation and crop residue turnover, with 92% of respondents adopting crop rotation and 86% of respondents adopting crop residue turnover. Some 53% of sampled farmers adopted DTWVs and used water-saving irrigation, whereas 26% adopted minimum tillage and 22% planted more than one wheat variety. The average adoption rate of crop–livestock systems was 17% and of organic soil amendments 16%.

The middle part of Table 18.1 summarizes the output and input variables used in the production function. The average farm revenue of wheat production was 799 yuan/mu (1 mu = 1/15 ha). Sampled farmers, on average, cultivated 9 mu for wheat production, and they spent 82 yuan/mu on seeds, 182 yuan/mu on fertilizers, 47 yuan/mu on pesticides and 145 yuan/mu on other inputs. Used labor, including family and hired laborers, was eight working days/mu.

The lower part of the table comprises the dependent and independent variables used in the 2SPS model. The variance was 0.83, skewness was −3.60 and kurtosis was 22.46. On average, farmers adopted 3.63 strategies to adapt to climate change. Regarding the control variables, the average age of the household heads was 56.5 years, and 60% were male. On average, a household had five members, 29% of whom participated in off-farm work. About 53% of the households adopted information and communication technologies (ICTs), 77% purchased insurance for wheat cultivation, 6% owned cooperative memberships and 18% owned a microwave oven. Here, microwave oven ownership is a proxy for household wealth. Households living in Shandong province account for 48%, those in Henan province for 26% and those in Anhui province for 26% of respondents. The average distance to the nearest input market was 2.33 km.

18.4 Empirical Results and Discussions

18.4.1 Determinants of the adoption of CCAs

Table 18.2 comprises the results obtained from the first stage of the 2SPS model, estimating the

Table 18.1. Variable definition and summary statistics. (Source: Authors' own work.)

Variables	Measurements	Mean	(SD)
Climate change adaptations adopted by wheat farmers			
DTWVs adoption	1 if farmers adopted drought-tolerant wheat varieties (DTWVs), 0 otherwise	0.53	(0.50)
Wheat diversification	1 if farmers planted more than one wheat variety, 0 otherwise	0.22	(0.42)
Crop rotation	1 if farmers adopted crop rotation, 0 otherwise	0.92	(0.28)
Water-saving irrigation	1 if farmers used water-saving irrigation technology, 0 otherwise	0.53	(0.50)
Organic soil amendments	1 if farmers adopted organic soil amendments (e.g. organic fertilizer or farmyard manure), 0 otherwise	0.16	(0.36)
Crop residue turnover	1 if farmers used crop residue turnover technology, 0 otherwise	0.86	(0.35)
Minimum tillage	1 if farmers used minimum tillage technology, 0 otherwise	0.26	(0.44)
Crop–livestock systems	1 if farmers adopted integrated crop–livestock systems, 0 otherwise	0.17	(0.37)
Output and inputs variables used in the production function			
Output variable			
Farm revenue	Gross revenue of wheat production (100 yuan/mu)[a,b]	7.99	(3.14)
Input variables			
Land	Total area of land for wheat production (100 mu)	0.09	(0.11)
Seeds	Expenditure on seeds (100 yuan/mu)	0.82	(0.49)
Fertilizers	Expenditure on fertilizers (100 yuan/mu)	1.82	(0.81)
Pesticides	Expenditure on pesticides (100 yuan/mu)	0.47	(0.45)
Labor	Working days of family and hired laborers (100 days/mu)	0.08	(0.10)
Other inputs	Expenditure on machinery and irrigation (100 yuan/mu)	1.45	(0.76)
Dependent and independent variables used in the 2SPS model			
Dependent variables			
Variance[c]	Variance of farm revenue, measured by the square term of the error term in the production function	0.83	(4.67)
Skewness	Skewness of farm revenue, measured by the third power of the error term in the production function	–3.60	(29.25)
Kurtosis	Kurtosis of farm revenue, measured by the fourth power of the error term in the production function	22.46	(184.62)
CCAs adoption	Number of climate change adaptations adopted by wheat farmers (0–8)	3.63	(1.33)
Independent variables			
Age	Age of household head (years)	56.50	(11.22)
Gender	1 if household head is male; 0 = otherwise	0.60	(0.49)
Education	Educational years of household head (years)	4.73	(3.84)
Family size	Number of family members (persons)	4.70	(2.45)
Off-farm-worker ratio	Ratio of off-farm workers to the family size (0–1)	0.29	(0.23)
Farm size	Total land area for wheat cultivation (mu)	9.07	(11.31)
ICT adoption	1 if household adopted information and communication technologies (ICTs), 0 otherwise	0.53	(0.50)
Insurance	1 if household purchased insurance for wheat cultivation, 0 otherwise	0.77	(0.42)
Cooperatives	1 if household participated in any agricultural cooperatives, 0 otherwise	0.06	(0.24)
Asset ownership	1 if household owns a micro oven, 0 otherwise	0.18	(0.39)
Shandong	1 if household is in Shandong province, 0 otherwise	0.48	(0.50)
Henan	1 if household is in Henan province, 0 otherwise	0.26	(0.44)
Anhui	1 if household is in Anhui province, 0 otherwise	0.26	(0.44)
IV	Distance to the nearest input market (km)	2.33	(3.57)
Observations		558	

Notes: SD refers to the standard deviation.
[a]1 mu = 1/15 ha; [b]yuan is Chinese currency (US$1 = 6.90 yuan in 2019); [c]Mean of the semi-variance indicator is not presented in the table because its calculation is not based on the full sample. IV: instrumental variable.

Table 18.2. Determinants of CCAs: Poisson model estimation. (Source: Authors' own work.)

Variables	Coefficients	
Age	0.001	(0.002)
Gender	0.010	(0.030)
Education	0.004	(0.004)
Family size	0.012	(0.006)**
Off-farm-worker ratio	0.163	(0.065)**
Farm size	0.001	(0.001)
ICT adoption	0.008	(0.038)
Insurance	0.164	(0.038)***
Cooperatives	0.007	(0.044)
Asset ownership	0.085	(0.037)**
Shandong	0.227	(0.038)***
Henan	0.041	(0.044)
IV	0.011	(0.004)***
Constant	1.163	(0.113)***
Sample size	558	

Notes: *$p < 0.10$, **$p < 0.05$ and ***$p < 0.01$.
The reference province is Anhui. Robust standard errors are presented in parentheses. IV: instrumental variable.

determinants of the number of CCAs farmers used. The negative and statistically significant coefficient of family size indicates that families with more members employ fewer of these adaptations. This result supports the findings of previous studies, including Aryal *et al.* (2020) for India and Martey and Kuwornu (2021) for Ghana. The result is reasonable as larger families have more labor available to them. Consequently, farmers may be more inclined to use family labor for wheat production rather than adopt capital-intensive CCAs. For example, they may move crop residue using labor instead of machinery. Also, to increase their earnings, some household members may work off-farm.

The coefficient of the off-farm-worker ratio is negative and statistically significant, implying that households with more members participating in off-farm work adopt fewer CCAs. More specifically, a unit increase in the off-farm-worker ratio is associated with a reduction in the number of adaptations used. This finding supports those of previous studies. For example, by analyzing household survey data from Ghana, Issahaku and Abdulai (2020) found that off-farm participation decreases the likelihood of adopting climate-smart practices. Aftab *et al.* (2021) showed that off-farm work negatively affects joint flood adaptation decisions in Pakistan.

We find that farmers who purchased crop insurance have a greater likelihood of adopting more CCAs than those who did not. This result supports the evidence provided by Zakaria *et al.* (2020), who showed that agricultural insurance increases the uptake of climate-smart agricultural practices in Ghana. Crop insurance helps farmers offset losses from crop failure, which may encourage them to invest in more CCAs. The positive and significant coefficient of asset ownership suggests that those farmers who own a microwave oven are more likely to adopt CCAs than non-owners. Adaptations such as water-saving irrigation and organic soil amendments are usually capital-intensive and require large investments. Asset ownership, an indicator of wealth, indicates that farmers have the wherewithal to adopt CCAs (Ma and Wang, 2020).

The results for different provinces vary. The number of CCAs for farmers in Shandong is higher than for those living in Anhui. In comparison, there is no significant difference in the number of adaptations for farmers located in Henan and Anhui. Finally, the coefficient of the IV is positive and statistically significant, signifying that the larger the distance to the input market, the greater the likelihood of adopting CCAs. This is in line with the findings of Bairagi *et al.* (2020). In their analysis of Cambodian farmers, distance to market was positively associated with adopting climate-resilient strategies. Longer distances to input markets increase transportation costs and time spent on purchasing inputs, resulting in higher production costs and loss of time that otherwise could be allocated to production activities. This may motivate farmers to adopt more technologies that reduce input requirements.

18.4.2 Determinants of risk exposure

In the second stage of the 2SPS model, risk exposure indicators are regressed on the predicted *CCAs* obtained from the first stage and all of the control variables. In the interest of brevity, the OLS results, which estimate the production function to calculate the risk exposure indicators, are not presented but are available upon request. An OLS model was employed to estimate Eqn 18.5, given the nature of the four risk exposure indicators. The results are presented in Table 18.3. In this section, we first focus on the impact of the key variable, CCAs, on risk exposure, followed by the effects of the other control variables on the same.

Does the number of CCAs affect farm revenue variability?

Our results suggest that the number of CCAs is negatively and significantly associated with the variance, indicating that these adaptations reduce the variability of farm revenue. This finding is consistent with the findings of Huang *et al.* (2015), who found that farmers' adaptation to extreme weather events significantly reduces the variance of rice yield in China. Furthermore, we also find that the number of adaptations has a negative and statistically significant relationship with the semi-variance. This finding suggests that CCAs are vitally important to those more vulnerable to climate shocks, as the semi-variance reflects losses below a specific benchmark (Finger *et al.*, 2018; Bozzola and Smale, 2020).

As shown in the fourth column of Table 18.3, the number of CCAs positively and significantly affects the skewness of farm revenue. Farmers who adopt more of these adaptations are less likely to be exposed to downside risks, and thus the probability of crop failure for these farmers is lower than for farmers with fewer adaptations. This is consistent with the results of Sarr *et al.* (2021), who showed that adopting climate-friendly agricultural practices is associated with a lower possibility of crop failure for smallholder farmers in Tanzania. The coefficient of the number of CCAs in the last column of Table 18.3 is negative and significant. As kurtosis reflects infrequent extreme deviations in farm revenue, this suggests that increasing the number of CCAs reduces the probability of obtaining extremely low or extremely high farm revenue from wheat production, providing adopting farmers with a more stable income source.

The inverse association between the number of adaptations and kurtosis is consistent with Shahzad and Abdulai (2020). They found that adaptation to extreme weather conditions significantly reduces kurtosis of net returns in rural Pakistan. In practice, smallholder farmers may adopt mixed CCAs since adopting any single adaptation or even two adaptations may be inadequate to tackle challenges from different types of climate shocks: heavy rainfall, drought, flooding, frost and hail. The findings mentioned above highlight that using more CCAs is a practical way to decrease farmers' exposure to risk in crop production.

Impacts of control variables

Turning to the effects of the control variables, the second column of Table 18.3 shows that the coefficient of gender is negative and significant

Table 18.3. Impact of CCAs on risk exposure: OLS estimations. (Source: Authors' own work.)

Variables	Variance		Semi-variance		Skewness		Kurtosis	
CCAs (predicted)	−1.169	(0.650)*	−4.929	(2.394)**	6.805	(3.920)*	−42.580	(23.631)*
Age	0.007	(0.015)	0.002	(0.045)	−0.029	(0.093)	0.158	(0.581)
Gender	−0.632	(0.056)	−1.498	(1.052)	3.674	(2.293)	−22.973	(14.157)
Education	0.036	(0.056)	0.111	(0.136)	−0.234	(0.359)	1.703	(2.313)
Family size	−0.149	(0.081)*	−0.307	(0.189)	0.925	(0.495)*	−5.821	(3.047)*
Off-farm-worker ratio	−1.692	(1.021)*	−3.923	(2.987)	9.754	(6.219)	−59.950	(37.950)
Farm size	−0.014	(0.013)	−0.087	(0.044)**	0.142	(0.068)**	−0.772	(0.448)*
ICT adoption	−0.756	(0.429)*	−1.940	(1.068)*	4.729	(2.724)*	−30.772	(17.467)*
Insurance	−1.167	(0.668)*	−0.470	(1.118)	7.091	(4.119)*	−41.846	(25.593)
Cooperatives	1.004	(1.213)	2.412	(5.083)	−5.674	(7.348)	34.624	(44.337)
Asset ownership	0.440	(0.504)	1.036	(2.314)	−2.268	(2.905)	13.244	(16.691)
Shandong	0.913	(0.964)	5.782	(3.631)	−5.301	(5.886)	35.676	(36.021)
Henan	−0.395	(0.570)	−0.731	(1.310)	2.119	(3.534)	−12.711	(22.253)
Constant	7.025	(2.890)**	22.394	(9.913)**	−41.471	(17.633)**	255.988	(108.099)**
Sample size	558		205		558		558	

Notes: *$p < 0.10$, **$p < 0.05$ and ***$p < 0.01$.
The reference province is Anhui. Robust standard errors are presented in parentheses.

at the 10% level, implying that farm revenues are likely to be more variable for female-headed households than for male-headed households. China is a patriarchal society. Female-headed households may not enjoy the same access to credit and input markets. Moreover, female household heads may be less educated than their male counterparts, less able to cope with some farm work requiring strength and endurance or both. These factors may undermine the capabilities of female-headed households to adapt to climate shocks and may increase the variability in farm revenues. Bangwayo-Skeete *et al.* (2012) also found that households with more female adults tend to have higher variability of crop output than those with fewer female adults in Ethiopia.

Family size reduces the variance and kurtosis but increases the skewness of farm revenue, suggesting that households with more members are less likely to be exposed to production risk than those with fewer members. This contradicts the findings of Mehraban *et al.* (2021), who showed that household size is positively associated with downside economic risk in Indonesia. The contradiction may be explained by differences in the structure and developmental stages of off-farm job markets in the two countries. The off-farm work participation rate in rural China has increased sharply in recent years (Bai *et al.*, 2022). The larger the household, the more labor it can allocate to off-farm work, *ceteris paribus*. With off-farm income at their disposal, households can invest in inputs and technologies to mitigate production risks. The negative and statistically significant coefficient of the off-farm-worker ratio in the second column of Table 18.3 indicates that households with more members working off-farm have a lower variability of farm revenue. The effects of household size and off-farm-worker ratio lend credence to one another. Although previous studies have discussed the influence of off-farm work participation in terms of 'lost-labor effect' and 'income-increasing effect' (Ma *et al.*, 2018; Shi, 2018; Zheng *et al.*, 2022), our results reveal that the latter seems to play a larger role in helping smallholder farmers reduce the variability of crop production.

Farm size has a negative and significant relationship with the semi-variance and kurtosis of farm revenue, while its association with skewness is positive and statistically significant. These findings suggest that households with larger farms have less exposure to production risk than those with smaller farms. Kumar *et al.* (2020) reported similar results for India. Large farms lend themselves well to implementing strategies designed to mitigate risk. For example, farmers with large farms may select various wheat varieties to create an 'optimal portfolio,' while those with small farms cannot do so, as they own limited land; large farms also help farmers achieve economies of scale, reducing production risks.

We show that adoption of information and communication technologies significantly reduces risk exposure by decreasing variance, semi-variance and kurtosis and by increasing the skewness of farm revenue. Adoption of such technologies increases productivity and lowers the cost of disseminating information to farmers (Birner *et al.*, 2021; Mohammed and Abdulai, 2022). Digital tools such as smartphones and computers enable farmers to access financial support, modern inputs, advanced farm management technologies and product markets. Farmers, therefore, can manage both production risk and market risk using information and communication devices. Lastly, having insurance negatively and significantly affects the variance, while its impact on skewness is positive and significant. Crop insurance helps farmers manage the downside risk of unexpected crop failure. Our result is in line with the findings of Di Falco *et al.* (2014). Their analysis of Italian farms showed that crop insurance is a practical strategy to reduce crop risk exposure.

18.5 Conclusions and Implications

Recently, there has been renewed interest in investigating the relationship between CCAs and farmers' risk exposure. However, little attention has been paid to studying farmers' behavior in adapting to climate change. This chapter assessed the role of CCAs in managing wheat farmers' risk exposure in China. Using household survey data on 558 smallholder farmers, we used a moment-based production function to measure farmers' risk exposure, focusing on variability, semi-variability, skewness and kurtosis of farm revenue. Then, to address the endogeneity of such adaptations, we used a two-step estimation method – a Poisson regression to obtain the predicted number of CCAs variable followed by an OLS regression to analyze the effects of the number of adaptations on indicators of risk exposure.

The findings in this chapter illuminate the association between CCAs and risk exposure. We found that such adaptations are associated with lower risk exposure of farmers, thereby contributing to the stability of farm revenue. Specifically, increasing the number of adaptations significantly reduces risk exposure by decreasing variability, semi-variability and kurtosis and by increasing farm revenue skewness. Additionally, we found that households with male heads, relatively large families, more off-farm workers and larger farms are less likely to be exposed to production risk. Adopting information and communication technologies and purchasing crop insurance are also associated with lower production risk.

In the first stage of the 2SPS model, we studied the determinants of CCAs. The results showed that the number of adaptations farmers adopt is positively associated with crop insurance and asset ownership, while large households and those with more off-farm workers tend to employ fewer adaptations. These findings call for practical interventions to encourage CCAs. First, governments and program officers should target large rural households with members working off-farm with awareness and educational initiatives to encourage them to adapt their farming practices to climate change. Second, as the number of CCAs largely relies on household wealth, policies that promote rural income growth may encourage farmers to use a greater variety of adaptations over time. Thus, income growth and CCAs would reinforce one another, creating a virtuous circle of prosperity and growth. Third, in our study, wheat farmers allocate their household labor to off-farm jobs to supplement farm income. Thus, more local off-farm work opportunities should be created and provided to improve rural income and help farmers balance on-farm and off-farm labor allocation. Although we find a positive link between insurance and the number of CCAs used by farmers, the result should be interpreted with caution. Insurance is costly to provide because of covariate risks. When one farmer has crop losses, typically, many others also suffer. Also, US experience shows that insurance can distort farm management decisions. Given the ground realities of providing insurance and the potential downsides of doing so, resolving the low uptake of CCAs via insurance policies may not yield the desired results. The policy implications of this study are important, not only for the wheat industry in China but also more generally, as climate change threatens crop production globally.

References

Abro, Z.A., Jaleta, M. and Teklewold, H. (2018) Does intensive tillage enhance productivity and reduce risk exposure? Panel data evidence from smallholders agriculture in Ethiopia. *Journal of Agricultural Economics* 69, 756–776. DOI: 10.1111/1477-9552.12262.

Aftab, A., Ahmed, A. and Scarpa, R. (2021) Farm households' perception of weather change and flood adaptations in northern Pakistan. *Ecological Economics* 182, 106882. DOI: 10.1016/j.ecolecon.2020.106882.

Ahmed, M.H. (2022) Impact of improved seed and inorganic fertilizer on maize yield and welfare: evidence from Eastern Ethiopia. *Journal of Agriculture and Food Research* 7, 100266. DOI: 10.1016/j.jafr.2021.100266.

Amadu, F.O., McNamara, P.E. and Miller, D.C. (2020) Understanding the adoption of climate-smart agriculture: a farm-level typology with empirical evidence from southern Malawi. *World Development* 126, 104692. DOI: 10.1016/j.worlddev.2019.104692.

Amondo, E., Simtowe, F., Rahut, D.B. and Erenstein, O. (2019) Productivity and production risk effects of adopting drought-tolerant maize varieties in Zambia. *International Journal of Climate Change Strategies and Management* 11, 570–591. DOI: 10.1108/IJCCSM-03-2018-0024.

Aryal, J.P., Farnworth, C.R., Khurana, R., Ray, S., Sapkota, T.B. *et al.* (2020) Does womens participation in agricultural technology adoption decisions affect the adoption of climate-smart agriculture? Insights from Indo-Gangetic Plains of India. *Review of Development Economics* 24, 973–990. DOI: 10.1111/rode.12670.

Bai, Y., Wang, W. and Zhang, L. (2022) Off-farm employment experience triggers heterogeneity of exiting part-time farming in rural China. *China Agricultural Economic Review* 14, 567–582 DOI: 10.1108/CAER-04-2020-0069.

Bairagi, S., Mishra, A.K. and Durand-morat, A. (2020) Climate risk management strategies and food security: evidence from Cambodian rice farmers. *Food Policy* 95, 101935. DOI: 10.1016/j.foodpol.2020.101935.

Bangwayo-Skeete, P.F., Bezabih, M. and Zikhali, P. (2012) Crop biodiversity, productivity and production risk: panel data micro-evidence from Ethiopia. *Natural Resources Forum* 36, 263–273. DOI: 10.1111/1477-8947.12000.

Birner, R., Daum, T. and Pray, C. (2021) Who drives the digital revolution in agriculture? A review of supply-side trends, players and challenges. *Applied Economic Perspectives and Policy* 43, 1260–1285. DOI: 10.1002/aepp13145.

Bora, K. (2022) Rainfall shocks and fertilizer use: a district level study of India. *Environment and Development Economics*, 27, 556–577. DOI: 10.1017/s1355770x21000413.

Bozzola, M. and Smale, M. (2020) The welfare effects of crop biodiversity as an adaptation to climate shocks in Kenya. *World Development* 135, 105065. DOI: 10.1016/j.worlddev.2020.105065.

Challinor, A.J., Watson, J., Lobell, D.B., Howden, S.M., Smith, D.R. *et al.* (2014) A meta-analysis of crop yield under climate change and adaptation. *Nature Climate Change* 4, 287–291. DOI: 10.1038/nclimate2153.

Cholo, T.C., Fleskens, L., Sietz, D. and Peerlings, J. (2019) Land fragmentation, climate change adaptation, and food security in the Gamo Highlands of Ethiopia. *Agricultural Economics* 50, 39–49. DOI: 10.1111/agec.12464.

Diallo, A., Donkor, E. and Owusu, V. (2020) Climate change adaptation strategies, productivity and sustainable food security in southern Mali. *Climatic Change* 159, 309–327. DOI: 10.1007/s10584-020-02684-8.

Di Falco, S. and Veronesi, M. (2013) How can African agriculture adapt to climate change? A counterfactual analysis from Ethiopia. *Land Economics* 89, 743–766. DOI: 10.3368/le.89.4.743.

Di Falco, S., Adinolfi, F., Bozzola, M. and Capitanio, F. (2014) Crop insurance as a strategy for adapting to climate change. *Journal of Agricultural Economics* 65, 485–504. DOI: 10.1111/1477-9552.12053.

Faisal, M., Abbas, A., Xia, C., Haseeb Raza, M., Akhtar, S. *et al.* (2021) Assessing small livestock herders' adaptation to climate variability and its impact on livestock losses and poverty. *Climate Risk Management* 34, 100358. DOI: 10.1016/j.crm.2021.100358.

FAOSTAT (2022) Value of Agricultural Production, Food And Agriculture Organization. FAO, Rome. Available at: http://www.fao.org/faostat/en/#data/QV (accessed May 11, 2022).

Fentie, A. and Beyene, A.D. (2019) Climate-smart agricultural practices and welfare of rural smallholders in Ethiopia: does planting method matter? *Land Use Policy* 85, 387–396. DOI: 10.1016/j.landusepol.2019.04.020.

Finger, R., Dalhaus, T., Allendorf, J. and Hirsch, S. (2018) Determinants of downside risk exposure of dairy farms. *European Review of Agricultural Economics* 45, 641–674. DOI: 10.1093/erae/jby012.

Gorst, A., Dehlavi, A. and Groom, B. (2018) Crop productivity and adaptation to climate change in Pakistan. *Environment and Development Economics* 23, 679–701. DOI: 10.1017/S1355770X18000232.

Gupta, R., Somanathan, E. and Dey, S. (2017) Global warming and local air pollution have reduced wheat yields in India. *Climatic Change* 140, 593–604. DOI: 10.1007/s10584-016-1878-8.

Ho, T.T. and Shimada, K. (2021) The effects of multiple climate change responses on economic performance of rice farms: evidence from the Mekong Delta of Vietnam. *Journal of Cleaner Production* 315, 128129. DOI: 10.1016/j.jclepro.2021.128129.

Huang, J., Wang, J. and Wang, Y. (2015) Farmers' adaptation to extreme weather events through farm management and its impacts on the mean and risk of rice yield in China. *American Journal of Agricultural Economics* 97, 602–617. DOI: 10.1093/ajae/aav005.

Issahaku, G. and Abdulai, A. (2020) Adoption of climate-smart practices and its impact on farm performance and risk exposure among smallholder farmers in Ghana. *Australian Journal of Agricultural and Resource Economics* 64, 396–420. DOI: 10.1111/1467-8489.12357.

Issahaku, G., Abdul-rahaman, A. and Amikuzuno, J. (2021) Climate change adaptation strategies, farm performance and poverty reduction among smallholder farming households in Ghana. *Climate and Development* 13, 736–747. DOI: 10.1080/17565529.2020.1860884.

Jamil, I., Jun, W., Mughal, B., Raza, M.H., Imran, M.A. *et al.* (2021) Does the adaptation of climate-smart agricultural practices increase farmers' resilience to climate change?. *Environmental Science and Pollution Research* 28, 27238–27249. DOI: 10.1007/s11356-021-12425-8.

Kimathi, S.M., Ayuya, O.I. and Mutai, B. (2021) Adoption of climate-resilient potato varieties under partial population exposure and its determinants: case of smallholder farmers in Meru County, Kenya. *Cogent Food and Agriculture* 7, 1860185. DOI: 10.1080/23311932.2020.1860185.

Kpadonou, R.A.B., Owiyo, T., Barbier, B., Denton, F., Rutabingwa, F. *et al.* (2017) Advancing climate-smart-agriculture in developing drylands: joint analysis of the adoption of multiple on-farm soil and water conservation technologies in West African Sahel. *Land Use Policy* 61, 196–207. DOI: 10.1016/j.landusepol.2016.10.050.

Kumar, S., Singh, D.R., Singh, A., Singh, N.P. and Jha, G.K. (2020) Does adoption of soil and water conservation practice enhance productivity and reduce risk exposure? Empirical evidence from semi-arid tropics (SAT), India. *Sustainability* 12, 6965. DOI: 10.3390/SU12176965.

Liu, M., Min, S., Ma, W. and Liu, T. (2021) The adoption and impact of e-commerce in rural China: application of an endogenous switching regression model. *Journal of Rural Studies* 83, 106–116. DOI: 10.1016/j.jrurstud.2021.02.021.

Ma, W. and Wang, X. (2020) Internet use, sustainable agricultural practices and rural incomes: evidence from China. *Australian Journal of Agricultural and Resource Economics* 64, 1087–1112. DOI: 10.1111/1467-8489.12390.

Ma, W., Abdulai, A. and Ma, C. (2018) The effects of off-farm work on fertilizer and pesticide expenditures in China. *Review of Development Economics* 22, 573–591.

Ma, W., Zheng, H. and Yuan, P. (2022) Impacts of cooperative membership on banana yield and risk exposure: insights from China. *Journal of Agricultural Economics* 73, 564–579. DOI: 10.1111/1477-9552.12465.

Martey, E. and Kuwornu, J.K.M. (2021) Perceptions of climate variability and soil fertility management choices among smallholder farmers in northern Ghana. *Ecological Economics* 180, 106870. DOI:10.1016/j.ecolecon.2020.106870.

Mehraban, N., Kubitza, C., Alamsyah, Z. and Qaim, M. (2021) Oil palm cultivation, household welfare, and exposure to economic risk in the Indonesian small farm sector. *Journal of Agricultural Economics* 72, 901–915. DOI: 10.1111/1477-9552.12433.

Mohammed, S. and Abdulai, A. (2022) Do ICT based extension services improve technology adoption and welfare? Empirical evidence from Ghana. *Applied Economics* 54, 2707–2726. DOI: 10.1080/00036846.2021.1998334.

Mukasa, A.N. (2018) Technology adoption and risk exposure among smallholder farmers: panel data evidence from Tanzania and Uganda. *World Development* 105, 299–309. DOI: 10.1016/j.worlddev.2017.12.006.

NBS (2020) *China Statistical Yearbook*. National Bureau of Statistics of China, Beijing.

Ojo, T.O. and Baiyegunhi, L.J.S. (2020) Determinants of climate change adaptation strategies and its impact on the net farm income of rice farmers in south-west Nigeria. *Land Use Policy* 95, 103946. DOI: 10.1016/j.landusepol.2019.04.007.

Ojo, T.O., Ogundeji, A.A. and Belle, J.A. (2021) Climate change perception and impact of on-farm demonstration on intensity of adoption of adaptation strategies among smallholder farmers in South Africa. *Technological Forecasting and Social Change* 172, 121031. DOI: 10.1016/j.techfore.2021.121031.

Onyeneke, R.U. (2021) Does climate change adaptation lead to increased productivity of rice production? Lessons from Ebonyi State, Nigeria. *Renewable Agriculture and Food Systems* 36, 54–68. DOI: 10.1017/S1742170519000486.

Oparinde, L.O. (2021) Fish farmers' welfare and climate change adaptation strategies in southwest, Nigeria: application of multinomial endogenous switching regression model. *Aquaculture Economics and Management* 25, 450–471. DOI: 10.1080/13657305.2021.1893863.

Pizer, S.D. (2016) Falsification testing of instrumental variables methods for comparative effectiveness research. *Health Services Research* 51, 790–811. DOI: 10.1111/1475-6773.12355.

Ritchie, H. (2021) Smallholders Produce One-Third of the Worlds Food, Less than Half of What Many Headlines Claim. Our World in Data, Oxford, UK. Available at: https://ourworldindata.org/smallholder-food-production (accessed May 12, 2023).

Sarr, M., Bezabih Ayele, M., Kimani, M.E. and Ruhinduka, R. (2021) Who benefits from climate-friendly agriculture? The marginal returns to a rainfed system of rice intensification in Tanzania. *World Development* 138, 105160. DOI: 10.1016/j.worlddev.2020.105160.

Setsoafia, E.D., Ma, W. and Renwick, A. (2022) Effects of sustainable agricultural practices on farm income and food security in northern Ghana. *Agricultural and Food Economics* 10, 9. DOI: 10.1186/s40100-022-00216-9.

Shahzad, M.F. and Abdulai, A. (2020) Adaptation to extreme weather conditions and farm performance in rural Pakistan. *Agricultural Systems* 180, 102772. DOI: 10.1016/j.agsy.2019.102772.

Shahzad, M.F., Abdulai, A. and Issahaku, G. (2021) Adaptation implications of climate-smart agriculture in rural Pakistan. *Sustainability* 13, 11702. DOI: 10.3390/su132111702.

Shi, X. (2018) Heterogeneous effects of rural-urban migration on agricultural productivity. *China Agricultural Economic Review* 10, 482–497. DOI: 10.1108/CAER-10-2017-0193.

Terza, J.V., Basu, A. and Rathouz, P.J. (2008) Two-stage residual inclusion estimation: addressing endogeneity in health econometric modeling. *Journal of Health Economics* 27, 531–543. DOI: 10.1016/j.jhealeco.2007.09.009.

Wang, Y., Huang, J., Wang, J. and Findlay, C. (2018) Mitigating rice production risks from drought through improving irrigation infrastructure and management in China. *Australian Journal of Agricultural and Resource Economics* 62, 161–176. DOI: 10.1111/1467-8489.12241.

Weerasekara, S., Wilson, C., Lee, B. and Hoang, V.-N. (2022) Impact of natural disasters on the efficiency of agricultural production: an exemplar from rice farming in Sri Lanka. *Climate and Development* 14, 133–146. DOI: 10.1080/17565529.2021.1893635.

Wongnaa, C.A. and Babu, S. (2020) Building resilience to shocks of climate change in Ghana's cocoa production and its effect on productivity and incomes. *Technology in Society* 62, 101288. DOI: 10.1016/j.techsoc.2020.101288.

Wooldridge, J. (2015) Control function methods in applied econometrics. *The Journal of Human Resources* 50, 420–445.

World Bank (2020) *Climate Change in APEC: Assessing Risks, Preparing Financial Markets, and Mobilizing Institutional Investors*. World Bank Group, Washington, DC. DOI: 10.1596/33423.

Ying, A., Xu, R. and Murphy, J. (2019) Two-stage residual inclusion for survival data and competing risks – an instrumental variable approach with application to SEER-Medicare linked data. *Statistics in Medicine* 38, 1775–1801. DOI: 10.1002/sim.8071.

Zakaria, A., Azumah, S.B., Appiah-Twumasi, M. and Dagunga, G. (2020) Adoption of climate-smart agricultural practices among farm households in Ghana: the role of farmer participation in training programmes. *Technology in Society* 63, 101338. DOI: 10.1016/j.techsoc.2020.101338.

Zheng, H., Ma, W. and Zhou, X. (2021) Renting-in cropland, machinery use intensity, and land productivity in rural China. *Applied Economics* 53, 5503–5517. DOI: 10.1080/00036846.2021.1923642.

Zheng, H., Ma, W., Guo, Y. and Zhou, X. (2022) Interactive relationship between non-farm employment and mechanization service expenditure in rural China. *China Agricultural Economic Review* 14, 84–105. DOI: 10.1108/CAER-10-2020-0251.

Zhu, Y., Yang, Q. and Zhang, C. (2021) Adaptation strategies and land productivity of banana farmers under climate change in China. *Climate Risk Management* 34, 100368. DOI: 10.1016/j.crm.2021.100368.

19 Cropping System Diversification for Higher Productivity and Resilience of Agriculture: Evidence from India

Jaweriah Hazrana[1]* and Pratap S. Birthal[2]

[1]*The Ohio State University, Columbus, USA;* [2]*ICAR-National Institute of Agricultural Economics and Policy Research, New Delhi, India*

Abstract

Using a spatially disaggregated district-level panel dataset and applying a moment-based specification of the stochastic production function, this chapter evaluates crop diversification's potential for improving system productivity and reducing its sensitivity to climate change. This investigation has three key findings. First, crop diversification helps improve productivity and reduce risks. Second, its productivity benefits outweigh the risk benefits, and both have slowed over time. Third, the effects of crop diversification differ across agro-climatic zones – both the productivity and risk benefits are more prominent in ecologically fragile arid environments. These findings suggest the need to design regionally differentiated strategies to exploit the multifunctionality of crop diversification to improve the productivity and resilience of agriculture.

19.1 Introduction

Over the past five decades, the technological change in agriculture supported by massive investments in irrigation, infrastructure, markets and institutions (i.e. extension services and credit), and economic incentives (i.e. input subsidies and output support prices) has spurred significant increases in agricultural productivity and food supplies in developing countries (Fuglie *et al.*, 2020), propelling many of them toward self-sufficiency in food and non-food agricultural commodities. In India, for instance, between 1996/97 and 2019/20 food grain production increased fourfold, from 75 million to 302 million tonnes, lifting the annual per capita net availability from 149 kg to 185 kg (GoI, 2020), despite rapid population growth.

Nonetheless, the quest for food self-sufficiency compelled a greater emphasis on increasing the production of staple cereals, i.e. rice and wheat. This triggered the development of their monocultures, causing degradation of natural resources (i.e. land, water and air) and compromising the long-term sustainability of agriculture and agriculture-based livelihoods (Gautam, 2015). Simultaneously, climate change emerged as a significant threat to the sustainable development of agriculture (Dell *et al.*, 2012; Loayza *et al.*, 2012; Birthal *et al.*, 2021; Ortiz-Bobea *et al.*, 2021). Ortiz-Bobea *et al.* (2021) have estimated that since 1960, climate

*E-mail: jaweriah.hazrana@gmail.com; hazrana.1@osu.edu

DOI: 10.1079/9781800622289.0019

change has slowed the productivity growth of world agriculture by 21%, with the rate being higher for developing countries. Birthal *et al.* (2021) provide similar evidence for India – a 25% decline in agricultural growth at the national level, and a greater reduction in it for poorer states that derived a larger share of their gross domestic product (GDP) from agriculture. Such climate-induced changes in agri-food production systems are a matter of grave concern for food and nutrition security and for poverty reduction (Hallegatte *et al.*, 2016).

Several technological and agronomic options can partially offset the adverse effects of climate change on agriculture and agriculture-based livelihoods. Often, in the absence of a market for modern risk management tools like crop insurance and hedging (Binswanger-Mkhize, 2012; Ali *et al.*, 2020), farmers rely on traditional risk-mitigating measures, including crop diversification (Jodha *et al.*, 2012; Altieri and Nicholls, 2017; van Zonneveld *et al.*, 2020). Since crops differ in their response to climate change, risk-averse farmers often choose a portfolio of crops with low-correlated returns (Di Falco and Chavas, 2009; Lybbert and Bell, 2010). If a crop does not perform well under a stressed condition, the loss in revenue from it can be offset to an extent by the gains in revenue from other crops that are more tolerant to stress.

Several studies have assessed the effects of cropping system diversification on farm productivity or income (e.g. Jayne and Govereh, 2003; Joshi *et al.*, 2004; Chhatre *et al.*, 2016), and most of these conclude that diversification into horticultural and commercial crops is an efficient means of improving farm income and farmers' welfare. Diversification of cropping systems toward high-value crops also has been identified as a key source of agricultural growth in South and South-east Asia (Birthal *et al.*, 2014, 2019a,b; Birthal and Pandey, 2020). It also acts as a cushion against excess temperatures and deficit rainfall (Birthal and Hazrana, 2019), helps optimize the use of scarce resources (i.e. land and water), helps manage the threats of insect pests and diseases, and enhances the efficiency and sustainability of cropping systems (Loreau and Hector, 2001; Altieri and Nicholls, 2017; Renard and Tilman, 2019; Tamburini *et al.*, 2020; van Zonneveld *et al.*, 2020).

Although crop diversification performs several economic, social and environmental functions, to the best of our understanding, hardly any study simultaneously evaluates its multifunctionality (Di Falco and Chavas, 2008, 2009). This lack of information on the relative utility of the different functions of crop diversification restricts the recommendations regarding the choice of a crop portfolio that best fits the goal of efficient, sustainable and climate-resilient agriculture (van Zonneveld *et al.*, 2020). Therefore, understanding the trade-offs between multiple functions of diversification is essential for devising policies and programs to enhance diversification of economic, social and environmental outcomes.

This chapter evaluates the effects of crop diversification on agricultural productivity and its resilience to climate change by applying a moment-based specification of the stochastic production function to a spatially disaggregated district-level panel dataset from India. Findings show crop diversification to be an efficient means of improving productivity and of cushioning it against climate risks. Nevertheless, crop diversification's productivity benefits outweigh the risk benefits, and both appear to have slowed. Further, the effects of crop diversification vary across agro-climatic conditions – both the productivity and risk benefits are more significant for ecologically fragile arid environments.

The rest of the chapter is structured as follows. Section 19.2 discusses data and provides descriptive statistics. Section 19.3 briefly discusses the empirical model. The productivity and resilience effects of crop diversification are discussed in Section 19.4. Concluding remarks are made in Section 19.5.

19.2 Data and Descriptive Statistics

This study uses a panel dataset on 311 districts (at their 1970 base) from 29 geographically contiguous states of India for 48 years, i.e. from 1966/67 to 2015/16. The primary source of data on agricultural variables is the ICRISAT database maintained by the International Crops Research Institute for Semi-Arid Tropics (ICRISAT), Patancheru, India. Hereafter, it is referred to as the 'ICRISAT database.' This database contains information on several aspects of Indian agriculture, including the area, production and farm harvest prices of major field crops; and agricultural inputs (i.e. land, soil quality, irrigation, farm workers, draught animals and tractors). The main crops for which data are available include rice, wheat, maize, sorghum, pearl millet, finger millet, barley, chickpeas, pigeon pea and other pulses, soybeans, sesame, groundnut, rapeseed mustard, sunflower, safflower, linseed, castor, cotton and sugarcane. This database also contains information on the area under horticultural crops (i.e. potato, onion and others) but not on their production and farm harvest prices. Hence, horticultural crops are excluded from the analysis.

Our variables of interest are: crop diversification and agricultural productivity and its higher-order moments, that is, the variance and skewness that capture the riskiness of agriculture. We estimate agricultural productivity at the district level as the sum of the values of outputs of crops (at 2004/05 constant prices) divided by the sum of their cropped area, and it is expressed in rupees per hectare.

$$Y_{it} = \frac{\sum_{j=1}^{N} F^{j}_{i,t=2004-05} \times P^{j}_{it}}{A_{it}} \tag{19.1}$$

where Y_{it} is agricultural productivity and A_{it} is the sum of the area under j crops in district i at time t. To calculate productivity, crop-specific outputs, P^{j}_{it}, are multiplied by their respective farm harvest price in 2004/05 ($F^{j}_{it=2004-05}$), and the sum of the monetary values of all j crops is divided by the sum of their cropped area, A_{it}. The crops included in the estimation of agricultural productivity occupy approximately 85% of the country's gross cropped area.

To derive moments of the agricultural productivity, we first estimate a mean productivity function, then use its residuals to estimate second (variance) and third (skewness) moments. The moment-based production function approach has been widely used in estimating downside risk exposure (Di Falco and Chavas, 2009; Huang *et al.*, 2015; Zhang and Antle, 2018), as any of the central moments around the mean is a proxy for risk exposure.

The degree of the cropping system's diversification is assessed using the Simpson index, SI = $\left(1-\sum_{i=1}^{N} A^{2}_{it}\right)$; where A_{it} is the share of a crop in the total cropped area at time t. The index is bounded by 0 and 1; a value closer to 1 implies a higher degree of diversification.

The agricultural data are combined with climate data obtained from the India Meteorological Department (IMD), Ministry of Earth Sciences, Government of India. The department provides gridded data on daily minimum and maximum temperatures (at 0.5 × 0.5 degree grid) and precipitation (at 0.25 × 0.25 degree grid). From this dataset, we generate district-level estimates of temperature and rainfall by assigning each grid to a district whose centroid is nearest to it, then taking the average of each variable recorded at all the grids assigned to that district. Then, we estimate climate anomalies as the standardized deviations of the annual mean temperature and rainfall from their respective long-term averages (1966/67 to 2015/16).

Table 19.1 provides the means and standard deviations of key variables. The mean value of the Simpson index is estimated at 0.696 (SD = 0.095). This indicates that India's cropping system is fairly diversified.

To have an idea about the changes in cropping system, we estimate kernel densities for the Simpson index and agricultural productivity for 1966–1990 and 1991–2015. The density function of the Simpson index for 1991/92 to 2015/16 is slightly skewed toward the right (Fig. 19.1), which implies a marginal improvement in the cropping system diversity. On the other hand, the density function of agricultural productivity for this period is highly skewed toward the right, which means a significant improvement in agricultural productivity. These distributions provide an indication of a positive association between cropping system diversification and agricultural productivity.

The association between crop diversification and agricultural productivity is probed further by estimating a non-parametric relationship between them (Fig. 19.2). The association appears highly positive – the productivity curve rises steeply, but beyond a threshold it starts declining, possibly due to structural and operational constraints on further diversification.

19.3 Empirical Framework

Farmers confront several production risks, hence they are assumed to operate on a stochastic

Table 19.1. Descriptive statistics. (Source: Based on ICRISAT database.)

Variable	Variable description	Mean	Std. dev.
Productivity	Rupees/ha of cropped area	14,841	(9,789)
Human labor	No. of workers/ha of cropped area	0.4780	(0.4711)
Animal labor	No. of adult cattle/ha cropped area	0.0258	(0.1445)
Irrigation	Proportion of gross cropped area	0.3706	(0.2809)
Soil quality	Proportion of net sown area under high moisture retention soils	0.4780	(0.4711)
Temperature	Degrees Celsius	25.228	(2.0333)
Rainfall	Millimeters	1,134.03	(656.01)
Farm size	Hectares	2.2844	(2.0968)
Road density	Km/1000 sq. km	5.1601	(5.9929)
Market density	No./1000 sq. km	289.01	(348.88)
Bank density	No./1000 population	180.67	(170.96)
Diversity	Simpson index of diversity	0.6962	(0.0952)

Note: Standard deviations are in parentheses.

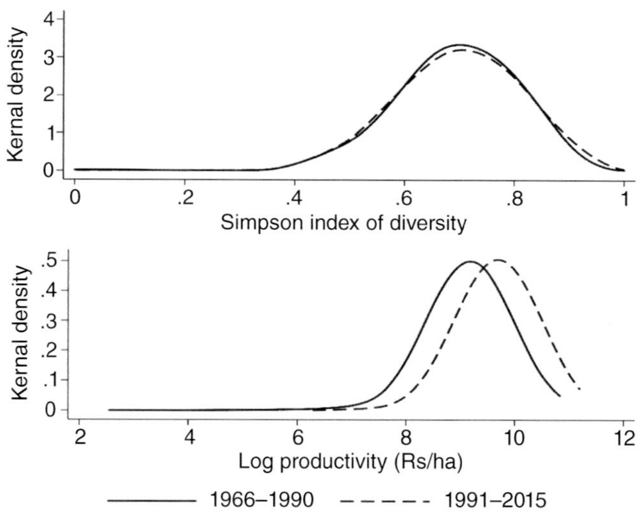

Source: Authors' calculation.

Fig. 19.1. Kernel density of Simpson index of diversity and agricultural productivity.

production function y = f(x, θ); where y is the system productivity, x is a vector of inputs and θ is a vector of random variables (e.g. climatic shocks and insect-pest infestations) that are beyond the farmers' control. Risk-averse farmers, thus, aim at maximizing productivity and minimizing output loss due to risks and uncertainty (Menezes *et al.*, 1980; Antle, 1983; Di Falco and Chavas, 2009; Zhang and Antle, 2018) by adopting technologies and agronomic practices.

In this chapter, our interest is in knowing the effects of crop diversification on agricultural productivity and its resilience to climate change. Crop diversification is an *ex-ante* means of adapting to climate change; hence, farmers' diversification decisions happen under uncertainty without prior knowledge of their outcomes. Following from the literature (Antle, 1983, 1987; Di Falco and Chavas, 2009; Huang *et al.*, 2015; Zhang and Antle, 2018), we apply a moment-based specification of the stochastic

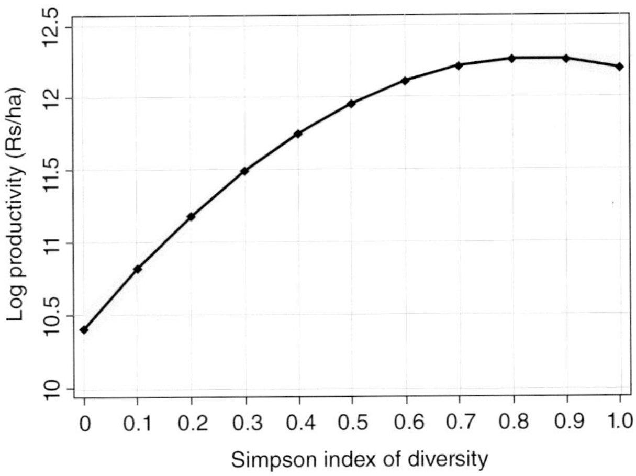

Source: Authors' calculation.

Fig. 19.2. Non-parametric relationship between the Simpson index of diversity and farm income.

production function to evaluate the effects of crop diversification:

$$f(x,\theta) = h_1(x,\delta_1) + u \quad (19.2)$$

Where $h_1(x, \delta_1) \equiv E[f(x, \theta)]$ is the mean of $f(x, \theta)$ and $u \equiv f(x, \theta) - h_1(x, \delta_1)$ is a random variable with zero mean. The higher-order moments of $f(x, \theta)$ can be specified as:

$$E\{[f(x,\theta) - h_1(x,\delta_1)]^k \mid x\} = h_k(x,\delta_k) \quad (19.3)$$

Equation 19.3 is a flexible representation of the moment functions and can capture the effect of diversification on the cropping system productivity under uncertainty. Equation 19.3 is also used to generate higher-order moments, i.e. variance (k = 2) and skewness (k = 3) of productivity.

A priori, the productivity function, $h_1(x, \delta_1)$, is expected to be increasing and concave in inputs and system diversity. However, the expected effects of inputs and crop diversification on the variance and skewness are ambiguous. For instance, the variance function can be increasing in inputs and diversification if $\frac{\partial h_2}{\partial x} > 0$, input- or diversification-neutral if $\frac{\partial h_2}{\partial x} = 0$ and decreasing in inputs and diversification if $\frac{\partial h_2}{\partial x} < 0$. The same is applicable to the skewness function.

The mean productivity function can be specified as:

$$y_{it} = \delta_0 + x_{it}\delta_1 + u_{it} \quad (19.4)$$

where, y_{it} denotes the agricultural productivity in district i at time t, x_{it} is a vector of the conditioning factors and u_{it} is the error term independently distributed with zero mean and finite variance.

The functional form of $h(\cdot)$ depends on the type of moment. To model the higher-order raw moments, Antle (1983) used an identity function, Schlenker and Roberts (2006) a natural logarithm function and Tack *et al.* (2012) a higher-power function. We use a natural logarithm function and write a higher-moment function as:

$$u_{it}^j = \delta_0 + x_{it}\delta + \epsilon_{jit} \quad E(\epsilon_{jit} \mid x_{it}) = 0 \text{ for } j = 2, 3 \quad (19.5)$$

Equations 19.4 and 19.5 are estimated applying the Prais–Winsten method with panel-corrected standard errors (PCSE). This method accounts for the contemporaneous correlation in panels and heteroskedasticity in the disturbances.

19.4 Results and Discussion

19.4.1 Specification tests

Time series of agricultural and climate variables often exhibit a random walk or non-stationarity, which may bias the estimates. We test for stationarity in the time series of important variables by applying Fisher-type unit root tests. Our null hypothesis is that the panel contains a unit root. The test statistics reported in Table A1 (in the appendix) reject the null hypothesis of a unit root; that is, the time series are stationary.

Further, we conduct several diagnostic tests (Table 19.2). First, we conduct a Wooldridge test for autocorrelation on the null hypothesis that there is no first-order autocorrelation. The p-value rejects the null hypothesis, suggesting estimation of the moment functions on the autocorrelation-transformed data. Next, we test for heteroskedasticity. A Wald test rejects the null hypothesis of group-wise homoskedasticity in the fixed effects regression; hence we estimate heteroskedasticity-consistent standard errors. Pesaran's test rejects the null hypothesis of cross-sectional independence, and therefore the standard errors are adjusted while estimating cross-sectional dependence. An F-test for the joint significance is highly significant, endorsing the use of non-linear terms in the regression equations.

19.4.2 Crop diversification and system productivity

We estimate three specifications of the mean productivity function (Table 19.3). Specification I includes the inputs (human and animal power and irrigation), infrastructure (roads, markets and banks), climate (rainfall and temperature) and farm size. Specification II extends Specification I by including the diversity index, its squared term and interactions with irrigation and time trend. Specification III, along with other variables, includes the interactions of the diversity index with irrigation, soils and climate variables.

We begin by examining the effects of crop diversification on system productivity. The diversity index and its squared term are positive and insignificant in Specification II. In Specification III, the coefficient on its linear term is highly positive and significant but not on its squared term. This means that crop diversification improves system productivity, but its effect becomes subdued after reaching a threshold. This is expected, as the greater the number of crops in a portfolio, the greater the competition for scarce resources, and a thin spread of resources may result in sub-optimal outcomes. Furthermore, the interaction of the diversity index with the time trend is negative, which means that the productivity gains of crop diversification have slowed over time.

Rainfall shocks have a positive and significant effect on productivity, but temperature shocks have an opposite effect. However, our interest is in knowing the significance of the interaction of the diversity index with climate variables. The coefficient on the interaction of the diversity index with temperature shocks is negative and statistically insignificant. In contrast, its interaction with rainfall shocks is significant.

Quality of soils (defined in terms of moisture-holding capacity) can also differentiate crop choices. In our regressions, the coefficient on soil quality is positive and significant in Specification III, but its interaction with the diversity index is negative. This means that the selection of a crop portfolio and its outcomes are sensitive to the soil quality. Good-quality soils are preferred for growing high-yielding crops, and low-quality soils for low-yielding, low-risk crops. The coefficient on irrigation is positive and significant in Specification I. Its interaction with the diversity index is also significantly positive in Specifications II and III. This means that irrigation enhances the productivity benefits of crop diversification.

Because of the confounding effects of interactions, the regression coefficients do not have an explicit connotation. Therefore, we estimate

Table 19.2. Model identification tests. (Source: Based on ICRISAT database.)

Test	Test statistic	p-value
Wooldridge (2002) test for autocorrelation F (1, 310)	4.080	0.0443
Modified Wald test of group-wise heteroskedasticity $Chi^2(290)$	18,759.97	0.0000
Pesaran's test of cross-section correlation	65.412	0.0000
Significance of interactions F (5, 11941)	9.530	0.0000

marginal effects (elasticities) of crop diversification and other variables at their respective means (Table 19.3). The marginal effects of diversification are positive and highly significant across all the specifications, confirming diversification's role in raising agricultural productivity.

The marginal effects of temperature are negative, while those of rainfall are positive and statistically significant. The marginal effects of irrigation are positive and significant. Notably, the marginal effects of all the variables, including diversity, remain stable across all the specifications.

Finally, we examine the influence of structural and infrastructural variables. There is a strong inverse relationship between farm size and agricultural productivity, and the relationship is stable across all the specifications. This is in line with earlier evidence of a negative size–productivity relationship in Indian agriculture (Chand *et al.*, 2011). As expected, marginal effects of roads, markets and banks are positive and significant, suggesting that improving markets and financial flow will help improve agricultural productivity.

19.4.3 Crop diversification and downside risk exposure

Table 19.4 presents the estimated effects of crop diversification on the variance and skewness of agricultural productivity. Let us first look at the estimates of the skewness function. The linear term of the diversity index is positive and significant, which means crop diversification reduces downside risk exposure. Note, a positive sign on any variable in the skewness function indicates its risk-reducing role. The coefficient on its squared term is negative and significant, indicating that risk benefits of crop diversification slow after reaching a threshold. The marginal effect of the diversity index (lower panel of Table 19.4) is positive and statistically significant. This suggests that, controlling for other covariates, crop diversification reduces agriculture's exposure to downside risk. Further, a negative and significant coefficient on the interaction of diversity with the time variable suggests a slowdown in the risk benefits of crop diversification.

Table 19.4 also shows that the marginal effects of temperature are positive for both the variance and skewness, and the marginal effects for rainfall are positive for skewness. This is possible as, over time, farmers learn to manage the adverse effects of climatic shocks. They may adopt stress-tolerant crop varieties, improve soil and water conservation practices, etc. The marginal effect of irrigation is positive but not significant. Infrastructure variables are found to positively skew productivity distribution, which means that infrastructure and financial institutions add to farmers' capacity to manage production risks. In contrast, farm size has a negative and significant sign, which means that larger farms have fewer requirements for managing risk.

The coefficient on the diversity index as well as its marginal effect derived from the variance function are the opposite of those estimated from the skewness function. This reinforces the risk-reducing role of crop diversification.

Contrary to expectations, climate shocks are observed to positively skew the productivity distribution. This is possible. As indicated earlier, farmers acquire more knowledge and information on novel technologies and practices that enhance their adaptive capacity. Infrastructure and financial institutions also positively skew the productivity distribution. Farm size, however, has a negative and significant effect on skewness, which confirms that smaller farms need more risk management. Diversity is found to reduce exposure to downside risk. However, the negative and significant coefficient on its interaction with time trend indicates a slowdown in diversity's risk benefits.

19.4.4 Spatial heterogeneity in diversification and its effects

There is considerable heterogeneity in India's agro-climatic conditions, giving rise to spatial differences in risk characteristics, cropping patterns and crop yields. From an agricultural perspective, India has four broad agro-climatic zones: humid, arid, semi-arid temperate and semi-arid tropics (Fig. 19.3). Their main characteristics are presented in Table 19.5. The growing period length exceeds 270 days in the humid zone, ranges between 70–180 days in the semi-arid temperate and semi-arid tropical zones, and is less than 75 days in the arid zone. The mean annual rainfall is the highest in the

Table 19.3. Regression estimates of the mean output function. (Source: Based on ICRISAT database.)

Variables	Specification I		Specification II		Specification III	
Regression coefficients						
Human labor	0.0336***	(0.0094)	0.0355***	(0.0092)	0.0337***	(0.0092)
Animal labor	–0.0200***	(0.0040)	–0.0207***	(0.0040)	–0.0197***	(0.0039)
Irrigation	1.0026***	(0.0308)	–0.1882	(0.1975)	0.2616	(0.2062)
Soil quality	–0.0463*	(0.0182)	–0.0203	(0.0190)	1.1410***	(0.1441)
Temperature shocks	–0.0191***	(0.0020)	–0.0187***	(0.0020)	–0.0175	(0.0254)
Rainfall shocks	0.0537***	(0.0018)	0.0538***	(0.0018)	0.1695***	(0.0227)
Farm size	–0.1190***	(0.0133)	–0.1419***	(0.0147)	–0.1359***	(0.0140)
Road density	0.0233***	(0.0058)	0.0232***	(0.0059)	0.0263***	(0.0059)
Market density	0.0686***	(0.0092)	0.0618***	(0.0091)	0.0688***	(0.0087)
Bank density	0.0470***	(0.0071)	0.0523***	(0.0075)	0.0471***	(0.0075)
Diversity			0.1613	(0.5752)	1.3237*	(0.5862)
Diversity × diversity			0.2713	(0.4210)	0.1332	(0.4142)
Diversity × time			–0.0277***	(0.0061)	–0.0246***	(0.0057)
Diversity × irrigation			1.7068***	(0.2816)	1.0484***	(0.2968)
Diversity × soil quality					–1.6366***	(0.1961)
Diversity × rainfall shocks					–0.1614***	(0.0311)
Diversity × temperature shocks					–0.0014	(0.0350)
Constant	8.3332***	(0.0516)	8.1014***	(0.2292)	7.3197***	(0.2599)
Time trend	Yes		Yes		Yes	
Marginal effects						
Temperature shocks	–0.0191***	(0.0020)	–0.0187***	(0.0020)	–0.0185***	(0.0021)
Rainfall shocks	0.0537***	(0.0018)	0.0538***	(0.0018)	0.0547***	(0.0021)
Diversity	–	–	0.5164***	(0.1022)	0.5492***	(0.1111)
Irrigation	1.0026***	(0.0308)	1.0258***	(0.0325)	1.0073***	(0.0315)
Number of observations	12,248		12,248		12,248	
Wald Chi-squared	7,197.39		6,921.46		7,235.43	

Notes: Heteroskedastic panel corrected standard errors (PCSEs) are in parentheses.
***, ** and * denote significance at 1%, 5% and 10%, respectively.

Table 19.4. Regression estimates of higher-order moments of the output function. (Source: Based on ICRISAT database.)

Variables	Variance		Skewness	
Regression coefficients				
Human labor	0.0023	(0.0012)	−0.0009	(0.0006)
Animal labor	0.0011	(0.0008)	−0.0001	(0.0003)
Irrigation	−0.2542***	(0.0502)	0.0679**	(0.0233)
Soil quality	−0.2451***	(0.0368)	0.0616***	(0.0166)
Temperature shocks	−0.0053	(0.0053)	0.0041	(0.0023)
Rainfall shocks	−0.0290***	(0.0048)	0.0089***	(0.0020)
Farm size	0.0204***	(0.0029)	−0.0056***	(0.0015)
Road density	−0.0008	(0.0008)	0.0009***	(0.0002)
Market density	−0.0062**	(0.0022)	0.0028*	(0.0013)
Bank density	0.0032*	(0.0016)	0.0006	(0.0007)
Diversity	−0.5410***	(0.1418)	0.2592***	(0.0665)
Diversity × diversity	−0.0061	(0.0607)	−0.0649**	(0.0227)
Diversity × time	0.0045**	(0.0015)	−0.0024***	(0.0007)
Diversity × irrigation	0.3306***	(0.0728)	−0.0936**	(0.0339)
Diversity × soil quality	0.3063***	(0.0498)	−0.0777***	(0.0220)
Diversity × rainfall shocks	0.0359***	(0.0065)	−0.0117***	(0.0027)
Diversity × temperature shocks	0.0086	(0.0072)	−0.0057	(0.0031)
Constant	0.4394***	(0.0857)	−0.1700***	(0.0422)
Time trend	Yes		Yes	
Marginal effects				
Temperature shocks	0.0009**	(0.0003)	0.0001	(0.0001)
Rainfall shocks	−0.0035***	(0.0003)	0.0005***	(0.0001)
Diversity	−0.1675***	(0.0233)	0.0357***	(0.0097)
Irrigation	−0.0190***	(0.0055)	0.0013	(0.0020)
Number of observations	12,248		12,248	
Wald Chi-squared	522.58***		102.22***	

Notes: Heteroskedastic panel corrected standard errors (PCSEs) are in parentheses.
*$p < 0.05$, **$p < 0.01$, ***$p < 0.001$.

humid zone and the lowest in the arid zone. Rainfall is highly erratic in the arid zone (CV = 62%). The semi-arid temperate zone also receives less rainfall than the humid and semi-arid tropics zones, but a considerable proportion of its agricultural land is equipped with irrigation.

The last row of Table 19.5 presents estimates of the Simpson index of diversity. The diversity indices for the semi-arid temperate and semi-arid tropical zones are similar, but they are higher than for other zones. The diversity index is the lowest for the humid zone. Rice, with a 40% share of the total cropped area, is a major crop in this zone (Table A2 in the appendix) due to higher precipitation. Wheat, pulses, oilseeds and coarse cereals are other important crops, each occupying between 7% and 9% of the total cropped area. In the semi-arid temperate zone, a higher level of irrigation favors wheat and rice cultivation, which respectively share about one-third and one-fifth of the total cropped area. Coarse cereals and pulses also are cultivated on a sizable area in this zone. Coarse grains, oilseeds and pulses dominate the cropping patterns in the arid and semi-arid tropics.

The differences in the crop diversification across agro-climatic zones are likely to differentiate these zones in their agricultural productivity and vulnerability to climate risks. To capture these, we estimate full moment functions separately for each agro-climatic zone, and the results are presented in Table A3 (in the appendix). For the sake of brevity, we discuss only the marginal effects or elasticities of some important variables (Table 19.6). Agricultural productivity responds positively and significantly to the diversity index in all the agro-climatic zones (except the humid zone). The response, however, is stronger in the

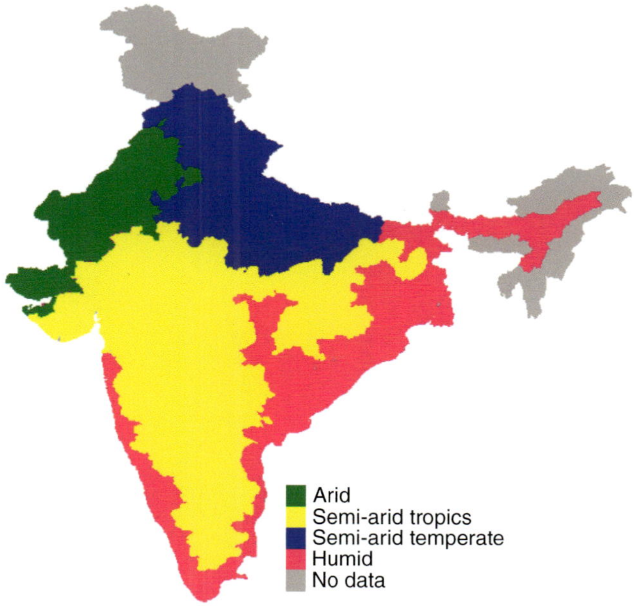

Source: Authors' own work.

Fig. 19.3. Agro-climatic zones of India.

Table 19.5. Means and standard deviations of key characteristics. (Source: Based on ICRISAT database.)

	Agro-climatic zone							
	Humid		Semi-arid temperate		Semi-arid tropics		Arid	
	Mean	SD	Mean	SD	Mean	SD	Mean	SD
Farm productivity (Rupees/ha)	16,628	9,775	17,998	10,490	11,816	8,392	11,869	10,038
Length of growing period (days)	>270		70–180		70–180		<75	
Temperature (degree Celsius)	24.61	2.51	24.56	1.72	26.15	1.23	24.92	1.86
Rainfall (mm)	1,700	768	894	428	911	352	464	289
Irrigated area (%)	0.29	0.23	0.62	0.27	0.28	0.21	0.36	0.34
Simpson index	0.41	0.25	0.69	0.11	0.68	0.16	0.57	0.23

arid zone. Rainfall shocks positively influence the cropping system productivity, but temperature shocks have a negative influence. Irrigation has an overall positive effect on the system productivity.

The effect of crop diversification on risk exposure also differs across agro-climatic zones. Diversification reduces the probability of output loss due to risks and uncertainty in the arid zone, but it leads to an increase in the probability of output loss in the humid zone. Diversification's effect in the semi-arid temperate and semi-arid tropical zones is not statistically significant. Rainfall shocks lead to an increase in the probability of output loss in the humid and semi-arid temperate zones but not in the arid and semi-arid tropical zones. Temperature shocks do not influence the probability of output loss, except in the semi-arid tropical zone. Irrigation plays a significant role in reducing risk in the semi-arid temperate and semi-arid tropical zones. Its effect in the humid zone is opposite. Most districts in this zone lie along the west and east coasts and in the north-eastern region, which are more prone to

Table 19.6. Marginal effects of diversity in different agro-climatic zones. (Source: Based on ICRISAT database.)

	Humid		Semi-arid temperate		Semi-arid tropics		Arid	
Mean								
Irrigation	0.5830***	(0.0555)	0.8778***	(0.0440)	1.4468***	(0.0561)	1.3273***	(0.1281)
Temperature shocks	−0.0112**	(0.0043)	−0.0099***	(0.0029)	−0.0337***	(0.0042)	−0.0283	(0.0154)
Rainfall shocks	0.0389***	(0.0040)	0.0425***	(0.0025)	0.0578***	(0.0035)	0.1655***	(0.0134)
Diversity	0.0912	(0.1518)	0.7491**	(0.2284)	0.3706*	(0.1820)	2.5731***	(0.4486)
Variance								
Irrigation	0.0074	(0.0047)	0.0074**	(0.0026)	−0.0306***	(0.0065)	0.0356	(0.0470)
Temperature shocks	0.0003	(0.0004)	0.0007***	(0.0002)	0.0018***	(0.0004)	0.0033	(0.0048)
Rainfall shocks	−0.0025***	(0.0004)	−0.0017***	(0.0002)	−0.0019***	(0.0004)	−0.0266***	(0.0041)
Diversity	−0.0357**	(0.0137)	0.0156	(0.0134)	−0.0273	(0.0181)	−0.5018**	(0.1548)
Skewness								
Irrigation	−0.0034***	(0.0006)	0.0009***	(0.0002)	0.0080***	(0.0014)	−0.0175	(0.0234)
Temperature shocks	0.0000	(0.0000)	0.0000	(0.0000)	−0.0004***	(0.0001)	0.0011	(0.0022)
Rainfall shocks	−0.0002***	(0.0000)	−0.0001***	(0.0000)	0.0002**	(0.0001)	0.0084***	(0.0019)
Diversity	−0.0037*	(0.0018)	−0.0007	(0.0012)	0.0006	(0.0039)	0.1314*	(0.0762)

Notes: Heteroskedastic panels corrected standard errors (PCSEs) are in parentheses. *p < 0.05, **p < 0.01, ***p < 0.001.

cyclones and floods. These differences in the productivity and risk benefits of cropping system diversification suggest the need to design regionally differentiated strategies to harness the multifunctionality of crop diversification.

19.4.5 Robustness test

To test the robustness of the results presented in Table 19.3, we estimate the Feasible Generalized Least Squares (FGLS) and Random Effects with Driscoll–Kraay standard errors. In FGLS, the disturbances are assumed to be independent and identically distributed instead of being heteroskedastic and contemporaneously correlated across panels. Beck and Katz (1995) have shown that Prais–Winsten estimates with PCSEs are closer to the nominal values compared to the full FGLS estimates. The results of these estimators are presented in Table 19.7. The diversity index is significant and positively associated with productivity in both models. The coefficients on its interactions and most of the control variables are similar in magnitude and direction to those presented in Table 19.3.

19.5 Conclusions

Employing a moment-based specification of the stochastic production function to the panel data on 311 districts of India for 1966/67 to 2015/16, this chapter has assessed the effects of crop diversification on agricultural productivity and downside risk exposure. This analysis has three important findings. One, crop diversification enhances agriculture's productivity and resilience

Table 19.7. Robustness tests. (Source: Based on ICRISAT database.)

	Feasible GLS		Random effects with Driscoll–Kraay	
Human labor	−0.0223**	(0.0083)	−0.0275	(0.0419)
Animal labor	−0.0099***	(0.0026)	−0.0212**	(0.0072)
Irrigation	0.0544	(0.1770)	1.6185***	(0.4023)
Soil quality	0.5429***	(0.1003)	1.5223***	(0.2638)
Temperature shocks	0.0128	(0.0194)	0.0151	(0.0562)
Rainfall shocks	0.0718***	(0.0169)	0.1064*	(0.0447)
Farm size	−0.0397***	(0.0102)	−0.0732*	(0.0326)
Road density	0.0378***	(0.0056)	0.0154	(0.0116)
Market density	0.0565***	(0.0060)	−0.0017	(0.0102)
Bank density	0.0586***	(0.0059)	0.0085	(0.0129)
Diversity	0.1743	(0.5194)	1.2905	(0.6442)
Diversity × diversity	0.1930	(0.3694)	1.2003**	(0.4208)
Diversity × time	−0.0075	(0.0048)	0.0024	(0.0073)
Diversity × irrigation	1.0694***	(0.2509)	−1.3295*	(0.5782)
Diversity × soil quality	−0.9782***	(0.1376)	−2.4218***	(0.3025)
Diversity × rainfall shocks	−0.0347	(0.0270)	−0.0516	(0.0746)
Diversity × temperature shocks	−0.0448	(0.0235)	−0.068	(0.0601)
Constant	8.0787***	(0.2044)	7.2409***	(0.3445)
Time trend	Yes		Yes	
Marginal effects				
Irrigation	0.8150***	(0.0228)	0.6728***	(0.0722)
Soil quality	0.1528***	(0.0141)	−0.2001*	(0.0900)
Temperature shocks	0.0118***	(0.0017)	−0.0216***	(0.0061)
Rainfall shocks	0.0399***	(0.0015)	0.0580***	(0.0056)
Diversity	0.2242**	(0.0782)	1.4252***	(0.1365)
Number of observations	12,248		12,248	
Wald Chi-squared	8,252.37***		10,673.91***	

Notes: Standard errors are in parentheses.
*p < 0.05, **p < 0.01, ***p < 0.001.

to climate risks. Two, its productivity benefits outweigh the risk benefits, and both have weakened over time. Three, the benefits of diversification are significantly influenced by agro-climatic conditions; the productivity and risk benefits of crop diversification are more prominent in the ecologically fragile arid environments than in semi-arid environments.

Given the predictions that extreme climatic events will be increasingly frequent in plausible future climate scenarios (IPCC, 2018), these findings serve as important feedback to farming communities and policymakers. At the farm level, the decisions on crop choices and acreage allocations happen under uncertainty without prior knowledge of their expected outcomes. Our findings on the effects of crop diversification on productivity and resilience are robust enough to inspire farmers to adopt it as an integral component of their cropping systems. At the policy level, these findings suggest the need for contemporary policies encompassing infrastructures, institutions and investments to exploit the multifunctionality of crop diversification for efficient and climate-resilient growth in agriculture.

References

Ali, W., Abdulai, A. and Mishra, A.K. (2020) Recent advances in the analyses of demand for agricultural insurance in developing and emerging countries. *Annual Review of Resource Economics* 12, 411–430.

Altieri, M.A. and Nicholls, C.I. (2017) The adaptation and mitigation potential of traditional agriculture in a changing climate. *Climatic Change* 140, 35–45.

Antle, J.M. (1983) Testing the stochastic structure of production: a flexible moment-based approach. *Journal of Business & Economic Statistics* 1, 192–201.

Antle, J.M. (1987) Econometric estimation of producers' risk attitudes. *American Journal of Agricultural Economics* 69, 509–522.

Beck, N. and Katz, J.N. (1995) What to do (and not to do) with time-series cross-section data. *American Political Science Review* 89, 634–647.

Binswanger-Mkhize, H.P. (2012) Is there too much hype about index-based agricultural insurance? *Journal of Development Studies* 48, 187–200.

Birthal, P.S. and Hazrana, J. (2019) Crop diversification and resilience of agriculture to climatic shocks: evidence from India. *Agricultural Systems* 173, 345–354.

Birthal, P.S. and Pandey, G. (2020) Structural transformation and sources of growth in South Asian agriculture. In: Mishra, A.K., Kumar, A. and Joshi P.K. (eds) *Transforming Agriculture in South Asia: The Role of Value Chains and Contract Farming*. Routledge, New York.

Birthal, P.S., Joshi, P.K., Negi, D.S. and Agrawal, S. (2014) Changing sources of growth in Indian agriculture: implications for regional priorities for accelerating agricultural growth. IFPRI Discussion Paper 1325. International Food Policy Research Institute (IFPRI), Washington, DC.

Birthal, P.S., Joshi, P.K., Roy, D. and Pandey, G. (2019a) Transformation and sources of growth in Southeast Asian agriculture. IFPRI Discussion Paper 1834, International Food Policy Research Institute (IFPRI), Washington, DC.

Birthal, P.S., Negi, D.S. and Hazrana, J. (2019b) Trade-off between risk and returns in farmers' choice of crops? Evidence from India. *Agricultural Economics Research Review* 32, 11–23.

Birthal, P.S., Hazrana, J. and Negi, D.S. (2021) Impacts of climatic hazards on agricultural growth in India. *Climate and Development*, 13, 895–908. DOI: 10.1080/17565529.2020.1867045.

Chand, R., Prasanna, L. and Singh, A. (2011) Farm size and productivity: understanding the strengths of smallholders and improving their livelihoods. *Economic and Political Weekly* 46, 5–11.

Chhatre, A., Devalkar, S. and Seshadri, S. (2016) Crop diversification and risk management in Indian agriculture. *Decision* 43, 167–179.

Dell, M., Jones, B.F. and Olken, B.A. (2012) Temperature shocks and economic growth: evidence from the last half century. *American Economic Journal: Macroeconomics* 4, 66–95.

Di Falco, S. and Chavas, J.P. (2008) Rainfall shocks, resilience and the dynamic effects of crop biodiversity on the productivity of the agroecosystem. *Land Economics* 84, 83–96.

Di Falco, S. and Chavas, J.P. (2009) On crop biodiversity, risk exposure and food security in the highlands of Ethiopia. *American Journal of Agricultural Economics* 91, 599–611.

Fuglie, K., Gautam, M., Goyal, A. and Maloney, W.F. (2020) *Harvesting Prosperity: Technology and Productivity Growth in Agriculture*. World Bank, Washington, DC.

Gautam, M. (2015) Agricultural subsidies: resurging interest in a perennial debate. *Indian Journal of Agricultural Economics* 70, 83–105.

GoI (Government of India) (2020) *Agricultural Statistics at a Glance*. Directorate of Economics and Statistics, Ministry of Agriculture and Farmers Welfare, New Delhi.

Hallegatte, S., Bangalore, M., Bonzanigo, L., Fay, M., Kane, T., Narloch, U. *et al.* (2016) *Shock Waves: Managing the Impacts of Climate Change on Poverty*. World Bank, Washington, DC.

Huang, J., Wang, Y. and Wang J. (2015) Farmers' adaptation to extreme weather events through farm management and its impacts on the mean and risk of rice yield in China. *American Journal of Agricultural Economics* 97, 602–617.

ICRISAT (undated) District Level Data for India (DLD). International Crops Research Institute for the Semi-Arid Tropics, Patancheru, India. Available at: http://data.icrisat.org/dld/ (accessed May 16, 2023).

IPCC (2018) *Global Warming of 1.5°C: An IPCC Special Report*. Intergovernmental Panel on Climate Change, Geneva, Switzerland. Available at: https://www.ipcc.ch/site/assets/uploads/sites/2/2019/06/SR15_Full_Report_High_Res.pdf (accessed May 16, 2023).

Jayne, T.S. and Govereh, J. (2003) Cash cropping and food crop productivity: synergies or trade-offs? *Agricultural Economics* 28, 39–50.

Jodha, N.S., Singh, N.P. and Bantilan, M.C.S. (2012) Enhancing farmers' adaptation to climate change in arid and semi-arid agriculture of India: evidences from indigenous practices. Working Paper Series No. 32, International Crops Research Institute for the Semi-Arid Tropics, Patancheru, India.

Joshi, P.K., Gulati, A., Birthal, P.S. and Tewari, L. (2004) Agriculture diversification in South Asia. *Economic and Political Weekly* 39, 2457–2467.

Loayza, V.N., Olaberria, E. and Rigolini, J. (2012) Natural disasters and growth: going beyond the averages. *World Development* 40, 1317–1336.

Loreau, M. and Hector, A. (2001) Partitioning selection and complementarity in biodiversity experiments. *Nature* 412, 72–76.

Lybbert, T.J. and Bell, A. (2010) Why drought tolerance is not the new Bt. *Nature Biotechnology* 28, 553–554.

Menezes, C., Geiss, C. and Tressler, J. (1980) Increasing downside risk. *American Economic Review* 70, 921–932.

Ortiz-Bobea, A., Ault, T.R., Carrillo, C.M., Chambers, R.G. and Lobell, D.B. (2021) Anthropogenic climate change has slowed global agricultural productivity growth. *Nature Climate Change* 11, 306–312.

Renard, D. and Tilman, D. (2019) National food production stabilized by crop diversity. *Nature* 571, 257–260.

Schlenker, W. and Roberts, M.J. (2006) Nonlinear effects of weather on corn yields. *Review of Agricultural Economics* 28, 391–398.

Tack, J., Harri, A. and Coble, K. (2012) More than mean effects: modeling the effect of climate on the higher order moments of crop yields. *American Journal of Agriculture Economics* 94, 1037–1054.

Tamburini, G., Bommarco, R., Wanger, T., Kremen, C., Heijden, M., Liebman, M. *et al.* (2020) Agricultural diversification promotes multiple ecosystem services without compromising yield. *Science Advances*, 6, eaba1715. DOI: 10.1126/sciadv.aba1715.

van Zonneveld, M., Turmel, M-S. and Hellin, J. (2020) Decision-making to diversify farm system for climate change adaptation. *Frontiers in Sustainable Food Systems* 4, 32. DOI: 10.3389/fsufs.2020.00032.

Wooldridge, J. (2002) *Econometric Analysis of Cross Section and Panel Data*. MIT Press, Cambridge, Massachusetts.

Zhang, H. and Antle, J.M. (2018) Weather, climate and production risk. IRENE Working Paper 18-01, Institute of Economic Research, University of Neuchatel, Switzerland.

Appendix

Table A1. Fisher-type panel unit root tests. (Source: Authors' own work.)

Variable	H0: All panels contain unit roots H1: At least one panel is stationary (lags 2)							
	Inverse Chi-squared (P)		Inverse normal (Z)		Inverse logit L		Modified inverse Chi-squared (Pm)	
Productivity	2959.36	(0.00)	−40.23	(0.00)	−45.87	(0.00)	66.27	(0.00)
Human labor	1562.69	(0.00)	−18.73	(0.00)	−20.27	(0.00)	26.67	(0.00)
Animal labor	6368.94	(0.00)	−69.91	(0.00)	−99.59	(0.00)	162.94	(0.00)
Irrigation	2028.22	(0.00)	−26.80	(0.00)	−28.79	(0.00)	39.87	(0.00)
Temperature	4469.37	(0.00)	−55.29	(0.00)	−69.86	(0.00)	109.08	(0.00)
Rainfall	4937.59	(0.00)	−58.90	(0.00)	−77.19	(0.00)	122.36	(0.00)
Farm size	1724.25	(0.00)	−20.64	(0.00)	−22.25	(0.00)	31.25	(0.00)
Road density	2448.79	(0.00)	−28.68	(0.00)	−34.44	(0.00)	51.79	(0.00)
Market density	2321.31	(0.00)	−32.35	(0.00)	−35.43	(0.00)	48.18	(0.00)
Bank density	4394.32	(0.00)	−48.89	(0.00)	−68.10	(0.00)	106.95	(0.00)
Diversity	2660.92	(0.00)	−34.39	(0.00)	−40.05	(0.00)	57.81	(0.00)

Notes: Figures in parentheses are p-values. ***, ** and * respectively indicate significance at 1%, 5% and 10%.

Table A2. Proportion of crops in the total cropped area. (Source: Authors' own work.)

	Humid		Semi-arid temperate		Semi-arid tropics		Arid	
	Mean	Std. dev.	Mean	Std. dev.	Mean	Std. dev.	Mean	Std. dev.
Rice	0.4013	(0.2358)	0.1896	(0.1514)	0.1237	(0.1769)	0.0592	(0.1001)
Wheat	0.0874	(0.1328)	0.3155	(0.1050)	0.0907	(0.1019)	0.1437	(0.1679)
Coarse cereals	0.0878	(0.1187)	0.145	(0.1172)	0.2431	(0.1682)	0.2682	(0.2021)
Pulses	0.073	(0.0879)	0.1287	(0.1222)	0.1498	(0.0937)	0.1439	(0.1191)
Oilseeds	0.0821	(0.0964)	0.066	(0.0909)	0.1752	(0.1515)	0.1074	(0.1446)
Sugarcane	0.0167	(0.0386)	0.0434	(0.0812)	0.0136	(0.0285)	0.0048	(0.0117)
Cotton	0.0031	(0.0120)	0.0032	(0.0117)	0.0764	(0.1176)	0.0593	(0.0804)
Others	0.0907	(0.1018)	0.0707	(0.0668)	0.0605	(0.0614)	0.1434	(0.1249)

Table A3. Regression estimates of the moment functions. (Source: Authors' own work.)

	Humid		Semi-arid temperate		Semi-arid tropics		Arid	
Mean								
Human labor	−0.0487**	(0.0161)	0.0191	(0.0178)	0.0441*	(0.0208)	0.0969*	(0.0407)
Animal labor	−0.0310**	(0.0103)	−0.0065	(0.0035)	−0.0554***	(0.0090)	−0.0207	(0.0134)
Irrigation	−0.1842	(0.3683)	0.3862	(0.4717)	−1.7207***	(0.4632)	6.8432***	(0.8604)
Soil quality	0.3398*	(0.1716)	0.2725	(0.3620)	0.7179**	(0.2612)	0.0000	(0.0000)
Rainfall shocks	0.0051	(0.0342)	−0.0025	(0.0413)	0.2742***	(0.0424)	1.2280***	(0.1548)
Temperature shocks	−0.0097	(0.0369)	0.0729	(0.0510)	−0.2568***	(0.0548)	−0.2237	(0.1749)
Farm size	0.0162	(0.0269)	0.0816***	(0.0176)	0.0007	(0.0289)	−0.3190***	(0.0561)
Road density	−0.0464***	(0.0140)	0.0558**	(0.0172)	0.0309***	(0.0079)	0.0722	(0.0575)
Market density	0.0682***	(0.0133)	0.0881***	(0.0100)	0.0028	(0.0134)	0.1315**	(0.0451)
Bank density	0.1058***	(0.0158)	0.0872***	(0.0112)	0.004	(0.0121)	−0.0038	(0.0378)
Diversity	−0.0356	(0.7709)	4.6429	(2.5238)	4.7222**	(1.5267)	0.0000	(0.0000)
Diversity × diversity	0.436	(0.6096)	−2.3327	(1.7722)	−2.1016*	(1.0252)	7.3735**	(2.6770)
Diversity × time	−0.013	(0.0094)	−0.0372**	(0.0140)	−0.0663***	(0.0111)	−0.0354	(0.0330)
Diversity × soil quality	−0.8859***	(0.2628)	−0.3859	(0.4896)	−1.1550***	(0.3386)	0.0000	(0.0000)
Diversity × irrigation	1.1663*	(0.5700)	0.6943	(0.6728)	4.2496***	(0.6317)	−7.8497***	(1.2062)
Diversity × rainfall shocks	0.0513	(0.0508)	0.0636	(0.0587)	−0.2904***	(0.0564)	−1.5120***	(0.2127)
Diversity × temperature shocks	−0.0022	(0.0552)	−0.1169	(0.0726)	0.2993***	(0.0724)	0.2781	(0.2387)
Constant	8.1127***	(0.3035)	5.9253***	(0.9264)	6.2302***	(0.6081)	0.0000	(0.0000)
Time trend	Yes		Yes		Yes		Yes	
Number of observations	2,953		3,340		5,165		790	
Wald Chi-squared	1,086.26***		2,939.01***		3,286.56***		313,428.02***	
Variance								
Human labor	−0.0008	(0.0012)	0.0035***	(0.0009)	0.0008	(0.0024)	0.0175	(0.0130)
Animal labor	0.0024*	(0.0011)	0.0003	(0.0002)	0.0005	(0.0011)	0.0011	(0.0041)
Irrigation	−0.1543***	(0.0319)	−0.0244	(0.0262)	0.0565	(0.0432)	0.1743	(0.2338)
Soil quality	−0.0717***	(0.0158)	−0.0043	(0.0199)	−0.0638*	(0.0272)	0.0000	(0.0000)
Rainfall shocks	−0.0006	(0.0035)	−0.0169***	(0.0028)	0.0076	(0.0044)	−0.4026***	(0.0576)
Temperature shocks	0.0018	(0.0037)	0.0061	(0.0033)	0.0053	(0.0055)	0.033	(0.0635)
Farm size	0.0164***	(0.0022)	0.0052***	(0.0010)	−0.0021	(0.0033)	0.0643***	(0.0145)
Road density	0.0007	(0.0011)	0.0020*	(0.0010)	−0.0002	(0.0006)	−0.0071	(0.0151)
Market density	−0.0009	(0.0010)	0.0005	(0.0006)	0.0005	(0.0014)	−0.0221	(0.0177)
Bank density	0.0031**	(0.0012)	0.0023***	(0.0006)	0.0084***	(0.0015)	0.0106	(0.0114)
Diversity	0.0462	(0.0816)	−0.2868*	(0.1369)	−0.1746	(0.1334)	0.0000	(0.0000)

Diversity × diversity	−0.1176	(0.0658)	0.2411*	(0.0995)	0.0492	(0.0903)	2.7001**	(0.8542)
Diversity × time	−0.0021**	(0.0008)	−0.0026***	(0.0008)	0.0027*	(0.0012)	0.0331*	(0.0133)
Diversity × soil quality	0.0893***	(0.0233)	0.0079	(0.0268)	0.0592	(0.0361)	0.0000	(0.0000)
Diversity × irrigation	0.2457***	(0.0490)	0.0449	(0.0377)	−0.1169	(0.0605)	−0.1975	(0.3395)
Diversity × rainfall shocks	−0.003	(0.0051)	0.0215***	(0.0040)	−0.0127*	(0.0059)	0.5350***	(0.0784)
Diversity × temperature shocks	−0.0023	(0.0055)	−0.0077	(0.0047)	−0.0047	(0.0074)	−0.0423	(0.0856)
Constant	0.0278	(0.0292)	0.083	(0.0484)	0.1149*	(0.0550)	0.0000	(0.0000)
Time trend	Yes		Yes		Yes		Yes	
Number of observations	2,953		3,340		5,165		790	
Wald Chi-squared	170.94***		373.78***		220.58***		159.97***	
Skewness								
Human labor	−0.0005**	(0.0002)	0.0002*	(0.0001)	−0.0002	(0.0006)	−0.0106	(0.0061)
Animal labor	−0.0001	(0.0001)	0.0001**	0.0000	(0.0000)	(0.0002)	0.0001	(0.0018)
Irrigation	−0.0154***	(0.0043)	0.0031	(0.0022)	0.0018	(0.0083)	−0.0135	(0.1081)
Soil quality	−0.0142***	(0.0022)	−0.0008	(0.0018)	0.0000	(0.0063)	0.0000	
Rainfall shocks	−0.0013**	(0.0004)	−0.0011***	(0.0003)	0.0000	(0.0007)	0.1607***	(0.0275)
Temperature shocks	−0.0002	(0.0005)	0.0009**	(0.0003)	−0.0039***	(0.0010)	0.014	(0.0301)
Farm size	0.0018***	(0.0003)	0.0007***	(0.0001)	0.0012	(0.0008)	−0.0195**	(0.0062)
Road density	−0.0003*	(0.0001)	0.0000	(0.0001)	0.0001	(0.0001)	0.0045	(0.0063)
Market density	−0.0001	(0.0001)	0.0000	(0.0000)	−0.0006	(0.0003)	0.0069	(0.0089)
Bank density	0.0006***	(0.0001)	0.0002***	(0.0001)	−0.0013***	(0.0003)	−0.0023	(0.0050)
Diversity	−0.0106	(0.0105)	−0.0133	(0.0116)	0.0266	(0.0273)	1.6983**	(0.6499)
Diversity × diversity	−0.0046	(0.0087)	0.0105	(0.0086)	−0.0151	(0.0184)	−0.9228*	(0.3929)
Diversity × time	−0.0001	(0.0001)	0.0000	(0.0001)	−0.0003	(0.0002)	−0.0103	(0.0067)
Diversity × soil quality	0.0177***	(0.0031)	0.0015	(0.0024)	0.0031	(0.0084)	0.0000	(0.0000)
Diversity × irrigation	0.0183**	(0.0066)	−0.0032	(0.0032)	0.0083	(0.0117)	−0.0056	(0.1595)
Diversity × rainfall shocks	0.0016*	(0.0006)	0.0014***	(0.0004)	0.0003	(0.0010)	−0.2167***	(0.0374)
Diversity × temperature shocks	0.0003	(0.0007)	−0.0013**	(0.0004)	0.0047***	(0.0013)	−0.0183	(0.0405)
Constant	0.0092*	(0.0037)	0.0032	(0.0040)	−0.0109	(0.0115)	0.0000	(0.0000)
Time trend	Yes		Yes		Yes		Yes	
Number of observations	2,953		3,340		5,165		790	
Wald Chi-squared	123.44***		354.26***		168.59***		59.62***	

Notes: Heteroskedastic panel-corrected standard errors (PCSEs) are in parentheses. *p < 0.05, **p < 0.01, ***p < 0.001.

20 Extreme Weather Events and Credit Demand

Puneet Vatsa,[1]* Wanglin Ma,[1] Hui Mao[2] and Xiaoheng Zhang[3]
[1]Lincoln University, Christchurch, New Zealand; [2]Shaanxi Normal University, Xi'an, China; [3]Nanjing University of Aeronautics and Astronautics, Nanjing, China

Abstract

Anthropogenic climate change has increased the intensity and frequency of extreme weather events. The agricultural sector is especially vulnerable, with smallholder farmers often lacking the wherewithal to cope with weather shocks. Thus, providing farmers access to credit is vital for building financial resilience in rural communities. Accordingly, this chapter analyzes the linkages between extreme weather events and apple farmers' demand for credit in six counties of China's Shaanxi province using a double-hurdle model. The results show that extreme weather events are associated with both an increase in the likelihood of farmers borrowing and in the amount borrowed. However, the demand for credit in response to different categories of weather events varies. Droughts do not affect credit demand, but frosts are associated with a rise in the amount borrowed and the probability of farmers borrowing. We find differences in credit demand across the six counties. Furthermore, older farmers are less likely to borrow than younger ones, and self-reported health status is negatively associated with the likelihood of borrowing – farmers who perceive themselves as healthy are less likely to borrow than those who do not.

20.1 Introduction

Greenhouse gases (GHG) emitted as a consequence of human activities have been responsible for approximately 1.1°C of global warming since 1850–1900 (IPCC, 2021). If these emissions remain unabated, global mean temperatures will continue to rise during the 21st century. The critical questions are by how much and how quickly they will do so.

The intensity and frequency of extreme weather events have risen due to anthropogenic climate change (Stott, 2016; Doherty *et al.*, 2021; Elahi *et al.*, 2021). A warmer climate raises the likelihood of extremely hot days and nights, leading to more frequent heavy snowfall and rainfall; it also causes more evaporation, thus intensifying droughts (Fischer and Knutti, 2015; Mukherjee *et al.*, 2018; Fontes *et al.*, 2021). Climate change is more apparent in some regions than in others. Temperatures in China, for example, have risen faster than the average temperature globally. Heat waves and severe rainfall have also occurred more frequently in the country (Zhao, 2020). The record-breaking heat and precipitation in southern China during 2020 can largely be attributed to climate change (Ye and Qian, 2021). Severe rainfall raises the risk of flooding, which has caused significant damage in China, killing more than 280,000

*E-mail: puneet.vatsa@lincoln.ac.nz

 Managing Risk in Agriculture: A Development Perspective (eds A.K. Mishra, S.C. Kumbhakar and G. Lien).

DOI: 10.1079/9781800622289.0020

people between 1950 and 2018. By the end of the 21st century, floods may affect 40 million people in the country, causing 110 billion euros of damage annually (Alfieri *et al.*, 2017; Guo *et al.*, 2020). These estimates are associated with an increase of 4°C in the mean global temperature above pre-industrial levels (the monetary damage is expressed in 2010 euros, adjusted for purchasing power parity).

Smallholder farmers are especially vulnerable to climate shocks (Harvey *et al.*, 2014, 2018; Fontes *et al.*, 2021). Volatile weather and adverse weather events increase production risks and reduce yields and farm profitability (Huang *et al.*, 2015). A comprehensive suite of risk management strategies should be developed and made available to farmers to help them adapt to and cope with such shocks. Technology, farm management, land-use policies and practices, agricultural regulations, insurance products, awareness and educational initiatives, financial incentives, and credit markets all have a role to play. Understanding how farmers interact with these aspects is crucial to developing practical solutions. This chapter investigates one of these aspects: credit markets. It is devoted to analyzing the association between extreme weather events and farmers' demand for credit.

A considerable body of research has investigated how farmers react and adapt to weather shocks and perceived weather risks (Acosta *et al.*, 2021; Elahi *et al.*, 2021; Fontes *et al.*, 2021; Bora, 2022). Analysts and policymakers have turned their attention to studying how information and communication technology can help farmers optimize agricultural practices (Vaughan *et al.*, 2016; Clarkson *et al.*, 2017; Dayamba *et al.*, 2018; Partey *et al.*, 2020). For example, Partey *et al.* (2020) analyzed the use of climate information services to manage climate risk in Ghana. They found that, although men are more likely to use these services than women, both find them helpful in making farming decisions. Smallholder farmers in Mali and Senegal viewed the Participatory Integrated Climate Services for Agriculture (PICSA) approach – which integrates historical climate records, participatory decision-making, and seasonal forecasts – positively and deemed it valuable (Dayamba *et al.*, 2018).

Owning livestock is common among farmers, who often rely on credit to acquire these animals due to their potentially high costs. Thus, access to credit is crucial to owning livestock. However, whether livestock ownership helps rural households cope with climate shocks is unclear. On the one hand, Hänke and Barkmann (2017) and Acosta *et al.* (2021) confirmed that livestock portfolios and herding provide a buffer against climate shocks and crop failures. On the other hand, Kazianga and Udry (2006) reported that livestock ownership was ineffective in helping rural households smooth their consumption during a severe drought in Burkina Faso.

Abbas *et al.* (2015), Arshad *et al.* (2016), and Aftab *et al.* (2021) studied the factors determining farmers' uptake of flood insurance in Pakistan. Arshad *et al.* (2016) found that only about 30% of the farmers were open to the idea of purchasing flood insurance to mitigate the risk from flooding; Abbas *et al.* (2015) suggested that financial constraints rather than perceived flood risk determine the uptake of flood insurance. Because credit markets can help farmers secure loans, these findings highlight the role of credit markets in helping farmers engage in risk management practices. In the event of extreme weather shocks, autonomous adaptation strategies, such as grain storage, plinth elevation, and participation in communal flood preparations, even when implemented simultaneously, may not suffice (Aftab *et al.*, 2021). Access to credit and insurance would help farmers navigate financial hardships caused by climate shocks.

The use of microcredit in response to climate change has also attracted considerable attention recently (Fenton *et al.*, 2017; Di Falco and Sharma, 2018; Johnson *et al.*, 2019; Jordan, 2021). According to Jordan (2021), communities exposed to climate shocks in Bangladesh benefited only incrementally from using microcredit as a coping strategy. The efficacy of microcredit was tempered by a lack of alternative sources of credit. Aryal *et al.* (2021) showed that borrowing and using savings were among the most common strategies farmers used to cope with climate risk.

Evidently, the importance of credit and finance to farmers' abilities to cope with climate shocks is widely acknowledged. At the same time, the literature on the effects of extreme weather events on farmers' credit demand is relatively scant. Focusing on smallholder apple farmers in Shaanxi, China, we bridge this gap in the literature. To this end, we use a double-hurdle model. The analysis proceeds in two stages. First, we ask how extreme weather events affect the probability of smallholder farmers accessing credit – a probit model is employed to answer this question. Next, we ask how these events affect the amount of credit obtained – this is answered using a truncated normal estimator. The results show that extreme weather events increase the probability of accessing credit and increase the amount of credit.

The rest of the chapter is structured as follows. Section 20.2 provides a brief background of apple production in China and briefly discusses how different extreme weather events affect apple crops and the strategies farmers employ to cope with these events. Section 20.3 details the methodological framework. Data are described in Section 20.4, and the empirical results are presented in Section 20.5. Section 20.6 concludes the chapter.

20.2 Apple Production in China

China is the world's largest apple producer, accounting for approximately half of global apple production. China's apple production and harvested area have grown rapidly in the past six decades. Specifically, the harvested area increased from 0.09 million hectares in 1961 to 1.91 million hectares in 2020, and production rose from 0.17 million tonnes in 1961 to 40.5 million tonnes in 2020 (Fig. 20.1). The harvested area has grown at a rate of 5.2% annually, and output has grown at a rate of 9.6% annually.

Of all the provinces, Shaanxi accounts for the largest proportion of apples produced in China. It is also the largest apple producer in China. Shaanxi's share of China's total apple production has remained above 20% for the past two decades, reaching 27% in 2020. (Fig. 20.2). Apples grow mainly on the sunny Weibel Loess Plateau, which is located at a high altitude, has deep layers of soil and experiences significant temperature variations between day and night. The apples grown in the province are brightly colored, crisp and moderately sweet, with a hint of tartness. With a relatively long shelf life, they lend themselves well to storage and transportation (Jin *et al.*, 2022; Wang *et al.*, 2022). The cities of Xianyang, Yan'an and Weinan account for almost 80% of the province's apple production.

20.3 Estimation Strategies

Broadly, acquiring loans to support farm production comprises two stages. First, farmers decide

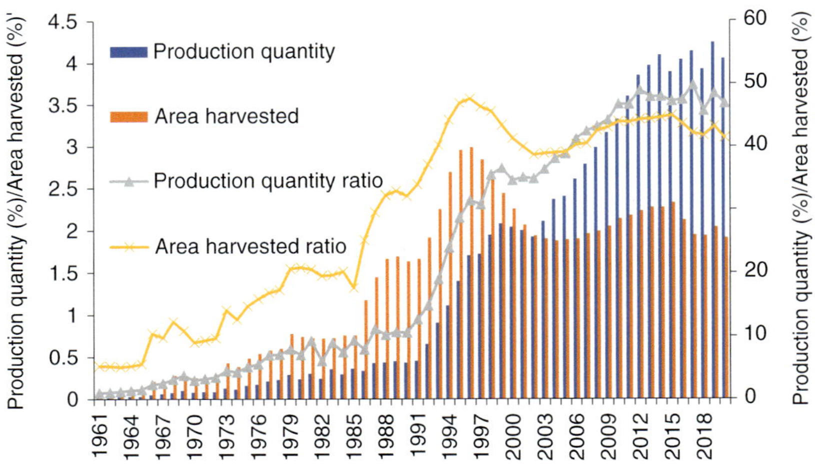

Notes: The area harvested is expressed in million hectares, and the output is expressed in 10 million tonnes.

Source: FAO (2022).

Fig. 20.1. Area harvested and output of apples in China (1961–2020).

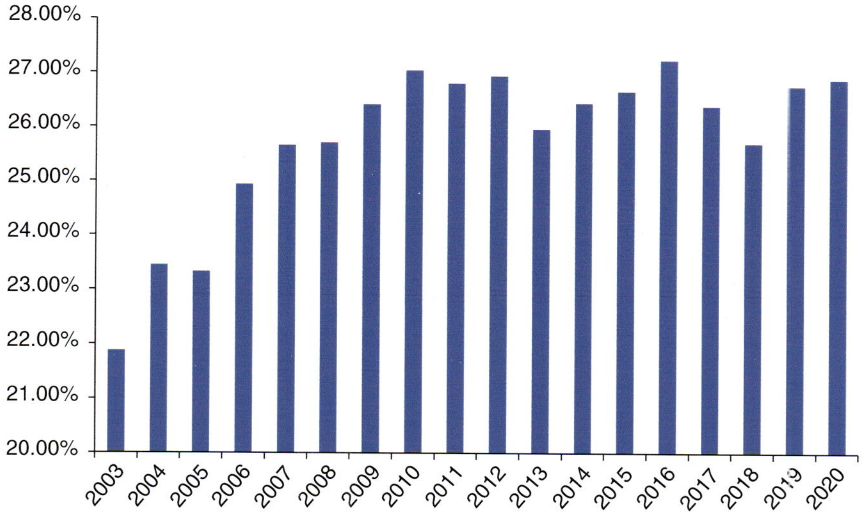

Source: FAO (2022).

Fig. 20.2. Shaanxi's share of China's apple production.

whether or not to participate in the credit market. To be clear, farmers may only explore borrowing possibilities to be better informed and may not, in fact, borrow any money. However, should the farmers decide to borrow, they must decide *how much* to borrow.

The double-hurdle model may be used in both of the above cases (García, 2013). This model has been widely applied in the literature (Mishra *et al.*, 2009; Adusah-Poku and Takeuchi, 2019; Mandikiana, 2021; Wu *et al.*, 2021; Leschewski and Sellnow, 2022). For example, Mandikiana (2021) estimated the household demand for private tutoring using a double-hurdle model and found that demographic, socioeconomic and student and school characteristics affected the decision to participate in private tutoring and the amount households spent on private tutoring.

This chapter uses a double-hurdle model to estimate the influence of climate shocks measured by the number of extreme weather events on the borrowing patterns of apple farmers. The double-hurdle model simultaneously captures the two decisions apple farmers make to acquire loans: whether to access credit (participation decision) and the amount of loan they should acquire (loan-amount decision). The two hurdle Eqns 20.1 and 20.2 can be expressed formally as follows:

$$\text{Participation decision: } C_i^{P*} = E_i\alpha_i + Z_i'\beta_i + \varepsilon_i, C_i^P = \begin{cases} 1, \textit{ if } C_i^{P*} > 0 \\ 0 \textit{ otherwise} \end{cases} \quad (20.1)$$

$$\text{Loan-amount decision: } A_i^* = E_i\gamma_i + Z_i'\delta_i + \eta_i,\ A_i = \begin{cases} A_i^* \textit{ if } A_i^* > 0 \textit{ and } C_i^P = 1 \\ 0 \textit{ otherwise} \end{cases} \quad (20.2)$$

where C_i^{P*} represents the probability that a farmer *i* chooses to participate in credit markets and it is observed by C_i^P. Specifically, $C_i^P = 1$ if farmers participate in credit markets, and $C_i^P = 0$ otherwise. A_i^* is a latent variable, while A_i measures the observed value of the loan amount for farmers participating in the credit markets; E_i measures the number of extreme weather events; Z_i' is a vector of control variables; α_i, β_i, γ_i and δ_i are parameters to be estimated; and ε_i and η_i are error terms that are assumed to be independent and distributed as $\nu_i \sim N(0, 1)$ and $\mu_i \sim N(0, \sigma^2)$.

Consistent estimates of the double-hurdle model can be obtained by maximizing the following likelihood Eqn 20.3:

$$LL = \sum \ln\left[1 - \Phi\left(E_i\alpha_i + Z_i'\beta_i\right)\Phi\left(\frac{E_i\gamma_i + Z_i'\delta_i}{\sigma_i}\right)\right] + \sum_{+} \ln\left[\Phi\left(E_i\alpha_i + Z_i'\beta_i\right)\frac{1}{\sigma_i}\phi\left(\frac{A_i - E_i\gamma_i - Z_i'\delta_i}{\sigma_i}\right)\right] \quad (20.3)$$

The results obtained from the maximum likelihood estimation are used to calculate the marginal effects of each regressor on the dependent variable. We also re-estimated the double-hurdle model by replacing the key independent variable, i.e. extreme weather events, with other variables that capture specific extreme weather events – frost, drought, strong winds, heavy rainfall and hail. Doing so provides a more detailed perspective on the associations between climate shocks and credit demand.

20.4 Data and Descriptive Statistics

20.4.1 Data collection

This chapter uses data collected in November 2021 from China's Shaanxi province. Shaanxi was chosen because it is the largest apple-producing province in China. In 2019, apple orchards constituted 615,000 ha in Shaanxi province, accounting for 31% of China's total area devoted to apple production (CRSY, 2020). Shaanxi is endowed with unique natural conditions conducive to producing high-quality apples – high altitude, sufficient sunlight, large temperature difference between day and night and deep soil layers. We used a multi-stage sampling procedure to select households for face-to-face interviews.

First, we selected two cities, Yan'an and Weinan, due to their geographical locations and regional economic development. Yan'an is located in northern Shaanxi on the Loess Plateau. It has a humid continental climate that borders on a steppe climate with cold, dry and moderately long winters and hot and somewhat humid summers. Weinan lies on the lower section of the Wei River confluence into the Yellow River. Second, we selected three top apple-producing counties (Luochuan, Fu and Yichuan) from the eight counties belonging to the city of Yan'an and another top three apple-producing counties (Baishui, Heyang and Chengcheng) from the six counties belonging to Weinan. The land area used in 2020 for apple production was 35,700 ha in Luochuan, 24,900 ha in Fu and 20,400 ha in Yichuan, producing appropriately 2 million tonnes of apples. Baishui, Heyang and Chengcheng produced 1.18 million tonnes of apples in 2020. Third, we randomly selected two towns from each county and then randomly selected two villages from each town. This resulted in a sample of 24 villages. Finally, we randomly selected 20 to 30 households in each village. Thus, we collected data from 700 rural households in total. We dropped the households with missing information and analyzed the 620 that remained.

The dependent variables used in the analysis were credit access and loan amount. Specifically, credit access was measured as a dichotomous variable. It was 1 if a household participated in credit markets and accessed credit from informal sources (e.g. neighbors, friends or relatives) or formal sources (e.g. banks or credit cooperatives) for supporting agricultural production, and 0 otherwise. The loan amount captured the actual amount of loan a household acquired. These definitions are consistent with previous studies (e.g. Twumasi *et al.*, 2020).

The key explanatory variable was the number of extreme weather events – frost, drought, strong winds, heavy rainfall and hail – reported by household heads for the period 2017–2021. To better understand the associations between extreme weather events and borrowing patterns, we also considered five explanatory variables measuring the areas of farmland affected by each type of extreme weather event. Drawing upon the literature on extreme weather events (e.g. Huang *et al.*, 2015; Shahzad and Abdulai, 2020; Zhao, 2020; Acosta *et al.*, 2021; Elahi *et al.*, 2021; Bora, 2022) and credit accessibility (e.g. Ankrah Twumasi *et al.*, 2019; Ma *et al.*, 2019; Xu *et al.*, 2019; Twumasi *et al.*, 2020; Li and Huo, 2021), we also selected a number of control variables. These comprised age, sex, education, health status and farming experience of the household head; variables denoting whether the family owns a farm, farm size, family size and whether the household encountered physical or financial hardship also were used. We also included county dummies to capture the unobserved heterogeneities related to socio-economic conditions, climate conditions and resource endowments.

20.4.2 Descriptive statistics

Table 20.1 reports the variable definitions and descriptive statistics. It shows that 35% of apple farmers in our sample accessed credit to support production activities. Farmers, on average, received 30,970 yuan in loans. The average number of extreme weather events reported by farmers was 3.56. Out of the five extreme weather events considered, hail affected the largest crop area: on average, 8.21 mu of apple orchards were affected by hail. In comparison, 2.9 mu of apple orchards were affected by strong winds. The average reported age of household heads was 54 years, and they had completed, on average, 8 years of formal education. Approximately 14% of households owned a family farm, and about 20% reported experiencing significant financial or physical hardship.

20.5 Empirical Results and Discussions

20.5.1 Impacts of extreme weather events on credit demand

The effects of extreme weather events on the probability of apple farmers' access to credit and the amount of credit obtained are presented in Table 20.2. Comparing the second and third columns reveals that the factors affecting farmers' decisions to borrow money were similar to those influencing how much they borrowed. Extreme weather events increased the likelihood of farmers borrowing money and increased the amount borrowed. More specifically, exposure to an extreme weather event increased the probability of apple farmers borrowing money by 2.3% and the amount borrowed by 2200 yuan. This result is consistent with the findings of Aryal *et al.* (2021) and Jordan (2021), who found that it is common for farmers to use credit to cope with climate shocks. Farm households in China usually face liquidity constraints (Yi *et al.*, 2015), which hinder their adaptation to extreme weather events. Access to credit markets can relax apple farmers' liquidity constraints, helping them cope with climate shocks. Should credit markets be accessible to apple farmers, they may use them to minimize and offset losses incurred due to extreme weather events, as revenue from apple sales accounts for a large proportion of their total income.

The effects of the control variables on whether and how much farmers borrow also are reported in Table 20.2. The results show that the age and health status of the household heads reduced the probability of farmers borrowing money and the amount borrowed. Noting that the average age of the household heads was 54 years, we surmise that the household heads participate in farming activities. More importantly, they hold sway in financial decisions. As farmers get older, they may become more risk-averse. Older farmers also have fewer remaining years of working life. The confluence of these factors may cause older farmers to doubt their abilities to repay the loans, instilling an unwillingness to borrow.

Farmers reporting good health were less likely to borrow than those reporting poor health. Notwithstanding technological advances, farming entails physical labor. Those in good health are better suited to addressing the physical demands of farming on their own without outsourcing, making them less likely to borrow to pay for outsourced tasks. Furthermore, farmers cultivating large areas were more likely to borrow money than those cultivating small areas. Specifically, a 1 mu increase in the cultivated area was associated with a 0.7% higher probability of farmers borrowing money. It was also associated with an increase of 800 yuan in the amount borrowed. This stands to reason. As larger farms produce more output, *ceteris paribus*, the losses due to extreme weather events are likely to be more substantial for farmers cultivating larger farms. Farmers may turn to credit markets to offset and cope with such losses. In addition, the borrowing behaviors of the farmers vary across different counties: this may be due to differences in the local governments' credit support policies.

20.5.2 Impacts of types of extreme weather on credit demand

As discussed above, we classify extreme weather events into different categories: frosts, droughts,

Table 20.1. Definitions and descriptive statistics. (Source: Authors' own work.)

Variables	Measurements	Mean	SD
Dependent variables			
Access to credit	1 if household has accessed credit for supporting agricultural production, 0 otherwise	0.350	0.478
Credit amount	The total amount of loan received (10,000 yuan)[a]	3.097	13.02
Key explanatory variables			
Extreme weather events	Self-reported extreme weather events in the last five years (times)	3.564	2.089
Frost	Area of farmland affected by frost (mu)[b]	5.959	8.442
Drought	Area of farmland affected by drought (mu)	7.383	9.196
Strong wind	Area of farmland affected by strong winds (mu)	2.904	5.853
Heavy rainfall	Area of farmland affected by heavy rainfall (mu)	5.133	8.432
Hail	Area of farmland affected by hail (mu)	8.214	9.073
Control variables			
Age	Age of household head (HH) (years)	54.12	9.565
Sex	Sex of HH: 1 if male, 0 otherwise	0.939	0.240
Education	Formal education received by HH (years)	7.993	4.263
Village cadres	1 if HH is a village cadre, 0 otherwise	0.106	0.308
Health status	HH's self-reported health status: from 1 = very bad to 4 = very good	3.143	0.910
Farming experience	HH's experience in apple farming (years)	24.72	8.983
Family farm	1 if household has a family farm, and 0 otherwise	0.138	0.345
Farm size	Total area of farmland cultivated (mu)[b]	16.49	15.99
Family size	Total number of people residing in a household (persons)	4.235	1.604
Physical or financial hardship	1 if the household encountered significant physical or financial hardship, 0 otherwise	0.203	0.402
Luochuan	1 if household resides in Luochuan county, 0 otherwise	0.184	0.388
Yichuan	1 if household resides in Yichuan county, 0 otherwise	0.147	0.354
Baishui	1 if household resides in Baishui county, 0 otherwise	0.185	0.389
Fu	1 if household resides in Fu county, 0 otherwise	0.148	0.356
Chengcheng	1 if household resides in Chengcheng county, 0 otherwise	0.165	0.371
Heyang	1 if household resides in Heyang county, 0 otherwise	0.171	0.377

Notes: [a]Yuan is the Chinese currency: US$1 was worth 6.34 yuan in 2022.
[b]1 mu = 1/15 ha.

strong winds, heavy rainfall and hail. This section explores the impacts of different types of extreme weather on farmers' borrowing decisions. The marginal effects of different types of weather events are reported in Table 20.3. Comparing the results reported in Tables 20.2 and 20.3 shows that demographics, households' characteristics and farm size had similar effects.

Frost increased the likelihood of farmers borrowing money and the amount they borrowed. A 1 mu increase in the area affected by frost was associated with a rise of 0.6% in the probability of farmers borrowing money, with 800 yuan being the average amount borrowed. Light frost is more common in Shaanxi province than hard frost. Temperatures do not stay below −1.6°C, the approximate freezing temperature of apples, for prolonged periods. Given that frost is not devastating for apple crops, farmers equipped with the appropriate tools and techniques can cope with frost without difficulty, making them more likely to borrow money to buy the requisite goods and services. For example, if needed, farmers may hire more workers to harvest apple crops before the onset of frost in autumn.

Although historically droughts have hindered apple production in Shaanxi province, we found that they did not significantly influence the borrowing patterns of apple farmers. One possible explanation is that Shaanxi province is located in the arid and semi-arid area of

Table 20.2. Impacts of extreme weather events on credit demand: marginal effects estimated using the double-hurdle model. (Source: Authors' own work.)

Variables	Hurdle 1 Access to credit Probit model	Hurdle 2 Credit amount Truncated normal estimator
Extreme weather events	0.023 (0.007)***	0.220 (0.092)**
Age	−0.010 (0.002)***	−0.090 (0.029)***
Sex	0.042 (0.146)	0.035 (1.675)
Education	0.002 (0.005)	−0.006 (0.067)
Health status	−0.040 (0.019)**	−0.484 (0.242)**
Farming experience	0.000 (0.002)	0.004 (0.024)
Farm size	0.007 (0.001)***	0.080 (0.016)***
Family size	0.006 (0.011)	0.037 (0.117)
Physical or financial hardship	0.082 (0.052)	0.505 (0.645)
Luochuan	0.208 (0.058)***	1.794 (0.568)***
Yichuan	0.257 (0.057)***	2.353 (0.559)***
Baishui	0.134 (0.053)**	1.387 (0.667)**
Fu	0.204 (0.059)***	1.892 (0.552)***
Chengcheng	0.097 (0.051)*	1.191 (0.800)
Sample size	620	620

Notes: Standard errors in parentheses; $*p < 0.1$, $**p < 0.05$, $***p < 0.01$.
Heyang is used as the reference county; credit amount is measured in 1000 yuan.

north-west China, and, over time, farmers have adapted to these conditions by installing irrigation systems and using drought-tolerant varieties. With the appropriate infrastructure and best practices already in place, farmers are well equipped to deal with droughts without incurring additional expenses, reducing the need to borrow money.

Extreme weather events in the form of strong winds did not influence the probability of farmers borrowing money; however, such weather events were associated with larger amounts borrowed. Specifically, a 1 mu increase in the area affected by strong winds was associated with an increase of 1190 yuan borrowed. Of course, this rise is seen among those who already have access to credit. In contrast, heavy rainfall increased the probability of apple farmers borrowing money but did not affect the amount borrowed. For an increase of 1 mu in the area affected by heavy rainfall, there was a 5% increase in the likelihood of farmers borrowing money. Hail did not affect the probability of farmers borrowing money. However, surprisingly, it was associated with a decrease of 850 yuan in the amount borrowed.

Hail can devastate apple crops, as it can severely injure the fruits, rendering them unfit for sale. Farmers may be unable to salvage these crops regardless of the resources at their disposal. Therefore, farmers may choose not to borrow money – to invest in ventures deemed futile – to cut their losses. Be that as it may, those with access to credit markets may borrow to offset lost income and smooth their consumption spending. However, they may borrow smaller sums, mindful of lost revenue and the need to repay loans. Investigating the mechanisms underpinning these borrowing patterns would be an interesting area for future research.

20.6 Conclusions and Policy Implications

The intensity and frequency of extreme weather events have risen during recent decades due to anthropogenic climate change, and China has been significantly affected. Although there has been a palpable shift in how people, policymakers and industry stakeholders view climate change, the progress toward carbon neutrality is slow. Climate change may intensify if greenhouse gas emissions remain unabated, and extreme weather events may become more catastrophic. The agricultural sector on the whole, and smallholder farmers in particular, are

Table 20.3. Impacts of different types of extreme weather events on credit demand: Marginal effects estimated using the double-hurdle model. (Source: Authors' own work.)

Variables	Hurdle 1 Credit access decision Probit model	Hurdle 2 Credit amount Truncated normal estimator
Frost	0.006 (0.003)**	0.080 (0.029)***
Drought	−0.002 (0.003)	−0.004 (0.033)
Strong wind	0.004 (0.003)	0.119 (0.044)***
Heavy rainfall	0.005 (0.003)*	0.072 (0.046)
Hail	0.002 (0.003)	−0.085 (0.049)*
Age	−0.009 (0.002)***	−0.039 (0.031)
Sex	0.235 (0.085)***	1.873 (1.253)
Education	0.001 (0.006)	−0.076 (0.083)
Health status	−0.042 (0.019)**	−0.412 (0.340)
Farming experience	−0.000 (0.002)	0.004 (0.027)
Farm size	0.008 (0.001)***	0.101 (0.013)***
Family size	−0.001 (0.010)	0.008 (0.119)
Physical or financial hardship	0.298 (36.061)	−1.430 (172.383)
Luochuan	0.206 (0.057)***	1.396 (0.514)***
Yichuan	0.275 (0.075)***	2.491 (0.661)***
Baishui	0.116 (0.056)**	1.378 (0.656)**
Fu	0.170 (0.057)***	3.197 (1.020)***
Chengcheng	0.064 (0.055)	1.299 (0.689)*
Sample size	620	620

Notes: Standard errors in parentheses.
$^{*}p < 0.1$, $^{**}p < 0.05$, $^{***}p < 0.01$.
Heyang is used as the reference county.
Credit amount is measured in 1000 yuan.

especially vulnerable to climate shocks – credit markets will play an important role in helping farmers manage risk and cope with climate shocks. Accordingly, this chapter examined the effects of extreme weather events on farmers' demand for credit. Using a double-hurdle model, we analyzed data collected from face-to-face interviews with farmers in China's Shaanxi province. The estimation proceeded in two stages: first, we determined how extreme weather events affected the likelihood of farmers borrowing money; second, we estimated how these events influenced the amount borrowed.

The results showed that extreme weather events increased the likelihood of farmers borrowing money by 2.3% and the amount they borrowed by 2200 yuan. The age of the household head was negatively associated with the probability of the household borrowing money and the amount borrowed. Furthermore, the better the household heads' self-reported health, the less likely the households were to borrow money. Those with larger farms were more likely to borrow than those with relatively small farms; the former also borrowed larger sums. We also found regional differences in borrowing patterns. The differences were quite apparent between farmers in Yichuan and Heyang counties – farm households in Yichuan county were 25.7% more likely to borrow than those in Heyang county; apropos the credit amount, Yichuan farm households borrowed 2353 yuan more than their Heyang counterparts.

We also examined the effects of specific types of extreme weather events on borrowing behaviors. The results showed that frost was associated with an increase both in the likelihood of borrowing and in the amount borrowed. Surprisingly, hail was associated with an increase in the probability of borrowing but a decrease in the amount borrowed. Drought did not affect borrowing. The other extreme weather events increased the likelihood of borrowing or the amount borrowed.

These links between apple farmers' credit demand and extreme weather events have important implications for policymakers, lending institutions, agricultural cooperatives and industry stakeholders. A positive association between extreme weather events and credit demand suggests that farmers use credit to cope with extreme weather. Policymakers can establish funds and programs to help farmers secure adequate short-term credit at reasonable interest rates in times of need. Timeliness is important. Therefore, rules and regulations related to determining creditworthiness should be simplified; they should facilitate quick access to funds to avoid delays in making credit available to farmers. That older farmers are less likely to access credit is suggestive. Their aversion to borrowing may stem from various factors: a lack of confidence in their ability to repay the loans, reluctance to pay interest, or unawareness of borrowing options. Regardless of the reasons, the evidence points to potential systematic financial exclusion. Credit institutions and policymakers should invest in outreach programs to help older farmers use credit markets to cope with climate shocks. Educational initiatives and subsidized lending may prove to be useful.

Farmers in Heyang and Chengcheng counties were less likely to borrow than those in Luochuan, Yichuan, Baishui and Fu counties. Differences in the borrowing patterns across counties may reflect different credit support policies implemented by local governments or different social ethos. Understanding why credit demand was relatively low in Heyang and Chengcheng counties may inform policies to ensure financial inclusion across the province. As noted above, different types of extreme weather events were variously associated with credit demand. Accordingly, credit forecast models should incorporate weather forecasts. Insurance providers, financial institutions, and weather services companies should work collaboratively to cater to the credit needs of rural farm households.

Although we have presented evidence on the links between credit demand and extreme weather events, we do not formally model the mechanisms underpinning these links. We propose that future research be devoted to this endeavor. This study focused on six counties in Shaanxi province and found significant differences in credit demand across them, suggesting the possibility of notable regional differences in credit demand. Thus, future research should undertake similar research for other regions. Insights into regional credit demand would help customize lending practices, regulations, and outreach initiatives.

References

Abbas, A., Amjath-Babu, T.S., Kächele, H. and Müller, K. (2015) Non-structural flood risk mitigation under developing country conditions: an analysis on the determinants of willingness to pay for flood insurance in rural Pakistan. *Natural Hazards* 75, 2119–2135.

Acosta, A., Nicolli, F. and Karfakis, P. (2021) Coping with climate shocks: the complex role of livestock portfolios. *World Development* 146, 105546.

Adusah-Poku, F. and Takeuchi, K. (2019) Household energy expenditure in Ghana: a double-hurdle model approach. *World Development* 117, 266–277.

Aftab, A., Ahmed, A. and Scarpa, R. (2021) Farm households' perception of weather change and flood adaptations in northern Pakistan. *Ecological Economics* 182, 106882.

Alfieri, L., Bisselink, B., Dottori, F., Naumann, G., de Roo, A. *et al.* (2017) Global projections of river flood risk in a warmer world. *Earth's Future* 5, 171–182.

Ankrah Twumasi, M., Jiang, Y., Osei Danquah, F., Chandio, A.A. and Agbenyo, W. (2019) The role of savings mobilization on access to credit: a case study of smallholder farmers in Ghana. *Agricultural Finance Review* 80, 289–304.

Arshad, M., Amjath-Babu, T.S., Kächele, H. and Müller, K. (2016) What drives the willingness to pay for crop insurance against extreme weather events (flood and drought) in Pakistan? A hypothetical market approach. *Climate and Development* 8, 234–244.

Aryal, J.P., Sapkota, T.B., Rahut, D.B., Marenya, P. and Stirling, C.M. (2021) Climate risks and adaptation strategies of farmers in East Africa and South Asia. *Scientific Reports* 11, 10489.

Bora, K. (2022) Rainfall shocks and fertilizer use: a district level study of India. *Environment and Development Economics* 27, 556—577.

Clarkson, G., Dorward, P., Kagabo, D.M. and Nsengiyumva, G. (2017) Climate services for agriculture in Rwanda: initial findings from PICSA monitoring and evaluation. CCAFS Info Note. CGIAR Research Program on Climate Change, Agriculture and Food Security (CCAFS), Wageningen, Netherlands.

CRSY (2020) *China Rural Statistical Yearbook*. Statistical Press, China.

Dayamba, D.S., Ky-Dembele, C., Bayala, J., Dorward, P., Clarkson, G. *et al.* (2018) Assessment of the use of participatory integrated climate services for agriculture (PICSA) approach by farmers to manage climate risk in Mali and Senegal. *Climate Services* 12, 27–35.

Di Falco, S. and Sharma, S. (2018) Investing in climate change adaptation: motivations and green incentives in the Fiji Islands. *Ecological Economics* 154, 394–408.

Doherty, E., Mellett, S., Norton, D., McDermott, T.K.J., Hora, D.O. *et al.* (2021) A discrete choice experiment exploring farmer preferences for insurance against extreme weather events. *Journal of Environmental Management* 290, 112607.

Elahi, E., Khalid, Z., Tauni, M.Z., Zhang, H. and Lirong, X. (2021) Extreme weather events risk to crop-production and the adaptation of innovative management strategies to mitigate the risk: a retrospective survey of rural Punjab, Pakistan. *Technovation* 117, 102255. DOI: 10.1016/j.technovation.2021.102255.

Fenton, A., Paavola, J. and Tallontire, A. (2017) The role of microfinance in household livelihood adaptation in Satkhira District, Southwest Bangladesh. *World Development* 92, 192–202.

Fischer, E.M. and Knutti, R. (2015) Anthropogenic contribution to global occurrence of heavy-precipitation and high-temperature extremes. *Nature Climate Change* 5, 560–564.

Fontes, F., Gorst, A. and Palmer, C. (2021) Threshold effects of extreme weather events on cereal yields in India. *Climatic Change* 165, 26.

García, B. (2013) Implementation of a double-hurdle model. *Stata Journal* 13, 776–794.

Guo, Y., Wu, Y., Wen, B., Huang, W., Ju, K., Gao, Y. *et al.* (2020) Floods in China, COVID-19, and climate change. *The Lancet Planetary Health* 4, e443-e444, DOI: 10.1016/S2542-5196(20)30203-5.

Hänke, H. and Barkmann, J. (2017) Insurance function of livestock: farmer's coping capacity with regional droughts in South-Western Madagascar. *World Development* 96, 264–275.

Harvey, C.A., Rakotobe, Z.L., Rao, N.S., Dave, R., Razafimahatratra, H. *et al.* (2014) Extreme vulnerability of smallholder farmers to agricultural risks and climate change in Madagascar. *Philosophical Transactions of the Royal Society B: Biological Sciences* 369, 20130089.

Harvey, C.A., Saborio-Rodríguez, M., Martinez-Rodríguez, M.R., Viguera, B., Chain-Guadarrama, A., Vignola, R. *et al.* (2018) Climate change impacts and adaptation among smallholder farmers in Central America. *Agriculture and Food Security* 7, 57.

Huang, J., Wang, J. and Wang, Y. (2015) Farmers' adaptation to extreme weather events through farm management and its impacts on the mean and risk of rice yield in China. *American Journal of Agricultural Economics* 97, 602–617.

IPCC (2021) Climate Change Widespread, Rapid, and Intensifying. Intergovernmental Panel on Climate Change. Available at: https://www.ipcc.ch/2021/08/09/ar6-wg1-20210809-pr/ (accessed May 17, 2023).

Jin, S., Li, W., Cao, Y., Jones, G., Chen, J. *et al.* (2022) Identifying barriers to sustainable apple production: a stakeholder perspective. *Journal of Environmental Management* 302(Pt B), 114082.

Johnson, B.A., Scheyvens, H., Baqui Khalily, M.A. and Onishi, A. (2019) Investigating the relationships between climate hazards and spatial accessibility to microfinance using geographically-weighted regression. *International Journal of Disaster Risk Reduction* 33, 122–130.

Jordan, J.C. (2021) Climate shocks and adaptation strategies in coastal Bangladesh: does microcredit have a part to play? *Climate and Development* 13, 454–466.

Kazianga, H. and Udry, C. (2006) Consumption smoothing? Livestock, insurance and drought in rural Burkina Faso. *Journal of Development Economics* 79, 413–446.

Leschewski, A. and Sellnow, C. (2022) Determinants of US household expenditures on fortified fruit juice. *International Food and Agribusiness Management Review* 25, 69–82.

Li, X. and Huo, X. (2021) Impacts of land market policies on formal credit accessibility and agricultural net income: evidence from China's apple growers. *Technological Forecasting and Social Change* 173, 121132.

Ma, S., Wu, X. and Gan, L. (2019) Credit accessibility, institutional deficiency and entrepreneurship in China. *China Economic Review* 54, 160–175.

Mandikiana, B.W. (2021) Choice and expenditure: a double hurdle model of private tutoring in Qatar. *Economic Analysis and Policy* 71, 1–15.

Mishra, A.K., Williams, R.P. and Detre, J.D. (2009) Internet access and internet purchasing patterns of farm households. *Agricultural and Resource Economics Review* 38, 240–257.

Mukherjee, S., Mishra, A. and Trenberth, K.E. (2018) Climate change and drought: a perspective on drought indices. *Current Climate Change Reports* 4, 145–163.

Partey, S.T., Dakorah, A.D., Zougmoré, R.B., Ouédraogo, M., Nyasimi, M., Nikoi, G.K. *et al.* (2020) Gender and climate risk management: evidence of climate information use in Ghana. *Climatic Change* 158, 61–75.

Shahzad, M.F. and Abdulai, A. (2020) Adaptation to extreme weather conditions and farm performance in rural Pakistan. *Agricultural Systems* 180, 102772.

Stott, P. (2016) How climate change affects extreme weather events. *Science* 352, 1517–1518.

Twumasi, M.A., Jiang, Y., Ameyaw, B., Danquah, F.O. and Acheampong, M.O. (2020) The impact of credit accessibility on rural households clean cooking energy consumption: the case of Ghana. *Energy Reports* 6, 974–983.

Vaughan, C., Buja, L., Kruczkiewicz, A. and Goddard, L. (2016) Identifying research priorities to advance climate services. *Climate Services* 4, 65–74.

Wang, F., Ge, S., Lyu, M., Liu, J., Li, M. *et al.* (2022) DMPP reduces nitrogen fertilizer application rate, improves fruit quality, and reduces environmental cost of intensive apple production in China. *Science of the Total Environment* 802, 149813.

Wu, C., Mai, F. and Li, X. (2021) The effect of content depth and deviation on online review helpfulness: evidence from double-hurdle model. *Information and Management* 58, 103408.

Xu, N., Shi, J., Rong, Z. and Yuan, Y. (2019) Financial literacy and formal credit accessibility: evidence from informal businesses in China. *Finance Research Letters* 36, 101327.

Ye, Y. and Qian, C. (2021) Conditional attribution of climate change and atmospheric circulation contributing to the record-breaking precipitation and temperature event of summer 2020 in southern China. *Environmental Research Letters* 16, 044058.

Yi, F., Sun, D. and Zhou, Y. (2015) Grain subsidy, liquidity constraints and food security-impact of the grain subsidy program on the grain-sown areas in China. *Food Policy* 50, 114–124.

Zhao, W. (2020) Extreme weather and climate events in China under changing climate. *National Science Review* 7, 938–943.

21 How Beliefs about Climate Change Adapt: A Natural Experiment with Evidence on Excess Sensitivity to a Weather Event

Seo Woo Lee,[1]* Hongli Feng[2] and David A. Hennessy[2]
[1]*Michigan State University, East Lansing, USA;* [2]*Iowa State University, Ames, USA*

Abstract
A prerequisite for adapting decisions to a new environment is the belief that the environment has changed. Agricultural field crop and grass-based cattle production involve long-term investments, including land conversion, buildings, drainage infrastructure and the acquisition of crop-specific skills. Agricultural practices evolve to match the prevailing climate, so climate change will have major effects on the sector, and timely adaptation is important for an efficient, economically sustainable production base. For a drought-prone region with variable weather and continental climate extremes, this chapter considers how a drought that occurred between two surveys of the same landowners affected responses to queries on viewpoints regarding changing weather patterns. Although drought is quite typical for the region and a single drought period is unlikely to be informative regarding climate change, we find that climate change beliefs were quite sensitive to the degree of drought experienced. The findings suggest that climate change perceptions are likely unstable, implying the need for caution when enacting policies intended to foster changing perspectives.

21.1 Introduction

Weather outcomes and the underlying climate enter many of the most important agricultural decisions. Despite the development of various agricultural technologies to cope with climate risk, climate remains a great influence on crop yields, and so on farmers' profits. For example, in 2019 heavy rainfall and flooding events prevented US farmers from planting crops on more than 14 million acres in a timely manner (USDA, 2019). As a result, farmers are well-motivated to seek a better understanding of changes in weather patterns relevant to them so as to mitigate damage (e.g. Abendroth *et al.*, 2021). In the process, they encounter various forms of information. For our purposes, we classify weather/climate information used to make production decisions into: (i) subjective climate information based on first-hand experiences and feelings; and (ii) objective data as might best be measured by official statistical sources.

These two forms of information receive different weights in decision-making, and the processes involved in assimilating subjectively obtained climate information can lead to bias. Individuals tend to base decisions on their perceptions (subjective information) as distinct from objective data (Akerlof and Dickens, 1982). Furthermore, statistical information is often

*E-mail: leeseo8@msu.edu

DOI: 10.1079/9781800622289.0021

recontextualized by decision makers based upon their own experiences (Marx *et al.*, 2007). Therefore, how actual data affect human perceptions should be taken into consideration when analyzing farmer behavior and adaptation to climate. Support for climate policies is related to individual perceptions about climate change (Leiserowitz, 2006). Knowing the relationship between projected future climate and a farmer's viewpoint on climate change will allow policymakers to set the appropriate scope of and targets for climate change adaptation and mitigation policy as they seek to engage farmers.

In this chapter, we investigate the relation between human perception and weather events. Specifically, we focus on how beliefs about climate change adapt after the experience of an extreme weather event. We will examine objective statistical data by considering the weather history in our study area so as to understand respondents' previous experiences. We will also examine subjective, first-hand data, obtained from surveys conducted before and after a particular extreme weather event. Then, using difference-in-difference (DID) analysis, we analyze how perceptions about climate change shifted after the experience of the extreme weather event. The results will allow us to infer the relation between extreme weather events and human perception.

This chapter makes the following contributions to the literature. First, to our knowledge, no previous research regarding weather perceptions has used farmer panel data that was collected both before and after an extreme weather event. The vast majority of relevant existing inquiries have used cross-sectional data that were collected only after an extreme weather event occurred (Spence *et al.*, 2011; Haden *et al.*, 2012; Niles *et al.*, 2013). Second, previous studies on the topic have focused principally on personal experiences among the general public (Joireman *et al.*, 2010; Li *et al.*, 2011; Goebbert *et al.*, 2012; Myers *et al.*, 2013; Shao, 2016; Howe *et al.*, 2019; Gärtner and Schoen, 2021). Only a few have focused on professionals whose business choices and performance outcomes are weather-sensitive (Carlton *et al.*, 2016). Our study subjects are farmers whose decisions are influenced by weather and, often, whose family livelihood depends on harvests that are dictated by weather.

Experienced professionals have been found to make more rational decisions than people who are less experienced in the decision-making contexts at hand, likely because experiences admit a better understanding of the subject and deeper appreciation of related consequences (List, 2004; List and Haigh, 2009). Third, this study is distinguished from previous studies in that the survey was conducted without using the term 'climate change.' Opinion on climate change can be heavily influenced by political and cultural background (McCright and Dunlap, 2011; Egan and Mullin, 2012; Myers *et al.*, 2013; Yazar *et al.*, 2021). Therefore, to prevent any possible bias the term 'climate change' might evoke, our survey questions directly asked about changes in weather patterns. Finally, our most general contribution to the literature regards information on the stability and sensitivity of beliefs. We provide evidence that climate domain beliefs, at least regarding drought, may be excessively sensitive to recent events. Such sensitivity has important policy implications.

The rest of this chapter is organized as follows. In Section 21.2, we review existing literatures on how beliefs about climate change, and more generally beliefs about stochastic laws, change. We then describe our data and study area and follow up with a description of approaches to data analysis and methodologies used. After presenting the estimation results, we conclude with a brief discussion.

21.2 Literature Review

How we learn is determined in part by what is to be learned. Some information can be established deterministically by a controlled experiment designed and implemented by an experimenter so that learning can take place discretely over time. This is not the case when what is to be learned regards a distribution shift. Then, repeated experiments are required to settle on the truth. Climate change refers to a change in climate at a location that occurs over many years and then persists for decades or longer (IPCC, 2021). Climate itself is an intricate assembly of stochastic events in which climate change can be viewed as a re-weighting of event probabilities for a location as well as the emergence of weather events new to the location. Therefore, climate change can only be experienced by samples with timing determined by nature (Swim *et al.*, 2011).

Furthermore, personal perceptions about climate change are subjective as each individual has different preferences and circumstances and may feel the impacts of climate change differently (van der Linden, 2014).

No consensus has yet emerged about whether a relationship exists between an extreme weather experience and perceptions about climate change. Many scholars have identified evidence that experiencing extreme weather events affects one's perception about climate change (Weber, 2013). However, not all studies agree that people change their perceptions after experiencing extreme weather events or how perceptions change. Whitmarsh (2008) has argued that there is no difference in responses to climate change between those who do and do not experience extreme floods. Brulle *et al.* (2012) found that none among high temperature, extreme precipitation and drought had a significant effect on public concerns about climate change. On the other hand, Spence *et al.* (2011) examined the linkage between flooding experience and perceptions about climate change to infer that experiencing a flood does promote both concern about and actions to mitigate climate change. In an analysis that connected the U.S. National Oceanic and Atmospheric Administration (NOAA) storm events database with public opinion surveys, Konisky *et al.* (2015) found a positive relationship between an experience of extreme weather events and expressions of concern about climate change. Working with U.S. crop loss data and survey information about federal farm advisor viewpoints, Niles *et al.* (2019) demonstrated that experiencing extreme drought alters perceptions about weather variability as distinct from perceptions about climate change.

Although the relationship between experiencing extreme weather events and climate change perceptions remains unclear, no literature has considered subjects' prior experiences when analyzing the relation between extreme weather experiences and climate change perceptions. Individuals perceive events based on their experience and personal values (Lazarus, 1984; Loewenstein *et al.*, 2001). Individuals who have frequently experienced certain weather events can increase their knowledge about these events and adapt their decisions on how to respond (Hansson *et al.*, 1982). As climate change is a progressive process of adjustments in average weather patterns over a long time period, in order to examine how extreme weather events affect climate change perceptions it is important to control for a respondent's entire prior weather event experiences.

Belief formation regarding climate change is an important strand in the larger literature on belief updating, a literature that a brief review cannot possibly do justice to (Benjamin, 2019). Shiller (1981) and De Bondt and Thaler (1985), for example, have demonstrated that beliefs about future stock returns are excessively sensitive to information and may even respond to information of no relevance for stock returns. Besides the novel challenge these observations pose for the efficient market hypothesis (Fama, 1970) and for neoclassical economics more generally, this finding is remarkable in the behavioral economics possibilities that it has posed and in making connections with disparate literatures. Among these are connections with experimental psychology and psychology more generally.

The earliest experiments on Bayesian updating were conducted by Ward Edwards and colleagues (Edwards and Phillips, 1968), who demonstrated conservatism in updating prior information and also the underweighting of prior beliefs (Kahneman and Tversky, 1973; Benjamin, 2019). This corpus also showed that more recent events are given extra weight in belief updates (Pitz and Reinhold, 1968; Grether, 1992; Benjamin, 2019), i.e. there are recency effects. Recency effects have long been detected in laboratory settings (Broadbent and Broadbent, 1981). They also are known to be present in learning to manage finances (Agarwal *et al.*, 2008) and popular assessments of performance (Page and Page, 2010). Perhaps the most relevant work on how weather events affect beliefs about climate change is due to Deryungina (2013), who used Gallup Poll surveys in the United States over the years 2003–2010. She found little evidence of recency effects and substantial evidence in favor of Bayesian updating.

21.3 Data

In this section we introduce the data used in our analysis. As we are interested in the relationship between human perceptions and actual weather events, these data consist of two sets. One includes responses to two surveys that were conducted in 2015 and 2018. This set provides us with perceptions about climate change as well as respondent and farm characteristics. The other set contains objective official weather data as provided by the US Drought Monitor (2021). The sets are linked by information on the location of respondents' residences. Details on each dataset are provided in separate sub-sections below.

21.3.1 Survey data

Two surveys were conducted in 2015 and 2018 as part of a larger endeavor to understand land-use adaptation and climate change. The first was conducted before a drought, and indeed before any reasonable indicators became available that a drought would occur in 2 years. In 2017 a drought hit North Dakota, South Dakota and Montana, decreasing agricultural production and resulting in more than US$1 billion in economic losses (Hoell *et al.*, 2019). The drought, which occurred unexpectedly during the rainy season (Otkin *et al.*, 2017), commenced in mid-May and spread through the northern Great Plains until July 25. The drought separated farmers into groups according to drought severity incidents in their localities, as measured objectively by government agencies through well-established procedures. The event provides a natural economic experiment in that it is a randomized treatment (Dunning, 2012).

Our study includes 20 counties in North Dakota and 37 counties in South Dakota, all east of the Missouri River (Fig. 21.1). In each county, the majority of all farmland acres are under cultivated annual crops. The survey targeted farm operators who operated more than 100 acres and had recently raised at least some wheat, corn, soybeans or grass/hay. The survey sample was purchased from a farm sampling frame provided by Survey Sampling International (now, after company rebranding, Dynata, self-claimed to be 'the world's largest first-party data company, with a global reach of nearly 70 million consumers and business professionals' (Dynata, 2023)). Sample selection was proportional by county so that more farmers were included from counties with comparatively more eligible farms. All survey mailing and data coding were

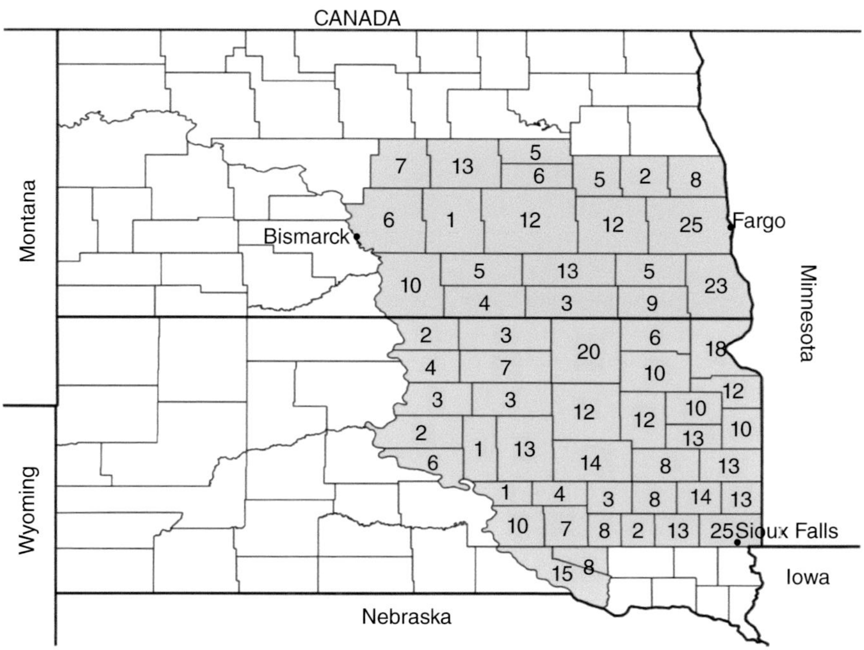

Source: Authors' own work.

Fig. 21.1. Useable responses by county level, 2018 survey.

handled by the Iowa State University Survey Research Center. The survey has been reported elsewhere and further details are available in Wang *et al.* (2017) and Wimberly *et al.* (2017), where studies examined issues related to land-use change.

Figure 21.2 describes the timeline for our belief adaptation natural experiment. For the first survey (Survey #1), conducted in 2015, 3000 surveys were mailed and 1026 completed surveys were returned (34.2% response rate). The drought, which we consider to be nature's treatment, arrived 2 years later. In 2018, a follow-up survey (Survey #2) was conducted for the 884 respondents who had completed the 2015 survey and were less than 70 years old at the time of the 2018 survey. We received 517 surveys back, a 58.5% response rate. Among Survey #2 responses, 506 were sufficiently complete for inclusion in our analysis. Figure 21.1 shows the number of sufficiently complete responses for both surveys across the counties studied.

For weather pattern perceptions about drought, the respondents were asked to indicate 'less,' 'same' or 'more' compared to the past 10 years. We also collected respondents' demographics and farm business characteristics such as whether they applied no-till cropping or used tile drainage, as well as land ownership status and soil characteristics, including slope and Land Capability Classes (LCC). LCC classification partitions land into eight categories based on soil and land attributes (Helms, 1992). Variable descriptions and summary statistics are presented in Table 21.1.

Given non-responses (41.5%) in the second survey, representativeness and sample selection issues are concerns. Table 21.2 compares responses given in the repeated samples with those given by subjects who took part only in Survey #1. Significant differences in mean drought pattern perceptions and farm ownership become apparent. Respondents who answered the follow-up survey in 2018 were less likely to own a comparatively higher percentage of farmland. The farm ownership variable is 0.17 lower in the repeated group. Also, when compared with those who completed Survey #1 only, Survey #2 respondents reported on average 0.38 units lower perceived changes in drought. This difference leads us to conjecture that there may be a selection bias. Except for these differences, the repeated group and the non-repeated group show no significant difference.

21.3.2 Drought data

As the survey data provide us with only how respondents perceived weather, we also need an objective weather metric to compare with. For this purpose, the Drought Index (DI) from the US Drought Monitor (USDM) as recorded on July 25 from the years 2000 through 2017 was collected. We chose the drought condition on July 25, 2017, because the drought was most widespread during that week. The US Drought Monitor is reported weekly, and its values are temporally stable, but to confirm robustness we also considered data recorded on July 18 and August 1. The magnitude of drought effect is smaller with these alternative dates, as the drought was most common in the week of July 25, 2017, but they had no material effect on our results. We applied DID analysis with the data 1 week before July 25 and 1 week after July 25 and both results show that the drought changed people's perceptions toward that there were more droughts. The data range 2000–2017 was chosen because Drought Index data are available from 2000, and 2017 represents the extent of data available to the 2018 survey respondents.

The Drought Index is produced jointly by the National Drought Mitigation Center

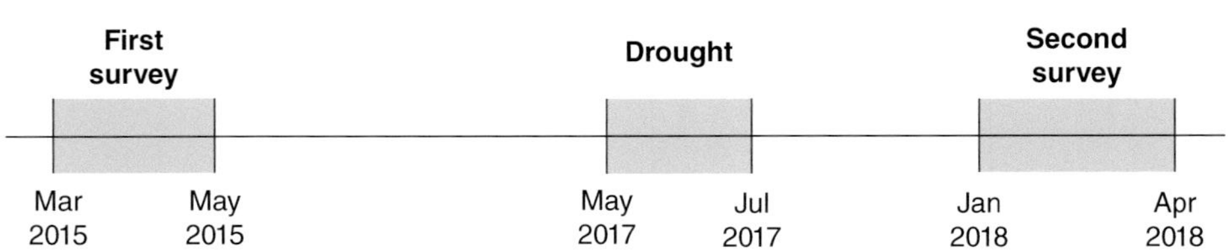

Source: Authors' own work.

Fig. 21.2. Natural experiment timeline.

Table 21.1. Variable descriptions and summary statistics. (Source: Authors' own work.)

Variable	Description	Mean (std. dev.)		Min	Max
drou15	Drought pattern perception in 2015: Less drought (= 1); Same (= 2); More drought (= 3) compared to the past 10 years.	1.89	(0.70)	1	3
drou18	Drought pattern perception in 2018: Less drought (= 1); Same (= 2); More drought (= 3) compared to the past 10 years.	2.16	(0.74)	1	3
Birth year	Birth year	1960	(9.93)	1929	1992
Education	Highest education level completed: Less than high school (1); high school (2); some college/technical school (3); 4-year college degree (4); advanced degree (5)	3.04	(0.84)	1	5
Earn	Level of annual gross farm/ranch sales: <$50K (1); $50K–$99.9K (2); $100K–$249.9K (3); $250K–$499.9K (4); $500K–$999.9K (5); $1 million + (6)	3.91	(1.31)	1	6
Ownership	Land ownership status: Own all acres farmed (1); own most acres farmed, renting the remainder (2); own and rent roughly equal number of farmland acres (3); rent most acres farmed, owning the remainder (4); rent all acres farmed (5)	2.79	(1.16)	1	5
No-till	= 1 if adopted or increased use of no-till between 2005 and 2015; = 0 otherwise.	0.51	(0.50)	0	1
Drainage	= 1 if adopted or increased drainage on cropland acres between 2005 and 2015; = 0 otherwise.	0.23	(0.42)	0	1
LCC4	Percentage of soils with Land Capability Classification (LCC) less than or equal to IV within 1 mile radius	95.41	(11.44)	0.02	100
Slope	Percentage of soils with slope less than or equal to 4 within 1 mile radius.	48.52	(36.73)	0	100
Latitude	Latitude of respondent's location	45.29	(1.31)	42.91	47.83
Longitude	Longitude of respondent's location	–98.00	(1.07)	– 100.78	– 96.47

(NDMC) at the University of Nebraska-Lincoln, the NOAA and the US Department of Agriculture (USDA) using satellite-based assessments and climatological indices such as the Palmer Drought Severity Index and the Keech–Byram Drought Index for fire (US Drought Monitor, 2021). This index identifies drought severity under five categories as described in Table 21.3. Unlike other indices such as the Palmer Drought Severity Index or the Standardized Precipitation Index, the Drought Index is not a statistical model. It aggregates a variety of contributory factors, including soil moisture, hydrological inputs, climatological inputs and local data such as grazing conditions or information about how drought is affecting people, to ensure index robustness and suitability for monitoring impact on agriculture (Svoboda *et al.*, 2002). The severity of drought experienced by each respondent was assumed to be that for the farm address as collated with Drought Index spatial maps.

We form H, for history, as our historical drought indicator by using 18 years (2000–2017) of drought data to arrive at the unconditional probability of experiencing any level of drought at a given location on July 25 as:

$$H \equiv \Pr(\text{Drought } 07/25) = \frac{\text{Years among 2000-'17 in which drought occurred on } 07/25}{18} \tag{21.1}$$

Table 21.2. 2015 Summary statistics for Survey #1 responses, broken down by those who answered both surveys and those who answered only Survey #1. (Source: Authors' own work.)

	Repeated (answered both surveys)		Non-repeated (i.e. answered Survey #1 only)	
	Mean	Std. dev.	Mean	Std. dev.
Drou15***	1.89	0.70	2.27	1.15
Drought severity	1.07	0.91	1.14	0.95
Age	3.03	0.85	2.98	0.94
Education	3.04	0.84	2.99	0.88
Annual income	3.91	1.31	3.82	1.39
Farm ownership**	2.79	1.16	2.96	1.23
LCC4	95.41	11.44	95.98	10.44
Slope	48.52	36.73	49.18	3841
Drainage	0.23	0.42	0.21	0.41
No-till	0.51	0.50	0.51	0.50

Note: *, **, *** indicate mean difference is significant at 10%, 5%, 1% levels, respectively.

Table 21.3. Description of Drought Monitor Index. (Source: Authors' own work.)

Drought category	Description of the category (US Drought Monitor, 2021)	N[a]
None	–	13
D0	Abnormally dry: when going into drought, short-term dryness slows growth of crops/pastures and when going out of drought, there are some lingering water deficits.	150
D1	Moderate drought: some damage to crop/pastures, some developing water shortages.	196
D2	Severe drought: crop/pasture losses are likely; water shortages are common.	108
D3	Extreme drought: major crop/pasture losses, widespread water shortages.	39
D4	Exceptional drought: exceptional and widespread crop/pasture losses and shortages of water creating water emergencies.	0

Note: [a]N represents the number of respondents who answered survey in both 2015 and 2018. Total N = 506.

Historical data, not reported due to space constraints, show that droughts are comparatively less frequent in North Dakota.

21.4 Study Area, Drought History and Hypothesis Development

The study area is part of the Prairie Pothole region in the Dakotas where corn, soybeans and wheat are currently the dominant cropland uses. Wheat is grown throughout the area, being most prevalent toward the north and west. Corn is dominant in the region's southern and eastern parts and soybean is planted most widely in the north (Alemu *et al.*, 2020). The region's agriculture is susceptible to drought damage because all crops, but especially corn, require water throughout growth in order to mature. This region has experienced significant land-use changes in the past two decades or so, as documented by several studies (e.g. Wright and Wimberly, 2013). Although crop prices and improved crop yields have been identified as the main drivers of the conversion of grassland to cropland in this region, climate and weather factors also have been proposed as an important driver (e.g. Rashford *et al.*, 2016). Thus, understanding climate change patterns and farmers' perceptions of these patterns will help better design policies intended to address grassland loss in the region.

Due to complex interactions among three air masses – continental polar, maritime tropical and maritime polar (Ahrens, 2007) – this region experiences inherently high weather variability as well as comparatively extreme precipitation and temperature events within and between years (Conant *et al.*, 2018). Most of the total annual precipitation occurs during spring months (April to June), but inter-annual variability in springtime precipitation is large (Knapp and Smith, 2001). A time series presentation of Drought Index measurements broken down by severity level but averaged over all areas in the two states is provided in Fig. 21.3. It can be seen that droughts are common in the region, and also that a severe drought occurred in 2012 for an extended period of time.

Figure 21.4 describes the study area's 1-year Standardized Precipitation Index (SPI; McKee *et al.*, 1993) by state. SPI is an index based on the distribution of precipitation over a recent time window. It is formed by taking the difference between precipitation and mean precipitation for a specified time, and then dividing that by the standard deviation, where mean and standard deviation are determined from the climatological record. A larger value indicates a more severe condition. Generally, 3-, 6-, 12-, 24- and 48-month time windows of SPI values are used depending on research purposes. Indices with shorter time windows describe short-term precipitation patterns, and these are appropriate precipitation indices for monsoon regions where conditions are generally wet during a given period. Longer-timescale SPI values reflect long-term precipitation patterns and tend toward zero unless there is a distinctive dry or wet trend. Also, while the U.S. Drought Monitor Drought Index became available in 2000, the Standardized Precipitation Index has been available since 1980.

We think the 1-year precipitation index is most appropriate for studying historical weather patterns. A positive index value on a date indicates that precipitation on this date exceeds the average precipitation over the given timescale, and a negative value indicates precipitation is less than average. According to SPI conventions, a drought occurs whenever the SPI value is below −1.0. Figure 21.4 illustrates extensive historical inter-annual fluctuations in precipitation, showing that the region has been prone to recurring droughts. In the region's meteorological records, most droughts have lasted less than 1 year, but every few decades prolonged drought periods (the 1930s Dust Bowl, 1955–1960 and 1988–1990) also occur (Jencso *et al.*, 2019). In general, northern Plains region droughts lasting 3–5 months occur approximately once every 5 years, and droughts lasting 6–8 months occur approximately once every 10 years (NIDIS/NOAA, 2018).

When we look at the historical Drought Index, according to Eqn 21.1, the average probability of experiencing a drought on July 25 during 2000–2017 was 0.35 across all respondents and the average severity level on July 25 during

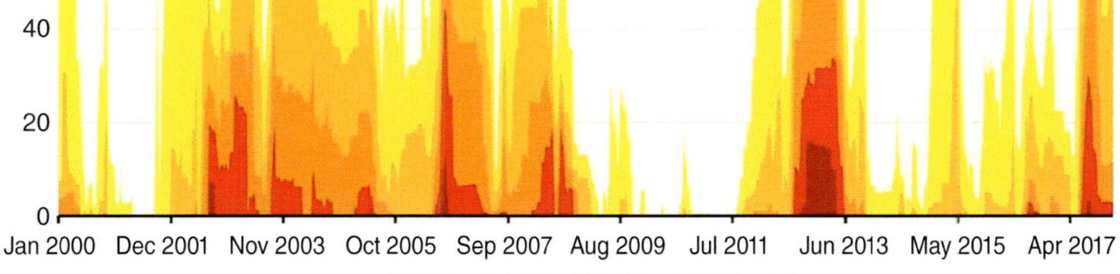

Source: US Drought Monitor.

Fig. 21.3. North Dakota and South Dakota cumulative percent area in USDM categories (%).

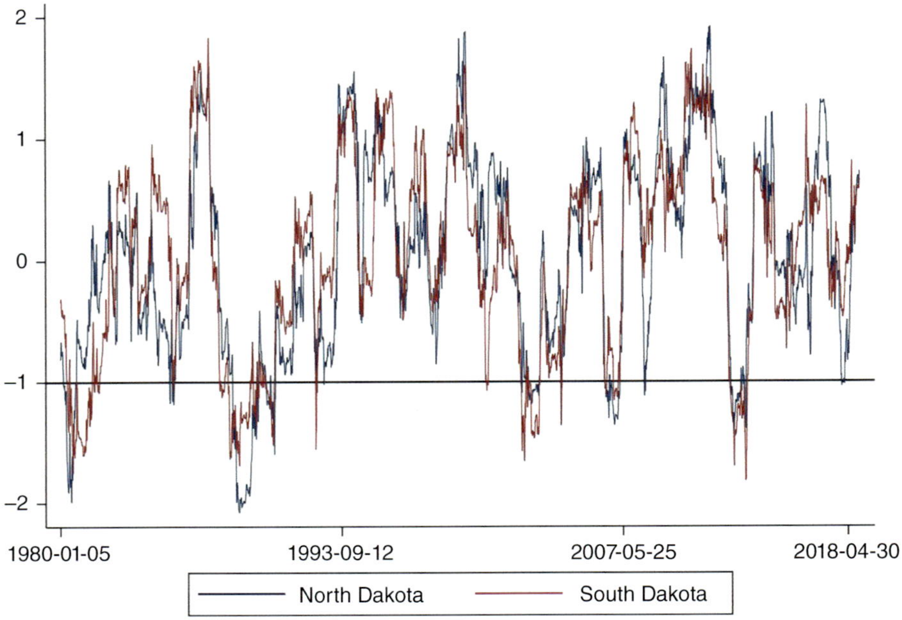

Source: gridMET; Abatzoglou (2013).

Fig. 21.4. One-year SPI from 1980 to 2018.

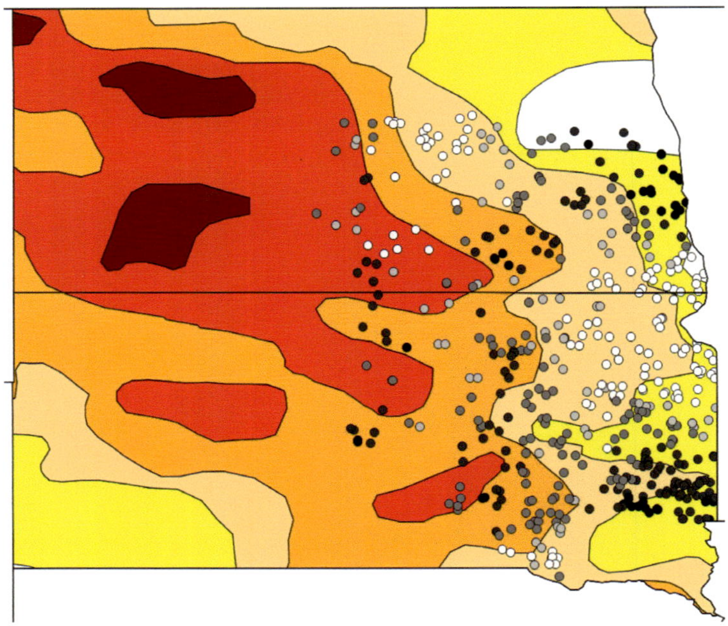

Note: Darker dot shades depict more severe average drought level while darker red area represents more severe drought on July 25, 2017.

Source: Authors' own work.

Fig. 21.5. Average severity of drought and severity of drought on July 25th, 2017 as measured by the SPI.

2000–2017 was 0.72, with a range between 0.17 and 2. Figure 21.5 describes the actual drought severity on July 25, 2017, with average drought severity on July 25 each year during 2000–2017. The drought on July 25, 2017, is presented in shaded areas, and the darker the shade, the more severe the drought. The drought severity in 2017 was highest in the west. The dots in gray scale describe the average severity level during 2000–2017, and the darker the dot, the more severe the drought. When we compare the historical average with the July 25, 2017, severity level, the Spearman correlation coefficient is 0.229 (p = 0.000), also indicating recurring droughts in the region. Thus, our main hypothesis is that the 2017 drought should not have been a surprise.

21.5 Methods

21.5.1 Conceptual considerations

The information content of the 2017 drought should, of course, depend very much on a location's drought history. In what follows, we take the standard approach to understand information updating, namely Bayesian analysis (Stone, 2013). We denote the 2017 drought event by I_{2017} and signify prior perceptions by $\Pr(M_t)$ where $t = 2015$ or 2018, and M_t represents the survey respondent's subjective assessment at time t, that droughts are becoming more common. The complementary expressed subjective probability, i.e. that they are not becoming more common, is given as $\Pr(N\text{-}M_t) = 1 - \Pr(M_t)$. From Bayes' theorem, when $\Pr(M_{2018})$ represents a respondent's 2018 survey posterior probability, then

$$\Pr(M_{2018}\,\text{and}\,I_{2017}) \equiv \overbrace{\Pr(M_{2018}\,|\,I_{2017})}^{\substack{\text{Posterior}\\ \text{given event } I_{2017}}} \times \Pr(I_{2017})$$
$$\equiv \overbrace{\Pr(I_{2017}\,|\,M_{2018}) \times \frac{\Pr(M_{2018})}{\Pr(M_{2015})}}^{\substack{\text{Update given}\\ \text{event } I_{2017}}} \times \overbrace{\Pr(M_{2015})}^{\text{Prior}} \tag{21.2}$$

so that

$$\frac{\Pr(M_{2018}\,|\,I_{2017})}{\Pr(M_{2015})} \equiv \frac{\Pr(M_{2018})}{\Pr(M_{2015})} \times \frac{\Pr(I_{2017}\,|\,M_{2018})}{\Pr(I_{2017})} \tag{21.3}$$

If the unconditional probabilities are the same, i.e. $\Pr(M_{2018}) = \Pr(M_{2015})$, then we may write the equation as

$$\frac{\Pr(M_{2018}\,|\,I_{2017})}{\Pr(M_{2015})} \equiv \frac{\Pr(I_{2017}\,|\,M_{2018})}{\Pr(I_{2017})} \tag{21.4}$$

as the relative drought-conditioned probability update regarding future droughtiness. The left-hand side in Eqn 21.4 determines how the drought event shapes beliefs about climate, and the right-hand side characterizes how beliefs about climate shape drought probabilities. If the drought is not a surprising event, then $\Pr(M_{2018}\,|\,I_{2017}) / \Pr(M_{2015}) \approx 1$ so that the extent to which $\Pr(M_{2015}) / \Pr(M_{2018}\,|\,I_{2017})$ is less than 1 measures surprise. Baldi and Itti (2009) use this conditioning observation to characterize surprise (or Wow) as, in our terminology,

$$W(I_{2017}, M) = -\ln\left(\frac{\Pr(M_{2018}\,|\,I_{2017})}{\Pr(M_{2015})}\right) \tag{21.5}$$

We write $S_u(k, D_i) \in [0,1]$ as the share of respondents in 2017 drought zone D_i who take the view in year $u \in \{2015, 2018\}$ that drought is becoming $k \in K \equiv \{\text{Less, Same, More}\}$ likely. We follow Baldi and Itti in using the entropy form to characterize zone-specific *ex-ante* average surprise as

$$W(D_i) = -\sum\nolimits_{k \in K} S_{2015}(k, D_i) \ln\left(\frac{S_{2018}(k, D_i)}{S_{2015}(k, D_i)}\right) \tag{21.6}$$

where no surprise can be depicted by $S_{2018}(k, D_i) = S_{2015}(k, D_i)\ \forall\ k \in K$ so that . $W(D_i) = 0$, the statistic's least value. Using our data, we test our main hypothesis that average surprise is zero. Given that surprise is registered, alternative reasonable hypotheses are that surprise should increase with drought severity, i.e. $W(D_{i+1}) \geq W(D_i)$ $\forall\ i \in \{0, 1, 2, 3\}$, and that the drought was more of a surprise for North Dakota than for South Dakota.

21.5.2 Empirical analysis

With the intent to further our understanding of how drought affects climate change perception, the methods used are of two types. The first involves descriptive statistics, including intertemporal comparisons of cumulative distributions, and comparisons among means. The second general approach is DID analysis. This analysis allows us to measure the treatment effect by examining differences between the average change over time for a treatment group, who experienced the drought on July 25, 2017, and the average change over time for a control group, who did not experience any drought on July 25, 2017.

Two identifying assumptions should be satisfied in order to apply DID analysis (Angrist and Pischke, 2009). The treatment should be mean-independent with respect to the error term. In other words, there should be no unobserved variables that affect both the farmers' perception and the severity of the drought that a farmer experienced. Because ours is a natural experiment and farmers had no control over which drought they experienced, this assumption is not violated. The second assumption, generally referred to as 'parallel trends,' is that the background trend over time should be the same across the treatment and control groups. A commonly used method to verify this second assumption is visual inspection, i.e. comparing plots for the treatment and control groups. A statistical test of the parallel trend assumption can be made by including a time regressor in the model to control for trends (Card and Krueger, 2000; Besley and Burgess, 2004; Hastings, 2004). However, as our data consists of only two periods, we can apply neither visual inspection nor time lags to validate this assumption. We assume that the parallel trends assumption holds. We consider this a reasonable assumption because our treatment is randomized by nature, and so is likely exogenous to respondents' characteristics (Cunningham, 2021).

The DID analysis allows us to measure the treatment effect by examining differences between estimated coefficients for the treatment group and those for the control group. Our DID model is specified as

$$\text{Perception}_{i,t} = \alpha_0 + \alpha_1 \text{Dtime} + \beta * \text{Tr} + \gamma * \text{Dtime} * \text{Tr} + \sum_i \lambda_i x_i + \varepsilon_{i,t} \quad (21.7)$$

where Dtime (0 = pre-treatment, 1 = post-treatment) and *Tr* (1 = experienced drought, 0 = did not experience drought) are dummy variables and $\varepsilon_{i,t}$ is the error term. The variables denoted by x_i are controls such as respondent demographics. The interaction term Dtime $*$ Tr is our primary object of interest. The interaction term coefficient should capture changes in respondents' perceptions between surveys.

When viewed collectively, prior drought history in the region (Figs 21.3 and 21.4), the 2017 drought map (Fig. 21.5) and Spearman correlations suggest that farmers in the region are well-accustomed to drought. For this reason, we frame our hypothesis as $\gamma = 0$ because farmers who experienced drought in 2017 should not alter their climate change perception as a result of the drought event.

21.6 Results

In this section, we first examine intertemporal cumulative distributions and compare how responses changed over the two surveys. Then, we calculate how much of a surprise the drought was to the respondents based on the Baldi and Itti (2009) entropy metric. Lastly, effects of the drought on perceptions are estimated using DID analysis, as the entropy analysis tells us only whether perceptions shifted significantly, not the direction of the shift.

21.6.1 Temporal changes in perception responses

In this sub-section, we use intertemporal cumulative distributions to examine how the perceptions shifted over the two surveys. Then, we calculate the measure of surprise. This analysis allows us to ascertain whether prior perceptions about weather held constant or changed after the experience of drought.

Figure 21.6 shows the cumulative percentage of views expressed in 2018 conditional on views expressed in 2015. From the graph, respondents who expressed the view in 2015 that there were more droughts are also more likely to indicate in the 2018 survey that there were more droughts. Similarly, respondents with the view in 2015 that there were fewer droughts are also more likely to provide this answer in the 2018 survey. Figure 21.7 focuses on those who

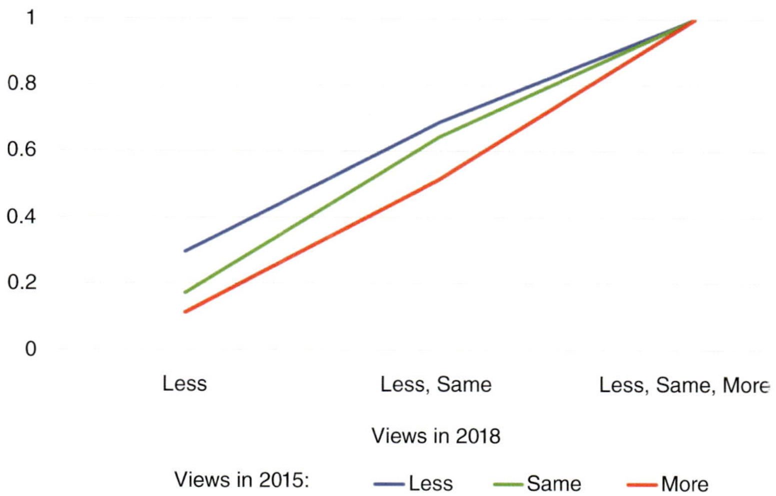

Source: Authors' own work.

Fig. 21.6. Cumulative percentage of views in 2018 by views in 2015.

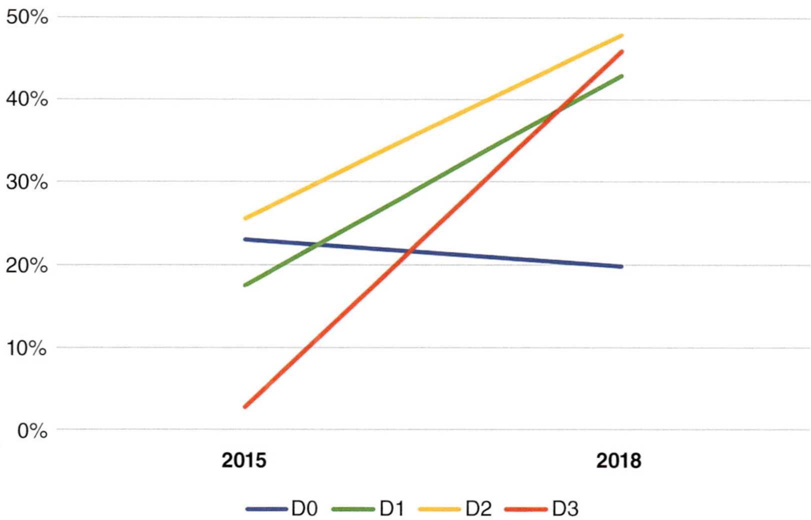

Source: Authors' own work.

Fig. 21.7. Percentage of 2018 survey respondents who answered that there were more droughts than in the past, compared with responses that these respondents gave in 2015, broken down by 2017 drought severity level.

answered that there were more droughts in 2018 and decomposes respondents by the severity of drought incurred in 2017. The figure shows that for D1-, D2- and D3-level droughts, the percentage of respondents who held that there had been more droughts increased between surveys. Among those who had experienced D0-level drought, the percentage decreased slightly. When we look at how individual views changed between the two surveys, in Fig. 21.8, a distinctive pattern emerges. The share of respondents who experienced the D0-level drought and changed their views toward more drought (from 'Less' to 'Same' or 'More' in union with from 'Same' to 'More') was smaller than the proportion who changed their views toward less drought (from 'More' to

'Same' or 'Less' in union with from 'Same' to 'Less'). However, for drought levels D1, D2 and D3, the fraction shift was reversed. Noteworthy is that the difference between 'change toward more drought' and 'change toward less drought' increases with the 2017 drought severity level, being −2% for level D0, 27% for level D1, 30% for level D2 and 54% for level D3.

The shifts in perception can be measured by how surprised the respondents were when they experienced the 2017 drought. According to Table 21.4, the average surprise increases with drought severity when the respondents experienced the drought in 2017. Not experiencing any drought was also a surprise and was more surprising than the level D2 drought. Comparing

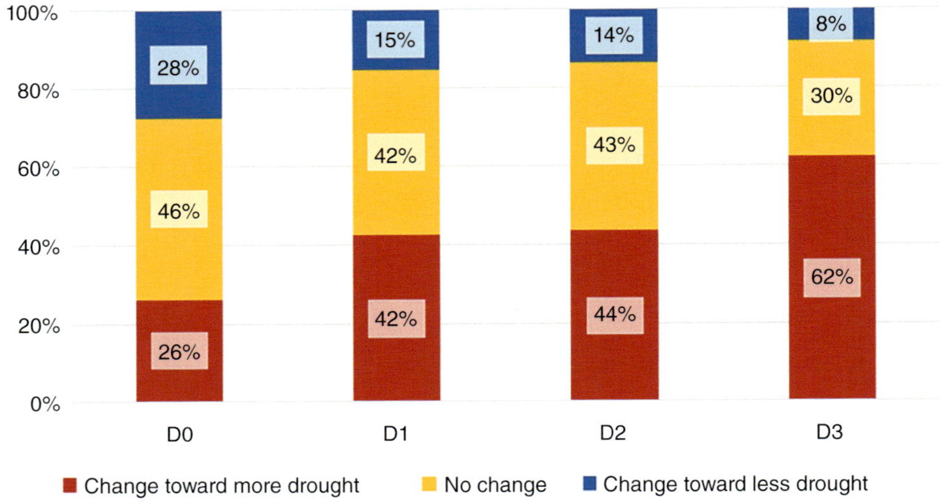

Source: Authors' own work.

Fig. 21.8. How the respondents change their views (%) between 2015 and 2018, broken down by 2017 drought severity level.

Table 21.4. Share of drought pattern perception and the measure of surprise by their drought experience in 2017 and state. (Source: Authors' own work.)

		Drought pattern perception			
Drought experience	Year	Less (%)	Same (%)	More (%)	Wow
No drought	2015	18	45	36	0.30
	2018	36	55	9	
D0	2015	30	47	23	0.003
	2018	31	49	20	
D1	2015	32	50	18	0.16
	2018	18	39	43	
D2	2015	23	51	26	0.18
	2018	7	45	48	
D3	2015	43	54	3	0.54
	2018	16	38	46	
State	Year	Less (%)	Same (%)	More (%)	Wow (%)
North Dakota	2015	40	49	12	0.15
	2018	21	46	33	
South Dakota	2015	25	50	25	0.05
	2018	19	42	39	

states, the drought was more surprising in North Dakota than in South Dakota. In short, the 2017 drought was a surprise to the respondents and shifted drought pattern perceptions.

21.6.2 Effect of the drought on the farmer's perception

Table 21.5 provides the DID estimates. The dependent variable is a perception about drought, and treatment is whether the respondents experienced any level of drought in 2017. Respondents who experienced no drought are considered untreated. Control variables include farmer demographics, farm characteristics and the probability of drought. The coefficient for the interaction term Dtime $*$ Tr, generated by multiplying the drought experience and time dummy variables, is 0.530 and is statistically significant ($p = 0.000$). It is larger than zero, indicating that the 2017 drought shifted perceptions toward there being more droughts. This is a large shift as the perception index has range [1, 3].

Table 21.5. DID analysis on the perception of drought pattern. (Source: Authors' own work.)

Variable	Coefficient (std. err.)	
Dtime (= 1 if after 2017 drought)	−0.073	(0.085)
Tr (= 1 if experienced drought)	−0.071	(0.081)
Dtime*Tr	0.530***	(0.101)
Birth year	0.009***	(0.003)
Education	0.054*	(0.028)
Income	−0.060***	(0.019)
Ownership	0.052***	(0.020)
Crop ratio	−0.028	(0.101)
Probability of drought	0.906***	(0.253)
LCC4	0.003*	(0.002)
Slope	0.001	(0.001)
Drainage	0.007	(0.064)
No-till	0.051	(0.049)
Constant	−17.079***	(4.907)

Note: *, **, *** indicate significance at 10%, 5%, 1% levels, respectively.

21.7 Assessment and Concluding Remarks

Figures 21.3 and 21.4 show that the 2017 drought was less severe than the one that occurred 3 years before the first survey. The 2017 drought lasted for 3 months, and only the 1-year Standardized Precipitation Index for the north central region in South Dakota went below −1. By contrast, the 2012 drought lasted for a year and was much more severe than the 2017 drought. Between 1980 and 2017, the Standardized Precipitation Index fell to −1 or lower on more than ten occasions. These past events show that the 2017 drought should not have been a surprisingly severe event for farmers in the area. Yet respondents were surprised.

A single survey comparison cannot allow for distinctions to be made between a shift in opinions that is fundamentally unstable and one that, although perhaps similar to prior shifts, turns out to be permanent. An approach other than ours, probably using regular polling, would be needed to better understand opinion stability. In any case, surveys alone will not convey how and why opinions harden. Information is likely not the only, or even dominant, factor in determining viewpoints because a taste for conformity, arising from fear or to gain acceptance or to internalize dissonant pressures (Kelman, 1958), may also matter, as may the tangibility of the weather event at issue (Asch, 1955).

A matter that remains unclear is whether the sensitivity of climate change perceptions to weather events can facilitate the management of climate change, including consent to policy adjustments. For better or worse, public and private responses that are conditioned on weather events will generate a more receptive citizenry so long as policies are put into effect quickly. However, those that take longer to implement or modify may miss the moment. The wisdom of policy enactment based on ephemeral, although genuine, support is another matter.

References

Abatzoglou, J.T. (2013) Development of gridded surface meteorological data for ecological applications and modelling. *International Journal of Climatology* 33, 121–131.

Abendroth, L.J., Miguez, F.E., Castellano, M.J., Carter, P.R., Messina, C.D. *et al.* (2021) Lengthening of maize maturity time is not a widespread climate change adaptation strategy in the US Midwest. *Global Change Biology* 27, 2426–2440.

Agarwal, S., Driscoll, J.C., Gabaix, X. and Laibson, D. (2008) Learning in the credit card market. *National Bureau of Economic Research Working Paper 13822*, National Bureau of Economic Research, Cambridge, Massachusetts.

Ahrens, C.D. (2007) *Meteorology Today: An Introduction to Weather, Climate, and the Environment*, 8th edn. Thompson, Brooks/Cole, Pacific Grove, California.

Akerlof, G.A. and Dickens, W.T. (1982) The economic consequences of cognitive dissonance. *The American Economic Review* 72, 307–319.

Alemu, W.G., Henebry, G.M. and Melesse, A.M. (2020) Land cover and land use change in the US Prairie pothole region using the USDA cropland data layer. *Land* 9, 166.

Angrist, J.D. and Pischke, J.S. (2009) *Mostly Harmless Econometrics: An Empiricist's Companion*. Princeton University Press, Princeton, New Jersey.

Asch, S.E. (1955) Opinions and social pressure. *Scientific American* 193, 31–35.

Baldi, P. and Itti, L. (2009) Bayesian surprise attracts human attention. *Vision Research* 49, 1295–1306.

Benjamin, D.J. (2019) Errors in probabilistic reasoning and judgment biases. In: Douglas Bernheim, B., DellaVigna, S. and Laibson, D. (eds) *Handbook of Behavioral Economics – Foundations and Applications, Vol.* 2. Elsevier, Kidlington, UK, pp. 69–186.

Besley, T. and Burgess, R. (2004) Can labor regulation hinder economic performance? Evidence from India. *Quarterly Journal of Economics* 119, 91–134.

Broadbent, D.E. and Broadbent, M.H.P. (1981) Recency events in visual memory. *Quarterly Journal of Experimental Psychology* 33, 1–15.

Brulle, R.J., Carmichael, J. and Jenkins, J.C. (2012) Shifting public opinion on climate change: an empirical assessment of factors influencing concern over climate change in the U.S., 2002–2010. *Climatic Change* 114, 169–188.

Card, D. and Krueger, A.B. (2000) Minimum wages and employment: a case study of the fast-food industry in New Jersey and Pennsylvania: reply. *American Economic Review* 90, 1397–1420.

Carlton, J.S., Mase, A.S., Knutson, C.L., Lemos, M.C., Haigh, T. *et al.* (2016) The effects of extreme drought on climate change beliefs, risk perceptions, and adaptation attitudes. *Climatic Change* 135, 211–226.

Conant, R.T., Kluck, D., Anderson, M., Badger, A., Boustead, B.M. *et al.* (2018) Northern Great Plains. In: Reidmiller, D.R., Avery, C.W., Easterling, D.R., Kunkel, K.E., Lewis, K.L.M. *et al.* (eds) *Impacts, Risks, and Adaptation in the United States: Fourth National Climate Assessment, Volume II*. U.S. Global Change Research Program, Washington, DC, pp. 941–986. DOI: 10.7930/NCA4.2018.CH22.

Cunningham, S. (2021) *Causal Inference*. Yale University Press, New Haven, Connecticut.

De Bondt, W.F.M. and Thaler, R. (1985) Does the stock market overreact? *Journal of Finance* 40, 793–805.

Deryugina, T. (2013) How do people update? The effects of local weather fluctuations on beliefs about global warming. *Climatic Change* 118, 397–416.

Dunning, T. (2012) *Natural Experiments in the Social Sciences: A Design-Based Approach*. Cambridge University Press, Cambridge, UK.

Dynata (2023) Dynata, Plano, Texas. https://www.dynata.com/ (accessed May 9, 2023).

Edwards, W. and Phillips, L.D. (1968) Man as transducer for probabilities in Bayesian command and control systems. In: Kleinmuntz, B. (ed.) *Human Judgements and Optimality*. Wiley, New York, pp. 360–401.

Egan, P.J. and Mullin, M. (2012) Turning personal experience into political Americans' perceptions about global warming. *Journal of Politics* 74, 796–809.

Fama, E.F. (1970) Efficient capital markets: a review of theory and empirical work. *Journal of Finance* 25, 383–417.

Gärtner, L. and Schoen, H. (2021) Experiencing climate change: revisiting the role of local weather in affecting climate change awareness and related policy preferences. *Climatic Change* 167, 1–20.

Goebbert, K., Jenkins-Smith, H.C., Klockow, K., Nowlin, M.C. and Silva, C.L. (2012) Weather, climate, and worldviews: the sources and consequences of public perceptions of changes in local weather patterns. *Weather, Climate, and Society* 4, 132–144.

Grether, D.M. (1992) Testing Bayes' rule and the representativeness heuristic: some experimental evidence. *Journal of Economic Behavior and Organization* 17, 31–57.

Haden, V.R., Niles, M.T., Lubell, M., Perlman, J. and Jackson, L.E. (2012) Global and local concerns: what attitudes and beliefs motivate farmers to mitigate and adapt to climate change? *PLoS ONE* 7, e52882.

Hansson, R.O., Noulles, D. and Bellovich, S.J. (1982) Knowledge, warning, and stress: a study of comparative roles in an urban floodplain. *Environment and Behavior* 14, 171–185.

Hastings, J.S. (2004) Vertical relationships and competition in retail gasoline markets: empirical evidence from contract changes in Southern California. *American Economic Review* 94, 317–328.

Helms, D. (1992) The development of the Land Capability Classification. In: *Readings in the History of the Soil Conservation Service*. Soil Conservation Service, Washington, DC.

Hoell, A., Perlwitz, J. and Eischeid, J. (2019) *The Causes, Predictability, and Historical Context of the 2017 U.S. Northern Great Plains Drought*. NOAA National Integrated Drought Information System, Washington, DC. Available at: https://repository.library.noaa.gov/view/noaa/23003 (accessed May 9, 2023).

Howe, P.D., Marlon, J.R., Mildenberger, M. and Shield, B.S. (2019) How will climate change shape climate opinion? *Environmental Research Letters* 14, 113001.

IPCC (2021) Annex VII: glossary. In: Masson-Delmotte, V., Zhai, P., Pirani, A., Connors, S.L., Péan, C., Berger, S. *et al.* (eds) *Climate Change 2021: The Physical Science Basis. Contribution of Working Group I to the Sixth Assessment Report of the Intergovernmental Panel on Climate Change*. Cambridge University Press, Cambridge, UK, pp. 2215–2256.

Jencso, K., Parker, B., Downey, M., Hadwen, T., Howell, A. *et al.* (2019) *Flash Drought: Lessons Learned from the 2017 Drought Across the U.S. Northern Plains and Canadian Prairies*. NOAA National Integrated Drought Information System, Washington, DC.

Joireman, J., Barnes, H. and Duell, B. (2010) Effect of outdoor temperature, heat primes and anchoring on belief in global warming. *Journal of Environmental Psychology* 30, 358–367.

Kahneman, D. and Tversky, A. (1973) On the psychology of prediction. *Psychological Review* 80, 237–251.

Kelman, H.C. (1958) Compliance, identification, and internalization: three processes of attitude change. *Journal of Conflict Resolution* 2, 51–60.

Knapp, A.K. and Smith, M.D. (2001) Variation among biomes in temporal dynamics of aboveground primary production. *Science* 291, 481–484.

Konisky, D.M., Hughes, L. and Kaylor, C.H. (2015) Extreme weather events and climate change concern. *Climatic Change* 134, 533–547.

Lazarus, R.S. (1984) On the primacy of cognition. *American Psychologist* 39, 124–129. DOI: 10.1037/0003-066X.39.2.124.

Leiserowitz, A. (2006) Climate change risk perception and policy preferences: the role of affect, imagery, and values. *Climatic Change* 77, 45–72.

Li, Y., Johnson, E.J. and Zaval, L. (2011) Local warming: daily temperature change influences belief in global warming. *Psychological Science* 22, 454–459.

List, J.A. (2004) Neoclassical theory versus prospect theory: evidence from the marketplace. *Econometrica* 72, 615–625.

List, J.A. and Haigh, M.S. (2009) Investment under uncertainty: testing the options model with professional traders. *Review of Economics and Statistics* 92, 974–984.

Loewenstein, G.F., Weber, E.U., Hsee, C.K. and Welch, N. (2001) Risk as feelings. *Psychological Bulletin* 127, 267.

Marx, S.M., Weber, E.U., Orlove, B.S., Leiserowitz, A., Krantz, D.H. *et al.* (2007) Communication and mental processes: experiential and analytic processing of uncertain climate information. *Global Environmental Change* 17, 47–58.

McCright, A.M. and Dunlap, R.E. (2011) Cool dudes: the denial of climate change among conservative white males in the United States. *Global Environmental Change* 21, 1163–1172. DOI: 10.1016/j.gloenvcha.2011.06.003.

McKee, T.B., Doesken, N.J. and Kleist, J. (1993) The relationship of drought frequency and duration to time scales. *Proceedings of the 8th Conference on Applied Climatology, 17–22 January, Anaheim, California*, pp. 179–183.

Myers, T.A., Maibach, E.W., Roser-Renouf, C., Akerlof, K. and Leiserowitz, A.A. (2013) The relationship between personal experience and belief in the reality of global warming. *Nature Climate Change* 3, 343–347.

NIDIS/NOAA (2018). *Historical Character of United States Northern Great Plains Drought*. NIDIS/NOAA, Washington, DC.

Niles, M.T., Lubell, M. and Haden, V.R. (2013) Perceptions and responses to climate policy risks among California farmers. *Global Environmental Change* 23, 1752–1760.

Niles, M.T., Wiener, S., Schattman, R.E., Roesch-McNally, G. and Reyes, J. (2019) Seeing is not always believing: crop loss and climate change perceptions among farm advisors. *Environmental Research Letters* 14, 044003.

Otkin, J.A., Svoboda, M., Hunt, E.D., Ford, T.W., Anderson, M.C. *et al.* (2017) Flash droughts: a review and assessment of the challenges imposed by rapid-onset droughts in the United States. *Bulletin of the American Meteorological Society* 99, 911–919.

Page, L. and Page, K. (2010) Last shall be first: a field study of biases in sequential performance evaluation on the idol series. *Journal of Economic Behavior and Organization* 73, 186–198.

Pitz, G.F. and Reinhold, H. (1968) Payoff effects in sequential decision-making. *Journal of Experimental Psychology* 77, 249–257.

Rashford, B.S., Adams, R.M., Wu, J., Voldseth, R.A., Guntenspergen, G.R. *et al.* (2016) Impacts of climate change on land-use and wetland productivity in the Prairie Pothole Region of North America. *Regional Environmental Change* 16, 515–526.

Shao, W. (2016) Are actual weather and perceived weather the same? Understanding perceptions of local weather and their effects on risk perceptions of global warming. *Journal of Risk Research* 19, 722–742. DOI: 10.1080/13669877.2014.1003956.

Shiller, R.J. (1981) Do stock prices move too much to be justified by subsequent changes in dividends? *American Economic Review* 71, 421–436.

Spence, A., Poortinga, W., Butler, C. and Pidgeon, N.F. (2011) Perceptions of climate change and willingness to save energy related to flood experience. *Nature Climate Change* 1, 46–49. DOI: 10.1038/nclimate1059.

Stone, J.V. (2013) *Bayes' Rule: A Tutorial Introduction to Bayesian Analysis*. Sebtel Press, Sheffield, UK.

Svoboda, M., Lecomte, D., Hayes, M., Heim, R., Gleason, K. *et al.* (2002) The Drought Monitor. *Bulletin of the American Meteorological Society* 83, 1181–1190.

Swim, J., Clayton, S., Doherty, T., Gifford, R., Howard, G. *et al.* (2011) *Physiology and Global Climate Change: Addressing a Multi-Faceted Phenomenon and Set of Challenges. A Report by the American Psychological Association's Task Force on the Interface Between Psychology and Global Climate Change*. American Psychological Association, Washington, DC.

USDA (2019) 2019 Acreage Data as of August 1, 2019. Available at: https://www.fsa.usda.gov/news-room/efoia/electronic-reading-room/frequently-requested-information/crop-acreage-data/index (accessed July 13, 2022).

US Drought Monitor (2021). What is the USDM. Available at: https://droughtmonitor.unl.edu/About/WhatistheUSDM.aspx (accessed 22 May 2021).

van der Linden, S. (2014) On the relationship between personal experience, affect and risk perception: the case of climate change. *European Journal of Social Psychology* 44, 430–440.

Wang, T., Luri, M., Janssen, L., Hennessy, D.A., Feng, H. *et al.* (2017) Determinants of motives for land use decisions at the margins of the Corn Belt. *Ecological Economics* 134, 227–237.

Weber, E.U. (2013) Seeing is believing. *Nature Publishing Group* 3, 1–2.

Whitmarsh, L. (2008) Are flood victims more concerned about climate change than other people? The role of direct experience in risk perception and behavioural response. *Journal of Risk Research* 11, 351–374.

Wimberly, M.C., Janssen, L.L., Hennessy, D.A., Luri, M., Chowdhury, N.M. *et al.* (2017) Cropland expansion and grassland loss in the eastern Dakotas: new insights from a farm-level survey. *Land Use Policy* 63, 160–173.

Wright, C.K. and Wimberly, M.C. (2013). Recent land use change in the Western Corn Belt threatens grasslands and wetlands. *Proceedings of the National Academy of Sciences* 110, 4134–4139.

Yazar, M., York, A. and Kyriakopoulos, G. (2021) Heat exposure and the climate change beliefs in a desert city: the case of phoenix metropolitan area. *Urban Climate* 36, 100769. DOI: 10.1016/j.uclim.2020.100769.

Index

Note: The page numbers in *italics* and **bold** represent figures and tables respectively.